Molecular Technology

Related Titles

Yamamoto, H., Kato, T. (eds.)

Molecular Technology

Volume 1: Energy Innovation

2018

ISBN: 978-3-527-34163-4

Yamamoto, H., Kato, T. (eds.)

Molecular Technology

Volume 2: Life Innovation

2018

ISBN: 978-3-527-34162-7

Yamamoto, H., Kato, T. (eds.)

Molecular Technology

Volume 4: Synthesis Innovation

2019

ISBN: 978-3-527-34588-5

Molecular Technology

Materials Innovation

Edited by Hisashi Yamamoto and Takashi Kato

Volume 3

WILEY-VCH

Editors

Hisashi Yamamoto
Chubu University
Molecular Catalyst Research Center
1200 Matsumoto
Kasugai
487-501 Aichi
Japan

Takashi Kato
University of Tokyo
Department of Chemistry and
Biotechnology
7-3-1 Hongo, Bunkyo-ku
113-8656 Tokyo
Japan

Cover
fotolia_VAlex and fotolia_monsitj

Library of Congress Card No.:
applied for

British Library Cataloguing-in-Publication Data
A catalogue record for this book is available from the British Library.

Bibliographic information published by the Deutsche Nationalbibliothek
The Deutsche Nationalbibliothek lists this publication in the Deutsche Nationalbibliografie; detailed bibliographic data are available on the Internet at <http://dnb.d-nb.de>.

Print ISBN: 978-3-527-34161-0
ePDF ISBN: 978-3-527-80273-9
ePub ISBN: 978-3-527-80272-2

Cover Design Adam-Design, Weinheim, Germany
Typesetting SPi Global, Chennai, India
Printing and Binding C.O.S. Printers Pte Ltd, Singapore

Printed on acid-free paper

10 9 8 7 6 5 4 3 2 1

Foreword by Dr. Hamaguchi

Molecular Technology is a newly developed research field supported through Japan Science and Technology Agency (JST) research funding programs. These programs aim to establish an innovative research field that harnesses the characteristics of molecules to enable new scientific and commercial applications. It is our great pleasure to publish this book, with the ambition that it will develop both an understanding of and further support for this new research field within the research and student community.

Molecular Technology as introduced in this book began in 2012 as a research area within JST's Strategic Basic Research Programs. JST is an advanced network-based research institution that promotes state-of-the-art R&D projects and leads the way in the co-creation of future innovation in tandem with wider society. JST develops a wide range of funding programs related to the promotion of scientific and technological innovation, which include strategy planning, target-driven basic research, and promotion of research and development.

Various research projects focused on Molecular Technology are currently underway within JST's Strategic Basic Research Programs, including

- the team-based research program " CREST (Core Research for Evolutionary Science and Technology)" and
- the individual research program " PRESTO (Precursory Research for Embryonic Science and Technology)".

Dr. Yamamoto (CREST) and Dr. Kato (PRESTO) manage the Molecular Technology research area as research supervisors.

In addition, JST's Strategic International Collaborative Research Program promotes research projects in the area of Molecular Technology, including ongoing cooperation with *L'Agence nationale de la recherche* (The French National Research Agency, ANR).

A wide range of researchers from the young to seniors across fields from green science, life science, and energy are participating in successful research aimed at establishing the new field of Molecular Technology. They are already producing excellent research results and it is our hope that these will develop into technologies capable of initiating a new era in energy and green and life sciences.

I encourage you to read not only researchers in related fields but also look more broadly to researchers working in other fields. Inspired by this book, I look forward to emerging new research fields and seeds toward future innovation.

Michinari Hamaguchi
Japan Science and Technology Agency
President

Foreword by Dr. Noyori

As an affiliated institution of the Japan Science and Technology Agency (JST), the Center for Research and Development Strategy (CRDS) navigates the latest global trends in science, technology, and innovation to aid the Japanese government in formulating its national strategies. *Molecular Technology* is the outcome of a research project born of a CRDS Strategic Proposal realized under the excellent editorial supervision of Hisashi Yamamoto and Takashi Kato. To them and to the scientists who have made major advances in molecular technology through their uninhibited research I extend my heartfelt congratulations and respect.

The significance of molecular science in all areas of scientific endeavor is certain to increase. Accurate understanding of molecular assemblies and molecular complexes is essential for comprehending the elaborate workings of natural phenomena and of the genesis and mechanisms of materials and life functions. Now, more than ever, science must be seen as a single entity, a comprehensive whole. Mathematical science and the most advanced technologies of observation and information help us explore the essence of materials and substances in a way that brings together all fields of science. It is the nature of molecular science to continually advance and expand. Using the metaphor of light, we can say that molecules behave in the manner of both "waves and particles."

The traditional separation of science into physics, chemistry, and biology no longer applies. Neither does it make any sense to maintain those seemingly self-contained subdivisions of organic chemistry, inorganic chemistry, physical chemistry, or polymer chemistry. So long as specialized groups and rigid educational systems cling to outdated perceptions, the more important it is to encourage an "anti-disciplinary" type of science in which diverse fields converge rather than conventional interdisciplinary or transdisciplinary attempts to link diverse fields.

Molecular technology, while firmly grounded in fundamental scientific knowledge, aims for practical applications within contemporary society. Johann Wolfgang von Goethe once said, "Knowing is not enough; we must apply. Willing is not enough; we must do." Technology with no practical application is meaningless to society. Researchers should not hesitate to set their own themes and topics of exploration in academia where self-determination holds strong and creativity wins the highest respect. Researchers must show ingenuity in the pursuit of their chosen mission even as they fulfill their duty to pursue science-based technology for society. Never forget that it is by no means advisable to function purely as a support for activities that industry should actually undertake on its own.

The creative outcomes of the Molecular Technology Project launched in 2013 in conjunction with new collaborations are certain to lead to a wide range of innovations and to make significant contribution to achieving the Sustainable Development Goals (SDGs) of the United Nations' 2030 Agenda.

Science is one; and the world is one. Those who will follow us have a responsibility to the world after 2030 and it is my hope that new generations will pioneer revolutionary molecular technology that will bring science and humanity ever closer together. Brain circulation and international collaboration are essential to achieving these goals. V. S. Naipaul, winner of the 2001 Nobel Prize in Literature, once noted that knowing what you wanted to write was three-quarters of the task of writing. Humanity's future is to be found in the unbounded imagination of the young and in their ability to support the challenges they undertake.

December 2017

Ryoji Noyori
Tokyo, Japan

Preface

Chemical science enables us to qualitatively change existing science and technology by purposefully designing and synthesizing molecules and creating the desired physical, chemical, and biological functions of materials and drugs at molecular level. In 2012, we have started a big funding project in Japan, "Molecular Technology" (Establishment of Molecular Technology toward the Creation of New Functions [CREST], and Molecular technology and new functions [PRESTO]), and numerous research groups in Japan have joined the project in diverse research areas. All of these are typical transdisciplinary research projects between chemistry and various other research areas. In other words, molecular technology is the brand-new scientific discipline. In principle, most of the proposed projects try to solve the big bridge between chemistry and other basic science and technology. We thus propose a good model for this bridge, which can make valuable contribution to human welfare.

Between JST and ANR we initiated a number of international collaboration projects in molecular technology in 2014. Since then, 12 new projects have been started. The projects provide quite unique opportunity for collaboration between Japan and France and active research groups have participated with very close discussions on molecular technology between the two countries. We are sure this project will provide close contacts between the research groups of Japan and France for otherwise impossible discoveries. Overall, this international collaboration will be a new entry for even more important discoveries in future.

In 2016, we started the discussion to bring out a new and comprehensive book on molecular technology for the benefit of all the researchers in the world to provide typical and leading examples in molecular technology. Overall, researchers of 15 CREST, 50 PRESTO, and 12 INTERNATIONAL groups have contributed to this book. This book covers extremely diverse areas of molecular technology, from materials to pharmaceuticals.

Hisashi Yamamoto
Chubu University,
Supervisor of CREST

Takashi Kato
The University of Tokyo,
Supervisor of PRESTO

Contents

1

Control of Electronic Property of C_{60} Fullerene via Polymerization

Nobuyuki Aoki

Chiba University, Graduate School of Science and Engineering, Department of Materials Science, 1-33 Yayoi, Inage, Chiba 263-8522, Japan

1.1 Introduction

1.1.1 History of Polymerization of C_{60} Fullerene

Fullerenes, the spherical cage molecules composed of carbon atoms, were discovered by Kuroto and Smalley coworkers in 1985 [1]. After the development of a large-scale synthesis method by Kraetschmer et al. in 1990 [2], they became usable not only in vacuum but also in atmospheric conditions. C_{60} is the most popular molecule in the fullerenes composed of 60 carbon atoms as shown in Figure 1.1a. It has semiconductor characteristics having a bandgap. The experimental values have been shown as typically 1.5–1.8 eV for the highest occupied molecular orbital–lowest unoccupied molecular orbital (HOMO–LUMO) gap [3, 4]; however, the value varies within the range of 1.43–2.35. The electrical properties of fullerene are based on the bandgap. It shows n-type semiconductor characteristics; the activation energy is less than half of the bandgap and is close to the half value only at high temperatures. At around room temperature, the activation energy is in the range of 100–200 meV, which relates to activation from the donner-like state within the optical gap [5]. The transport properties are strongly affected also by the orientation state of the crystalline fullerene. On decreasing the temperature from room temperature, the first kink of the conductivity is observed at around 260 K related to the restriction of the orientation angle of rotation of C_{60} molecules [6]. The second kink at 90 K relates to the glass transition of the orientation angle.

The conductivity drastically falls by several orders of magnitude, mostly becoming insulating due to absorption of oxygen molecules [7]. This is due to the formation of deep level trap sites lying 0.7 eV below the bottom of the conduction band. Therefore, most of the transport measurements of the semiconducting properties are done in vacuum conditions. An inert atmosphere such as argon, nitrogen, or helium also helps maintain the conductivity. If the sample was exposed in air once, heating at 160–180 °C in vacuum is necessary to recover the conductivity by desorbing oxygen from the thin film of fullerene [8].

Molecular Technology: Materials Innovation, Volume 3, First Edition.
Edited by Hisashi Yamamoto and Takashi Kato.

Crystalline C_{60} fullerene exhibits face centred cubic (fcc) structure due to van der Waals interaction. Intermolecular interactions change the electrical and optical properties. A possibility of intermolecular coupling was observed first by photo-irradiation using laser light [9]. UV–visible light illumination read the photopolymerization of the C_{60} molecules in oxygen-free condition since oxygen hinders the reaction by forming photoexcited triplets [10]. Such a process occurs only above 260 K since a random orientation of the rotation is essential for the polymerization process.

Photo-transformation takes place by [2+2] cycloaddition reaction mechanism [11], where faced double molecular bonds are broken and a four-member ring is formed as shown in Figure 1.1a.

Such a dimer structure is called "dumbbell type" polymerization. For the occurrence of the polymerization reaction, the following requirements can be summarized:

(i) The C_{60} molecules should be situated close enough to each other (an application of pressure assists this situation).
(ii) They must be rotating freely.
(iii) Their double bonds must be faced in parallel.
(iv) Certain external energy that opens the double bond must be applied (photoexcitation, thermal agitation, plasma, electron beam [EB] absorption, pressure application, etc.).
(v) A four-atom carbon ring is formed.

After forming the intermolecular bonding, the mean intermolecular length, typically 1.0 nm in fcc structure, shortens to 0.01–0.03 nm. And then, the free rotation of the molecule stops, and solubility in polar organic solvents such as toluene, xylene, hexane, and so on is lost. For the photochemical reaction between C_{60} molecules, the following reaction scheme is proposed [11]. Although the [2+2] cycloaddition of neutral C_{60} molecules is thermally forbidden due to the Woodward–Hoffmann rules, this type of reaction is photochemically allowed between an excited and a ground state molecule. The interaction of the singly occupied π^*-orbital of the photoexcited molecule with the unoccupied π^*-orbital of the ground state molecule, as well as the interaction of the singly and doubly occupied π-orbitals gives rise to a symmetrically allowed and energetically favorable transition state. The simplified orbital interactions of two C=C double

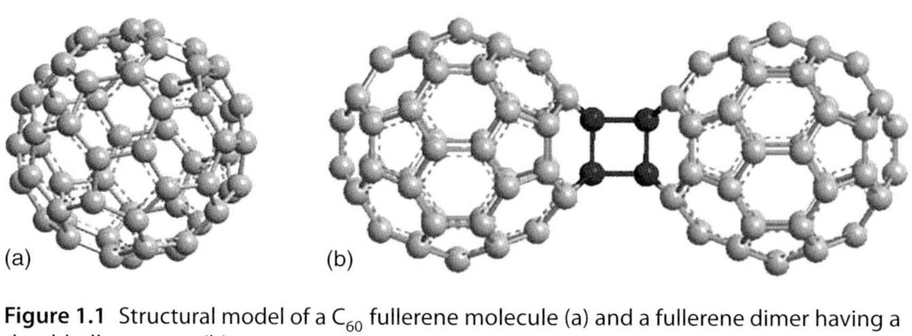

Figure 1.1 Structural model of a C_{60} fullerene molecule (a) and a fullerene dimer having a dumbbell structure (b).

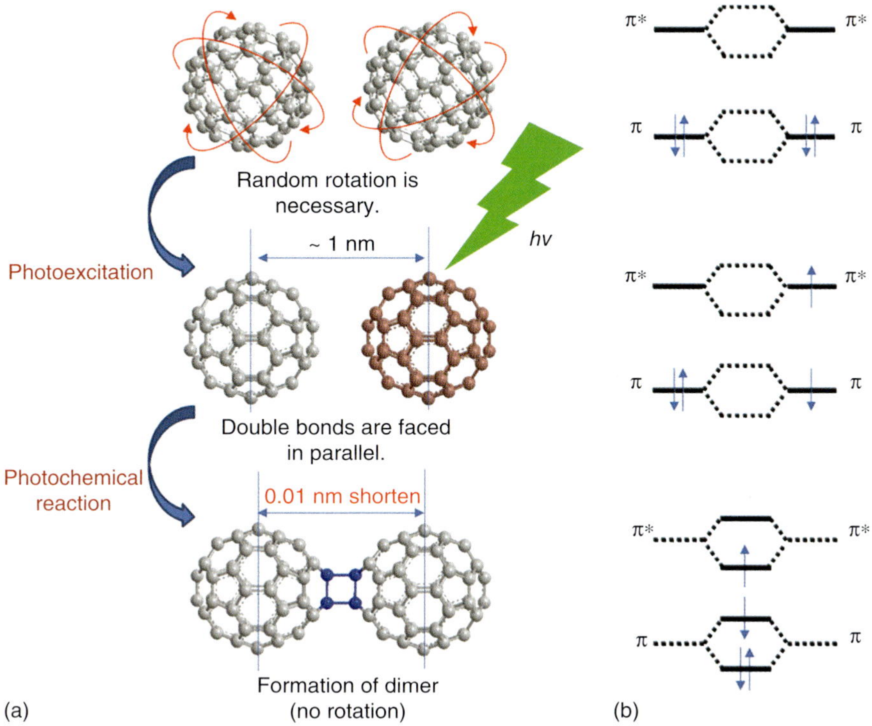

Figure 1.2 Schematic of the photochemical reaction (a) and energy diagrams of frontier orbital interactions of reactants in [2+2] cycloaddition reactions (b).

bonds are illustrated in Figure 1.2. Such a polymerization process in a solid or a thin film of C_{60} can also take place by other means: application of high pressure at high temperature, intercalation of alkali metals, plasma treatment, EB irradiation, and so on. If a negative ion reacts with a neutral ground state molecule, the interaction of the singly occupied and the vacant π^*-orbitals results in a lower energy transition state, similarly to the photochemical mechanism.

For the intermolecular bond formation, two types of possibilities are proposed. In one configuration, the two C atoms shared by the two adjacent hexagons in a C_{60} are covalently bonded to the C atoms that are shared by the two hexagons in the adjacent C_{60} (66/66 bond); in another, the two C atoms shared by a hexagon and a pentagon in a C_{60} are bonded to the adjacent C_{60} (66/65 bond) shown in Figure 1.3b. It is known that the 66/66 bond is more stable thermodynamically than the 66/65 one; therefore, the 66/65 one must be a very rare case [12, 13]. However, existence of a C_{60} polymer composed of 66/65 bonds (Figure 1.3b) is the only model to explain the metallic property of the polymer having a two-dimensional rhombohedral (rh) structure realized by high temperature and high pressure application [14].

Hence, we can control the electrical properties of C_{60} fullerene depending on the polymerized structure. In this chapter, after showing the basic electrical properties of a pristine C_{60} fullerene, various kinds of polymerization regimes and the electrical properties will be introduced.

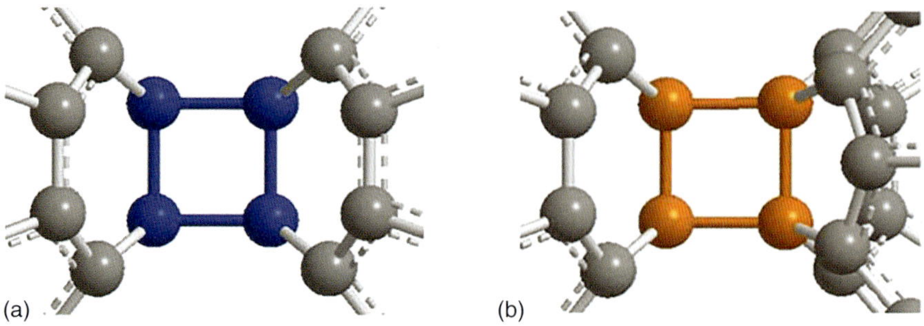

Figure 1.3 (a) 66/66 and (b) 66/65 bonding structures.

1.1.2 Electronic Property of Pristine C_{60} and n-Type FET Action

The first C_{60} field effect transistor (FET) was reported by Paloheimo et al. [15] in 1993 and developed by Haddon et al. [16] in 1995. They clarified that C_{60} FET works as an n-type transistor and exhibits fairly good carrier mobility around 0.1 $cm^2\ V^{-1}\ s^{-1}$. The highest mobility is more than 10 $cm^2\ V^{-1}\ s^{-1}$ as reported by Li et al. in 2012 [17]. The properties are severely dependent on the environment. After being exposed in air or oxygen, the conductance suddenly drops and it becomes almost insulating [7]. Therefore, most of the transport experiments are performed in vacuum conditions. The conductance also depends on temperature, so that there might be a large amount of charge trapping states, in other words donor-like states, in the pseudo gap of C_{60} FET [18]. In general, these states usually come from structural disorders at crystalline defect or grain boundary, as well as from polaronic disorders introduced by guest impurities. The latter would be an effect of oxygen adsorption, by which the conductance decreases drastically. Therefore, transport measurements of a C_{60} FET require an oxygen-free environment such as vacuum or inert gas, or covering by a passivation layer [19]. Figure 1.4 shows typical current–voltage (I–V) curve and transfer curves of a C_{60} thin film FET. The FET structure is fabricated on a SiO_2 layer of heavily doped Si wafer. The electrical contacts are performed by Au. The channel length and width are 5 and 100 μm, respectively. The field effect mobility, μ, in the low field region can be estimated from the following equation [20]:

$$\mu = \frac{I_D L}{C_{ox} W \left\{ V_{DS}(V_{GS}-V_T)-\frac{1}{2}{V_{DS}}^2 \right\}}$$

where I_D is the drain current, L is the channel length, W is the channel width, C_{OX} is the gate capacitance of oxide layer, V_{DS} is the drain voltage, V_{GS} is the gate voltage, and V_T is the threshold voltage. Considering the threshold voltage to be 15 V, the mobility is estimated as 0.1 $cm^2\ V^{-1}\ s^{-1}$.

The temperature dependence of a C_{60} FET shows nearest neighbor hopping, which depends on T^{-1} at temperatures lower than room temperature. However, below 100 K, variable range hopping (VRH) has been observed in the conductance of C_{60} FET [21]. In this transport regime, the conductance drops with decreasing temperature depending on the following relation:

$$\sigma \propto \exp(-T^{-1/4})$$

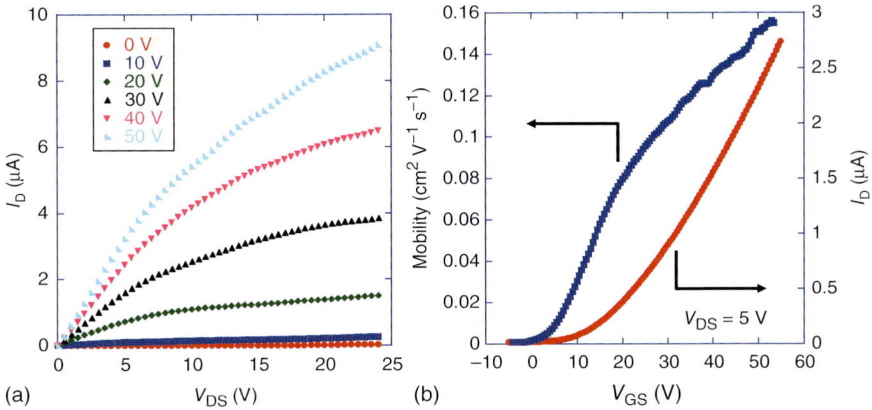

Figure 1.4 Typical transistor curves of a C_{60} thin film FET (a). Transfer curve and estimated mobility (b).

This relationship indicates that the transport is three dimensional in the thin film of C_{60} since the suffix of T can be obtained from $1/(d+1)$ where d is the dimensionality of electron transport of the system. In general, conduction carriers in VRH can transport by hopping among electronic states at Fermi level (E_F), which enables us to estimate the density of charge trapping states in the pseudo gap of C_{60} thin film.

1.2 Polymerization of C_{60} Fullerene

1.2.1 Photo-irradiation

Photopolymerization of C_{60} thin film under visible or UV irradiation was confirmed first by Rao et al. in 1993 [9]. The reaction was confirmed using UV–visible light from a Hg arc lamp or an Ar ion laser having a wavelength of 488 nm on C_{60} thin films of thickness of several hundred nanometers on a Si substrate. One typical evidence of the photopolymerization is a peak shift of $A_g(2)$ mode in the Raman scattering spectrum, which corresponds to the pentagonal pinch mode of the C_{60} molecule [22]. The peak occurs at 1469 cm^{-1} before the light irradiation and shifts to 1460 cm^{-1} after the irradiation. The mechanism of photo-transformation is based on [2+2] cycloaddition reaction [11]. A 1D polymer chain of C_{60} molecules having an orthorhombic crystal structure is proposed for this photopolymer. The polymerization reaction occurs within a temperature range between 260 and 400 K without applying pressure. The lower limit is due to the necessity of free rotation of C_{60} molecules. On the other hand, the upper limit is restricted by a depolymerization process due to an increase in thermal vibrations. Therefore, the intermolecular bonding is broken by heating to more than 473 K so that the photopolymer reversibly returns to the monomer [23]. The activation energy for the depolymerization process is approximately 1.25 eV [22]. For the photopolymerization, light having photon energy greater than the bandgap energy of the C_{60} molecules is necessary. Existence of long chains of more than 20 C_{60} molecules was confirmed by laser desorption mass spectrum from C_{60} thin film after sufficient UV–visible light irradiation [9]. Figure 1.5a shows typical Raman shift peaks of $A_g(2)$ mode of a C_{60} thin film. The peak occurs at 1469 cm^{-1} before irradiation; however, it shifts to 1460 cm^{-1} after green laser irradiation for 10 minutes. On the other hand, no significant peak shift can be confirmed; Raman peaks of a solid single crystal before and after laser light irradiation are shown in Figure 1.5b. In the case of the thin film, the peak shift occurs very easily even during the irradiation for Raman spectroscopy. In other words, it is very difficult to obtain a Raman peak of the pristine C_{60} thin film at 1469 cm^{-1}. In the case of photopolymerization, it is basically difficult to obtain long chain polymers. The polymer chain is mainly of short-range oligomers such as dimer or tetramers. The $A_g(2)$ peak position shifts to further lower frequency through progress of the polymerization reaction of C_{60}. The shoulder at 1453 cm^{-1} can be assumed as the peak corresponding to a polymer higher than a dimer (e.g. trimer). Generally, no long-range order can

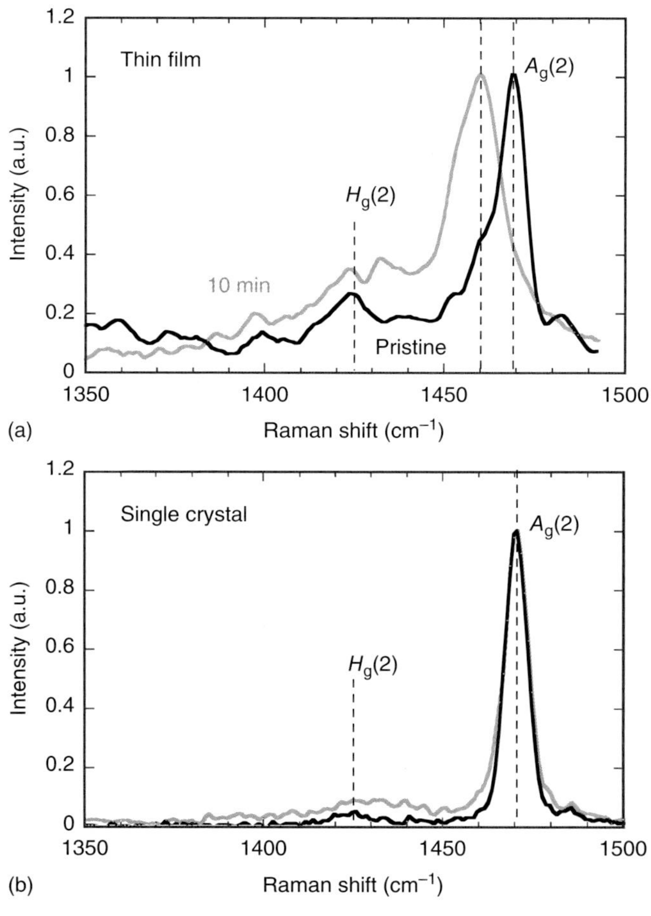

Figure 1.5 Typical peak shifts of $A_g(2)$ mode before (black) and after (gray) laser irradiation of a thin film (a) and a single crystal (b).

be observed in the photopolymer of the thin film unlike in pressure-induced polymerization.

A C_{60} photopolymer still shows semiconducting electric property [24]. The evolution of gate voltage dependences of a C_{60} thin film FET with laser light irradiation is shown in Figure 1.6. The FET sample has a C_{60} thin film of 50 nm thickness and the electric contacts occur underneath the thin film, forming a bottom contact structure. In the case of such an FET, the ON current of the FET decreases with increasing dose of the green laser beam. Consequently, a decrease in the mobility was confirmed. In this sample structure, since a laser beam was irradiated from the top of the thin film, photopolymerization takes place from the surface of the film. Therefore, at the bottom of the film, the molecules must be polymerized partially and then such a disorder affects the decrease of the mobility.

Basically, polymerization by photo-irradiation is never perfect in the bulk region of a C_{60} thin film and a solid of a single crystal since the dimerization of two C_{60} molecules hinders successive, long-chained, photo-polymerization. However, polymerization at the surface of a thin film shows different behavior. 2D-rhombohedral rinks of C_{60} molecules were confirmed on a thin film after irradiation with UV–visible light (photon energy: 2–4 eV, light intensity: 3–4 W cm^{-2}) for 400 hours at room temperature in vacuum [25]. The resistivity was of the order of 10^3 Ω cm. More recently, we also observed the formation of two-dimensional polymer films synthesized on a C_{60} thin film after sufficient irradiation with optical vortex (OV) laser light [26]. Figure 1.7 shows the atomic force microscope image of the surface of the irradiated region. A film having a thickness of 1 nm can be confirmed on the grains in a C_{60} thin film. The coverage on the surface depends on the dose of the laser light. Such a film reminds us of a 2D polymer phase having te- and rh-structures. When we placed two electrodes on the irradiated region after the irradiation, we confirmed no gate voltage dependence of current through the sample, which suggested a metallic property (sufficient amount of carriers exist in a sample without applying a gate voltage). Surprisingly, the current can be confirmed even in air since a C_{60} thin film becomes insulating immediately after being exposed in air. The mechanism is still under discussion; however, the experimental results suggest that even

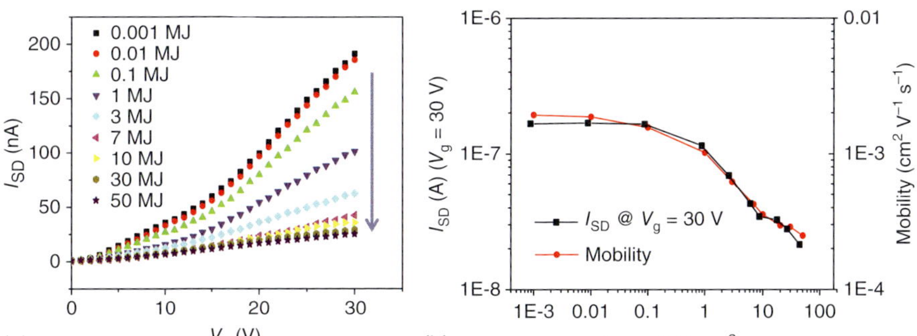

Figure 1.6 Typical effect of photopolymerization on transfer curves of a C_{60} FET (a). The mobility decreases as well as the ON current incrementing the dose (b).

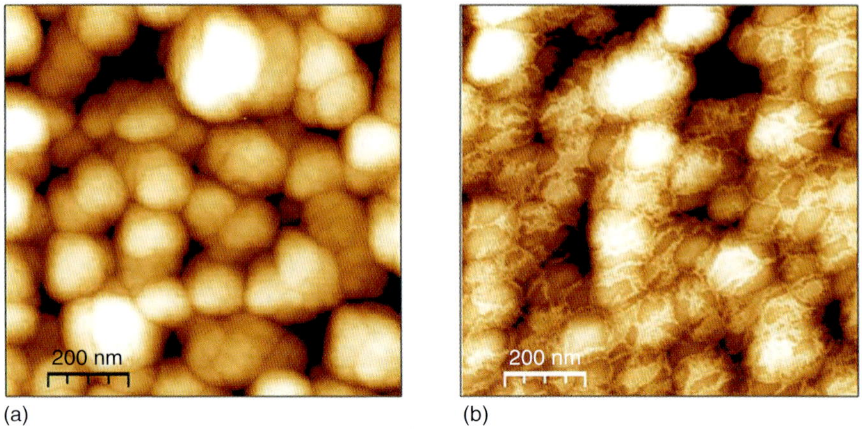

Figure 1.7 AFM image of C_{60} thin film out of (a) and within (b) the OV irradiated region.

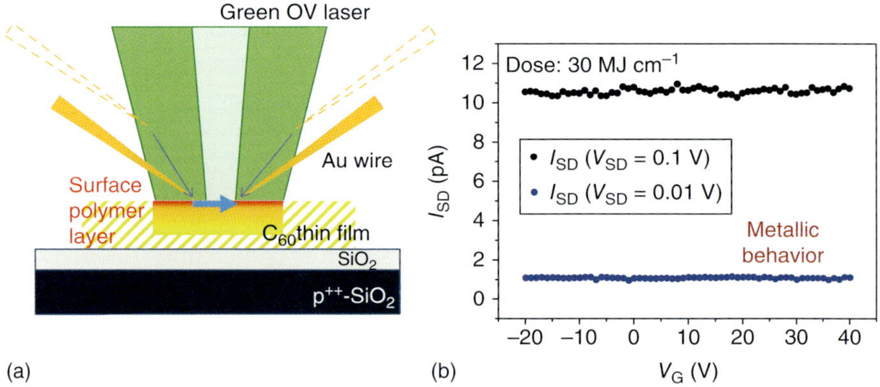

Figure 1.8 Schematic view of OV irradiation on a C_{60} thin film and the electrical measurement within the irradiated region using Au wires after the irradiation (a). Two terminal transfer curves at different bias voltages (b). No gate voltage dependence was observed.

photo-polymerization may provide a metallic 2D polymer at the surface of thin films after sufficient irradiation (Figure 1.8).

1.2.2 Doping Effect Using Alkali Metal and Superconductivity

Doping into C_{60} fullerene molecules is accompanied by 3-D intercalation of alkali, alkali earth, and rare earth metals forming compounds [27]. Owing to the intercalation, electrons can be delivered into the unoccupied band whose lattice can store up to six electrons, without forming the chemical bond unlike the conventional covalently-bonded semiconductors such as Si. A_XC_{60}, where X can be changed from one to six in stoichiometric ratio, which has different characteristics depending on the number of X. A_1C_{60} and A_3C_{60} show metallic property while the other structures have semiconducting property. Especially, A_3C_{60} compound is well known to exhibit superconducting property at low temperature. The critical temperatures of K_3C_{60} and Rb_3C_{60} are 19.8 and 30.2 K at normal pressure, respectively [28]. On the other hand, the A_1C_{60} (A = K, Rb, or Cs) compound shows different behavior. It is reported that the crystal has an orthorhombic structure and the C_{60} molecules have a covalent bond with each other one dimensionally. X-ray diffraction pattern of the material shows good agreement with the calculated atomic position of orthorhombic phase with [2+2] cycloaddition below 350 K [29]. The mechanism of the formation of intermolecular bonds is explained by ion-induced [2+2] cycloaddition [30]. Such a structure can be obtained via the following process; alkali ions such as K^+, Rb^+, and Cs^+, which are larger than the tetrahedral space of fcc C_{60} crystal, can form a doped structure with one ion per octahedral position. The compound takes a NaCl (fcc) structure at high temperatures (e.g. more than 420 K for KC_{60}). It can be balanced by Coulomb interaction between A^+ and C_{60} and a slight distortion of the crystal. The structure undergoes first-order phase transition into a body-centered orthorhombic structure. The structure is formed by the distortion of the fcc lattice without diffusing the metal ion out from the crystal. And then, [2+2] cycloaddition takes place and the one-dimensional polymer is formed. This polymerization scheme is basically similar to a process in high-pressure application; however, the dimensionality of the polymerized crystal is different. The fcc structure is not stable at room temperature and transits into an orthorhombic polymer phase. A single crystal of $(KC_{60})_N$ was grown; it was a few millimeters long and the degree of polymerization exceeded 100 000 [31]. At low temperatures such as liquid nitrogen temperature, the crystal forms a superstructure by dimerization within the 1D crystal.

1.2.3 High-Pressure and High-Temperature Application

As mentioned above, depolymerization process occurs at more than 200 °C in normal pressure; however, application of pressure with heating brings about the other phases of polymerization. Iwasa et al. reported the first polymerization with a rhombohedral (rh-phase) structure in 1994 [32]. They heated C_{60} crystals up to 1073 K by applying a pressure of 4–8 GPa. The other phases, orthorhombic (or-phase) and tetragonal (te-phase), were confirmed in latter experiments [33]. Their crystalline structures have been determined by X-ray diffraction patterns. Successive transition was observed from fcc structure to 1D- (or-phase) or 2D- (te- and rh-phase) polymerized phase by increasing the temperature and applying pressure. A mixture of rh- and te- phases is formed at 2–4 GPa and 673–1073 K, and a pure rh-phase at 4–8 GPa and the same temperatures [34]. By applying pressure, the intermolecular distance is decreased and the regularity of rotation angle is restricted; therefore, the crystalline structure changes from fcc to simple cubic (sc) phase, which is a more restricted structure. Since such a restriction of rotation angle hinders the formation of intermolecular bond by [2+2] cycloaddition process, polymerization does not progress in the sc phase. However, by increasing the temperature, the transition in the sc phase is suppressed and it returns to the fcc phase again, and then the angular disorder of the rotation also reappears. Therefore, higher order polymerization such as the te- and rh-phase can be obtained at high pressure and high temperature conditions.

Typically, more than 5 GPa and 973 K are necessary for the formation of the rh-phase [34].

Since rh-polymer phase has 2D and a hexagonal symmetry, the crystal exhibits a hexagonal or triangular shape. On the other hand, the te-phase polymer shows a square-shaped crystal. Makarova et al. show the transport properties of each polymer [35]. Both te- and rh-polymers show semiconducting characteristics; the resistance increases with decreasing temperature. However, an rh-phase sample synthesized at 7–8 GPa and 973–1073 K shows metallic property; the resistance decreases with decreasing temperature (Figure 1.9).

Okada and Oshiyama showed the results of the first principles calculation of band structures of the rh-polymer phase [36], which is composed of normal 66/66 bonds and has a bandgap at the Fermi energy as shown in Figure 1.10. Therefore, there is still room for discussion on the mechanism; however, the only clear mechanism is that of an rh-phase polymer having 66/65 bonds. Metallic band structures are suggested in both cases of ABA and ABC staking of rh-phase polymer layers of 66/65 bonds.

If a C_{60} crystal is heated at over 1273 K in 5 GPa, the cages of C_{60} molecules are broken and become amorphous carbon [37].

The peak shifts of $A_g(2)$ mode in Raman spectrum in the different polymer structures are summarized by Rao et al. [38]. However, the suggested peak assignments of the Raman shift in each phase include ambiguity. It is difficult to obtain a pure sample having a single polymer phase, as multiple phases are usually involved in it. Therefore, in the early stages of the experiments, mixed peaks were occasionally misassigned in the spectrum. The most reliable experimental result shown by Meletov et al. is that a peak wave number of $A_g(2)$ mode in pure rh-phase of C_{60} crystal appears at 1408 cm^{-1} as shown in Figure 1.11 [39]. They prepared the single crystal of rh-phase sample by applying a high pressure of 5 GPa at 773 K. By heat treatment at more than 523 K for 0.5 hour, the peak height at 1408 cm^{-1} decreased and then peaks corresponding to the 2D-te-phase and the 1D-or-phase began to appear as shown in Figure 1.11. After thermal treatment at 548 K for 0.5 hour, these peaks also disappeared and only a peak of 1469 cm^{-1} corresponding to the C_{60} monomer finally remained. Even at

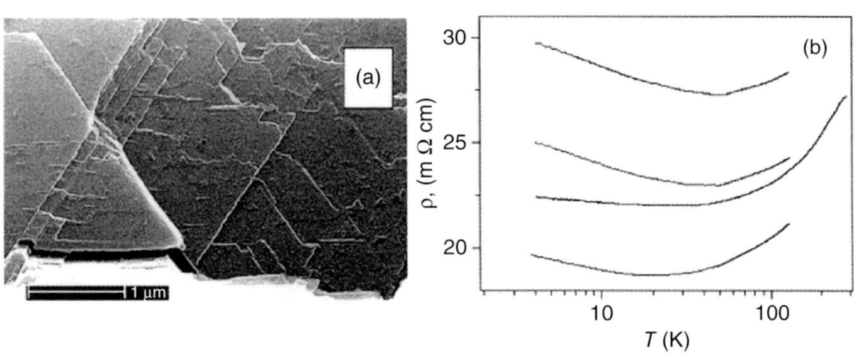

Figure 1.9 Scanning electron microscope image of polymeric C_{60} crystal having rh-phase (a) and temperature dependences of the resistivity (b) [35].

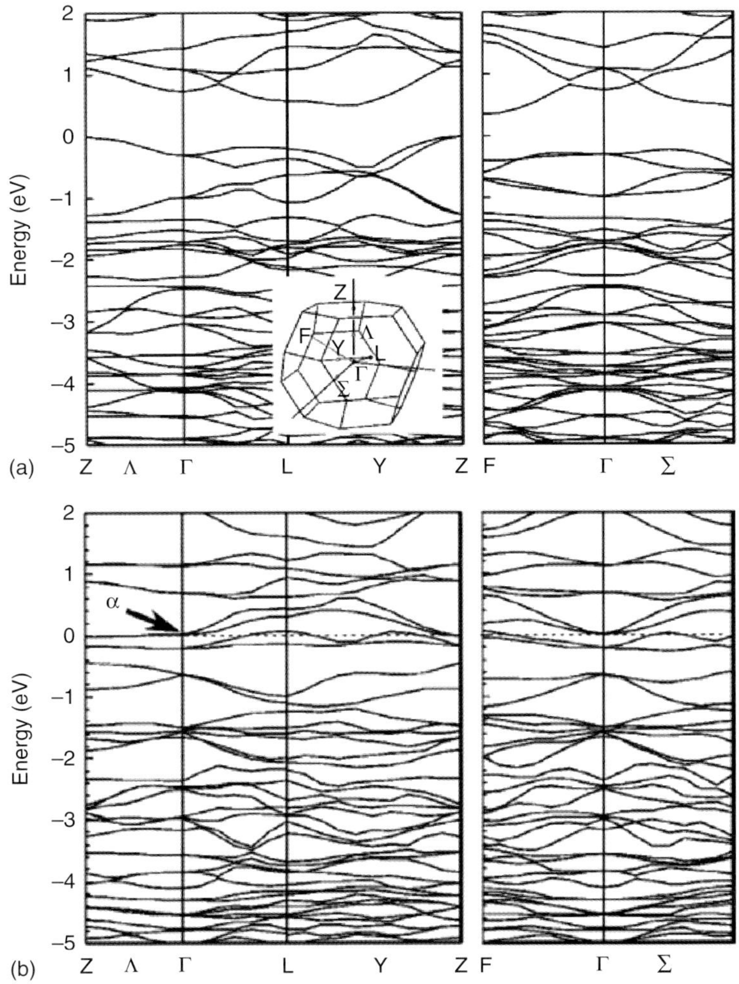

Figure 1.10 Electronic energy bands of the ABC-stacking of rh-phase C_{60} polymer crystals composed of 66/66 bond (a) and 66/65 bond (b) [36].

513 K, the polymer returned to the monomer phase when the temperature was maintained for two hours.

1.2.4 Plasma and EB Irradiation

Polymerization of C_{60} molecules occurs also by applying plasma and irradiating EB on the thin film. Maruyama sublimated C_{60} powder in argon plasma [40]. Owing to the energy of plasma, polymerization occurred. It was confirmed by a shift of $A_g(2)$ peak from 1469 cm^{-1} to a lower frequency; however, the intensity decreased with plasma power and almost diminished after being treated at 50 W. Higher order of merging of the C_{60} molecules such as peanut-shaped inter-C_{60} crosslink structures is proposed in this process.

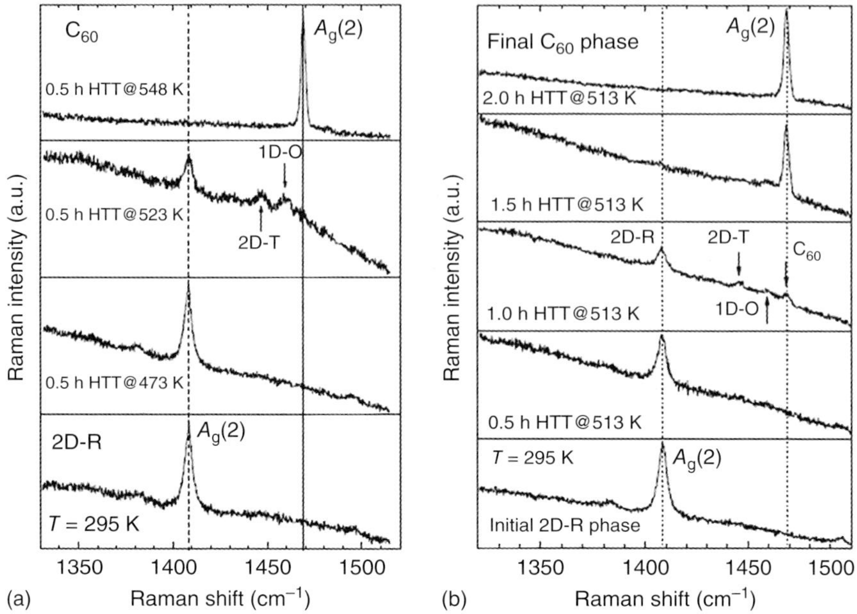

Figure 1.11 Raman spectra of $A_g(2)$ peak in the pure rhombohedral polymer measured at room temperature after heat treatment at various temperatures (a). Time dependence of the peak shift at 513 K (b) [39].

Especially, since such an EB can be focused and irradiated in a selected region, this method is proposed to use C_{60} molecules as a resist material for EB lithography. Tada et al. reported that C_{60} thin film acted as a negative e-beam resist with a sensitivity of 1×10^{-2} C cm^{-2} [41]. Although it is several orders of magnitude worse than the usual resist materials in sensitivity, it shows higher dry-tech durability than conventional novolac resists. However, the solubility in organic solvent becomes almost disappear after in the EB treatment. Therefore, how to remove the polymer in the process is still remained as one of the problems in the practical use.

1.2.5 Low-Energy EB Irradiation

Although sufficient EB irradiation may break a cage of C_{60} fullerene, low-energy EB can provide a different type of C_{60} polymer. The formation of a peanuts-type C_{60} dimer (C_{120} structure) as shown in Figure 1.12a is proposed by Onoe et al. and then a 1D chain of such a structure would be formed by EB irradiation [42]. The intermolecular bonding scheme is different from that of the usual [2+2] cycloaddition reaction. The fusion of adjacent C_{60} molecules via Stone–Wales (S–W) transition is assisted by EB irradiation. Since the S–W transition is not unique, many kinds of allotropes of peanuts-type structures are shown by first principles density functional theory (DFT) calculation with their band structures by Ohno et al. [43]. According to their results, some of the structures do not have a bandgap at the Fermi level and then they may have a metallic property.

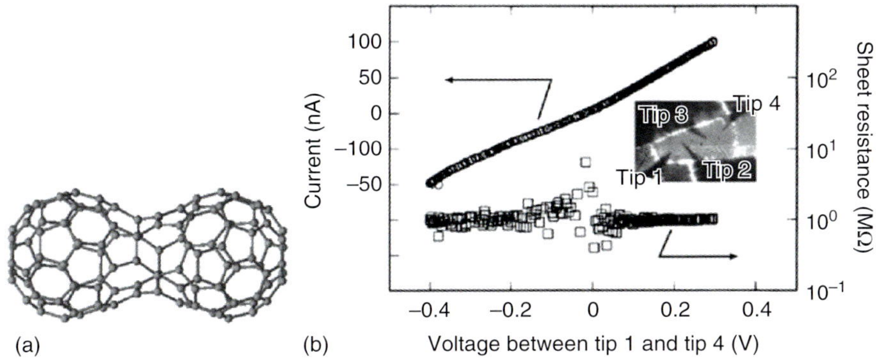

Figure 1.12 (a) Schematic illustration of peanuts-type C_{60} dimer (C_{120} structure). (b) Current–voltage curve and sheet resistance of electron-beam-irradiated sample for 20 hours [42].

Onoe et al. irradiated EB with an acceleration voltage of less than 3 keV on a C_{60} thin film in ultrahigh vacuum conditions. The polymer showed metallic transport properties, and can be explained as a pi-conjugated system. The resistivity of the EB-irradiated sample for 20 hours showed 7 Ω cm by a four-probe method at room temperature. An almost linear current–voltage curve was obtained even in air. No trace of transition into an amorphous phase was observed in the infrared spectra of the sample even after 20 hours of EB irradiation [44]. In order to confirm the electron structure of the polymer around the Fermi energy, they observed *in situ* high-resolution ultraviolet photoemission spectroscopy (UPS). The irradiation time dependence of the UPS spectrum shows increases in the density of states of the HOMO band and monotonically close to the Fermi level as well as the broadening of HOMO peaks with increase in the irradiation time (Figure 1.13). After five hours of irradiation, the density of states exists almost

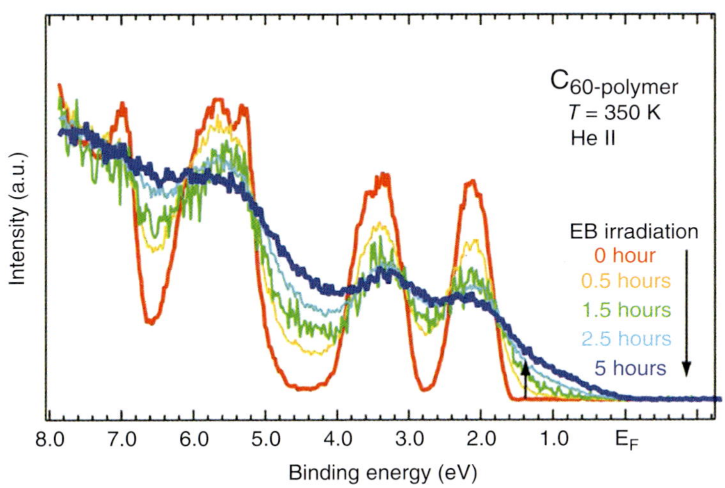

Figure 1.13 Evolution of valence photoelectron spectra of a C_{60} thin film by irradiating with low-energy electron beam as a function of irradiation time [44].

close to the Fermi level. In addition, a finite density of states at the Fermi level has been observed in the UPS spectrum after 12 hours of irradiation, which suggests a metallic property. Unfortunately, the temperature dependence of the resistance of the EB-irradiated sample increased with decreasing temperature, showing a thermal activation-type transport property. This indicates that the system is highly disordered even in the metallic state. However, the activation energy is reported as approximately 100 meV, which is almost half of that of a pristine C_{60} thin film. Therefore, a low-energy EB-irradiated thin film is more conductive than the pristine one. They confirmed that the conductivity can be maintained even in air (Figure 1.12b).

1.3 Summary

Various kinds of polymerization regimes have been studied in C_{60} fullerene. Although C_{60} fullerene thin film and crystal have an fcc structure, one-dimensional or-phase and two-dimensional te- and rh-phases have been confirmed in the polymerized structures. The intermolecular bonding is realized by [2+2] cycloaddition mechanism; therefore, the bonding can be broken by heating it over 573 K and the polymer can reversibly return to the monomer. Such an intermolecular bonding can be realized by photo-irradiation, doping with alkali or alkali earth metals, and plasma pressure application at high temperature. Although the pristine thin film FET shows semiconducting properties, it becomes insulating after exposure to air or EB irradiation. In addition, metallic properties have been observed in a C_{60} polymer synthesized by high pressure and high temperature application. Recently, such a metallic property has been confirmed on C_{60} thin films after sufficient photo-irradiation using optical vortex laser or low-energy electron beam irradiation. Very interesting experimental results have been confirmed and reported individually so far. Combining such unique characteristics, flexible integration circuits or a new class of opto-electric device might be realized by using fullerene molecules.

Acknowledgments

I thank Prof. T. Omatsu, Prof. J. P. Bird, Y. Ochiai, and Prof. T. Kato for the kind support and discussion. This work was supported by the JST-PRESTO program "Molecular technology and creation of new functions" and JSPS KAKENHI Grants No. 24656009.

References

1 Kuroto, H.W., Heath, J.R., O'Brien, C.S. et al. (1985). *Nature* 318: 162.
2 Kraetschmer, W., Lamb, L.D., Fostiropoulos, K., and Huffman, D.R. (1990). *Nature* 347: 354.
3 Hosoya, M., Ichimura, K., Wang, Z.H. et al. (1994). *Phys. Rev. B* 49: 4981.

4 Hora, J., Pánek, P., Navrátil, K. et al. (1996). *Phys. Rev. B* 54: 5106.
5 Mishori, B., Shapira, Y., Belu-Marian, A. et al. (1997). *Chem. Phys. Lett.* 264: 163.
6 Pintschovius, L., Chaplot, S.L., Roth, G., and Heger, G. (1995). *Phys. Rev. Lett.* 75: 2843.
7 Lee, C.H., Yu, G., Kraabel, B. et al. (1994). *Phys. Rev. B* 49: 10572.
8 Hamed, A., Sun, Y.Y., Tao, Y.K. et al. (1993). *Phys. Rev. B* 47: 10873.
9 Rao, A.M., Zhou, P., Wang, K.A. et al. (1993). *Science* 259: 955.
10 Eklund, P.C., Rao, A.M., Zhou, P. et al. (1995). *Thin Solid Films* 257: 185.
11 Pekker, S., Forro, L., Mihaly, L., and Janossy, A. (1994). *Solid State Commun.* 90: 349.
12 Adams, G.B., Page, J.B., Sankey, O.F., and O'Keeffe, M. (1994). *Phys. Rev. B* 50: 17471.
13 Strout, D.L., Murry, R.L., Xu, C.H. et al. (1993). *Chem. Phys. Lett.* 214: 576.
14 Okada, S. and Saito, S. (1997). *Phys. Rev. B* 55: 4039.
15 Paloheimo, J., Isotalo, H., Kastner, J., and Kuzmany, H. (1993). *Synth. Met.* 55–57: 3185.
16 Haddon, R.C., Perel, A.S., Morris, R.C. et al. (1995). *Appl. Phys. Lett.* 67: 121.
17 Li, H., Tee, B., C.-K., Cha, J.J. et al. (2012). *J. Am. Chem.Soc.* 134: 2760.
18 Horiuchi, K., Uchino, S., Nakada, K. et al. (2003). *Physica B* 329–333: 1538.
19 Horiuchi, K., Nakada, S., Uchino, S. et al. (2002). *Appl. Phys. Lett.* 81: 1911.
20 Sze, S.M. and Ng, K.K. (2007). *Physics of Semiconductor Devices*, 3e. Wiley.
21 Horiuchi, K., Uchino, S., Hashii, S. et al. (2004). *Appl. Phys. Lett.* 85: 1987.
22 Zhou, P., Rao, A.M., Wang, K.-A. et al. (1992). *Appl. Phys. Lett.* 50: 2871.
23 Iwasa, Y., Tanoue, K., Mitani, T., and Yagi, T. (1998). *Phys. Rev. B* 58: 16 374.
24 Dzwilewski, A., Wågberg, T., and Edman, L. (2007). *Phys. Rev. B* 75: 75203.
25 Onoe, J., Nakayama, T., Aono, M., and Hara, T. (2004). *J. Appl. Phys.* 96: 443.
26 Aoki, N., Omatsu, T. et al., to be published in elsewhere.
27 (a) Gunnarson, O. (1997). *Rev. Mod. Phys.* 69: 575. (b) Zhao, W.B., Chen, J., Wu, K. et al. (1994). *J. Phys.: Condens. Matter* 6: 631.
28 Xiang, X.-D. et al. (1992). *Science* 256: 1190.
29 Chauvet, O., Oszlanyi, G., Forro, L. et al. (1994). *Phys. Rev. Lett.* 72: 2721.
30 Pakker, S. et al. (1994). *Solid State Comm.* 90: 349.
31 Pekker, S., Janossy, A., Mihaly, L. et al. (1994). *Science* 265: 1077.
32 Iwasa, Y., Arima, T., Fleming, R.M. et al. (1994). *Science* 264: 1570.
33 Núñez-Regueiro, M., Marques, L., Hodeau, J.-L. et al. (1995). *Phys. Rev. Lett.* 74: 278.
34 Davydov, V.A., Kashevarova, L.S., Rakhmanina, A.V. et al. (1997). *Carbon* 35: 735.
35 Makarova, T.L., Sundqvist, B., Scharff, P. et al. (2001). *Carbon* 39: 2203.
36 Okada, S. and Oshiyama, A. (2003). *Phys. Rev. B* 68: 235402.
37 Rao, C.N.R., Govindaraj, A., Aiyer, H.N., and Seshadri, R. (1995). *J. Phys. Chem.* 99: 16814.
38 Rao, A.M., Eklund, P.C., Hodeau, J.-L. et al. (1997). *Phys. Rev. B* 55: 4766.
39 Meletov, K.P., Arvanitidis, J., Christofilos, D. et al. (2010). *Cabon* 48: 2974.
40 Maruyama, R. (2007). *J. Molec. Struct.* 10–17: 831.
41 Tada, T. and Kanayama, T. (1997). *J. Photopolymer Sci. and Technol.* 10: 647.

42 Onoe, J., Nakayama, T., Aono, M., and Hara, T. (2003). *Appl. Phys. Lett.* 82: 595.
43 Ohno, K., Noguchi, Y., Ueda, S., and Onoe, J. (2007). *Eur. Phys. D* 43: 137.
44 Onoe, J., Ito, T., and Kimura, S. (2008). *J. Appl. Phys.* 104: 103706.

2

Flapping Molecules for Photofunctional Materials

Shohei Saito

Kyoto University, Department of Chemistry, Graduate School of Science, Kitashirakawa-Oiwake, Sakyo, Kyoto 606-8502, Japan

2.1 Introduction

2.1.1 Motivation

2.1.1.1 Hybridization of Rigidity and Flexibility

π-Conjugated molecules have been utilized in a wide range of applications. Starting with dyes, pigments, and aromatic polymers, in recent years π-conjugated molecules have become indispensable as building blocks for organic electronics materials, such as organic field-effect transistors (OFETs) [1, 2], organic photovoltaics (OPVs) [3, 4], and organic light-emitting diodes (OLEDs) [5, 6]. In addition, π-conjugated units are important components of fluorescent probes [7, 8], molecular machines [9, 10], self-assembled supramolecules [11–13], and porous frameworks [14–17]. Most of these applications rely on the merit of the *rigidity of skeleton* originally possessed by π-conjugated molecules (Figure 2.1). For example, since porphyrin and fullerene have small reorientation energies due to their structural rigidity [18], these molecules are often used for construction of electron-transfer systems. Perylene shows bright luminescence with high fluorescence (FL) quantum yield because its molecular vibration is suppressed by the polycyclic structure, and nonradiative relaxation from the photoexcited state is slow [19]. Likewise, triphenylene and hexabenzocoronene have a flat and circular shape; therefore these compounds are easily stacked [20–22], and triphenylene has been actually used in industry for discotic liquid crystal (LC) materials [23]. Moreover, in the construction of molecular assembly structures by metal coordination and hydrogen bonds, aromatic units are useful because a variety of established cross-coupling reactions are available for the synthesis of the building blocks [24] and because their conformational rigidity leads to structurally defined molecular assembly [25, 26]. However, from another point of view, the inherent rigidity of π-skeletons suggests difficulty in producing convertible photophysical properties derived from the flexibility of molecular structures. If conformational flexibility is effectively introduced to rigid π-conjugated molecules consisting of aromatic rings, new "molecular technologies" would be exploited by using dynamic motions of the synthetic π-skeletons.

Molecular Technology: Materials Innovation, Volume 3, First Edition.
Edited by Hisashi Yamamoto and Takashi Kato.

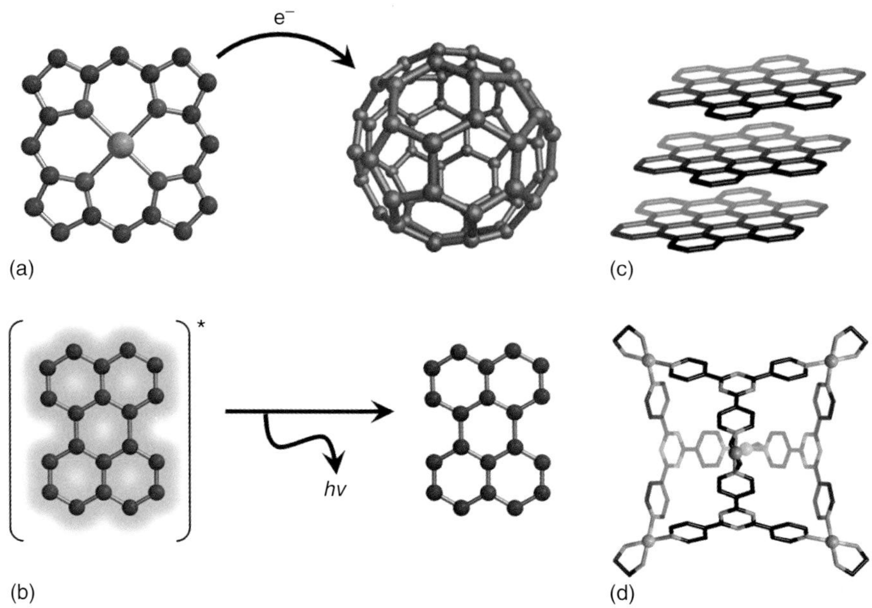

Figure 2.1 Merits of molecular rigidity in materials science: (a) small reorientation energy in electron transfer, (b) high fluorescence quantum yield owing to slow nonradiative decay process, (c) stacking ability in construction of columnar liquid crystals and supramolecules, and (d) easily synthesized building units for construction of structurally defined molecular assembly.

In other words, construction of a hybrid π-electron system that combines rigidity and flexibility will explore new functional molecules and materials, for example, sophisticated systems featuring luminescent properties and molecular stacking ability derived from the *structural rigidity* and, at the same time, convertible photophysical properties originating from the *conformational flexibility.*

2.1.2 Background

2.1.2.1 How to Change Photophysical Properties by Changing Conformation of Molecules

Conformational flexibility is the key to giving rise to dynamic functions; however, not every flexible π-conjugated molecule shows convertible photophysical properties. In general, conformational change of π-conjugated molecules does not necessarily result in significant conversion of photophysical properties. To create a characteristic system in which photophysical properties change in conjunction with the conformational change, some sort of ingenuity is necessary in molecular design. Photochromic compounds such as azobenzene [27–31], diarylethene [32, 33], and spiropyran [34] have been reported as successful examples to switch photophysical properties and to induce mechanical motions in films and crystals. These photoresponses originate from photoisomerization reactions involving recombination of chemical bonds. A well-known example of the conformationally flexible functional system is the twisted intramolecular charge transfer (TICT) molecules that are composed of electron donor and acceptor units connected with a rotatable single bond [35, 36]. Photoexcitation of TICT molecules induces intramolecular electron transfer from the donor to the acceptor.

In this chapter, these photochromic compounds and TICT systems are not highlighted, since other reviews can be referred [27–36]. On the other hand, conformational flexibility of parent π-conjugated molecules without bond breaking or bond forming is focused on, which can be found in the following representative examples (Figure 2.2): (i) inversion of tub-shaped cyclooctatetraene (COT) [37, 38]; (ii) inversion of bowl-shaped π-systems such as corannulene [39, 40] and sumanene [41, 42], which are substructures of the fullerene skeleton; (iii) flipping of helical π-systems such as helicenes [43] and twisted perylene bisimides [44]; and (iv) Hückel–Möbius topological switch of expanded porphyrins [45–47], which is a flexible macrocyclic π-conjugated system. Practical application of these conformational flexibilities to produce useful functions is still challenging. In particular, COT and expanded porphyrins are characteristic π-electron systems among them, in which the electronic structure significantly depends on their molecular conformation because of the emergence of specific electronic states such as $4n\pi$ Hückel antiaromaticity [48–51] and $4n\pi$ Möbius aromaticity [52–54] in the ground state, and $4n\pi$ Baird aromaticity in the excited state [55–59].

2.1.3 Flapping Fluorophore

With the background mentioned above, a new hybrid π-system bearing rigidity and flexibility has been designed, where two luminescent anthraceneimide moieties are fused as rigid wings on opposite sides of a flexible COT ring (Figure 2.3). A series of the rigid–flexible hybrid π-systems are named FLAP

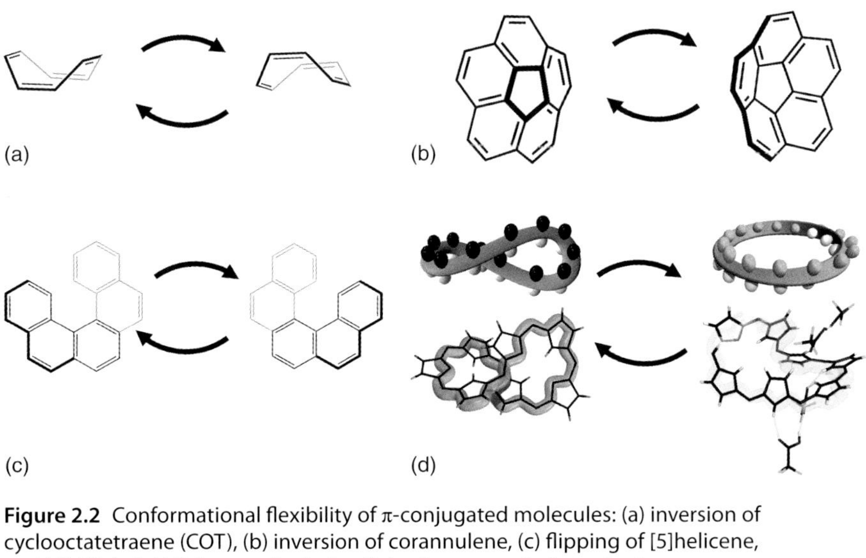

Figure 2.2 Conformational flexibility of π-conjugated molecules: (a) inversion of cyclooctatetraene (COT), (b) inversion of corannulene, (c) flipping of [5]helicene, and (d) Hückel–Möbius topological switch of an expanded porphyrin.

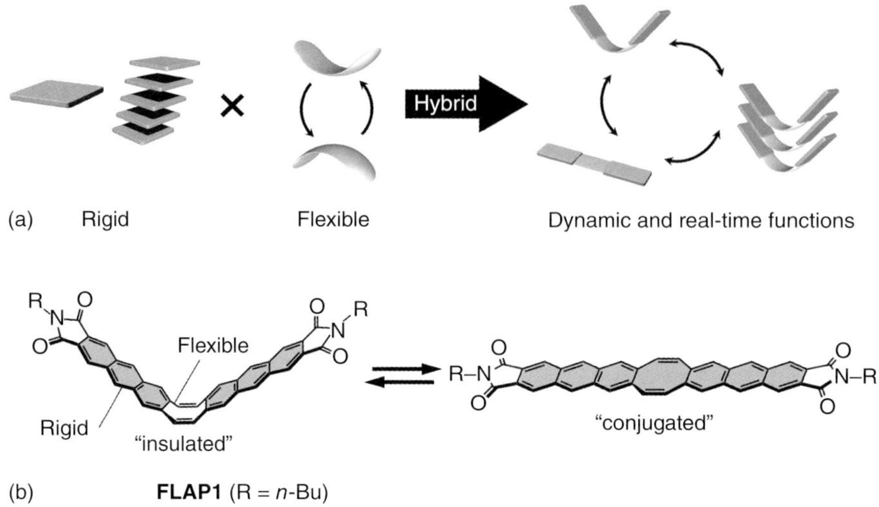

Figure 2.3 (a) Hybridization of rigidity and flexibility to design FLAP molecular systems and (b) an example of FLAP molecule (**FLAP1**) composed of rigid anthraceneimide wings and a flexible cyclooctatetraene joint.

(that is an abbreviation of *flexible aromatic photofunctional* systems) [60–63]. **FLAP1** is the first fluorophore that can emit an environment-dependent RGB (red, green, and blue) luminescence as a single component without changing the excitation wavelength [60]. That is, it shows green fluorescence (FL) in solution, blue FL in polymer thin film, and red FL in crystal (Figure 2.4). Although there are examples of multiluminescent materials combining multiple luminophores [64] and of single-component organic materials that can emit three-color luminescence by an equilibrium among fluorescent species [65] or by careful control of intermolecular interactions [66–70], **FLAP1** has a different mechanism of multiple emission. Conventional mechanisms of luminescence color changes are typically based on an ON/OFF switching of the following photochemical processes by external stimuli: [71–73] intramolecular charge transfer including TICT mechanism [35, 36], excited-state intramolecular proton transfer (ESIPT) [74, 75], fluorescence resonance energy transfer (FRET) [76, 77], excimer/exciplex formation [78, 79], J-aggregation [80], aggregation-induced emission (AIE) by restriction of intramolecular rotation [81], phosphorescence through intersystem crossing [82, 83], and *d*-orbital interaction of metal complexes especially with Au···Au [84, 85] and Pt···Pt [86, 87] contacts in the excited state. Recently, thermally activated delayed fluorescence (TADF) [6, 88] and photon upconversion fluorescence via triplet–triplet annihilation (TTA-UC) [89, 90] have also attracted attention as unique FL mechanisms. In contrast, the mechanism of the dual fluorescence of **FLAP1** is different from these cases. The dual fluorescence of **FLAP1** arises from a conformational change of a single π system that is not accompanied by charge transfer, the excited-state dynamics of which is rather similar to that of dibenzo[*b*,*f*]oxepine [91] and *N*,*N*′-disubstituted dihydrodibenzo[*a*,*c*]phenazines [92, 93].

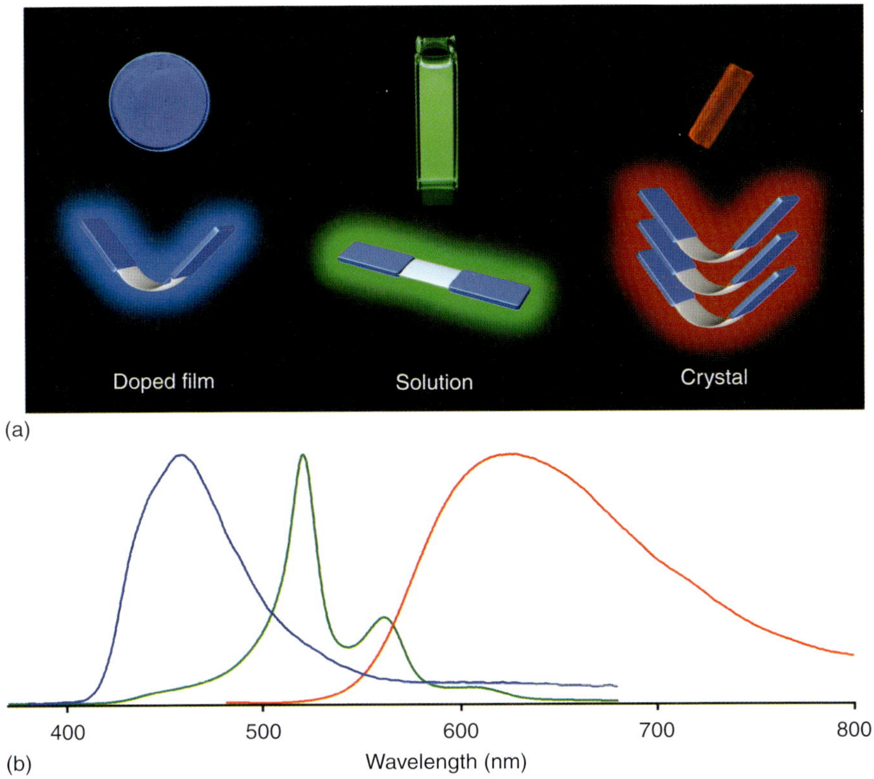

Figure 2.4 Single-component environment-dependent RGB luminescence of **FLAP1**. (a) Photographs and emissive molecular structures. (b) Fluorescence spectra of **FLAP1** in each environment. Blue fluorescence in a doped polymer film (left), green fluorescence in organic solvents (middle), and red fluorescence of single crystals (right).

FLAP1 is most stable in the V- and Λ-shaped conformations at the S_0 ground state, and the energy barrier of the conformational inversion is estimated to be lower than 10 kcal mol^{-1}; therefore, it undergoes fast flapping behavior at room temperature. In the planar conformation, π-electron delocalization is effectively extended and the $S_0 \rightarrow S_1$ excitation energy becomes significantly smaller than that in the V-shaped conformation. Consequently, the planar conformation is energetically more stable in S_1 (Figure 2.5). An excited molecule of **FLAP1** changes its conformation in solution from a V-shaped form into a planar form in S_1, emitting green fluorescence. In contrast, the molecule cannot relax its conformation in S_1 under highly viscous media such as in polymer matrix and in frozen glass, and blue fluorescence is emitted from the almost retained V-shaped conformation. In other words, **FLAP1** shows viscosity-dependent dual fluorescence in dispersed conditions (see Section 2.2) [63]. On the other hand, this molecule can form a twofold columnar stacking structure in spite of its nonplanar structure (Figure 2.6). In the crystalline phase, **FLAP1** shows excimer-like red FL because of intermolecular interaction in the excited state. Even in the kinetically formed amorphous phase, the solid-state FL is observed

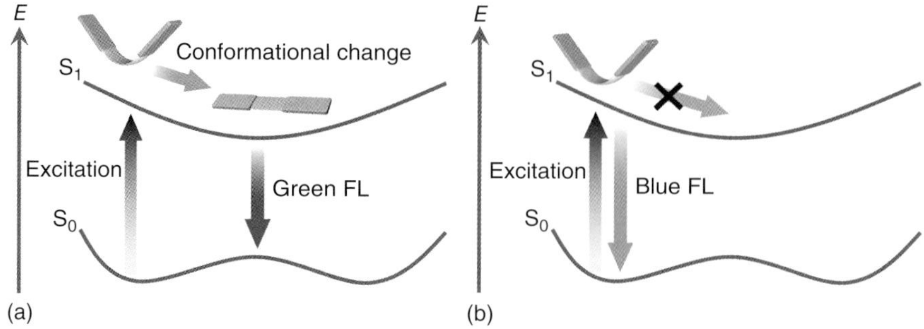

Figure 2.5 Rough sketch of the energy diagram of **FLAP1**. (a) Green fluorescence in low-viscosity solution and (b) blue fluorescence in highly viscous media.

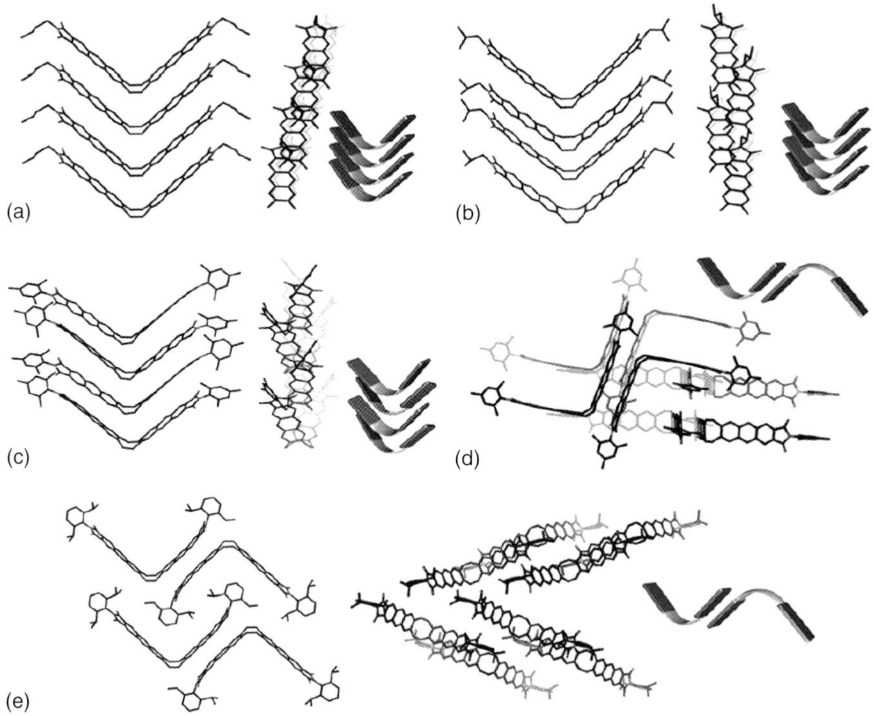

Figure 2.6 A variety of crystal-packing structures of **FLAP1** and its derivatives with different substituents. (a) Crystal packing of **FLAP1** (R = *n*-butyl) emitting an excimer-like red fluorescence, and of FLAP derivatives with the substituent of (b) R = isobutyl, (c) R = mesityl containing no solvent molecules in the unit cell, (d) R = mesityl containing octane molecules in the unit cell, and (e) R = 2,6-diisopropylphenyl (**FLAP2**) containing THF and *i*-PrOH molecules in the unit cell.

as a yellow to orange emission; hence, aggregation can be instantly identified by the FL color change. Depending on the bulkiness of the terminal substituents, derivatives of **FLAP1** form a variety of crystal packing structures. On the basis of FL spectral analysis on these crystals, it has turned out that the remarkable red shift observed in crystals of **FLAP1** bearing *n*-butyl substituents (λ_{em} = 625 nm, showing a 166-nm red shift from its blue monomer emission) originates from the unique twofold columnar stacking structure of the V-shaped molecules [61].

2.2 Viscosity Imaging Technique

Viscosity imaging using designed fluorescent probes is a useful technique because it can map the local viscosity of heterogeneous media. Although the local viscosity estimated with a molecular probe does not simply correspond to bulk viscosity measured by viscometer and rheometer, these properties are significantly related in homogeneous solvents, especially Newtonian fluids. Spatial heterogeneity mapping is an important advantage for molecular viscosity probes compared to these instruments [94–102].

2.2.1 Molecular Design of Chemical Viscosity Probes

Molecular rotors have long been recognized as representative viscosity probes [103–106]. In highly viscous media, the dynamic internal rotation in S_1 is suppressed, so that the fluorescence quantum yield of fluorogenic molecular rotors increases (Figure 2.7). These fluorophores have been applied to the study of microenvironments in polymeric materials [101, 102, 107–113], artificial membranes, and cell organelles [114–120], and recently in an interface of

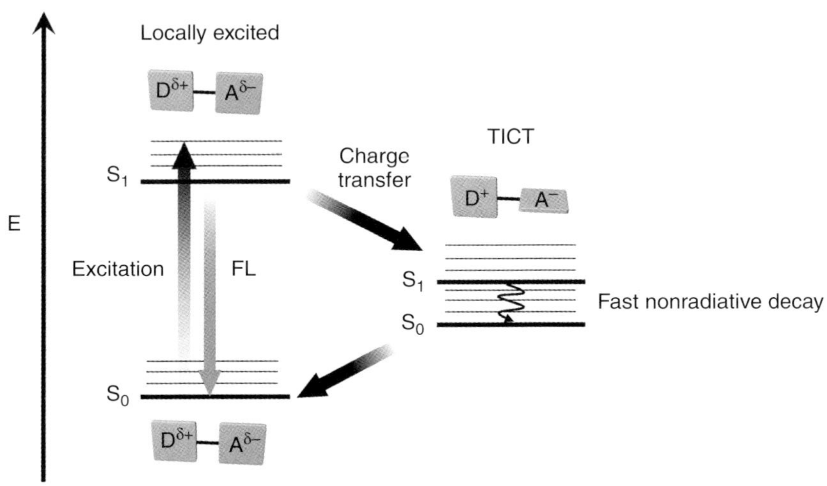

Figure 2.7 Excited-state dynamics of fluorogenic viscosity probe. Fluorescence quenching by fast nonradiative decay in a twisted conformation, which is suppressed in highly viscous media with enhancement of the fluorescence quantum yield.

objects [121]. Later, quantitative measurements of the local viscosity have been developed using the *ratiometric fluorescence technique* [122–124], which can estimate the local viscosity based on an intensity ratio of two FL bands of different wavelengths. In recent years, *fluorescence lifetime imaging* (FLIM) [125–135] has also become an important alternative to quantitative viscosity analysis in which viscosity-dependent FL lifetime of probe molecules is measured to map cellular and materials heterogeneity. In addition to the advancement of viscosity imaging techniques, various sophisticated probe molecules showing viscosity-sensitive FL with low polarity dependence [136, 137] have been developed and widely used, such as probe molecules based on CVJ (cyanovinyljulolidine), BODIPY (boron-dipyrromethene), and Cy (cyanine) structures (Figure 2.8). However, the structural design of these viscosity probes still relies on molecular rotors characterized by intramolecular rotational dynamics in S_1, including bond twists resulting in trans–cis photoisomerization [138–144]. Flapping fluorophores **FLAP2** and **FLAP4** in Figure 2.10 show polarity-independent ratiometric FL characteristics as a series of novel viscosity probes [63]. The molecular design is different from the molecular rotor, but it is based on the hybridization of a flexible eight-membered ring (COT) and rigid fluorescent aromatic wings. It is worth noting that viscosity probes with different structural motifs have also been reported recently, for example, phosphorescent molecular butterfly bearing a Pt–Pt contact [145] and amine-functionalized benzoborole showing B—N bond-cleavage-induced intramolecular charge transfer (BICT) [146].

2.2.2 Flapping Viscosity Probe

2.2.2.1 Synthesis

Outlines for the synthesis of the FLAP series are shown in Figures 2.9 and 2.10 [63]. The key step is the reaction to form an eight-membered ring (COT) at the center of the FLAP molecules. In an earlier scheme, photochemical rearrangement of 9,10-dihydro-9,10-ethenoanthracene [147] was used for this purpose, but large-scale synthesis is hampered because the reaction concentration cannot be increased for the photoreaction. Furthermore, starting from the precursor

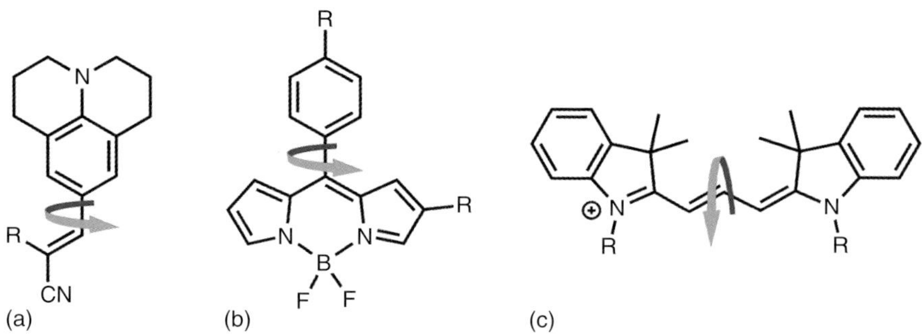

Figure 2.8 Molecular rotors based on (a) CVJ (cyanovinyljulolidine), (b) BODIPY (boron-dipyrromethene), and (c) Cy (cyanine) structures. These molecular systems act as viscosity probes by sensing the local environment through the internal rotation indicated by arrows.

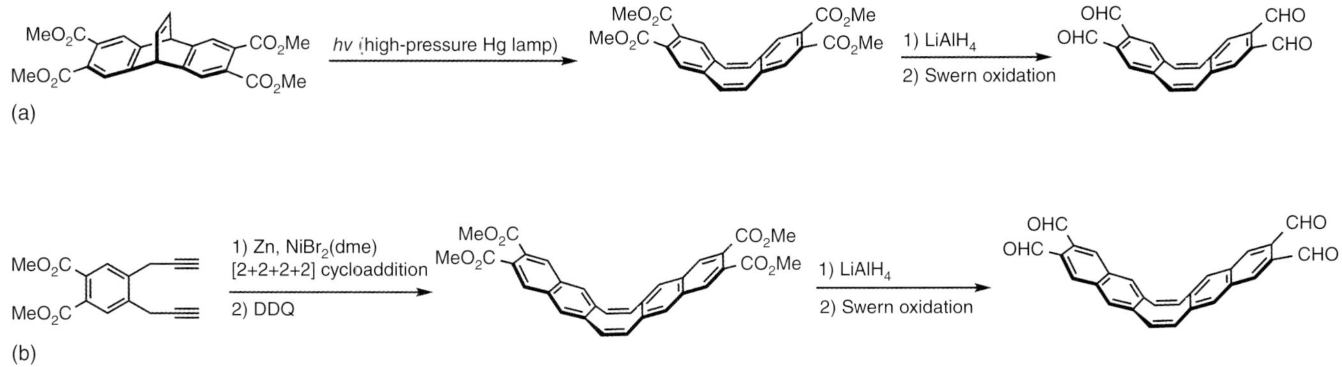

Figure 2.9 Synthetic approaches to form the COT moiety (8-membered ring) in the FLAP derivatives. (a) Photochemical rearrangement of substituted 9,10-dihydro-9,10-ethenoanthracene and (b) Ni(0)-catalyzed [2+2+2+2]cycloaddition of substituted 1,2-di(2-propynyl)benzene.

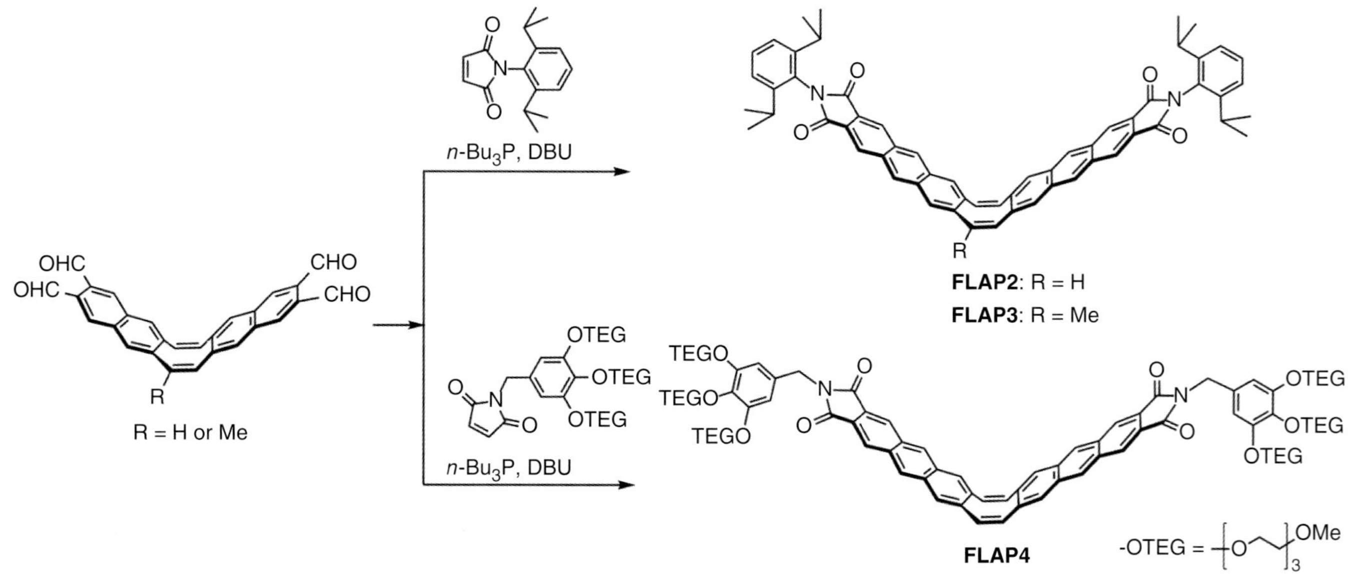

Figure 2.10 Synthesis of anthraceneimide-based FLAP compounds with a variety of terminal substituents.

obtained, subsequent acene elongation reaction [148, 149] must be performed two times to form the anthracene wings of the FLAP derivatives [61]. Then, a new synthetic scheme has been established using nickel-catalyzed [2+2+2+2] cycloaddition [150, 151] in the step of the COT formation. Through this reaction, tetraaldehyde of a COT-fused naphthalene dimer, a useful precursor for the synthesis of the FLAP series, can be obtained in gram scale (Figure 2.9b). Acene elongation reaction using this precursor affords FLAP compounds with a variety of terminal substituents (Figures 2.6 and 2.10).

2.2.2.2 Fluorescence and Excited-State Dynamics

FLAP2 shows intense green FL at around 520, 560, and 610 nm in common organic solvents. Fluorescence quantum yield (Φ_f) is acceptable, for example, $\Phi_f = 0.34$ in tetrahydrofuran (THF). A large structural change of **FLAP2** in S_1 is suggested by a large Stokes shift (4580 cm^{-1}). A weak broad FL band is also observed in the blue region (450–500 nm) of the FL spectrum at room temperature. In a frozen glass of 2-methyl tetrahydrofuran (MTHF) media at 77 K, the green emission band disappeared and the blue emission bands increased its intensity (Figure 2.11). Importantly, **FLAP3**, which has an extra methyl group at the COT ring, shows a fluorescence behavior distinct from **FLAP2**. Namely, **FLAP3** displays blue FL at around 440 and 465 nm ($\Phi_f = 0.28$ in THF). Stokes shift of **FLAP3** (1200 cm^{-1}) is much smaller than that of **FLAP2**, indicating that the bent-to-planar conformational change in S_1 is remarkably suppressed by the small steric hindrance on the central COT ring. In other words, excited-state

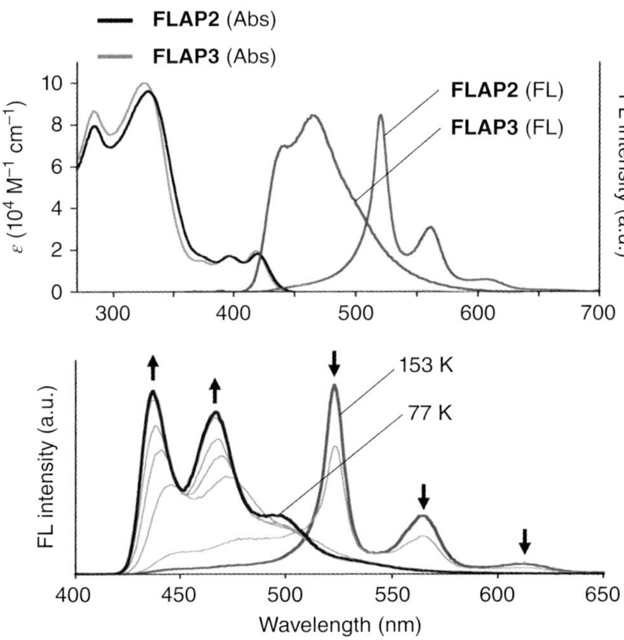

Figure 2.11 (a) Absorption and fluorescence spectra of **FLAP2** and **FLAP3** and (b) the spectral change of **FLAP2** in fluorescence at lower temperature.

dynamics of the FLAP systems are based on the conformational flexibility of the COT ring. The calculated energy profiles are reasonable for an explanation of the distinct FL properties of **FLAP2** and **FLAP3**. The corresponding model structures without terminal bulky substituents are optimized using density functional theory (DFT) for S_0 and time-dependent (TD) DFT for S_1. As mentioned above, V-shaped conformation is the most stable geometry in S_0, and the structural features are consistent with those of X-ray crystal structures. Owing to the bent conformation, bond alternation around the *cis*-olefins of the COT ring is explicit, in which the two anthraceneimide wings are not efficiently π conjugated. DFT calculations indicate that the S_0 energy level of both **FLAP2** and **FLAP3** gets higher as the conformation becomes more planar, while higher inversion behavior in S_0 is estimated for **FLAP3** between V- and Λ-shaped conformers. S_1 energy diagrams of **FLAP2** delineated using TD-DFT constrained optimization suggested two different energy minimum points with a shallow V-shaped conformation and a planar one. The energy barrier between them is small enough to get over at room temperature, and the planar conformation is slightly more stable in S_1 (Figure 2.12).

In this context, the intense green FL at 520 nm of **FLAP2** can be assigned to the emission from the planar conformation, while the weak blue FL observed in 450–500 nm can be assigned to the emission from the shallow V-shaped form. On the basis of the molecular orbitals analysis, less effective π conjugation through the COT joint is confirmed in the shallow V-shaped conformation, while effective π-electron delocalization is expected in the planar conformation. Electronic configuration of S_1 has changed during conformational planarization, which accounts for the green FL. In sharp contrast, the S_1 energy diagram of

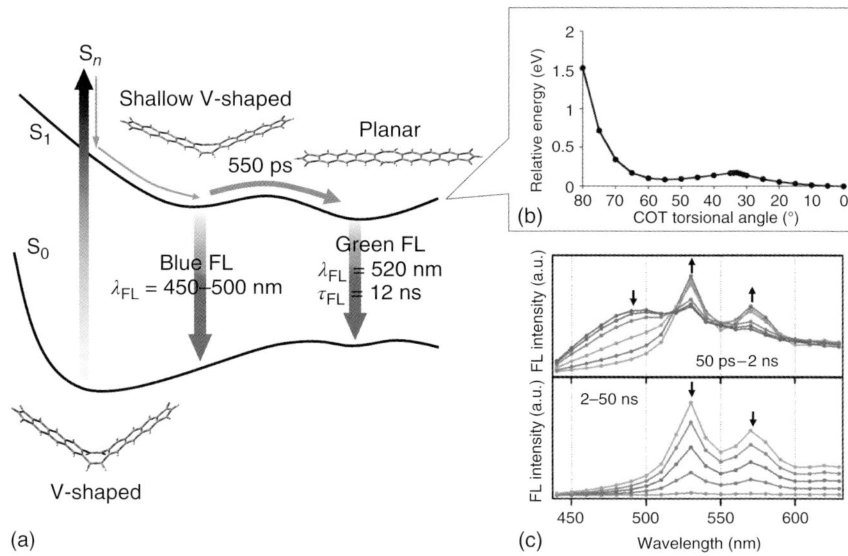

Figure 2.12 Energy diagram of **FLAP2** in S_0 and S_1. (a) Relaxed scan profile of **FLAP2** in S_1 at the cam-B3LYP/6-31+G* level, in which the terminal substituents are replaced by hydrogen atoms and (b) time-resolved fluorescence spectroscopy of **FLAP2**.

FLAP3 provided a single minimum with a V-shaped conformation, suggesting that the methyl group on the COT ring induces a steric hindrance for the conformational planarization in S_1 (as well as in S_0). It is reasonable that the blue FL of **FLAP3** with smaller Stokes shift is assigned to the emission from the V-shaped structure. It should be noted that the FL behavior of **FLAP2** is similar to that of dibenzo[*b,f*]oxepin [91], which was focused on for its possible excited-state 8π aromaticity in S_1 [55].

Conformational dynamics of **FLAP2** in S_1 has been supported by time-resolved FL and IR spectroscopies [63]. Time-resolved FL spectra of **FLAP2** in solution show a broad emission band around 450–500 nm from the shallow V-shaped conformation immediately after the pulse excitation (Figure 2.12c). This broad band gradually decreases with the time constant of 550 ps, and the intense emission bands from the planar conformation concomitantly increase around 530 and 570 nm, whose decay constant has been observed with 12.5 ns. Time-resolved IR spectroscopy of **FLAP2** has been also reported to show good agreement with this result. Through these time-resolved spectroscopies, the dynamics of **FLAP2** in S_1 has been elucidated as shown in Figure 2.12. The time constant of 550 ps is longer than the time constants of structural changes in most of photoresponsive molecules, for example, trans–cis isomerization of stilbene ($\tau \sim 1$ ps) [152]. In other words, the conformational change of **FLAP2** in S_1 is relatively slower, which is consistent with the presence of an energy barrier in the process of the structural change from V-shaped to planar conformation in S_1.

2.2.2.3 Polarity-Independent Viscochromism

Quantitative study of the viscochromism in the FLAP system has been demonstrated (Figure 2.13). A mixed solvent of dimethyl sulfoxide (DMSO)/glycerol supplies suitable Newtonian fluid for this purpose, that is, the viscosity of media is independent of the shear rate [153]. By changing the ratio of DMSO/glycerol, the media viscosity can be converted without changing the polarity. Owing to the poor solubility of **FLAP2** in DMSO/glycerol, hydrophilic **FLAP4** has been used for this measurement. Importantly, FL spectra of **FLAP4** as well as **FLAP2** are independent of media polarity. The green FL band shows a negligible shift within 520–526 nm in nonpolar to polar organic solvents such as toluene, THF, dichloromethane (DCM), dimethyl formamide (DMF), acetonitrile (MeCN), and DMSO. Using these solvents, a wide range of relative dielectric constants ε_r is covered from 2.4 to 46.7 at room temperature, while media viscosity is confined within a small range of 0.4–2.2 cP. The FL spectral features of **FLAP4** are well preserved despite the wide range of relative dielectric constants, suggesting that the emissive structures in S_1 are not charge polarized. The difference between excited-state and ground-state dipoles at the vertical geometry ($\Delta\mu$) of **FLAP2** is 1.30 D in the planar conformation and 1.10 D in the shallow V-shaped conformation at the S_1 energy minima. These $\Delta\mu$ values are significantly small when compared to the typical dual fluorescent systems having electron donor and acceptor moieties.

On the other hand, a remarkable viscosity dependence of **FLAP4** in FL has been confirmed simply by changing the volume ratio of DMSO/glycerol. With increasing viscosity, the relative intensity of the blue FL band (450–500 nm)

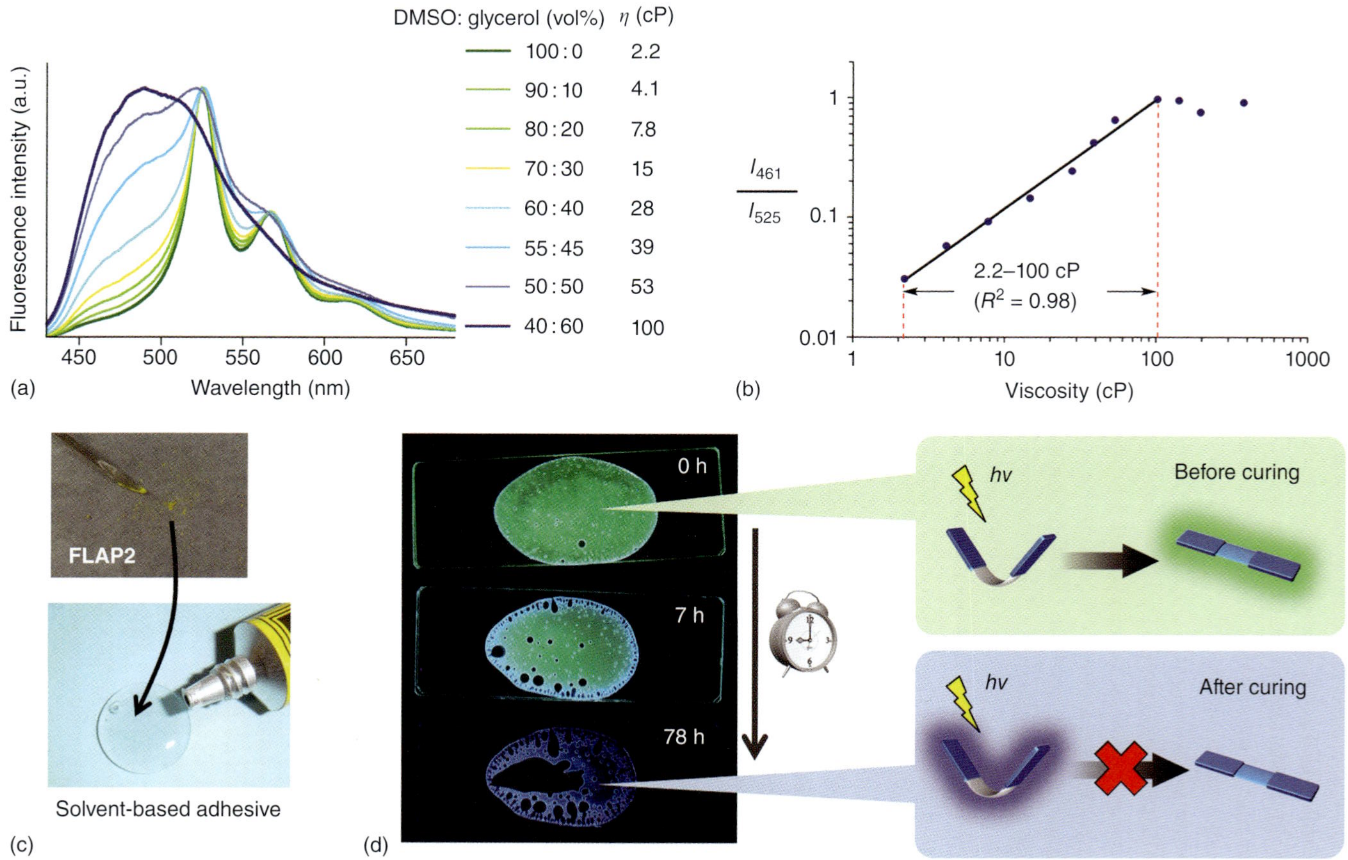

Figure 2.13 Viscochromism of **FLAP2**. (a) Viscosity-dependent fluorescence spectra of **FLAP2**. (b) Double logarithmic plot of FL intensity ratio I_{461}/I_{525} vs. viscosity measured by a viscometer. I_{461} and I_{525} mean the FL intensity at 461 and 525 nm, respectively. (c) Doping **FLAP2** into a solvent-based adhesive. (d) Real-time monitoring of the curing process of a solvent-based adhesive.

becomes higher compared with the green FL band (525 nm). According to the Förster–Hoffmann rule [94, 95, 106], the fluorescence intensity ratio (I_{461}/I_{525}) of the blue emission band (461 nm) and the green emission band (525 nm) is plotted with the media viscosity in a double logarithmic plot, and a linear relationship has been demonstrated to provide a calibration line in the viscosity range of 2.2–100 cP. In addition, the FLAP has a higher sensitivity in this viscosity range than the conventional viscosity probes based on molecular rotors. That is, the FL spectral shape is more largely dependent on the viscosity, probably because the flapping motion of **FLAP4** requires a larger volume to displace the surrounding solvent molecules as compared with the rotational motion of typical molecular rotors [154]. The polarity-independent ratiometric viscochromism of the FLAP fluorophores enables the real-time visualization of local viscosity change without sensing the local polarity environment.

2.2.2.4 Monitoring the Epoxy Resin Curing

Monitoring of epoxy resin curing has been realized in a real-time manner using the flapping viscosity probe **FLAP2** [63]. In the process of epoxy resin curing, both the local viscosity and polarity increase because the reaction of epoxy groups with curing agents generates polar hydroxy groups and because the microviscosity increases with decrease in free volume [111]. To confirm the efficacy of the FLAP probes, an epoxy prepolymer of bisphenol A diglycidyl ether (BADGE), pentaerythritol tetrakis (2-mercaptoacetate) (PETM), and n-Bu_3N have been used as an epoxy agent, a hardener, and a catalytic accelerator, respectively (Figure 2.14). These epoxy resin reagents are cured in two steps: first, the BADGE epoxy prepolymer is mixed with the PETM hardener, and second, a catalytic amount of n-Bu_3N accelerates the cross-linking polymerization between the epoxy prepolymer and the hardener. Before curing, a trace amount of **FLAP2** (0.1 wt% ratio) has been doped into the BADGE epoxy prepolymer. A fluid of the BADGE epoxy prepolymer doped with **FLAP2** shows green FL at 60 °C. After the addition of the PETM hardener to the BADGE epoxy prepolymer, the green FL has been still maintained, but the FL color changes into blue after the addition of n-Bu_3N, which indicates that the conformational planarization of **FLAP2** is eventually suppressed in S_1. Using another epoxy agent, ethylene glycol diglycidyl ether (EGDE), spectroscopic monitoring at room temperature has been demonstrated, and it has turned out that the curing process takes 60 minutes (Figure 2.14). In this experiment, the performance of typical dual fluorescent molecules with the mechanisms of TICT [35, 36] and excited- ESIPT [74, 75] has been also investigated. However, the TICT-type dye, 4-(dimethylamino) benzonitrile, does not work as a viscosity probe but rather a polarity probe because the charge polarization is explicit in S_1. Regarding the ESIPT-type dye, 2-(2′-hydroxyphenyl) oxazole, the structural change of the ESIPT process is too small to be suppressed by adhesive curing, and the FL band with a large Stokes shift remains even after the epoxy curing is completed. As a result, the competence of the FLAP systems has been demonstrated as a polarity-independent ratiometric fluorescent probe.

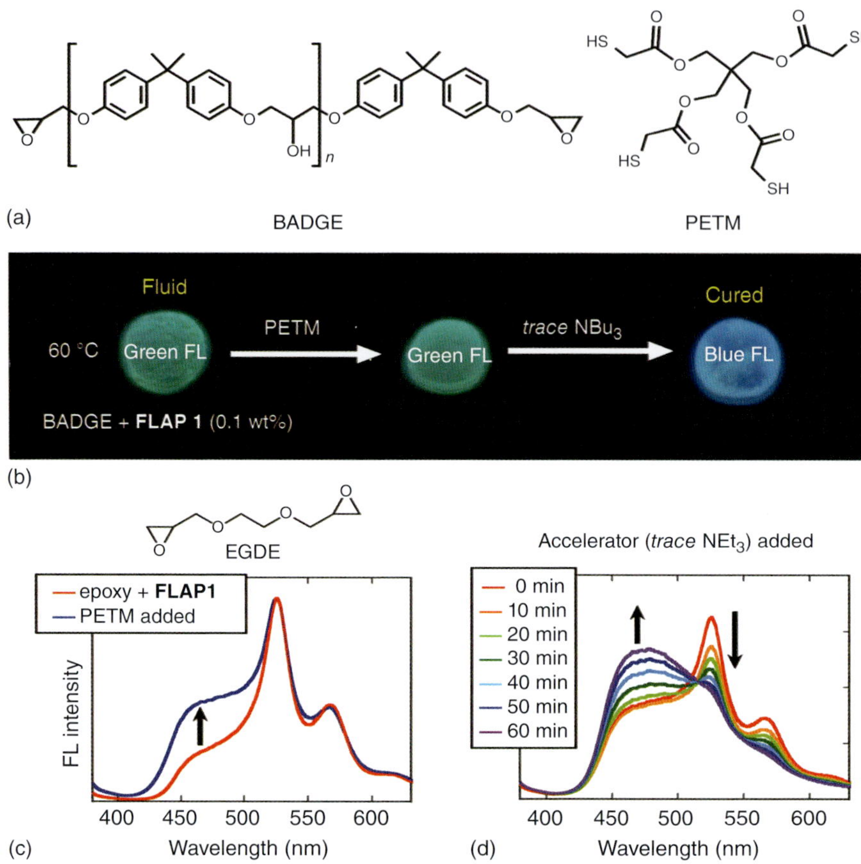

Figure 2.14 (a) Chemical structures of BADGE and PETM. (b) Fluorescence color change of curing polymer resin (BADGE + PETM) doped with **FLAP1**. (c) Fluorescence spectral change of curing polymer resin (EGDE + PETM) doped with **FLAP1**.

2.3 Light-Removable Adhesive

Photoinduced phase transition has been studied using materials containing isomerization dyes such as azobenzene [27–31], diarylethene [32, 33], and spiropyran [34]. It has attracted attention in materials science because optical properties and mechanical motion of the materials can be controlled. In industry, photocurable resins that harden by light irradiation are widely used for adhesion, coating, and sealing. One such example is dicing tape [155], which is used to temporarily attach semiconductor wafers during the dicing process (Figure 2.15). Dicing tape is commercialized to induce detachment by photocuring. However, there are limited reports on "rigid materials that dissolve [156–160] or soften [161–163] by light irradiation"; thus, functional adhesives developed with a new light-removable mechanism have potential applications. In the following sections, recently reported functional materials contributing to this field are introduced, and the light-melt adhesive based on the FLAP molecule is described in detail.

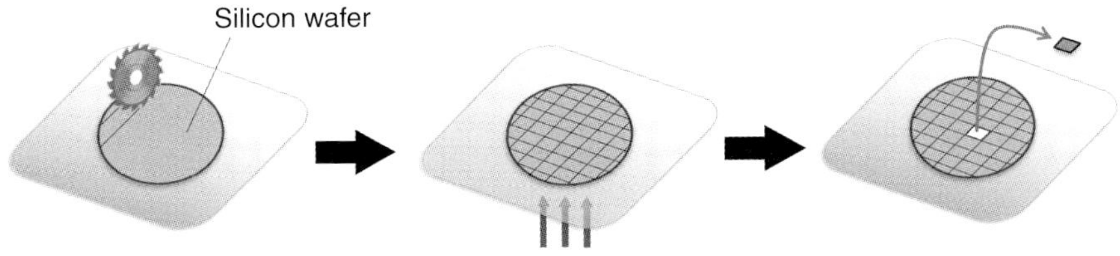

Figure 2.15 Dicing tape for temporary attachment of a silicon wafer. After dicing the wafer, detachment of the silicon pieces is induced by UV irradiation.

2.3.1 Polymer and Supramolecular Approach

Some polymer materials are developed as photo-deactivatable resins based on various operation principles, such as photoinduced cross-linking [164, 165], photoacid-catalyzed chain modification [166], photochemical cleavage of (supra) polymer chains [167, 168], and thermal cleavage of supramolecular chains by light irradiation [169]. For example, photodimerization of coumarin moiety can be used for cross-linking of polymer chains, inducing deactivation of a pressure-sensitive adhesive (PSA) (Figure 2.16a) [165]. A diarylethene-introduced polymer material has been designed to show Diels–Alder activity when the diarylethene moiety takes an open form (Figure 2.16b) [168]. Dissociation of the polymer chains is induced by visible light irradiation of diarylethene (closed form)-containing polymers at high temperature, which results in decrease of adhesive strength. Adhesion switching has also been realized using a host–guest complex of cyclodextrin (CD) and azobenzene at interfaces of polymer gels (Figure 2.16c) [167]. When the azobenzene moiety takes a trans form, host–guest interaction works for bonding two gels. By UV irradiation, azobenzene changes into a cis form, which is released from the CD host. As a result, two gels are separated spontaneously because the host–guest interaction at the interface should disappear.

2.3.2 Liquid Crystal Approach

LC is a suitable phase of materials to make use of structural changes of molecules in a condensed phase. Amplification of the structural changes enables a variety of technologies. Applications of the LC materials are not limited to LC displays, but cover lasers, OPVs, OLEDs, OFETs, nonlinear optics, biosensors, switchable windows, and nanophotonics [170–172]. Photochemical control of LC materials, as well as polymers and crystals, has long been an attractive research subject owing to the promising applications, for example, optical switching and image storage [173–177], photoalignment technique [178–184], and photoresponsive actuators [185–193]. In these studies, disorganized/ordered phase is spontaneously amplified in the LCs. Typically, the ordered alignment of rod-shaped LC molecules is destroyed by doping a guest molecule with a bent shape [28]. If the guest molecule generates upon *in situ* photoisomerization, as often observed in LC azobenzene analogues, an instant isothermal photoinduced phase transformation of the LC material takes place, depending on the concentration of the guest molecule. In

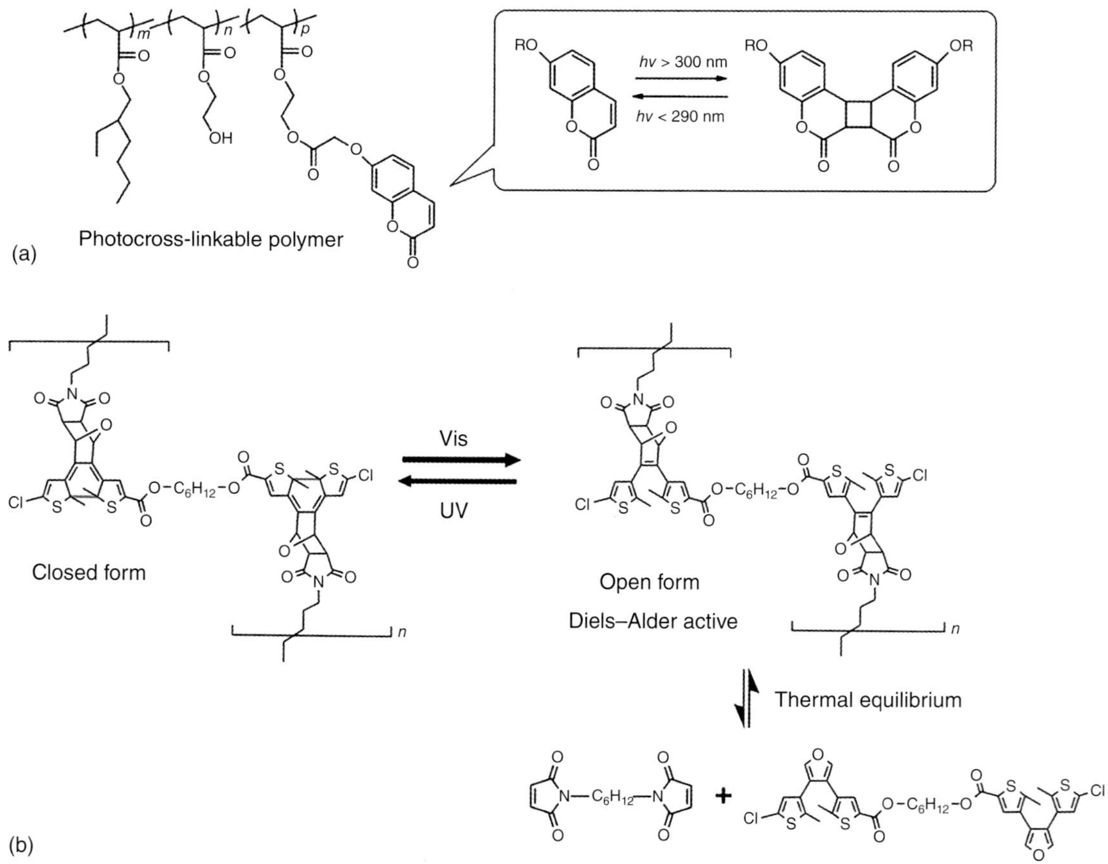

Figure 2.16 Photo-deactivatable polymers. (a) Photocross-linkable pressure-sensitive adhesive (PSA), (b) diarylethene-based Diels–Alder active polymer, and (c) host–guest complex gel composed of cyclodextrin (CD) and azobenzene (Azo) moieties.

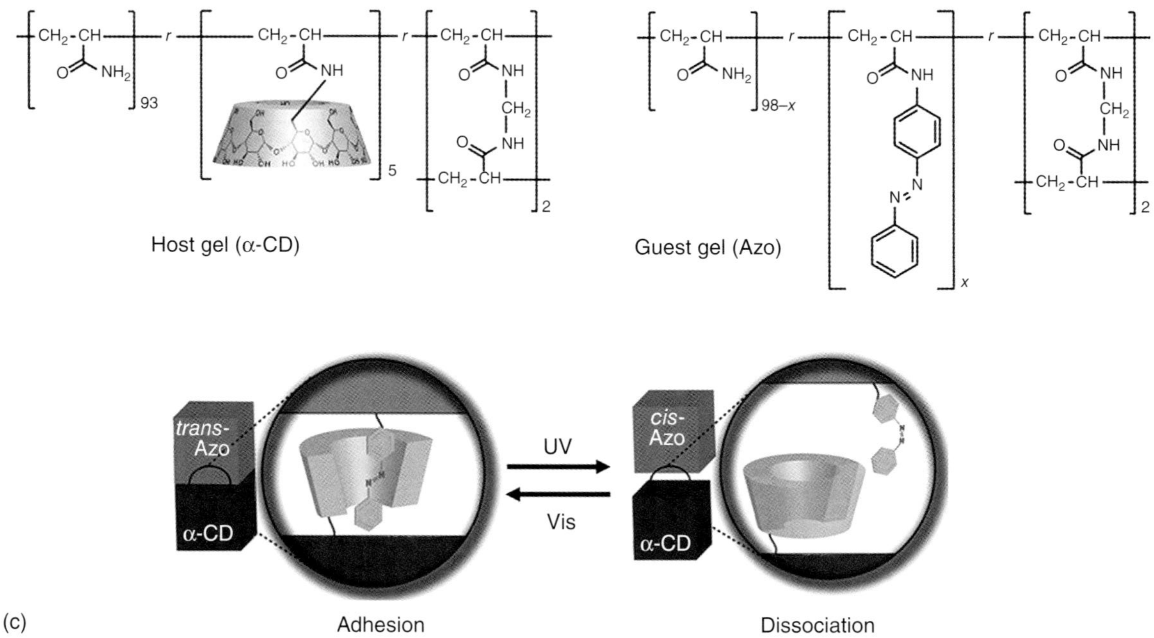

(c)

Figure 2.16 (*Continued*)

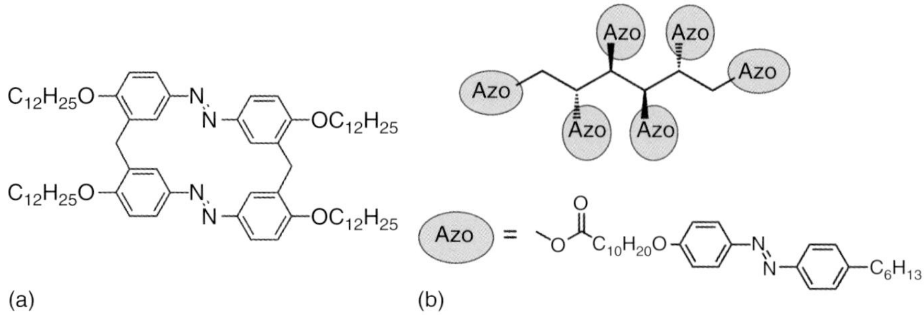

Figure 2.17 Early examples of photomelting small-molecule compounds. (a) Cyclic azobenzene dimer and (b) azobenzene oligomer.

this context, photoresponsive LC is the most suitable platform for realizing a quick photoinduced isothermal phase transformation into a fluid mixture (that is, photomelting function), while some crystals are reported to display a slow photomelting behavior [157, 159, 160] (Figure 2.17). Until recently, however, an application of LC materials for removable adhesives has not been explored [194, 195]. The difficulty arises from the requirement for satisfying both sufficient bonding strength and its rapid disappearance by photoirradiation.

2.3.3 Light-Melt Adhesive

A photomelting columnar LC material based on a FLAP mesogen unit is expected to meet the requirements for the practical application of "light-melt adhesive." V-shaped π-systems of FLAP are stacked in a columnar manner to display sufficient bonding strength even under high temperature, but its bonding ability is lost by a few seconds of UV irradiation leading to photodimerization of the FLAP wings.

2.3.3.1 Requirements for Applications

In industry, hot-melt adhesives are widely used for temporary bonding, which can be separated by heating. These adhesives are composed of polymers that melt at high temperature. Therefore, high-temperature-resistant bonding is a challenging issue for temporary adhesives. In this context, the light-melting function offers a new manufacturing technique if the following essential requirements are satisfied. First, adequate strength for a temporary bond is required even under heating conditions. Although permanent bonding in some applications needs more than 10 MPa adhesive strength, standard temporary bonding only needs 1 MPa strength. Second, significant reduction of the bonding strengths must be induced by light irradiation. It will be better if the bonding strength become smaller and smaller during the irradiation. Third, quick photoresponse, ideally within a few seconds, is necessary for efficient detachment technology in a manufacturing process. The wavelength of light required is dependent on each application, but instruments for 365-nm UV irradiation are widely equipped in the laboratories and factories where they use UV curable resins.

The abovementioned requisites have been fulfilled by the light-melt adhesive described below: (i) a shear strength over 1 MPa up to 110 °C for bonding glass plates, (ii) an 85% reduction of the strength by UV irradiation, and (iii) an instant photomelting of the LC film in a few seconds have been realized (Figure 2.18). In addition, the phase transition is reversible, and therefore the light-melt adhesive acts as a reworkable adhesive. The photoinduced isothermal phase transformation between the LC and melted phases is associated with a color change in

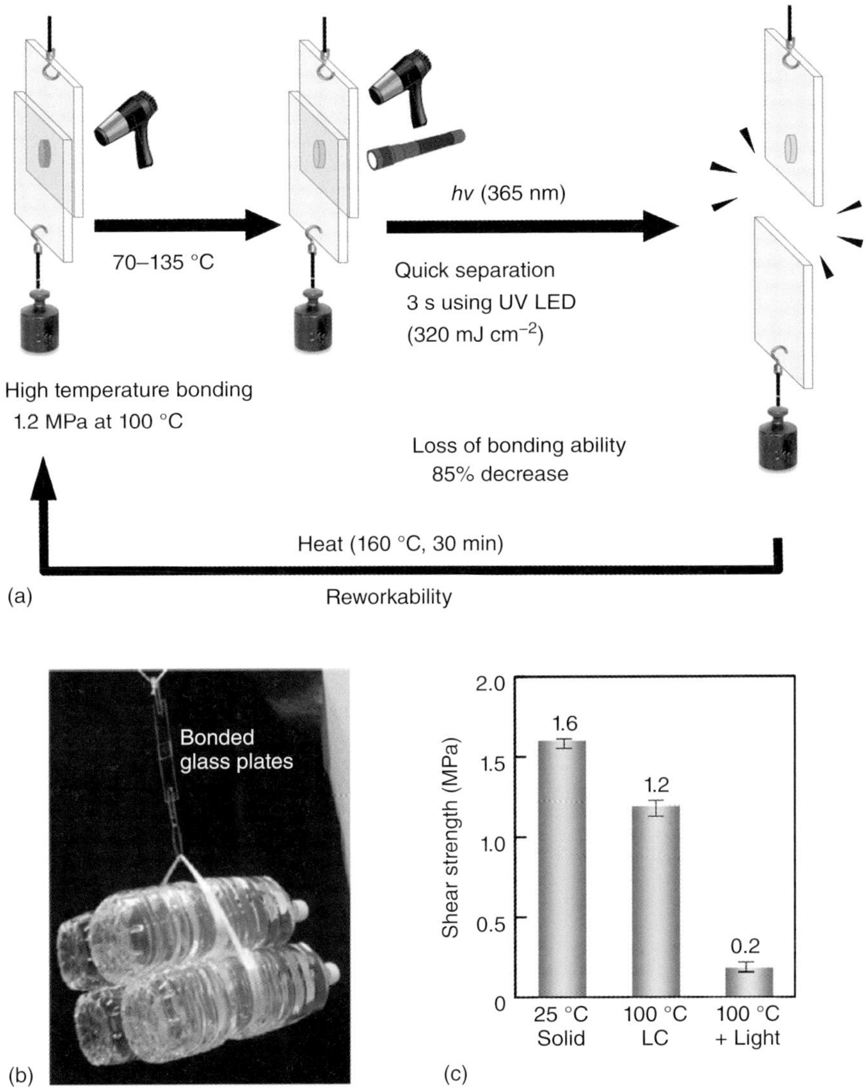

Figure 2.18 (a) Performance of the light-melt adhesive **FLAP5**. (b) Demonstration of strong bonding properties, in which a few milligrams of the adhesive is sandwiched between two glass plates. (c) Ultimate shear strength of the sample specimen in each condition.

fluorescence, by which the bonding/nonbonding phase can be distinguished in a contactless manner.

2.3.3.2 Materials Design

A light-melt adhesive is composed of the LC material **FLAP5**. Molecular design is based on the FLAP mesogen with typical dendritic carbon chains [197, 198] (Figure 2.19). The hybrid design of rigid anthracene and flexible COT gives two important features to this material. First, the rigid anthracene wings of the V-shaped FLAP show strong stacking ability to form a columnar array in the condensed phases, which results in high cohesive force of the LC material for realizing high-temperature-resistant bonding. Second, the flexible COT ring changes the conformation of FLAP into a planar geometry by photoexcitation in the LC phase, leading to the photodimerization of the anthracene moiety [199]. This photoreaction gives rise to the photomelting function, accompanied by the separation of bonded glass plates.

FLAP5 has a rectangular columnar LC phase between 65 and 140 °C, in which the V-shaped π-systems align on top of each other and the stacked arrays are located *side by side*. The columnar stacking is supported in a single crystal X-ray structure of the corresponding analogue that has no peripheral chains. The interfacial distance $d(\pi{-}\pi)$ between the stacked anthracene wings is 3.50 Å, indicating the formation of typical slipped π stacking at both wings. The strong intermolecular interaction and high cohesive strength of **FLAP5** are also confirmed by the large enthalpy change in the phase transition between the columnar LC and isotropic liquid (Figure 2.20). The enthalpy change over 30 kJ mol^{-1} is one of the largest values among LCs that do not contain hydrogen bonds [200].

2.3.3.3 Adhesive Performance

A thin film of **FLAP5** gets melted between glass plates by 365-nm UV-LED irradiation. Observation by polarized optical microscopy and thermography demonstrates a photoinduced isothermal phase transformation into an isotropic liquid. The nature of LC is important to induce the photomelting function, because this

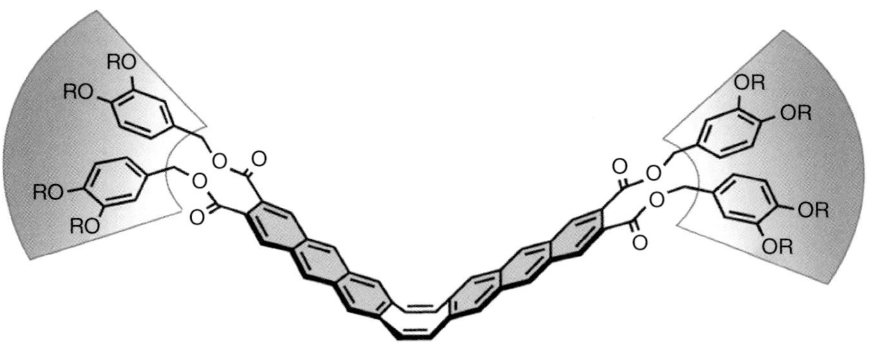

FLAP5 (R = $C_{12}H_{25}$)
FLAP6 (R = CH_3)

Figure 2.19 Molecular structures of **FLAP5** and its analogue **FLAP6**.

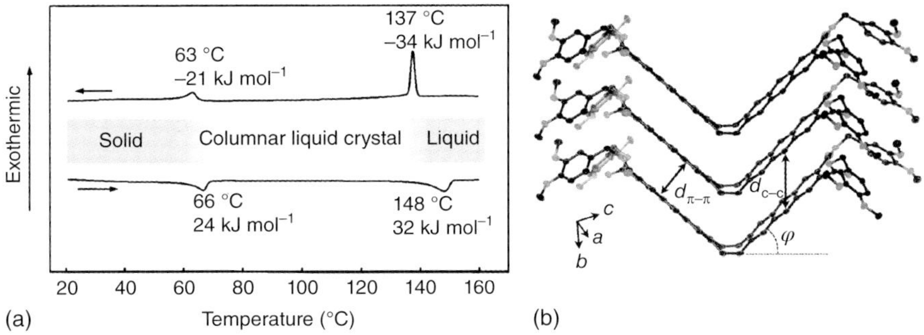

Figure 2.20 (a) Thermal analysis of **FLAP5** and (b) crystal packing of **FLAP6**.

transformation is not observed in the solid phase. The resulting isotropic fluid includes the dimer and trace of oligomers of **FLAP5** produced by UV irradiation, while the unreacted monomer also remains in the melted mixture. It turns out that, in a 130-μm-thick adhesive thin film, half of the monomer molecules are photodimerized until glass plates are separated by light. Complete consumption of all monomers is not necessary for the photomelting event because the destruction of the columnar LC structure is spontaneously induced by amplification of a disorganized phase. A less symmetric molecular structure of the isolated photodimer suggests a photomelting mechanism in which *in situ* photodimer generation destabilizes the LC phase of **FLAP5** due to a shape unsuitable for columnar packing (Figure 2.21).

The light-melt adhesive has a practical bonding strength (over 1 MPa) for temporary bonding. While the adhesive film shows high-temperature-resistant bonding, the glass plates are separated in a few seconds by UV-LED irradiation. Ultimate shear strengths are 1.6 MPa at 25 °C, 1.2 MPa at 100 °C, 1.1 MPa at 110 °C, and 0.9 MPa at 120 °C, when a 130-μm-thick film of **FLAP5** is sandwiched by non-treated glass plates.

In general, adhesive strength is discussed with cohesive force and adhesion force. The cohesive force derives from the internal strength of the adhesive material, which is closely related to the intermolecular interaction, particularly in the case of small molecules. On the other hand, the adhesion force is the interaction between the adhesive material and the substrate surface. If $F_{\text{cohesion}} > F_{\text{adhesion}}$, the bonding strength is determined by the adhesion force and thus largely depends on the surface conditions of the substrates such as hydrophilicity. When $F_{\text{cohesion}} < F_{\text{adhesion}}$, in contrast, the bonding strength is determined by the cohesive force regardless of the surface conditions.

Here, the shear strength is independent of the hydrophilicity of the glass surface. Therefore, the cohesive force, closely related to the intermolecular interaction of **FLAP5**, plays a key role in determining the shear strength of this test specimen rather than the adhesion force, the interaction between the glass surface and the adhesive material (Figure 2.22) [196].

The shear strength remarkably decreases to 0.2 MPa after 320 mJ cm^{-2} UV exposure at 100 °C. The total dose of UV irradiation is comparable to that required for UV curing of typical photoactive resins. Quick detachment has

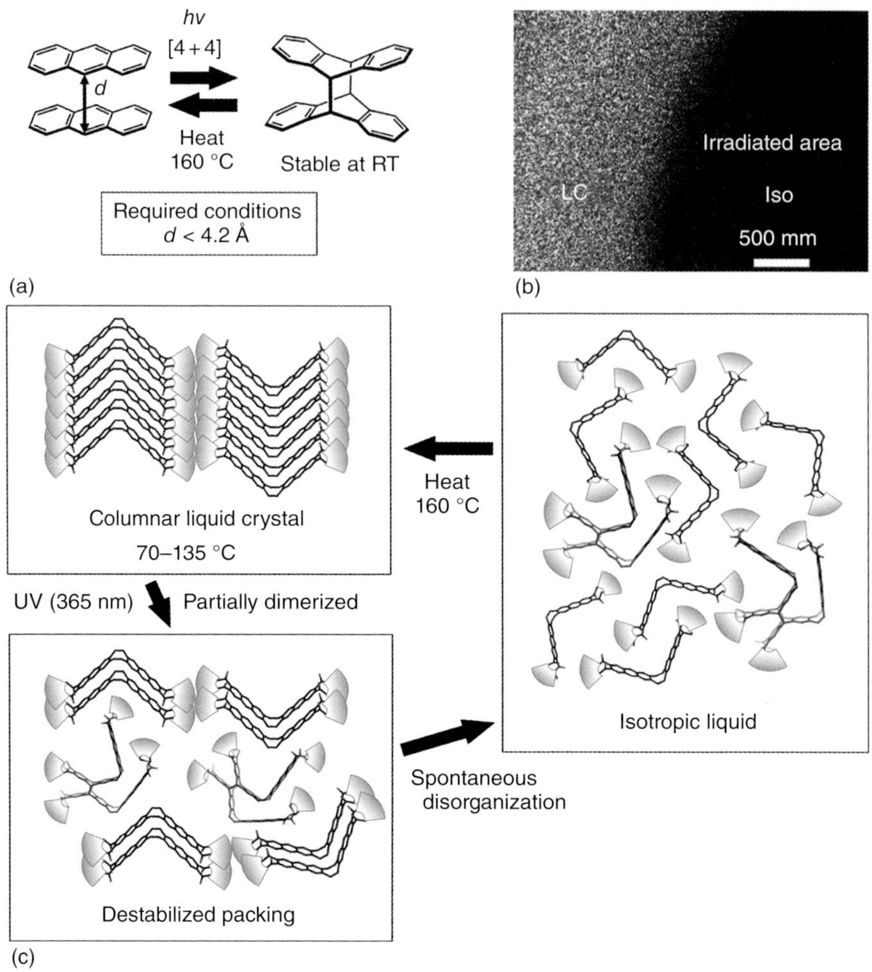

Figure 2.21 (a) Photodimerization of anthracene. (b) POM image of photomelting LC composed of **FLAP5**. (c) Photomelting mechanism of **FLAP5**.

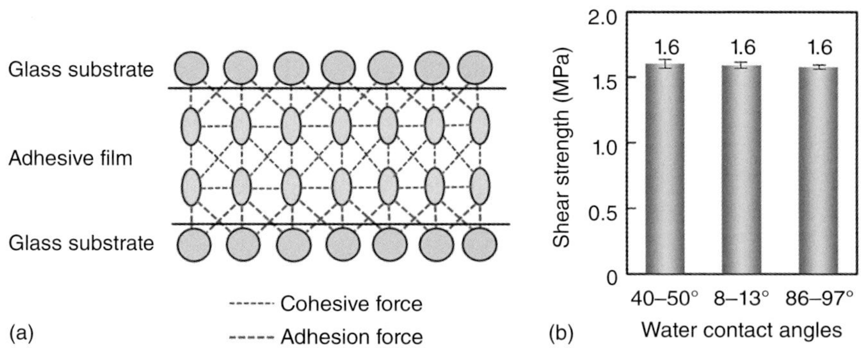

Figure 2.22 (a) Cohesive force and adhesion force in a bonded specimen. (b) Constant shear strength regardless of hydrophilicity of the glass surface, measured for the bonded specimens using **FLAP5**.

been achieved in a few seconds by using 365-nm UV-LED with 160 mW cm^{-2} irradiance. Since the condensed molecules of **FLAP5** efficiently absorb UV light, the photoinduced detachment takes place near the interface between the adhesive film and the irradiated glass plate. The light transmittance exponentially decreases depending on the thickness of the films of **FLAP5**. More than 95% of 365-nm light is absorbed within 3 μm depth from the interface. The total irradiation dose required for glass separation has been almost constant regardless of the film thickness over 5 μm (Figure 2.23). Therefore, it is not necessary for users to care about the adhesive thickness in practical applications and, more importantly, the amount of the adhesive residue on the irradiated glass plate is always small. Moreover, the small adhesive residue can be removed by common organic solvents because the solubility of **FLAP5** is high owing to the long alkyl chains introduced at the FLAP wings.

The thin film of **FLAP5** can act as a reworkable adhesive. When the irradiated mixture fluid is heated at 160 °C for 15 minutes, both the shear strength around 1.5 MPa at 25 °C and the quick photomelting function at 100 °C recover by at least four times. In addition, a fluorescence color change gives information about bonding/nonbonding phases. If the material is recovered by heating and the monomer of **FLAP5** is obtained as a single component, green fluorescence can be observed in the LC phase as a sign of the adhesive recovery. On the other hand, blue fluorescence is observed when the adhesive film gets melted (Figure 2.24).

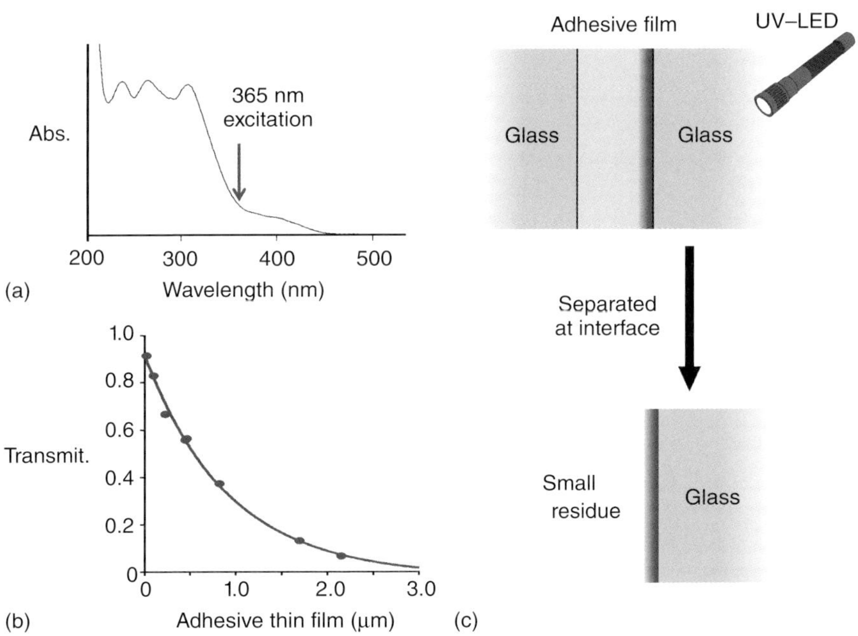

Figure 2.23 (a) Absorption spectrum of the LC film composed of **FLAP5**. (b) Transmittance of the 365-nm excitation light depending on the film thickness. (c) Photoinduced detachment near the interface between the adhesive and the glass plate.

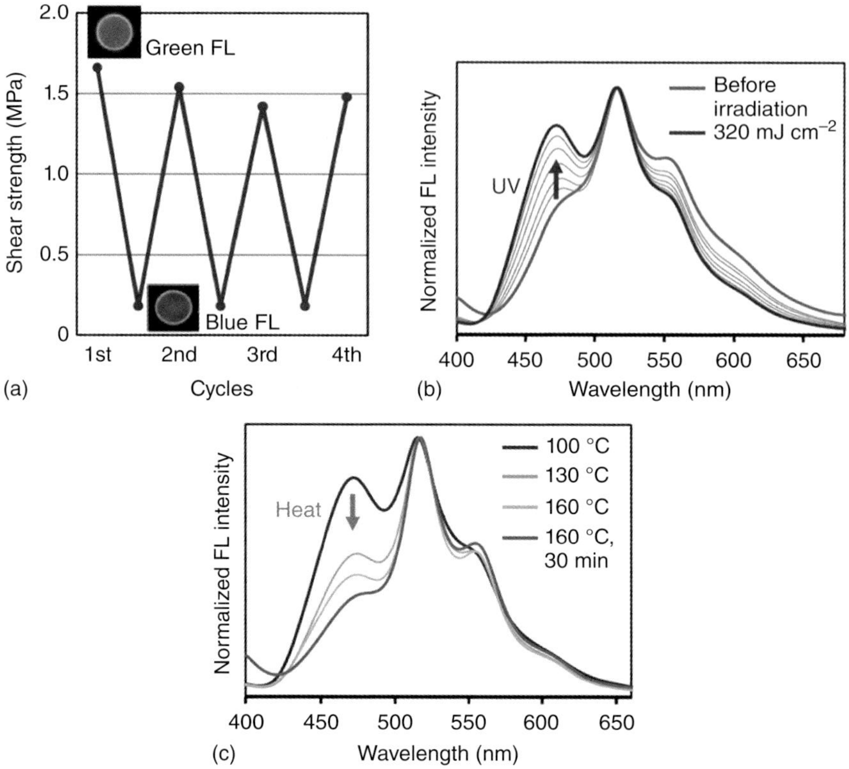

Figure 2.24 (a) Repeated cycles of the photomelting and the thermal recovery. FL spectral change of the **FLAP5** LC film (b) during UV irradiation and (c) thermal back reaction.

2.3.3.4 Working Mechanism

The photomelting mechanism has been demonstrated in consideration of the dynamics of **FLAP5** in the singlet excited state (S_1) (Figure 2.25). Photodimerization of anthracene moieties takes place in S_1 [201] if the reactive sites can approach each other in the condensed phase. In the literature, the required distance d(C—C) is reported to be within 4.2 Å in order to form new C—C bonds between the central carbon atoms of the anthracene units [202]. In spite of the tight π stacking of the anthracene wings with $d(\pi{-}\pi) \sim 3.5$ Å, the corresponding distance d(C—C) in the columnar packing of **FLAP5** is estimated to be c. 4.7 Å, which originates from the slipped alignment of the anthracene wings. As a result of the longer d(C—C) distance, photodimerization is not allowed at room temperature, in which the conformation of **FLAP5** is fixed in the solid phase. In contrast, the columnar LC structure can be perturbed by the photoexcitation of **FLAP5** because the conformational planarization in S_1 described above is allowed in the LC phase, and dimerization occurs when the d(C—C) distance gets closer. The dynamic conformational change of **FLAP5** in the LC phase has been elucidated by fluorescence spectroscopy of the LC film. The LC film of **FLAP5** at 100 °C shows green fluorescence with a large Stokes shift, which is the same as that observed in its solution phase. The overall mechanism of

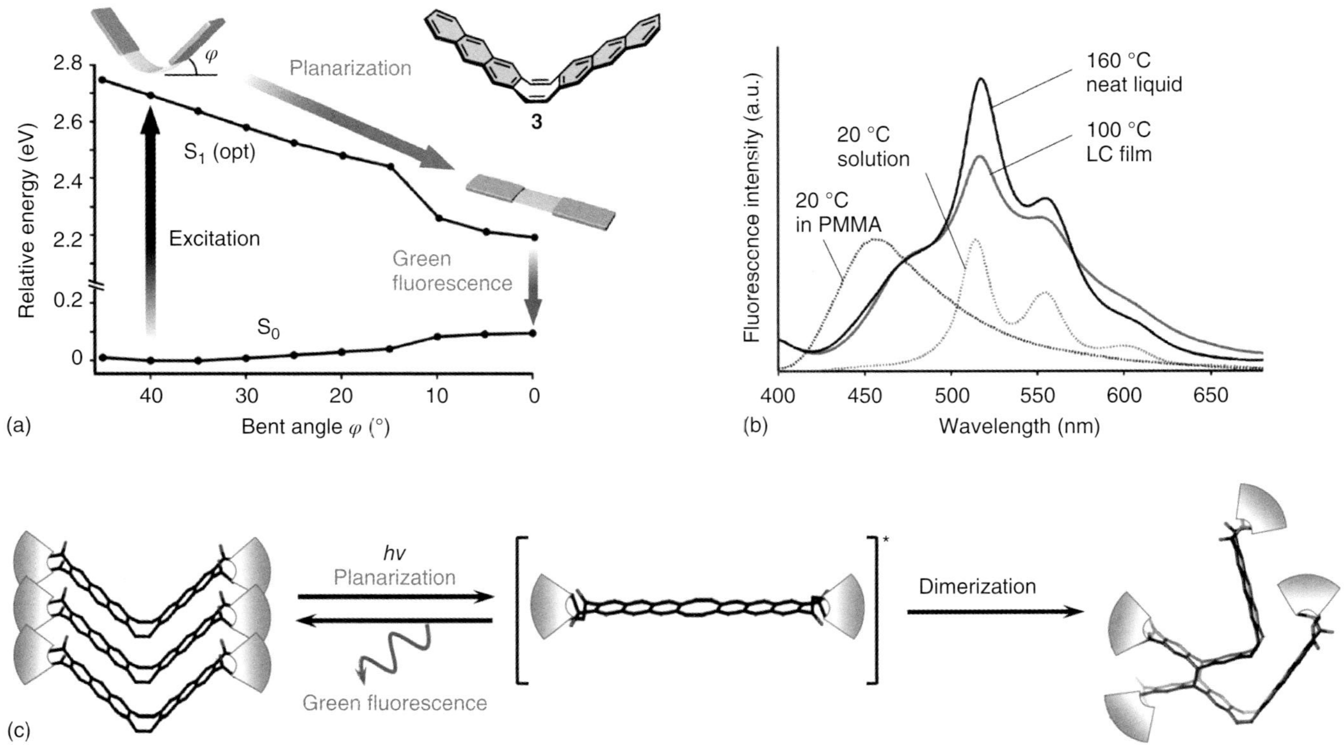

Figure 2.25 Photoresponse mechanism of **FLAP5** in the columnar LC phase. (a) Calculated energy profile of the parent structure **FLAP6**. (b) Fluorescence spectra of **FLAP5** in each phase of materials. (c) Conformational planarization in S_1 and subsequent dimerization of **FLAP5**.

the photomelting process has been interpreted as follows: Near the interface between the LC film of **FLAP5** and the glass substrate, the V-shaped molecules are photoexcited to induce the conformational planarization. Consequently, the intermolecular distance *d*(C—C) between the anthracene reactive sites is significantly perturbed. Photochemical dimerization occurs if an excited molecule in S_1 gets into pairs with a neighboring molecule in S_0, while other excited molecules relax back to S_0 accompanied by the green FL emitted from the planar conformation. The photoproducts work as "impurities" to induce disorganization of the columnar LC phase, leading to the photomelting event accompanied by a remarkable decrease in the bonding strength.

In summary, a light-melt adhesive based on a photoresponsive columnar LC material has been developed. Tight π-stacking of the V-shaped FLAP molecules resulted in sufficient bonding properties of the adhesive film even at high temperature. Remarkable photoinduced decrease in shear strength was realized by the *in situ* dimerization of the LC molecule and the following disorganization of the columnar structure. Fast melting response achieved a quick separation of bonded glass plates, leaving a small adhesive residue. We envision that composite materials with the light-melt function will further improve the performance in manufacturing processes, which will accelerate on-demand photoseparation technology complementary to the other switchable adhesion approaches [203].

2.4 Conclusion

Most of the applications using π-conjugated systems rely on the conformational rigidity of those molecular frames. It is still challenging to make use of conformational flexibility of molecules in materials science. Here, the molecular design, basic properties, and applications of conformationally flexible flapping molecules (FLAP) have been introduced. Flapping fluorophores enable polarity-independent ratiometric viscosity sensing, and a photoresponsive columnar LC based on the flapping structure exhibits high performance as a light-melt adhesive. A series of FLAPs bearing a rigid–flexible hybrid molecular design will help develop new functional materials that show convertible photophysical and electronic properties in the condensed phase.

References

1 Dong, H., Fu, X., Liu, J. et al. (2013). *Adv. Mater.* 25: 6158–6183.
2 Takimiya, K., Shinamura, S., Osaka, I., and Miyazaki, E. (2011). *Adv. Mater.* 23: 4347–4370.
3 Thompson, B.C. and Fréchet, J.M.J. (2008). *Angew. Chem. Int. Ed.* 47: 58–77.
4 Cheng, Y.-J., Yang, S.-H., and Hsu, C.-S. (2009). *Chem. Rev.* 109: 5868–5923.
5 Kulkarni, A.P., Tonzola, C.J., Babel, A., and Jenekhe, S.A. (2004). *Chem. Mater.* 16: 4556–4573.
6 Tao, Y., Yuan, K., Chen, T. et al. (2014). *Adv. Mater.* 26: 7931–7958.

7 Chan, J., Dodani, S.C., and Chang, C.J. (2012). *Nat. Chem.* 4: 793–984.
8 Lavis, L.D. and Raines, R.T. (2008). *ACS Chem. Biol.* 3: 142–155.
9 Balzani, V., Credi, A., Raymo, F.M., and Stoddard, J.F. (2000). *Angew. Chem. Int. Ed.* 39: 3348–3391.
10 Erbas-Cakmak, S., Leigh, D.A., McTernan, C.T., and Nussbaumer, A.L. (2015). *Chem. Rev.* 115: 10081–10206.
11 Swiegers, G.F. and Malefetse, T.J. (2000). *Chem. Rev.* 100: 3483–3537.
12 Leininger, S., Olenyuk, B., and Stang, P.J. (2000). *Chem. Rev.* 100: 853–908.
13 Yoshizawa, M., Klosterman, J.K., and Fujita, M. (2009). *Angew. Chem. Int. Ed.* 48: 3418–3438.
14 Furukawa, H., Cordova, K.E., O'Keeffe, M., and Yaghi, O.M. (2013). *Science* 341: 1230444.
15 Kitagawa, S., Kitaura, R., and Noro, S. (2004). *Angew. Chem. Int. Ed.* 43: 2334–2375.
16 Waller, P.J., Gándara, F., and Yaghi, O.M. (2015). *Acc. Chem. Res.* 48: 3053–3063.
17 Huang, N., Wang, P., and Jiang, D. (2016). *Nat. Rev. Mater.* 1: 16068.
18 Imahori, H., Tkachenko, N.V., Vehmanen, V. et al. (2001). *J. Phys. Chem. A* 105: 1750–1756.
19 Kowaka, Y., Suganuma, Y., Ashizawa, N. et al. (2010). *J. Mol. Spectrosc.* 260: 72–76.
20 Wöhrle, T., Wurzbach, I., Kirres, J. et al. (2016). *Chem. Rev.* 116: 1139–1241.
21 Hill, J.P., Jin, W., Kosaka, A. et al. (2004). *Science* 304: 1481–1483.
22 Feng, X., Marcon, V., Pisula, W. et al. (2009). *Nat. Mater.* 8: 421–426.
23 Mori, H., Itoh, Y., Nishiura, Y. et al. (1999). *Jpn. J. Appl. Phys.* 38: 2837–2844.
24 Seechurn, C.C.C.J., Kitching, M.O., Colacot, T.J., and Snieckus, V. (2012). *Angew. Chem. Int. Ed.* 51: 5062–5085.
25 Fujita, D., Ueda, Y., Sato, S. et al. (2016). *Nature* 540: 563.
26 Fujita, D., Ueda, Y., Sato, S. et al. (2016). *Chem* 1: 91–101.
27 Kumar, G.S. and Neckers, D.C. (1989). *Chem. Rev.* 89: 1915–1925.
28 Ikeda, T. (2003). *J. Mater. Chem.* 13: 2037–2057.
29 Browne, W.R. and Feringa, B.L. (2006). *Nat. Nanotechnol.* 1: 25–35.
30 Zhao, Y. and Ikeda, T. (eds.) (2009). *Smart Light-Responsive Materials.* Wiley.
31 White, T.J. and Broer, D.J. (2015). *Nat. Mater.* 14: 1087–1098.
32 Irie, M., Fukaminato, T., Matsuda, K., and Kobatake, S. (2014). *Chem. Rev.* 114: 12174–12277.
33 Kobatake, S., Takami, S., Muto, H. et al. (2007). *Nature* 446: 778–781.
34 Klajn, R. (2014). *Chem. Soc. Rev.* 43: 148.
35 Grabowski, Z.R., Rotkiewicz, K., and Rettig, W. (2003). *Chem. Rev.* 103: 3899.
36 Sasaki, S., Drummen, G.P.C., and Konishi, G. (2016). *J. Mater. Chem. C* 4: 2731–2743.
37 Anet, F.A.L. (1962). *J. Am. Chem. Soc.* 84: 671.
38 Paquette, L.A. (1993). *Acc. Chem. Res.* 26: 57–62.
39 Seiders, T.J., Baldridge, K.K., Grube, G.H., and Siegel, J.S. (2001). *J. Am. Chem. Soc.* 123: 517.
40 Wu, Y.-T. and Siegel, J.S. (2006). *Chem. Rev.* 106: 4843.

41 Sakurai, H., Daiko, T., and Hirao, T. (2003). *Science* 301: 1878.
42 Amaya, T., Sakane, H., Muneishi, T., and Hirao, T. (2008). *Chem. Commun.* 765.
43 Shen, Y. and Chen, C.-F. (2012). *Chem. Rev.* 112: 1463–1535.
44 Würthner, F. (2006). *Pure Appl. Chem.* 78: 2341.
45 Stępień, M., Sprutta, N., and Latos-Grażyński, L. (2011). *Angew. Chem. Int. Ed.* 50: 4288–4340.
46 Saito, S. and Osuka, A. (2011). *Angew. Chem. Int. Ed.* 50: 4342–4373.
47 Sessler, J.L. and Seidel, D. (2003). *Angew. Chem. Int. Ed.* 42: 5134–5175.
48 Breslow, R. (1973). *Acc. Chem. Res.* 6: 393.
49 Klärner, F.G. (2001). *Angew. Chem. Int. Ed.* 40: 3977–3981.
50 Fowler, P.W., Havenith, R.W.A., Jenneskens, L.W. et al. (2002). *Angew. Chem. Int. Ed.* 41: 1558–1560.
51 Nishinaga, T., Ohmae, T., and Iyoda, M. (2010). *Symmetry* 2: 76.
52 Rzepa, H.S. (2005). *Chem. Rev.* 105: 3697–3715.
53 Herges, R. (2006). *Chem. Rev.* 106: 4820–4842.
54 Osuka, A. and Saito, S. (2011). *Chem. Commun.* 47: 4330–4339.
55 Rosenberg, M., Dahlstrand, C., Kilså, K., and Ottosson, H. (2014). *Chem. Rev.* 114: 5379–5425.
56 Sung, Y.M., Yoon, M.-C., Lim, J.M. et al. (2015). *Nat. Chem.* 7: 418–422.
57 Sung, Y.M., Oh, J., Naoda, K. et al. (2016). *Angew. Chem. Int. Ed.* 55: 11930–11934.
58 Ueda, M., Jorner, K., Sung, Y.M. et al. (2017). *Nat. Commun.* 8: 346.
59 Hada, M., Saito, S., Tanaka, S. et al. (2017). *J. Am. Chem. Soc.* https://doi.org/10.1021/jacs.7b08021.
60 Yuan, C., Saito, S., Camacho, C. et al. (2013). *J. Am. Chem. Soc.* 135: 8842.
61 Yuan, C., Saito, S., Camacho, C. et al. (2014). *Chem. Eur. J.* 20: 2193–2200.
62 Saito, S., Nobusue, S., Hara, M. et al. (2016). *Nat. Commun.* 7: 12094.
63 Kotani, R., Sotome, H., Okajima, H. et al. (2017). *J. Mater. Chem. C* 5: 5248.
64 He, G., Guo, D., He, C. et al. (2009). *Angew. Chem. Int. Ed.* 48: 6132.
65 Yang, Y., Lowry, M., Schowalter, C.M. et al. (2006). *J. Am. Chem. Soc.* 128: 14081.
66 Tong, H., Hong, Y., Dong, Y. et al. (2007). *J. Phys. Chem. B* 111: 2000.
67 Zhao, Y., Gao, H., Fan, Y. et al. (2009). *Adv. Mater.* 21: 3165.
68 Ruff, Y., Buhler, E., Candau, S.-J. et al. (2010). *J. Am. Chem. Soc.* 132: 2573–2584.
69 Sagara, Y. and Kato, T. (2011). *Angew. Chem. Int. Ed.* 50: 9128.
70 Dong, Y., Xu, B., Zhang, J. et al. (2012). *Angew. Chem. Int. Ed.* 51: 10782.
71 Demchenko, A.P. (ed.) (2010). *Advanced Fluorescence Reporters in Chemistry and Biology I, Fundamentals and Molecular Design*, Springer Series on Fluorescence, vol. 8. Berlin, Heidelberg: Springer-Verlag.
72 Callan, J.F., Silva, A.P., and Magri, D.C. (2005). *Tetrahedron* 61: 8551–8588.
73 Wu, J., Liu, W., Ge, J. et al. (2011). *Chem. Soc. Rev.* 40: 3483–3495.
74 Kwon, J.E. and Park, S.Y. (2011). *Adv. Mater.* 23: 3615–3642.
75 Padalkar, V.S. and Seki, S. (2016). *Chem. Soc. Rev.* 45: 169–202.
76 Fan, J., Hu, M., Zhan, P., and Peng, X. (2013). *Chem. Soc. Rev.* 42: 29–43.

77 Yuan, L., Lin, W., Zheng, K., and Zhu, S. (2013). *Acc. Chem. Res.* 46: 1462–1473.
78 Saigusa, H. and Lim, E.C. (1996). *Acc. Chem. Res.* 29: 171–178.
79 Jenekhe, S.A. and Osaheni, J.A. (1994). *Science* 265: 765–768.
80 Würthner, F., Kaiser, T.E., and Saha-Möller, C.R. (2011). *Angew. Chem. Int. Ed.* 50: 3376–3410.
81 Mei, J., Leung, N.L.C., Kwok, R.T.K. et al. (2015). *Chem. Rev.* 115: 11718–11940.
82 Zhao, Q., Li, F., and Huang, C. (2010). *Chem. Soc. Rev.* 39: 3007–3030.
83 Han, M., Tian, Y., Yuan, Z. et al. (2014). *Angew. Chem. Int. Ed.* 53: 10908–10912.
84 He, X. and Yam, V.W.-W. (2011). *Coord. Chem. Rev.* 255: 2111–2123.
85 Seki, T. and Ito, H. (2016). *Chem. Eur. J.* 22: 4322–4329.
86 Kato, M. (2007). *Bull. Chem. Soc. Jpn.* 80: 287–294.
87 Ma, B., Li, J., Djurovich, P.I. et al. (2005). *J. Am. Chem. Soc.* 127: 28–29.
88 Uoyama, H., Goushi, K., Shizu, K. et al. (2012). *Nature* 492: 235–240.
89 Singh-Rachford, T.N. and Castellano, F.N. (2010). *Coord. Chem. Rev.* 254: 2560–2573.
90 Ye, C., Zhou, L., Wang, X., and Liang, Z. (2016). *Phys. Chem. Chem. Phys.* 18: 10818.
91 Shukla, D. and Wan, P. (1993). *J. Am. Chem. Soc.* 115: 2990–2991.
92 Zhang, Z., Wu, Y.-S., Tang, K.-C. et al. (2015). *J. Am. Chem. Soc.* 137: 8509–8520.
93 Chen, W., Chen, C.-L., Zhang, Z. et al. (2017). *J. Am. Chem. Soc.* 139: 1636–1644.
94 Haidekker, M.A. and Theodorakis, E.A. (2016). *J. Mater. Chem. C* 4: 2707–2718.
95 Kuimova, M.K. (2012). *Phys. Chem. Chem. Phys.* 14: 12671–12686.
96 Haidekker, M.A. and Theodorakis, E.A. (2007). *Org. Biomol. Chem.* 5: 1669–1678.
97 Klymchenko, A.S. (2017). *Acc. Chem. Res.* 50: 366–375.
98 Klymchenko, A.S. and Kreder, R. (2014). *Chem. Biol.* 21: 97–113.
99 Yang, Z., Cao, J., He, Y. et al. (2014). *Chem. Soc. Rev.* 43: 4563–4601.
100 Kowada, T., Maeda, H., and Kikuchi, K. (2015). *Chem. Soc. Rev.* 44: 4953–4972.
101 Itagaki, H., Horie, K., and Mita, I. (1990). *Prog. Polym. Sci.* 15: 361–424.
102 Strehmel, B., Strehmel, V., and Younes, M. (1999). *J. Polym. Sci., Part B: Polym. Phys.* 37: 1367–1386.
103 Stark, J. and Lipp, P. (1913). *Z. Phys. Chem.* 86: 36–50.
104 Schmidt, G.C. (1921). *Ann. Phys.* 65: 247–256.
105 Oster, G. and Nishijima, Y. (1956). *J. Am. Chem. Soc.* 78: 1581–1584.
106 Förster, T. and Hoffmann, G. (1971). *Z. Phys. Chem.* 75: 63–69.
107 Loutfy, R.O. (1981). *Macromol.* 14: 270–275.
108 Wang, F.W., Lowry, R.E., and Fanconi, B.M. (1986). *Polymer* 27: 1529–1532.
109 Sung, C.S.P. and Mathisen, R. (1987). *Polymer* 28: 941–945.
110 Pączkowski, J. and Neckers, D.C. (1991). *Macromol.* 24: 3013–3016.
111 Younes, M., Wartewig, S., Lellinger, D. et al. (1994). *Polymer* 35: 5269–5278.

112 Ghiggino, K.P., Hutchison, J.A., Langford, S.J. et al. (2007). *Adv. Funct. Mater.* 17: 805–813.
113 Zhu, L.-L., Qu, D.-H., Zhang, D. et al. (2010). *Tetrahedron* 66: 1254–1260.
114 Lukac, S. (1984). *J. Am. Chem. Soc.* 106: 4386–4392.
115 Kung, C.E. and Reed, J.K. (1986). *Biochemistry* 25: 6114–6121.
116 Furuno, T., Isoda, R., Inagaki, K. et al. (1992). *Immunol. Lett.* 33: 285–288.
117 Wandelt, B., Cywinski, P., Darling, G.D., and Stranix, B.R. (2005). *Biosens. Bioelectron.* 20: 1728–1736.
118 Yasuhara, K., Sasaki, Y., and Kikuchi, J. (2008). *Colloids Surf., B* 67: 145.
119 Hawe, A., Filipe, V., and Jiskoot, W. (2010). *Pharm. Res.* 27: 314–326.
120 Molin, M.D., Verolet, Q., Colom, A. et al. (2015). *J. Am. Chem. Soc.* 137: 568–571.
121 Suhina, T., Weber, B., Carpentier, C.E. et al. (2015). *Angew. Chem. Int. Ed.* 54: 3688–3691.
122 Haidekker, M.A., Brady, T.P., Lichlyter, D., and Theodorakis, E.A. (2006). *J. Am. Chem. Soc.* 128: 398–399.
123 Fischer, D., Theodorakis, E.A., and Haidekker, M.A. (2007). *Nat. Protoc.* 2: 227–236.
124 Dakanali, M., Do, T.H., Horn, A. et al. (2012). *Bioorg. Med. Chem.* 20: 4443–4450.
125 Suhling, K., Hirvonen, L.M., Levitt, J.A. et al. (2015). *Med. Photo.* 27: 3–40.
126 Kuimova, M.K., Yahioglu, G., Levitt, J.A., and Suhling, K. (2008). *J. Am. Chem. Soc.* 130: 6672–6673.
127 Kuimova, M.K., Botchway, S.W., Parker, A.W. et al. (2009). *Nat. Chem.* 1: 69–73.
128 Peng, X., Yang, Z., Wang, J. et al. (2011). *J. Am. Chem. Soc.* 133: 6626–6635.
129 Yang, Z., He, Y., Lee, J.-H. et al. (2013). *J. Am. Chem. Soc.* 135: 9181–9185.
130 Gatzogiannis, E., Chen, Z., Wei, L. et al. (2012). *Chem. Commun.* 48: 8694–8696.
131 Wang, L., Xiao, Y., Tian, W., and Deng, L. (2013). *J. Am. Chem. Soc.* 135: 2903–2906.
132 López-Duarte, I., Vu, T.T., Izquierdo, M.A. et al. (2014). *Chem. Commun.* 50: 5282–5284.
133 Hosny, N.A., Fitzgerald, C., Vyšniauskas, A. et al. (2016). *Chem. Sci.* 7: 1357–1367.
134 Lee, S.-C., Heo, J., Ryu, J.-W. et al. (2016). *Chem. Commun.* 52: 13695–13698.
135 Shimolina, L.E., Izquierdo, M.A., López-Duarte, I. et al. (2017). *Sci. Rep.* 7: 41097–41106.
136 Filarowski, A., Kluba, M., Cieślik-Boczula, K. et al. (2010). *Photochem. Photobiol. Sci.* 9: 996–1008.
137 Levitt, J.A., Chung, P.-H., Kuimova, M.K. et al. (2011). *ChemPhysChem* 12: 662–672.
138 Allen, B.D., Benniston, A.C., Harriman, A. et al. (2005). *Phys. Chem. Chem. Phys.* 7: 3035–3040.
139 Swalina, C. and Maroncelli, M. (2010). *J. Phys. Chem. C* 114: 5602–5610.
140 Yang, S. and Han, K. (2016). *J. Phys. Chem. A* 120: 4961–4965.

141 Jiao, L., Yu, C., Wang, J. et al. (2015). *RSC Adv.* 5: 89375–89388.
142 Prlj, A., Fabrizio, A., and Corminboeuf, C. (2016). *Phys. Chem. Chem. Phys.* 18: 32668–32672.
143 Aramendia, P.F., Negri, R.M., and Roman, E.S. (1994). *J. Phys. Chem.* 98: 3165–3173.
144 Sanborn, M.E., Connolly, B.K., Gurunathan, K., and Levitus, M. (2007). *J. Phys. Chem. B* 111: 11064–11074.
145 Zhou, C., Yuan, L., Yuan, Z. et al. (2016). *Inorg. Chem.* 55: 8564–8569.
146 Matsumoto, T., Takamine, H., Tanaka, K., and Chujo, Y. (2017). *Mater. Chem. Front.* https://doi.org/10.1039/c7qm00350a.
147 Rabideau, P.W., Hamilton, J.B., and Friedman, L. (1968). *J. Am. Chem. Soc.* 90: 4465–4466.
148 Lin, C.-H., Lin, K.-H., Pal, B., and Tsou, L.-D. (2009). *Chem. Commun.* 803–805.
149 Lin, Y.-C., Lin, C.-H., Chen, C.-Y. et al. (2011). *Org. Biomol. Chem.* 9: 4507–4517.
150 Wender, P.A. and Christy, J.P. (2007). *J. Am. Chem. Soc.* 129: 13402–13403.
151 Wender, P.A., Lesser, A.B., and Sirois, L.E. (2012). *Angew. Chem. Int. Ed.* 51: 2736–2740.
152 Takeuchi, S., Ruhman, S., Tsuneda, T. et al. (2008). *Science* 332: 1073–1077.
153 Angulo, G., Brucka, M., Gerecke, M. et al. (2016). *Phys. Chem. Chem. Phys.* 18: 18460–18469.
154 Chen, J., Kistemaker, J.C.M., Robertus, J., and Feringa, B.L. (2014). *J. Am. Chem. Soc.* 136: 14924–14932.
155 Ebe, K., Seno, H., and Horigome, K. (2003). *J. Appl. Polym. Sci.* 90: 436–441.
156 Uchida, K., Izumi, N., Sukata, S. et al. (2006). *Angew. Chem. Int. Ed.* 45: 6470–6473.
157 Norikane, Y., Hirai, Y., and Yoshida, M. (2011). *Chem. Commun.* 47: 1770–1772.
158 Akiyama, H. and Yoshida, M. (2012). *Adv. Mater.* 24: 2353–2356.
159 Hoshino, M., Uchida, E., Norikane, Y. et al. (2014). *J. Am. Chem. Soc.* 136: 9158–9164.
160 Uchida, E., Azumi, R., and Norikane, Y. (2015). *Nat. Commun.* 6: 7310.
161 Shimamura, A., Priimagi, A., Mamiya, J. et al. (2011). *ACS Appl. Mater. Interfaces* 3: 4190–4196.
162 Takeshima, T., Liao, W., Nagashima, Y. et al. (2015). *Macromolecules* 48: 6378–6384.
163 Zhou, H., Xue, C., Weis, P. et al. (2017). *Nat. Chem.* 9: 145–151.
164 Boyne, J.M., Millan, E.J., and Webster, I. (2001). *Int. J. Adhes. Adhes.* 21: 49–53.
165 Trenor, S.R., Long, T.E., and Love, B.J. (2005). *J. Adhes.* 81: 213–229.
166 Inui, T., Sato, E., and Matsumoto, A. (2012). *ACS Appl. Mater. Interfaces* 4: 2124–2132.
167 Yamaguchi, H., Kobayashi, Y., Kobayashi, R. et al. (2012). *Nat. Commun.* 3: 603.
168 Asadirad, A.M., Boutault, S., Erno, Z., and Branda, N.R. (2014). *J. Am. Chem. Soc.* 136: 3024–3027.

169 Heinzmann, C., Coulibaly, S., Roulin, A. et al. (2014). *ACS Appl. Mater. Interfaces* 6: 4713–4719.
170 Goodby, J.W., Collings, P.J., Kato, T. et al. (eds.) (2014). *Handbook of Liquid Crystals*. Wiley-VCH.
171 Li, Q. (ed.) (2012). *Liquid Crystals Beyond Displays*. Wiley.
172 Fleischmann, E.K. and Zentel, R. (2013). *Angew. Chem. Int. Ed.* 52: 8810–8827.
173 Ikeda, T. and Tsutsumi, O. (1995). *Science* 268: 1873–1875.
174 Tazuke, S., Kurihara, S., Yamaguchi, H., and Ikeda, T. (1987). *Chem. Lett.* 16: 911–914.
175 Huck, N.P.M., Jager, W.F., Lange, B., and Feringa, B.L. (1996). *Science* 273: 1686–1688.
176 Hrozhyk, U.A., Serak, S.V., Tabiryan, N.V., and Bunning, T.J. (2007). *Adv. Funct. Mater.* 17: 1735–1742.
177 White, T.J., McConney, M.E., and Bunning, T.J. (2010). *J. Mater. Chem.* 20: 9832–9847.
178 Ichimura, K., Suzuki, Y., Seki, T. et al. (1988). *Langmuir* 4: 1214–1216.
179 Gibbons, W.M., Shannon, P.J., Sun, S.-T., and Swetlin, B.J. (1991). *Nature* 351: 49–50.
180 Schadt, M., Schmitt, K., Nozinkov, V., and Chigrinov, V. (1992). *Jpn. J. Appl. Phys.* 31: 2155–2164.
181 Ichimura, K. (2000). *Chem. Rev.* 100: 1847–1873.
182 Yaroshchuk, O. and Reznikov, Y. (2012). *J. Mater. Chem.* 22: 286–300.
183 Seki, T., Nagano, S., and Hara, M. (2013). *Polymer* 54: 6053–6072.
184 Fukuhara, K., Nagano, S., Hara, M., and Seki, T. (2014). *Nat. Commun.* 5: 3320.
185 Finkelmann, H., Nishikawa, E., Pereira, G.G., and Warner, M.A. (2001). *Phys. Rev. Lett.* 87: 015501.
186 Yu, Y., Nakano, M., and Ikeda, T. (2003). *Nature* 425: 145.
187 Li, M.-H., Keller, P., Li, B. et al. (2003). *Adv. Mater.* 15: 569–572.
188 Camacho-Lopez, M., Finkelmann, H., Palffy-Muhoray, P., and Shelley, M. (2004). *Nat. Mater.* 3: 307–310.
189 Eelkema, R., Pollard, M.M., Vicario, J. et al. (2006). *Nature* 440: 163.
190 Yamada, M., Kondo, M., Mamiya, J. et al. (2008). *Angew. Chem. Int. Ed.* 47: 4986–4988.
191 Oosten, C.L., Bastiaansen, C.W.M., and Broer, D.J. (2009). *Nat. Mater.* 8: 677–682.
192 Iamsaard, S., Aßhoff, S.J., Matt, B. et al. (2014). *Nat. Chem.* 6: 229–235.
193 Priimagi, A., Barrett, C.J., and Shishido, A. (2014). *J. Mater. Chem. C* 2: 7155–7162.
194 Akiyama, H., Kanazawa, S., Okuyama, Y. et al. (2014). *ACS Appl. Mater. Interfaces* 6: 7933–7941.
195 Akiyama, H., Fukata, T., Yamashita, A. et al. (2017). *J. Adhes.* 93: 823–830.
196 Silva, L.F.M., Öchsner, A., and Adams, R.D. (eds.) (2011). *Handbook of Adhesion Technology*. Springer-Verlag.
197 Kato, T., Mizoshita, N., and Kishimoto, K. (2006). *Angew. Chem. Int. Ed.* 45: 38–68.

198 Rosen, B.M., Wilson, C.J., Wilson, D.A. et al. (2009). *Chem. Rev.* 109: 6275–6540.
199 Breton, G.W. and Vang, X. (1998). *J. Chem. Educ.* 75: 81–82.
200 Acree, W.E. Jr., and Chickos, J.S. (2006). *J. Phys. Chem. Ref. Data* 35: 1051–1330.
201 Okuda, M. and Katayama, K. (2008). *J. Phys. Chem. A* 112: 4545–4549.
202 Bhola, R., Bhola, R., Payamyar, P. et al. (2013). *J. Am. Chem. Soc.* 135: 14134–14141.
203 Kamperman, M. and Synytska, A. (2012). *J. Mater. Chem.* 22: 19390–19401.

3

Catechol-Containing Polymers: A Biomimetic Approach for Creating Novel Adhesive and Reducing Polymers

Hiroshi Yabu

Tohoku University, WPI-Advanced Institute for Materials Research (AIMR), 2-1-1, Katahira, Aoba-Ku, Sendai 980-8577, Japan

3.1 Background

3.1.1 Adhesive Proteins of Mussels

Mussels (*Mytilus edulis*) adhere strongly to the surface of rocks, plastics, and other materials under seawater. In 1981, Waite and coworkers reported that the phenolic compound L-3,4-hydroxyphenyl alanine (DOPA) was present in mussel adhesive proteins [1, 2]. In the 1990s, surface the interspecific protein sequences of mussels cultured in different regions were compared [3] and their genomes were analyzed [4]. Deming reported that DOPA moieties have multiple intra- and intermolecular reactivities based on the phenolic hydroxyl groups and end amino groups [5]. Because of the phenolic hydroxyl groups of catechol chelate metal ions, DOPA adheres to a wide variety of metals [6] (Figure 3.1). Oxidation of DOPA forms quinone derivatives that can react with amines and form imines. Radical generation at the hydroxyl or phenyl groups under basic conditions results in coupling of two DOPA molecules or polymerization of DOPA, which forms polydopamine (PDA). Additionally, hydroxyl groups can form hydrogen bonds among polar hydrophilic materials, and π–π interactions among phenyl groups allow adhesion to hydrophobic materials [7]. These multiple reactivities of DOPA result in the excellent adhesive properties of mussels to a wide variety of materials underwater. Lee et al. revealed the interactions between a single molecule catechol moiety and a surface by using atomic force microscopy and found that quinone formation through oxidation of the catechol decreased the adhesion force of catechol to a solid substrate [8].

3.1.2 Bio-Based Catechol-Containing Polymers

Bio-based catechol monomers that mimic the reactivity and adhesive properties of mussel adhesive proteins have been reported for adhesive polymers. DOPA is widely found in monomer units of adhesive proteins of marine adhesive organisms. One of the most well-known bio-based catechol-containing polymers is PDA (Figure 3.2). Under basic conditions, self-cyclization of dopamine

Molecular Technology: Materials Innovation, Volume 3, First Edition.
Edited by Hisashi Yamamoto and Takashi Kato.

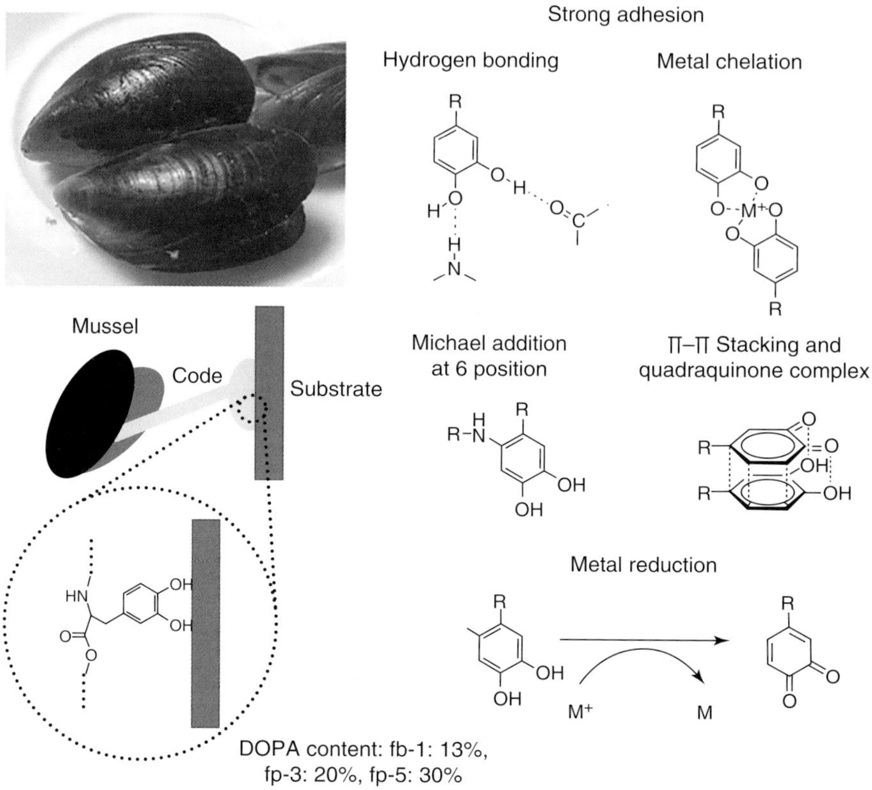

Figure 3.1 Photograph of mussels (top left) and schematic illustration of adhesion of mussels onto substrate with code and various properties of catechol moieties.

forms PDA in aqueous solution. When PDA formation occurs at a liquid–solid interface, the solid substrate surface is modified with a PDA membrane. Lee et al. reported that the PDA can modify a variety of inorganic substrates, including metal, metal oxides, and semiconductors, and organic substrates, including polytetrafluoroethylene (PTFE) despite its low surface free energy [9]. The 5-position of the PDA monomer unit is highly reactive because it is electron-poor; thus, the surface properties of the PDA membrane can be changed by nucleophilic substitution with thiol and amine derivatives. Furthermore, owing to the reducing properties of catechol moieties, metal plating of the PDA-modified surface has been reported. Norepinephrine, which is a low molecular weight catechol derivative, also polymerizes under alkaline conditions to form uniform nanolayers on a wide variety of materials. Based on this surface modification process, a poly(caprolactone) grafted layer was formed by ring-opening polymerization using the modified layer as an initiator.

Bio-based monomers from botanical sources also form catechol-containing polymers. Lignin, which is a phenolic extract from wood, contains a variety of catechol derivatives, including 3,4-dihydroxyhydrocinnamic acid (caffeic acid). Various catechol derivatives, such as caffeic acid and 3-(3-hydroxyphenyl)

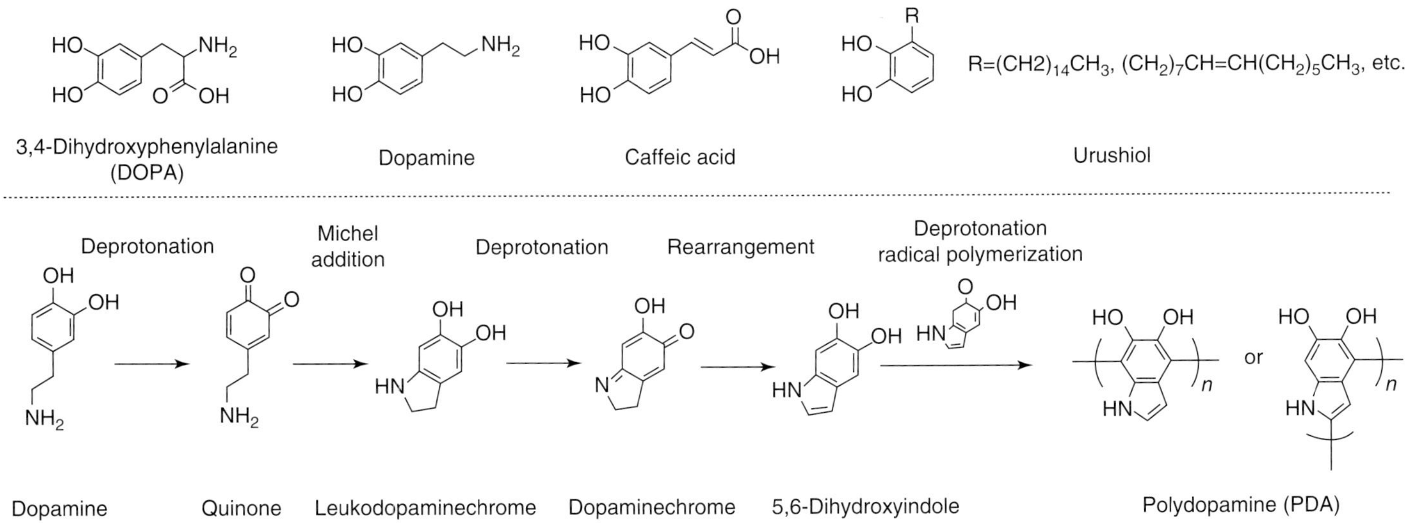

Figure 3.2 Bio-based catechol monomers and formation process of PDA.

propionic acid, have been reported. These can also be used to create adhesive polymers based on catechol moieties. Caffeic acid has both a double bond and a catechol moiety. Polycondensation of caffeic acid and 4-hydroxycinnamic acid results in a highly adhesive resin [10].

A well-known bio-based material containing a catechol moiety is urushiol, which is the main constituent of oriental lacquer and is used for coating wood [11]. Urushiol comprises a mixture of 3-position-substituted catechols and forms robust resins through oxidation and cross-linking under ambient conditions. These properties are used in coating traditional Japanese wooden dishes with lacquer, which consists of cross-linked urushiols. Recently, Watanabe et al. reported that sprayed urushiol solutions formed physically and chemically robust thin films on various kinds of surfaces [12].

3.1.3 Synthetic Polymers Containing Catechol Moieties

In addition to bio-based catechol polymers, a variety of synthetic polymers with catechol moieties on their side chains or end groups have been reported (Table 3.1). Catechol-containing synthetic polymers can be classified based on their structures and functionalized regions. Because synthetic polymers can be designed arbitrarily, polymers containing linear, branched [13], multiarm [14], cyclic [15], and cross-linked [16] main chains can be created depending on their applications (Figure 3.3). The number and position (end-functionalized or introduced as a side chain) of catechol moieties can be controlled.

Many surface modification applications of catechol moieties to adhesives have also been reported. Huang et al. reported polyethylene oxide and polypropylene oxide bioadhesives with DOPA end groups [17], and subsequently, various kinds of synthetic polymers with catechol moieties have been reported. Lee et al. demonstrated reversible underwater and dry adhesion to various substrates using a gecko-like hairy surface coated with poly(dopamine methacrylamide-*co*-methoxy ethyl acrylate) (P(DMA-*co*-MEA)) [18]. They showed that random side-chain copolymers with catechol moieties show strong adhesion underwater and reversible attachment and detachment due to the hairy surface structure. Guvendiren et al. reported hydrogel adhesives comprising poly(methyl methacrylate), polyacrylic acid, and dopamine methacrylamide [19]. In this system, the catechol moieties function as a molecular adhesive on the surface and as cross-linking points to guarantee mechanical strength.

Owing to their strong adhesive properties, catechol-containing polymers have been also been used as surface modifiers for solid substrates. Dalsin et al. reported that polyethylene glycols (PEGs) with catechol-functionalized end groups formed stable monolayers on a gold substrate and showed good antifouling properties [20]. Subsequently, various types of polymers with catechol-modified end groups have been reported. Zürcher et al. also reported surface modification with PEGs containing a metal chelator as an antifouling moiety [21].

This "grafting to" approach with end-functionalized polymers is a powerful method for functionalizing inorganic nanoparticles stably dispersed in organic solvents. Amstad et al. fabricated PEGs containing catechol, other phenolic moieties, and carboxylic acids as end groups for surface modification of iron

Table 3.1 Synthetic polymers containing catechol moieties.

Polymer structure	Catechol unit	Synthesis method	Yielded polymers	Applications
End-functionalized		Condensation surface-initiated polymerization	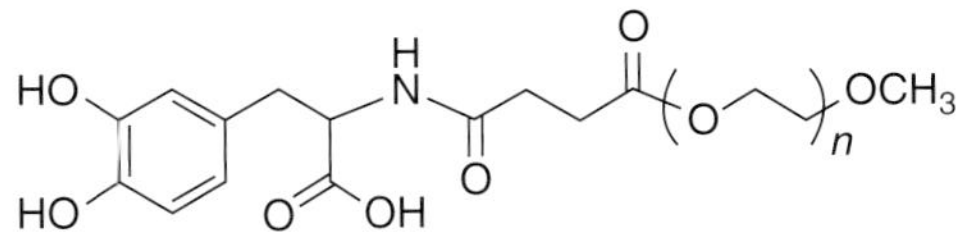	Surface coatings
Random	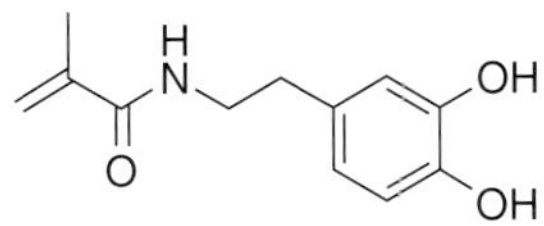	Free radical polymerization	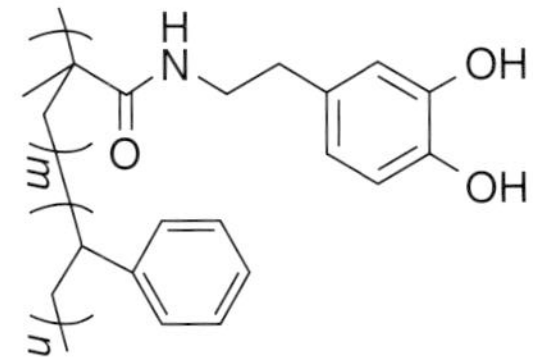	Surface coatings cross-linkers adhesives
	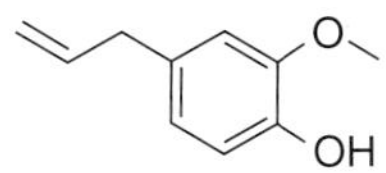	Click chemistry	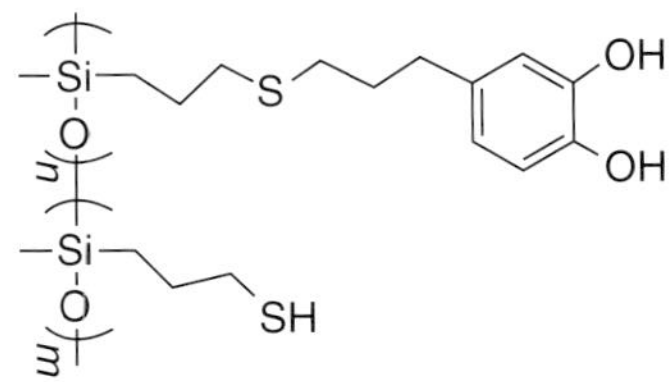	Surface coatings adhesives
	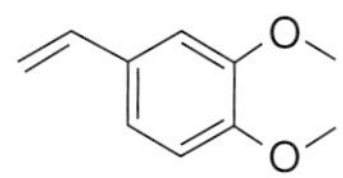	Free radical or anionic polymerization	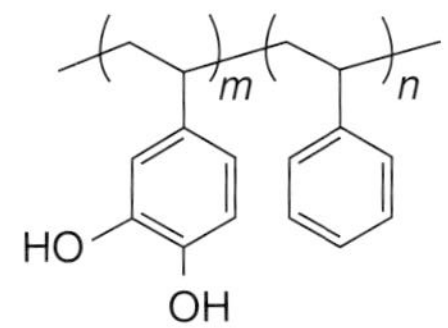	Surface coatings adhesives

(*continued*)

Table 3.1 (Continued)

Polymer structure	Catechol unit	Synthesis method	Yielded polymers	Applications
Block	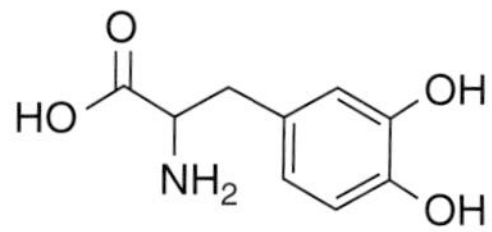	Condensation	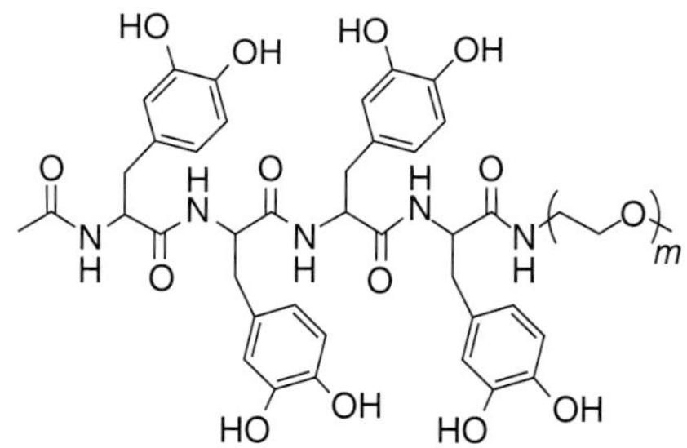	Reductants
	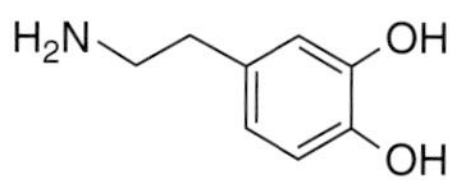	Condensation	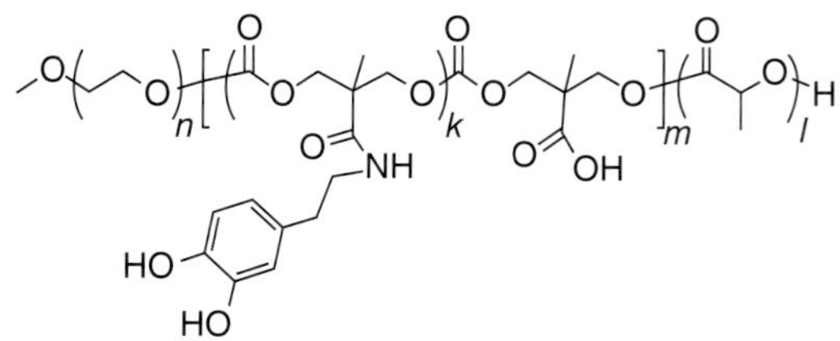	Cross-linkers
	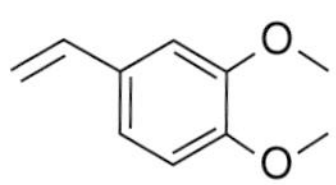	RAFT polymerization	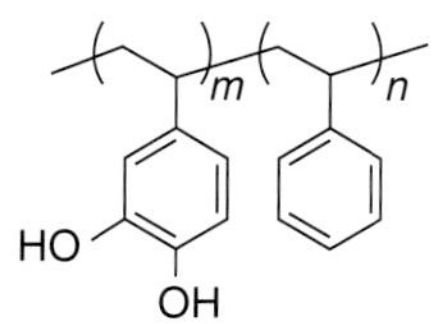	Surface coatings reductants

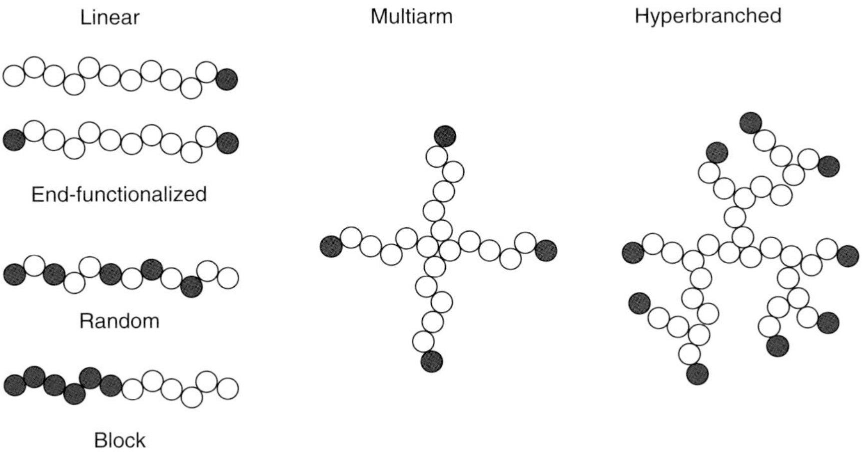

Figure 3.3 Schematic illustration of polymer architectures having catechol moieties. White and dark circles show non-catecholic and catecholic monomers.

oxide nanoparticles by ligand exchange with lipids, and they estimated the ligand exchange efficiency by elemental analysis [22]. Davis et al. also measured the stability of ^{14}C-labeled catechol-containing lipids on iron oxide nanoparticles [23]. These studies showed that lipids containing DOPA and nitro-DOPA, in which a nitro group was substituted at the 6-position of DOPA, have high stability on iron oxide particles.

Side-chain-functionalized polymers have also been used to modify the surfaces of materials. Saltz et al. synthesized an antifouling coating based on PEG and polypeptide block copolymers, in which catechol moieties were immobilized as side chains. Amphiphilic copolymers comprising dopamine methacrylamide and hydrophobic monomers, such as styrene, have also been reported as surface modifiers for inorganic nanoparticles.

In contrast to the "grafting to" approach, surface-initiated polymerization, which uses catechol-containing initiators, has been reported as a "grafting from" approach. Because atom-transfer living radical polymerization (ATRP) requires initiators containing halogen atoms, DOPA derivatives with bromide groups were immobilized onto a metal substrate, and then oligoethylene glycol methyl ether methacrylate was polymerized by ATRP [24]. Because the DOPA catechol groups reacted with the surface hydroxyl groups of the metal oxide layers and chelated the metal atoms, bromide-containing initiators were immobilized as self-assembled monolayers, and polymer brush layers were eventually formed on the surface of the metal substrate.

3.1.4 Toward Biomimetic Molecular Technology

Based on our understanding of the fundamental adhesive and surface modification properties of mussel-inspired polymers and their syntheses, we discuss the practical applications of these functional materials. In this section, molecular technologies based on biomimetic mussel-inspired polymers are discussed.

We provide an overview of the development of advanced adhesives and surface modifiers, which are suitable for a wide variety of adhesion and functionalization of materials. Furthermore, we discuss electrochemical applications of mussel-inspired polymers as reducing agents, ion conductors, and templates for carbon materials.

3.2 Advanced Adhesives and Surface Modification Agents

The fundamental adhesive properties of catechol-containing polymers have been discussed in various review articles; therefore, we focus on recent advances in adhesives and surface modifiers based on the adhesive properties of catechol groups.

Since Lee et al. reported underwater adhesives comprising copolymers of dopamine methacrylamide and ethylene oxide methacrylate, the strong adhesion properties of catechol groups have been used to create strong adhesive polymers, especially for underwater conditions [18, 25]. Recent methods for creating strong adhesives are classified as the addition of functional monomers and moieties to obtain stronger adhesion properties, and as simplifying systems to understand the adhesive properties of catechol-containing polymers. As an example of the addition method, Clancy et al. reported strong underwater adhesive polymers comprising four different monomers, in which anionic, cationic, hydrophobic, and catechol-containing monomers were polymerized [26]. Mussel adhesive proteins also containing multiple hydrophobic and ionic groups are a representative polymer for applying the concept of mussel adhesive proteins to synthetic polymers. In addition to linear polymers, other structural variations have been reported. Originally, end-functionalized four-arm polymers were used to create adhesive thin layer-by-layer (LbL) membranes [27]. Recently, hyperbranched polymers with all their end groups functionalized with catechol groups have been synthesized and used as tissue adhesives. Adhesion to biological tissues is difficult owing to their high content of water [13]. As an example of the second method of simplifying the structure of adhesive molecules, Ahn et al. reported that low molecular weight zwitterionic surfactants consisting of catechol, anionic, cationic, and hydrophobic moieties showed underwater adhesive properties. These properties arose from the adhesion of catechol moieties to the surface of materials, cross-linking of catechol moieties, and hydrophobic interactions [28]. These results showed that the hydrophobic moiety plays a significant role in adhesion properties. In a copolymer system, Wilker and coworker synthesized random copolymers of 3,4-dihydroxy styrene and styrene and examined their adhesion [29]. Because 3,4-dihydroxy styrene is the simplest monomer unit with a catechol moiety, the series of copolymers shows the effect of the catechol moieties on the adhesion properties.

The most common application of catechol-containing materials is surface modification. Compared with the strong interaction between gold and thiols and related moieties, the catechol moiety has the advantage of adhering stably

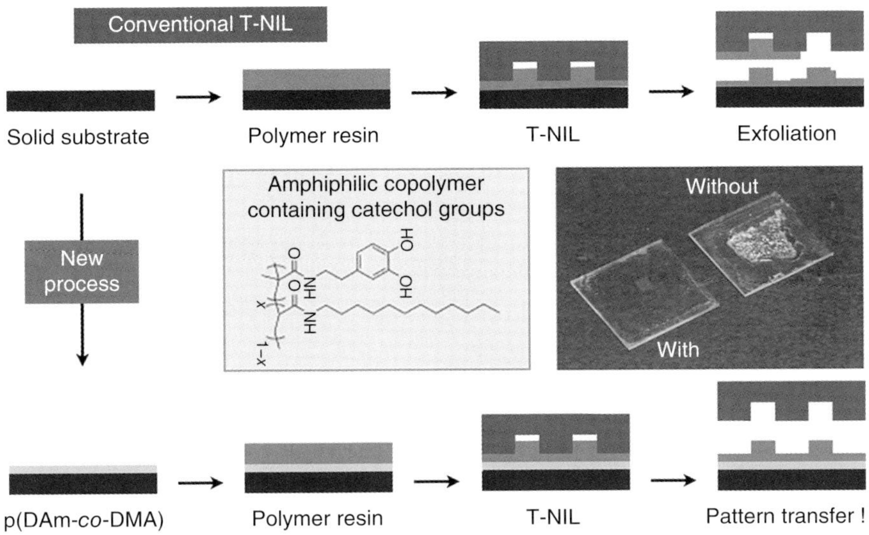

Figure 3.4 Application of mussel-inspired catechol-containing polymers for thermal nanoimprint lithography.

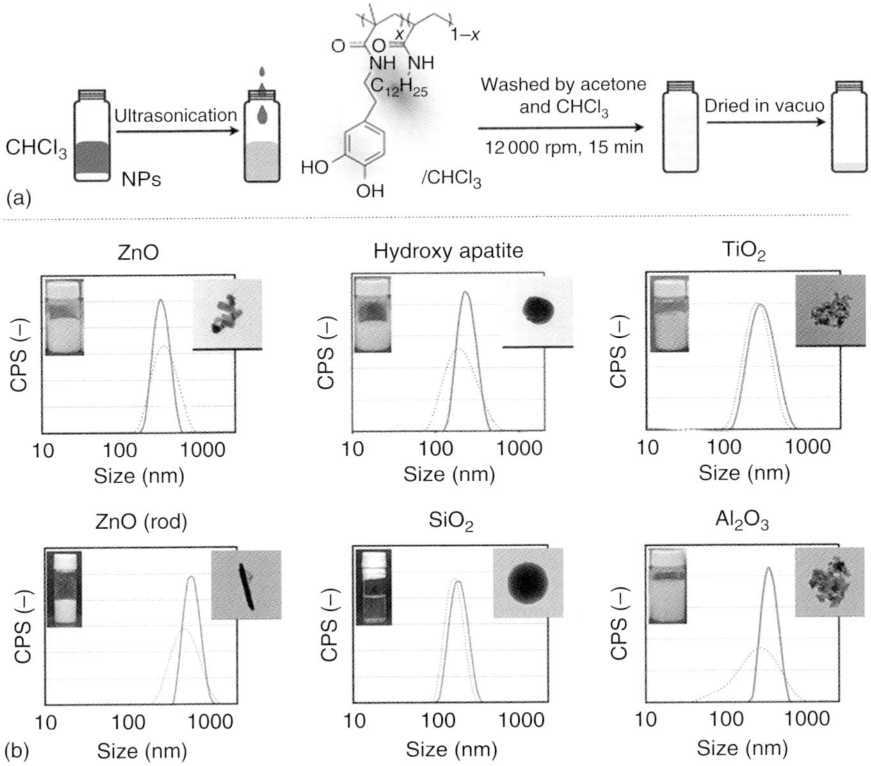

Figure 3.5 Schematic illustration of nanoparticle stabilization with mussel-inspired catechol-containing polymers (a) and TEM and size distribution of nanoparticles dispersed in chloroform (b). Inset photographs show dispersions of nanoparticles in bottles.

to a wide variety of materials. PDA films have been used as surface modification agents because they form thin flat films that adhere strongly to a variety of solid materials [9]. Recently, liquid marbles stabilized with PDA membranes have been fabricated by using liquid–air interface. The liquid droplets stabilized with PDA membranes act like solid particles and by selective surface modification of PDA at the liquid–air interface, Janus particles can be formed. Amphiphilic copolymers with catechol moieties can be used as surface modification agents for material and nanoparticle surfaces (Figure 3.4). Hydrophobic organic solvent dispersions of various inorganic nanoparticles can be obtained by immobilizing copolymers comprising hydrophobic and catechol-containing monomers on their surfaces (Figure 3.5).

3.3 Reducing Agents for Creating Nanoscale Metallic Structures

The catechol moiety has reducing properties arising from its two phenolic hydroxyl groups. Polyphenols, including tannic acid, extracted from various plant sources, have been used as reducing agents for noble metal ions [30]. Because phenol, catechol, and resorcinol groups have lower pK_a values (9–10) than alcohols, they function as reducing agents for noble metal ions. The catechol group is particularly useful for noble metal nanoparticle formation because of its reducing properties and strong surface adhesion. Various types of catechol-containing polymeric materials have been used to form noble metal nanoparticles. Lee et al. reported LbL assembly of a catechol-containing polyethyleneimine (PEI) derivative and a hyaluronic acid (HA) derivative, and the formation of silver nanoparticles embedded in the LbL films [27]. Dihydroxy phenyl propionic acid was conjugated with the PEI amino groups, and dopamine was conjugated with the hydroxyl groups of HA. A PTFE substrate, which has very low surface free energy, was alternately immersed in these two functionalized polymers to prepare the LbL films of the two polymers. After immersion in an aqueous solution of silver nitrate, silver(0) nanoparticles were formed inside the films. The silver composite films exhibited good antibacterial properties. A pseudomelanin membrane was also used as a platform for metal nanoparticle formation [31]. Dopamine oxidation forms thin films comprising pseudomelanin on flat surfaces. The pseudomelanin film changes the surface properties by adhering strongly to the whole surface owing to the strong adhesive properties of the catechol moieties. There are residue catechol moieties in the pseudomelanin films that do not contribute to adhesion and these moieties act as reducing agents. Ball et al. [31] reported the formation of silver nanoparticles on the surface of pseudomelanin film prepared on a polystyrene dish or a silicon substrate. Nonwoven fibers were also coated with noble metal nanoparticles by reducing silver ions with catechol moieties. Nonwoven fiber sheets consisting of catechol-grafted poly(vinyl alcohol) were prepared by electrospinning, and then were immersed in metal ion solution forming metal nanoparticle arrays on the surface of fibers. These techniques demonstrate the preparation and functionalization of noble metal nanoparticles on the surface of various shaped substrates and their potential use as antibacterial coatings or catalysts. However, the size of nanoparticles could not be controlled.

Block copolymer micelles, which are formed by self-assembly of amphiphilic block copolymers in a solvent, have uniform sizes that depend on their molecular weight. These micelles have been used as templates for inorganic nanoparticle formation [32]. Many kinds of block copolymers, including poly(styrene-*block*-vinyl pyridine) [33], poly(styrene-*block*-acrylic acid) [34], and so on [35, 36], have been used to form inorganic and metal nanoparticles with a reducing agent, such as hydrazine or sodium borohydride. Because catechol moieties have reducing properties, block copolymers containing catechol moieties as hydrophilic groups can form uniform-sized metal nanoparticles without using reducing agents. Black et al. synthesized hydrophilic block copolymers comprising methoxy-capped PEG and poly(3,4-dihydroxy phenyl alanine) (PDOPA) and used them to form silver and gold nanoparticles [37]. PDOPA, which is an adhesive fragment observed in mussel adhesive proteins, acted as a reducing moiety. The nanoparticle size was controlled at between 10 and 68 nm and the nanoparticles consisted of assemblies of smaller nanoparticles. The nanoparticles were dispersed in water.

Recently, examples of amphiphilic and synthetic block copolymers containing catechol moieties have been reported. Because catechol moieties usually prevent the propagation of radical polymerization, it is difficult to synthesize catechol-containing block copolymers directly by conventional living radical polymerization. There are two ways to synthesize catechol-containing block copolymers: substitution of side chains of pre-synthesized block copolymers with catechol moieties; and living polymerization of protected catechol moieties of monomers followed by deprotection. The synthesis of poly(oligo-ethoxy methacrylate-random-pentafluorobenzene methacrylate) by reversible-addition fragmentation transfer (RAFT) polymerization and substituting the pentafluorobenzene moiety with dopamine has been reported [38]. Hagberg et al. reported introduction of catechol moieties into side-chain vinyl groups by the en-thiol reaction [39]. This is not a block copolymer; however, the combination of living radical polymerization and the polymer reaction could be used to synthesize catechol-containing block copolymers. Protected catechol-containing monomers could also be used. Patil et al. reported the polymerization of acetonide-protected dopamine methacrylamide by RAFT polymerization and synthesis of diblock copolymers with catechol moieties followed by deprotection of acetonide groups with trifluoroacetic acid [40].

In contrast to these poly(methacrylamide) copolymers, copolymers containing poly(3,4-dihydroxy styrene), which is the simplest catechol-containing polymer – in other words, polymers containing poly(vinyl catechol) (PVCa) segments – have been developed by using living radical polymerization of 3,4-dimethoxy styrene (DMSt) followed by deprotection of methoxy groups (Figure 3.6). Controlled polymerization of styrene and DMSt has been reported using 2,2,6,6-tetramethylpiperidine 1-oxyl (TEMPO); however, the lowest polydispersity index (PDI) was 1.56, which is not sufficient for uniform self-assembly of block copolymers [41]. Isakova et al. reported RAFT polymerization of DMSt and deprotection of methoxy groups with BBr_3 to obtain hydroxyl groups [42],

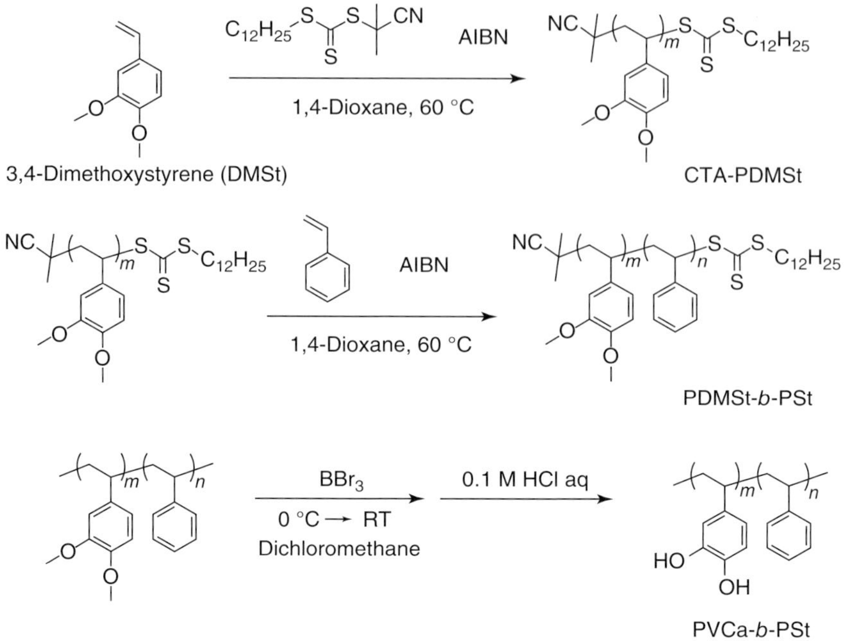

Figure 3.6 Synthetic scheme of PVCa-*b*-PSt.

and a narrow PDI of 1.1–1.3 was realized. Saito and Yabu were the first to synthesize block copolymers of styrene and DMSt by RAFT polymerization and prepare PVCa-*b*-PSt by deprotecting the methoxy groups [43]. In organic solvents, PVCa-*b*-PSt forms inverse micelles in which the PVCa block creates reducing hydrophilic cores. When metal ions, such as silver, gold, or other noble metal ions, are added to the solution, the metal ions are spontaneously reduced and form metal nanoparticles stabilized with PVCa-*b*-PSt micelles. Because the PVCa segments function as a reducing agent and as adhesive moieties for metal nanoparticles, the metal nanoparticles are stably dispersed in organic solvents. This bifunctionality of the PVCa segment is unique among block copolymer micelle systems. The size of nanoparticles can be controlled by changing the molecular weight of the PVCa segments (Figure 3.7).

Block copolymers form various phase-separated structures in the solution phase and in the bulk state. Changing the molecular weight, copolymerization ratio, and immiscibility of the block copolymers controls the phase-separated structures. PVCa-*b*-PSt also forms phase-separated structures in its bulk film [44]. The inner structures of the PVCa-*b*-PSt films were changed from spherical to cylindrical as the proportion of the PVCa component increased, and finally, a lamellar phase formed at a 1 : 1 block ratio of PVCa to polystyrene. Figure 3.8 shows transmission electron tomography (ET) images of films comprising bicontinuous and lamellar phase-separated structures. In the figure, large nanoparticles (>5 nm) and nonspecifically attached nanoparticles on the film surface are red, and smaller nanoparticles in the polymer are yellow. Green

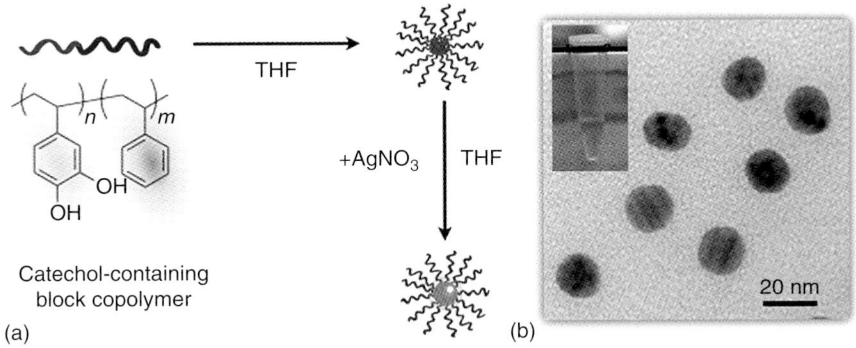

Figure 3.7 Schematic illustration of nanomicelle formation from PVCa-*b*-PSt in tetrahydrofuran (THF) (a) and TEM image of synthesized Ag nanoparticles (b). The inset image shows dispersion of Ag nanoparticles stabilized with PVCa-*b*-PSt.

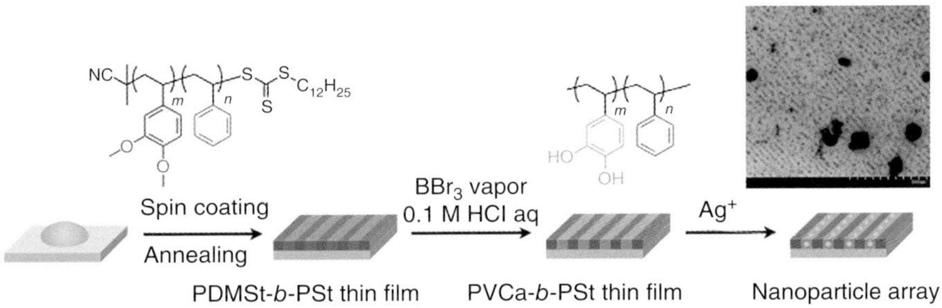

Figure 3.8 Schematic illustration of preparation of PVCa-*b*-PSt film having Ag nanoparticles.

regions indicate PVCa segments stained with OsO_4. This distribution of silver nanoparticles forms as follows. Silver ions diffuse from the edges of PVCa domains located at the surface of the film. The silver ions are spontaneously reduced by catechol moieties in the PVCa domains and form the nuclei of the nanoparticles. It has been reported that the morphologies and volumes of phase-separated structures strongly affect the size of the nanoparticles produced by reduction in block copolymer phase separation owing to limitations of the metal source and diffusion. Silver nanoparticles gradually grow, although the size of the nanoparticles in the PVCa domains is limited because the quantity of silver ions is dependent on the volume of the PVCa domains. Larger silver nanoparticles grow at the junction points where the volume of PVCa domains is larger than that in the cylinder phase. The ET observations showed that no silver nanoparticles were formed at the bottommost PVCa layer that adhered to the substrate surface. This PVCa layer is homogenously wetted and adhered to the substrate surface; therefore, it is impossible for silver ions to diffuse into this layer. Because silver ions dissolved in an aqueous solution, the solution can swell only in the PVCa phase. During the diffusion, silver ions are reduced to silver nanoparticles by the catechol moieties. Thus, the concentration of silver nanoparticles gradually decreases.

3.4 Application as Proton-Conductive Thin Films

Proton-conductive polymer electrolytes are essential materials for polymer electrolyte fuel cells. Commercialized polyelectrolytes, including Nafion membranes, have nanochannels with high sulfonic acid group content as proton carriers to transport protons [45]. Nanochannel structures induced by self-assembly of molecules are key materials for creating efficient proton- and ion-conductive membranes. Highly conductive liquid crystalline proton and ion conductors, which form unique self-assembled nanochannel structures [46], and aromatic ionomers [47] have been reported. Amphiphilic block copolymers, which form various phase-separated structures in their bulk state, are promising materials for high proton conductivity because they spontaneously self-assemble into nanoscale channel structures [48]. The PVCa-*b*-PSt block copolymer (Section 3.3) forms a variety of microphase-separated structures, including spheres, cylinders, bicontinuous phases, and lamellae, depending on the copolymerization ratio of PVCa and polystyrene. Furthermore, the reducing properties of the two phenolic groups in catechol mean that size-controlled metal nanoparticles were spontaneously formed when using the diblock copolymers as reductants and templates for nanoparticles. The proton conductivity of the PVCa-*b*-PSt film was increased 10-fold by the addition of silver nanoparticles into the proton conduction channels filled with catechol moieties [49]. There are several reports of the incorporation of inorganic nanoparticles into the nanochannels of ionic conductive materials to the ionic conductivity, although the nanoparticles were used as aligners for the nanochannels.

3.5 Templates for Carbon Materials

PDA films are dark brown owing to the long π conjugation length, which indicates that PDA films are good candidates as sources for carbon materials. N-doped carbon materials, which have wide electrical windows and can be applied to a wide variety of sensing devices, have been fabricated by pyrolysis of PDA films in vacuo. Li et al. reported wafer-scale production of N-doped carbon films by pyrolysis of PDA films prepared on copper substrates because carbon films are easily exfoliated from the copper substrate for transfer onto other substrates [50]. Because N-doped carbon has a unique electrochemical window, it can be used in a wide variety of applications including catalysis [51], capacitors [52], and fuel cells [53].

3.6 Summary

Mussel-inspired polymers with strong adhesion and reduction properties are promising for a wide variety of applications. Various chemical reactions occur with the catechol moiety. This reactivity for wide variety of materials is useful for functionalizing surfaces, adhering to different materials, and forming solid metals on the surface of organic materials. Some unique applications, including proton-conductive films and templates for carbon materials, have emerged. These are good examples of molecular biomimicry.

References

1 Waite, J.H. and Tanzer, M.L. (1981). Polyphenolic substance of Mytilus edulis: novel adhesive containing L-DOPA and hydroxyproline. *Science* 212: 1038–1040.

2 Waite, J.H. (1983). Evidence for a repeating 3,4-dihydroxyphenylalanine- and hydroxyproline-containing decapeptide in the adhesive protein of the mussel, Mytilus edulis L. *J. Biol. Chem.* 258: 2911–2915.

3 Inoue, K., Waite, J.H., Matsuoka, M. et al. (1995). Interspecific variations in adhesive protein sequences of *Mytilus edulis, M. galloprovincialis,* and *M. trossulus. Biol. Bull.* 189: 370–375.

4 Inoue, K., Takeuchi, Y., Miki, D., and Odo, S. (1995). Mussel adhesive plaque protein gene is a novel member of epidermal growth factor-like gene family. *J. Biol. Chem.* 270: 6698–6701.

5 Deming, T.J. (1999). Mussel byssus and biomolecular materials. *Curr. Opin. Chem. Biol.* 3: 100–105.

6 Yamamoto, H., Yoshihiro Sakai, A., and Ohkawa, K. (2000). Synthesis and wettability characteristics of model adhesive protein sequences inspired by a marine mussel. *Biomacromolecules* 1: 543–551.

7 Baty, A.M., Suci, P.A., Tyler, B.J., and Geesey, G.G. (1996). Investigation of mussel adhesive protein adsorption on polystyrene and poly(octadecyl methacrylate) using angle dependent XPS, ATR-FTIR, and AFM. *J. Colloid Interface Sci.* 177: 307–315.

8 Lee, H., Scherer, N.F., and Messersmith, P.B. (2006). Single-molecule mechanics of mussel adhesion. *Proc. Natl. Acad. Sci. U.S.A.* 103: 12999–13003.

9 Lee, H., Dellatore, S.M., Miller, W.M., and Messersmith, P.B. (2007). Mussel-inspired surface chemistry for multifunctional coatings. *Science* 318: 426–430.

10 Kaneko, D., Wang, S., Matsumoto, K. et al. (2011). Mussel-mimetic strong adhesive resin from bio-base polycoumarates. *Polym. J.* 43: 855–858.

11 Watanabe, H., Fujimoto, A., and Takahara, A. (2013). Characterization of catechol-containing natural thermosetting polymer 'urushiol' thin film. *J. Polym. Sci., Part A: Polym. Chem.* 51: 3688–3692.

12 Watanabe, H., Fujimoto, A., and Takahara, A. (2015). Spray-assisted nanocoating of the biobased material urushiol. *Langmuir* 31: 2360–2365.

13 Zhang, H., Bré, L.P., Zhao, T. et al. (2014). Mussel-inspired hyperbranched poly(amino ester) polymer as strong wet tissue adhesive. *Biomaterials* 35: 711–719.

14 Sedó, J., Saiz-Poseu, J., Busqué, F., and Ruiz-Molina, D. (2013). Catechol-based biomimetic functional materials. *Adv. Mater.* 25: 653–701.

15 Zieger, M.M., Pop-Georgievski, O., de los Santos Pereira, A. et al. (2017). Ultrathin monomolecular films and robust assemblies based on cyclic catechols. *Langmuir* 33: 670–679.

16 Nishida, J., Kobayashi, M., and Takahara, A. (2012). Gelation and adhesion behavior of mussel adhesive protein mimetic polymer. *J. Polym. Sci., Part A: Polym. Chem.* 51: 1058–1065.

17 Huang, K., Lee, B.P., Ingram, D.R., and Messersmith, P.B. (2002). Synthesis and characterization of self-assembling block copolymers containing bioadhesive end groups. *Biomacromolecules* 3: 397–406.

18 Lee, H., Lee, B.P., and Messersmith, P.B. (2007). A reversible wet/dry adhesive inspired by mussels and geckos. *Nature* 448: 338–341.

19 Guvendiren, M., Messersmith, P.B., and Shull, K.R. (2007). Self-assembly and adhesion of DOPA-modified methacrylic triblock hydrogels. *Biomacromolecules* 9: 122–128.

20 Dalsin, J.L., Hu, B.H., Lee, B.P., and Messersmith, P.B. (2003). Mussel adhesive protein mimetic polymers for the preparation of nonfouling surfaces. *J. Am. Chem. Soc.* 125: 4253–4258.

21 Zürcher, S., Wäckerlin, D., Bethuel, Y. et al. (2006). Biomimetic surface modifications based on the cyanobacterial iron chelator anachelin. *J. Am. Chem. Soc.* 128: 1064–1065.

22 Amstad, E., Gehring, A.U., Fischer, H. et al. (2011). Influence of electronegative substituents on the binding affinity of catechol-derived anchors to Fe_3O_4 nanoparticles. *J. Phys. Chem. C* 115: 683–691.

23 Davis, K., Qi, B., Witmer, M. et al. (2014). Quantitative measurement of ligand exchange on iron oxides via radiolabeled oleic acid. *Langmuir* 30: 10918–10925.

24 Fan, X., Lin, L., Dalsin, J.L., and Messersmith, P.B. (2005). Biomimetic anchor for surface-initiated polymerization from metal substrates. *J. Am. Chem. Soc.* 127: 15843–15847.

25 Lee, B.P., Messersmith, P.B., Israelachvili, J.N., and Waite, J.H. (2011). Mussel-inspired adhesives and coatings. *Annu. Rev. Mater. Res.* 41: 99–132.

26 Clancy, S.K., Sodano, A., Cunningham, D.J. et al. (2016). Marine bioinspired underwater contact adhesion. *Biomacromolecules* 17: 1869–1874.

27 Lee, H., Lee, Y., Statz, A.R. et al. (2008). Substrate-independent layer-by-layer assembly by using mussel-adhesive-inspired polymers. *Adv. Mater.* 20: 1619–1623.

28 Ahn, B.K., Das, S., Linstadt, R. et al. (2015). High-performance mussel-inspired adhesives of reduced complexity. *Nat. Commun.* 6: 1–7.

29 White, J.D. and Wilker, J.J. (2011). Underwater bonding with charged polymer mimics of marine mussel adhesive proteins. *Macromolecules* 44: 5085–5088.

30 Lee, Y. and Park, T.G. (2011). Facile fabrication of branched gold nanoparticles by reductive hydroxyphenol derivatives. *Langmuir* 27: 2965–2971.

31 Ball, V., Nguyen, I., Haupt, M. et al. (2011). *J. Colloid Interface Sci.* 364: 359–365.

32 Möller, M., Spatz, J.P., and Roescher, A. (1996). Gold nanoparticles in micellar poly (styrene)-b-poly (ethylene oxide) films—size and interparticle distance control in monoparticulate films. *Adv. Mater.* 8: 337–340.

33 Spatz, J.P., Mössmer, S., Hartmann, C. et al. (2000). Ordered deposition of inorganic clusters from micellar block copolymer films. *Langmuir* 16: 407–415.
34 Du, B., Zhao, B., Tao, P. et al. (2008). Amphiphilic multiblock copolymer stabilized Au nanoparticles. *Colloids Surf., A* 317: 194–205.
35 Kang, Y. and Taton, T.A. (2005). Core/shell gold nanoparticles by self-assembly and crosslinking of micellar, block-copolymer shells. *Angew. Chem. Int. Ed.* 117: 413–416.
36 Sakai, T. and Alexandridis, P. (2005). Mechanism of gold metal ion reduction, nanoparticle growth and size control in aqueous amphiphilic block copolymer solutions at ambient conditions. *J. Phys. Chem. B* 109: 7766–7777.
37 Black, K.C.L., Liu, Z., and Messersmith, P.B. (2011). Catechol redox induced formation of metal core–polymer shell nanoparticles. *Chem. Mater.* 23: 1130–1135.
38 Wang, C.X., Braendle, A., Menyo, M.S. et al. (2015). Catechol-based layer-by-layer assembly of composite coatings: a versatile platform to hierarchical nano-materials. *Soft Matter* https://doi.org/10.1039/C5SM01374G.
39 Hagberg, E.C., Malkoch, M., Ling, Y. et al. (2007). Effects of modulus and surface chemistry of thiol-ene photopolymers in nanoimprinting. *Nano Lett.* 7: 233–237.
40 Patil, N., Falentin-Daudré, C., Jérôme, C., and Detrembleur, C. (2015). Mussel-inspired protein-repelling ambivalent block copolymers: controlled synthesis and characterization. *Polym. Chem.* 6: 2919–2933.
41 Gravert, D.J., Datta, A., Wentworth, P., and Janda, K.D. (1998). Soluble supports tailored for organic synthesis: parallel polymer synthesis via sequential normal/living free radical processes. *J. Am. Chem. Soc.* 120: 9481–9495.
42 Isakova, A., Topham, P.D., and Sutherland, A.J. (2014). Controlled RAFT polymerization and zinc binding performance of catechol-inspired homopolymers. *Macromolecules* 47: 2561–2568.
43 Saito, Y. and Yabu, H. (2015). Synthesis of poly(dihydroxystyrene-block-styrene) (PDHSt-b-PSt) by the RAFT process and preparation of organic-solvent-dispersive Ag NPs by automatic reduction of metal ions in the presence of PDHSt-b-PSt. *Chem. Commun.* 51: 3743–3746.
44 Saito, Y., Higuchi, T., Jinnai, H. et al. (2016). Silver nanoparticle arrays prepared by in situ automatic reduction of silver ions in mussel-inspired block copolymer films. *Macromol. Chem. Phys.* 217: 726–734.
45 Kreuer, K.-D. (2014). Ion conducting membranes for fuel cells and other electrochemical devices. *Chem. Mater.* 26: 361–380.
46 Ichikawa, T., Kato, T., and Ohno, H. (2012). 3D continuous water nanosheet as a gyroid minimal surface formed by bicontinuous cubic liquid-crystalline zwitterions. *J. Am. Chem. Soc.* 134: 11354–11357.
47 Miyake, J., Watanabe, M., and Miyatake, K. (2014). Ammonium-functionalized poly(arylene ether)s as anion-exchange membranes. *Polym. J.* 46: 656–663.
48 Miranda, D.F., Versek, C., Tuominen, M.T. et al. (2013). Cross-linked block copolymer/ionic liquid self-assembled blends for polymer gel electrolytes

with high ionic conductivity and mechanical strength. *Macromolecules* 46: 9313–9323.

49 Yabu, H., Matsui, J., Hara, M. et al. (2016). Proton conductivities of lamellae-forming bioinspired block copolymer thin films containing silver nanoparticles. *Langmuir* 32: 9484–9491.

50 Li, R., Parvez, K., Hinkel, F. et al. (2013). Bioinspired wafer-scale production of highly stretchable carbon films for transparent conductive electrodes. *Angew. Chem. Int. Ed.* 125: 5645–5648.

51 Cong, H.-P., Wang, P., Gong, M., and Yu, S.-H. (2014). Facile synthesis of mesoporous nitrogen-doped graphene: an efficient methanol–tolerant cathodic catalyst for oxygen reduction reaction. *Nano Energy* 3: 55–63.

52 Yuan, L., Lu, X.-H., Xiao, X. et al. (2012). Flexible solid-state supercapacitors based on carbon nanoparticles/MnO_2 nanorods hybrid structure. *ACS Nano* 6: 656–661.

53 He, Y., Wang, J., Zhang, H. et al. (2014). Polydopamine-modified graphene oxide nanocomposite membrane for proton exchange membrane fuel cell under anhydrous conditions. *J. Mater. Chem. A* 2: 9548–9558.

4

Development of Ultra-microfabricating Polymeric Materials and Its Self-assembly Technology

Teruaki Hayakawa

Tokyo Institute of Technology, Department of Materials Science and Engineering, School of Materials and Chemical Technology, 2-12-1-S8-36 Ookayama, Meguro-ku, Tokyo 152-8552, Japan

4.1 Introduction

The convenience and comfort afforded by personal computers and smartphones today is remarkable compared to that experienced 10–15 years ago. This is attributed to progress in electronics technology and especially improved performance of integrated circuits that fully utilize semiconductor microfabrication technology. Decreased circuit pattern width is key to realizing high integration. Previously, the continuous development of microfabrication technology via photolithography technology reduced the minimum processing dimension of a circuit pattern several times. However, the search for a size in which the half pitch (width of the convex section only) of the pattern width is below 10 nm (sub-10 nm) has commenced, although this poses challenges in terms of both technology and cost [1, 2].

Under these circumstances, lithographic technology that utilizes the self-assembly phenomenon and directed self-assembly (DSA) technology of polymers and "block copolymers lithography (BCP lithography)" have garnered increasing attention and expectation [3–8]. BCP lithography corresponds to microfabrication technology in which spherical, cylindrical, or lamellar microphase-separated domains with size ranging from several to dozens of nanometers are formed by microphase separation of BCP and are arranged in order as a thin film on a substrate. When a domain is removed, the remaining domain components are used to resist the structure transfer on the substrate [9–12]. The pattern width of BCP lithography is closely related to the periodic length of microphase-separated domains. Additionally, the pattern shape is determined by the shape of microphase-separated domains and orientation direction relative to the substrate surface. Thus, the role of higher order structure of BCP thin films is significant, and optimization of molecular structural design and self-assembly of BCP in response to the required characteristics of materials is important.

Furthermore, BCP materials that realize a pattern width of sub-10 nm require polymers that form a clear microphase-separated structure with a strong

Molecular Technology: Materials Innovation, Volume 3, First Edition.
Edited by Hisashi Yamamoto and Takashi Kato.

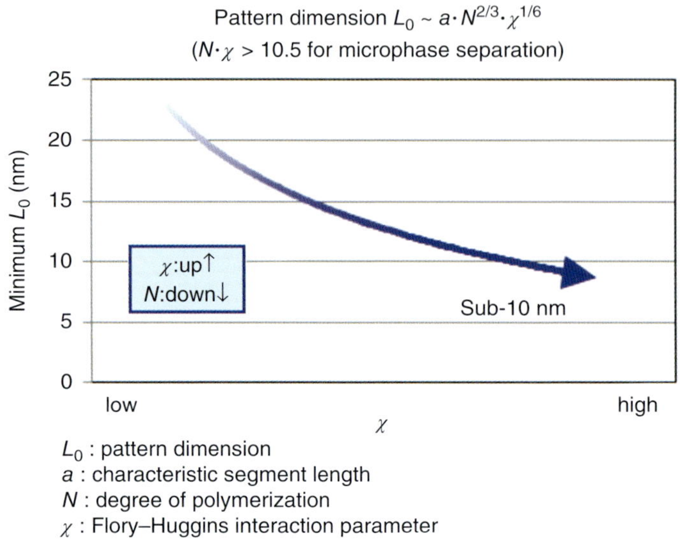

Figure 4.1 Relationship between pattern dimension and Flory–Huggins interaction parameter.

repulsive force between polymers while corresponding to a substance with relatively low molecular weight ranging from several thousands to c. 20 000. Generally, this is termed as high-χBCP since it corresponds to BCP with relatively high value of interaction parameter χ in Flory–Huggins solution theory (Figure 4.1). Recently, significant efforts have been focused on the material development of high-χBCP. Thus, a certain outlook is established as a goal for the sub-10 nm pattern. Conversely, major challenges exist with respect to the perpendicular orientation control of microphase-separated domains that are directly connected to circuit pattern formation. In this study, based on the authors' research background [13–24], we summarize the latest information on the development of new high-χBCP corresponding to perpendicular orientation control of microphase-separated domains in BCP lithography.

4.2 Perpendicular Orientation of High-χBCP Microphase-Separated Domains

4.2.1 Challenges in Perpendicular Orientation of High-χBCP

In the molecular structural design of high-χBCP, a strong repulsive force between polymers constituting BCP is required. As shown in Figure 4.2, for polymers on one side, polymers with molecular structure with relatively high intramolecular polarizability including polar functional groups, such as a hydroxyl group and heteroelements, are selected. The combination of these polymers and hydrocarbon-based polymers with relatively low polarizability forms the

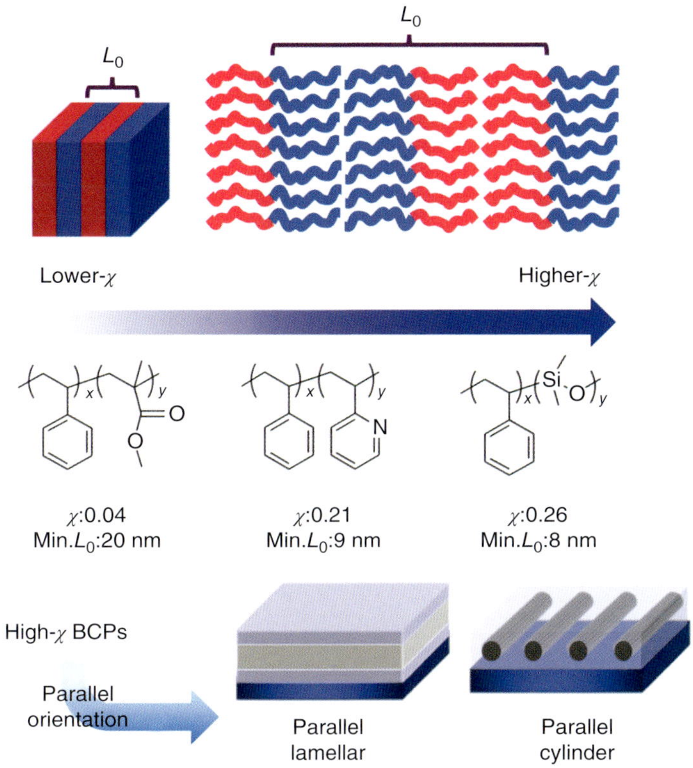

Figure 4.2 Typical high-χ block copolymers. L_0 describes bulk lamellar period of block copolymer.

high-χBCP. In contrast, several reports on high-χBCPs use silicon-containing polymer with superior dry-etching resistance as the component of BCP. These high-χBCPs are observed to clearly form microphase-separated domains with a periodic length of 20 nm or less, and this allows microfabrication of sub-10 nm patterns.

However, several high-χBCP are observed to possess a microphase-separated domain that is segregated to the surface or to the substrate interface during film formation. Therefore, the microphase separation interface is oriented parallel to the substrate surface. For example, a thin film termed as a "parallel cylinder" or "parallel lamella" forms inevitably. This is assumed to correspond to the difference in polarity on the molecular structure, and this is a factor that creates high repulsive interaction with a strong impact on the surface tension of polymers that constitute the BCP.

Each BCP constituting the polymer exhibits equal affinity to the neutralizing film, and thus the neutralizing film is used in BCP lithography with the objective of perpendicular orientation control of microphase-separated domains [3, 4]. For example, with respect to BCP (PS-*b*-PMMA), which consists of polystyrene (PS) and polymethyl methacrylate (PMMA) that is representative of BCP lithography material, microphase-separated domains of PS and PMMA are vertically oriented by using the substrate grafted on the surface of a random copolymer in which the composition ratio of PS and PMMA is appropriately adjusted (PS-*ran*-PMMA). In the case of PS-*b*-PMMA, the surface tensions of PS and PMMA are almost identical, and this is important for the perpendicular orientation of the domain from the thin film surface to the substrate interface. Based on these findings, solvent annealing method and top-coat method were proposed as methods to adjust the neutralizing environment of the film surface of high-χBCP with significantly different surface tension to attempt the perpendicular orientation of microphase-separated domains.

4.2.2 Solvent Annealing Method

In the solvent annealing method (Figure 4.3a), the BCP thin film under solvent vapor is exposed to incorporate solvent molecules into BCP to induce self-assembly [16, 24, 25]. At this time, the BCP thin film surface is placed under the environment in contact with solvent molecules. Therefore, the selection of an appropriate solvent in which the affinity of each polymer constituting high-χBCP becomes equal neutralizing conditions are provided on the BCP film interface. With respect to the high-χBCP, the preparation of this type of an environment achieves the perpendicular orientation of microphase-separated domains. Ross and coworkers [25] attempted solvent annealing of high-χBCP (PS-*b*-PDMS), which consisted of PS and polydimethylsiloxane (PDMS), and successfully achieved the perpendicular orientation of cylinder and lamellar domains. With respect to the solvent, acetone or chloroform was used. Annealing was performed until all solvents were volatilized, and the time for this process corresponded to approximately four hours. The PS-*b*-PDMS with superior dry-etching resistance was expected to correspond to high-χBCP, which allows for sub-10 nm pattern formation from the early stage. However, it involved major issues with perpendicular orientation control of microphase-separated domains. According to a report by Ross and coworkers perpendicular orientation of PS-*b*-PDMS is mostly achieved through solvent annealing while a parallel

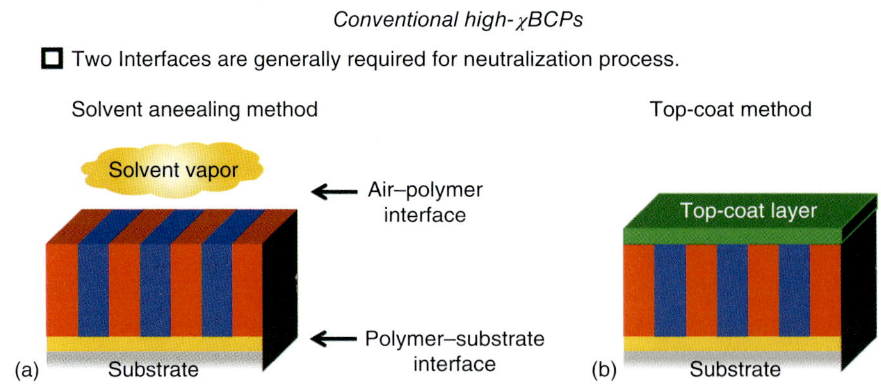

Figure 4.3 (a) Solvent annealing method and (b) top-coat method for neutralization process forming perpendicularly orientated nanostructures.

cylinder domain is formed on the surface layer. Therefore, a perpendicular orientation structure is obtained after removing the surface layer domain with dry etching. While it is easy to perform experiments with the solvent annealing method, it is generally known that optimization of solvent type search, annealing time, and BCP composition ratio is not easy.

4.2.3 Top-Coat Method

Willson and coworkers developed a top-coat method (Figure 4.3b) in which a neutralized film is placed on the substrate interface and on the film surface to achieve the perpendicular orientation of high-χBCP [26]. With respect to the polymers that constitute the top-coat film, random copolymers consisting of three components, namely, a combination of maleic anhydride, norbornene or its derivative, and styrene derivative, are used. The incorporation of maleic anhydride permits coating with a basic aqueous solution that does not dissolve the BCP thin film. Norbornene and styrene derivative are introduced to adjust the thermal stability and polarity of the film. When the BCP film is spin coated with this random copolymer and thermally treated, cyclization occurs due to the dehydration of maleic acid. This leads to reformation and switch of polarity by the anhydride, and it functions as a neutralizing film. Furthermore, BCP consisting of a silicon-containing polystyrene derivative in which perpendicular orientation does not only occur by thermal treatment was reported to form a perpendicular orientation pattern with a width of 9 nm via thermal treatment after the top-coat. Recently, the perpendicular orientation of various high-χBCPs was successful and is increasingly established as a reliable method. However, an issue related to the increased number of processes due to coating and removal of the film persists.

4.2.4 Perpendicular Orientation by Molecular Structure Design

4.2.4.1 Development of Perpendicular Orientation High-χBCP Using Silicon-Containing Polymer

Both the solvent annealing method and the top-coat method exhibit good records of perpendicular orientation of microphase-separated domains and are considered effective. However, it is ultimately desirable to achieve a vertical orientation through simple annealing, such as thermal treatment of BCP, without using these methods. Recently, we reviewed the molecular structure design of high-χBCP to control the perpendicular orientation of microphase-separated domains as shown in Figure 4.4.

Previous research on the development of high-χBCP materials involved examining polyhedral oligomeric silsesquioxane (POSS)-containing polymethacrylate BCP (polymethyl methacrylate-*b*-poly(POSS methacrylate) [PMMA-*b*-PMAPOSS]) [13–15]). Although PMMA-*b*-PMAPOSS allowed easy formation of the sub-10 nm pattern width and this was similar to several other high-χBCPs, the perpendicular orientation of domains was difficult. Subsequent experiments demonstrated that this was due to the high difference in surface tensions of PMMA and PMAPOSS, namely components of

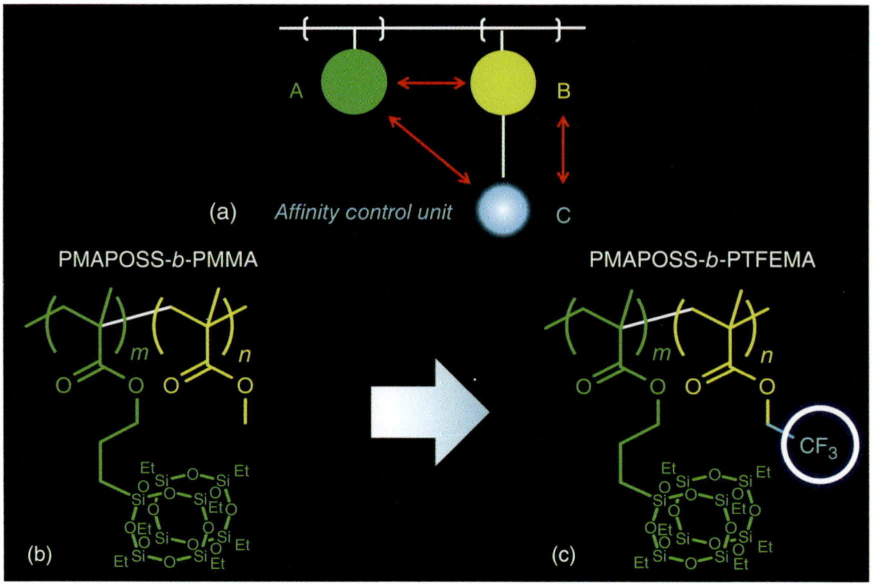

Figure 4.4 (a) Design concept of high-χ block copolymer materials with affinity control unit for perpendicular orientation domains, (b) chemical structure of poly(methyl methacrylate)-*b*-poly(polyhedral oligomeric silsesquioxane (POSS)-containing methacrylate) (PMMA-*b*-PMAPOSS), and (c) chemical structure of PMAPOSS-*b*-poly(trifluoroethyl methacrylate) (PMAPOSS-*b*-PTFEMA).

BCP. Therefore, as opposed to PMMA (47.2 mJ m^{-2}), we synthesized BCP (PMAPOSS-*b*-poly(trifluoroethyl methacrylate) [PMAPOSS-*b*-PTFEMA]) [18–20] combined with fluorine-containing polymethacrylate [16] (PTFEMA, 25.1 mJ m^{-2}) with a trifluoroethyl group with surface tension similar to that of PMAPOSS (28.7 mJ m^{-2}) via molecular design and anionic polymerization (all surface tensions corresponded to empirical values under the same measurement conditions in the laboratory). (Figure 4.4 shows the perpendicular orientation in PMAPOSS-*b*-PTFEMA).

We also employed the reversible addition–fragmentation chain transfer (RAFT) polymerization method to synthesize PTFEMA homopolymer and its BCPs [19, 20]. The conditions for the homopolymerization of TFEMA, such as controlling the types of solvents used and the concentration of reactants in solution, were varied. The findings indicated that when 1,1,1,3,3,3-hexafluoro-2-propanol (HFIP) was used as a solvent and the weight fraction (W_f) of TFEMA was 80 wt%, the molecular weight (M_n) of PTFEMAs was controlled in the range of 10–95k. The gel permeation chromatograms (GPCs) of the resulting polymers exhibited high symmetry, and the polymer dispersity indexes (PDI = M_w/M_n) were approximately 1.2. The optimization for MAPOSS polymerization was demonstrated in addition to controlling the polymerization conditions of TFEMA. When 30 wt% MAPOSS was stirred in toluene, the PMAPOSS obtained had a narrow PDI of 1.12 (M_n: 4000). The results revealed that PMAPOSS-*b*-PTFEMA was successfully synthesized with a mixed solvent of HFIP and toluene by using the PMAPOSS as a macro-RAFT agent. The desired PMAPOSS$_5$-*b*-PTFEMA$_{98}$ was obtained with M_n: 20 500 M_w/M_n: 1.10, and this indicated that the RAFT polymerizations were reasonably controlled.

The morphology of PMAPOSS$_5$-*b*-PTFEMA$_{98}$ was characterized from the bulk sample by using small angle X-ray scattering (SAXS). The bulk sample was prepared by solvent evaporation from PMAPOSS$_5$-*b*- PTFEMA$_{98}$ solution in chloroform. The SAXS profile exhibited two scattering peaks with relative peak position ratios of 1 : 2, and this is indicative of a lamellar morphology. The domain spacing (*d*-spacing) of the periodic structure was estimated from the position of first-order peak as 20.4 nm. In order to examine the phase behavior in thin films, a thin film of PMAPOSS$_5$-*b*-PTFEMA$_{98}$ with a thickness of 40 nm was prepared by spin-casting, and this was followed by thermal annealing under ambient conditions. The PMAPOSS$_5$-*b*-PTFEMA$_{98}$ was observed to form a well-organized line structure in the thin film with a periodicity of 20.4 nm by thermal annealing at 150 °C for three minutes in an air atmosphere. The periodic length matched with the *d*-spacing of PMAPOSS$_5$-*b*-PTFEMA$_{98}$ in the bulk sample, and this suggested that a perpendicularly oriented lamellar structure was formed in the thin film. Furthermore, oblique incidence SAXS structural analysis, oxygen plasma etching, and scanning electron microscope (SEM) observation of the thin film cross section showed that lamellar structure is perpendicular and that the pattern width after etching was 10 nm. A similar result was obtained with one minute of thermal treatment.

In order to examine structural change in thin film microphase-separated domains before and after thermal treatment, a high-speed scanning atomic force microscope (AFM) (ULTRA Speed AFM/JPK Instruments) was used to follow microphase separation behavior in the thermal annealing process. With a rate of temperature increase of 1.0 °C s^{-1}, we observed the film surface between 70 and 130 °C as summarized in Figure 4.5. At approximately 75 °C, only point-like structures were observed in some parts. However, when the temperature increased, point-like structure domains began connecting with each other to form short linear domains. Heating furthered the diffusion of BCP and microphase separation. Concentration on the film surface of PMAPOSS and PTFEMA with similar surface tensions commenced simultaneously. When the temperature exceeded 110 °C, long range correlation of linear domains increased, and fingerprint-like microphase-separated domains ultimately formed across the entire film. The SEM structural observation of film cross section revealed that it corresponded to a lamellar perpendicular oriented structure. The BCP is revolutionary and interesting since it does not require a neutralizing film, and this was essential in the past. With the short duration of thermal treatment under atmospheric conditions, the perpendicular orientation of microphase-separated domains was achieved.

4.2.4.2 Development of a Perpendicular Oriented High-χBCP by Using a Polysiloxane Derivative

With respect to the PS-*b*-PDMS that was introduced earlier, polarized concentration of PDMS with relatively low surface tension occurs on the thin

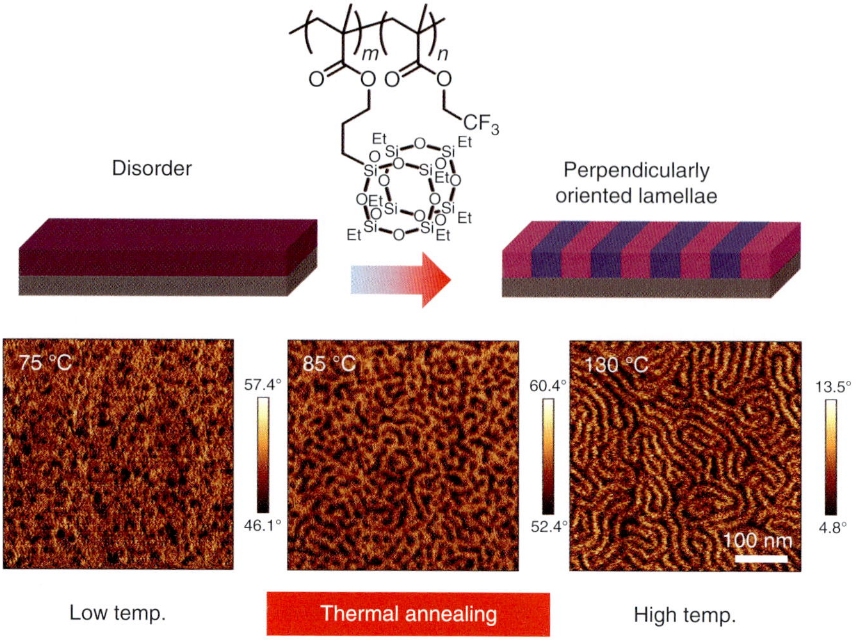

Figure 4.5 High-speed scanning atomic force microscopy images of microphase-separated nanostructures in thin films of PMAPOSS-*b*-PTFEMA at 75 °C, 85 °C, and 130 °C. Perpendicular oriented lamellae were obtained by thermal annealing at 130 °C for one minute.

film surface. The present study focused on the development of BCP with the aim of suppressing the surface polarization of polysiloxane while maintaining high phase separation [21–23]. In order to adjust surface tension, we used polystyrene-*block*-poly(methyl vinyl siloxane) (PS-*b*-PMVS) that uses polymethylvinylsiloxane to allow for the introduction of a side-chain group as a component, as shown in the synthetic scheme in Figure 4.6. With respect to the thiolene reaction, various functional groups were quantitatively introduced to the side chain. The PS_{130}-*b*-$PMHxS_{44}$ with alkyl group in the side chain did not clearly exhibit microphase-separated domains. However,

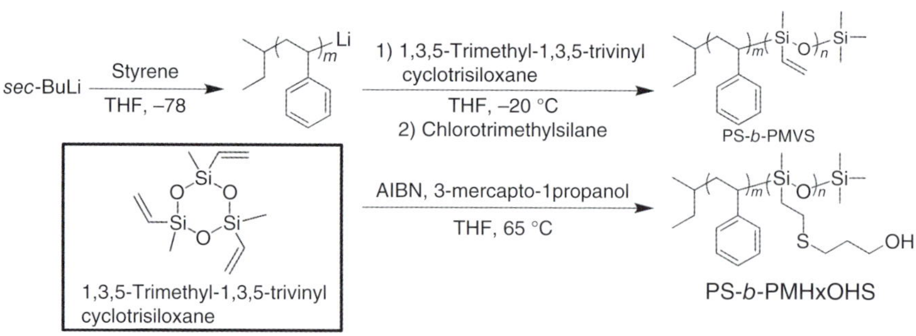

Figure 4.6 Synthetic scheme of polysiloxane-based high-χ block copolymer (PS-*b*-PMHxOHS) prepared by living anionic polymerization and post-functionalization of thiolene reaction.

PS_{130}-*b*-$PMHxOHS_{44}$ and PS_{100}-*b*-$PMHxOHS_{25}$ with alkyl alcohol group formed bulk lamellar structures. Therefore, we examined orientation control of the lamellar microphase-separated domains in a thin film by using PS_{100}-*b*-$PMHxOHS_{25}$.

The preliminary aim involved elucidating the impact of differences in primary structures including molecular structure and composition ratio of the base material on the perpendicular orientation of the PS-*b*-PMHxOHS film domain. With respect to the PS-*b*-PMHxOHS that contains polysiloxane with high oxygen plasma etching resistance in perpendicular orientation control of PS-*b*-PMMA, random copolymers of its constituting polymers cannot be used as neutralizing polymers. Therefore, we synthesized a polymer with surface free energy similar to that of poly(methyl methacrylate-*random*-methylacrylate) (PMMA-*ran*-PMA) [22] wherein successful perpendicular orientation was already reported and examined the impact of its molecular structure and surface free energy. Synthesis was performed by using radical polymerization with 2,2′-azo-bis-isobutyrylnitrile (AIBN) as the initiator. In order to fix a thin film on the silicon substrate through cross-linking, 2-hydroxyethyl methacrylate (HEMA) was added to all the polymers for polymerization. For monomers other than HEMA, we used styrene (S), methyl methacrylate (MMA), 2-vinylpridine (2VP), and 4-vinyl pyridine (4VP).

A film of the base material was formed on the silicon substrate through spin casting. Subsequently, heating resulted in furthering a condensation reaction between PHEMA contained in the base material and the hydroxyl group (—OH) on the substrate surface. The fixing was followed by washing with a solvent to remove base material that was not involved with the cross-linking. This was used as the base film to calculate surface free energy through contact angle measurement. Subsequently, a thin film of PS-*b*-PMHxOHS was created on the base film through spin casting. In the atmospheric condition, thermal annealing was performed for three minutes. We then analyzed the structure of the thin film surface with AFM.

The results indicated that the perpendicular orientation of the lamellar structure is strongly affected by the composition ratio of the base material. A comparison of each AFM imaging shows that poly(2-vinylpyrridine-*random*-hexylmethylacrylate) (P2VP-*ran*-PHEMA) film exhibited more point-like structures as opposed to the poly(methyl methacrylate-*random*-hexylmethylacrylate) (PMMA-*ran*-PHEMA) film (Figure 4.7). This only holds when P2VP-*ran*-PHEMA is used as the base material with a composition ratio of 95 : 5 mol%, which is identical to the composition ratio of 95 : 5 mol% of PMMA-*ran*-PHEMA. We calculated the surface free energy of the substrate with base materials, namely, PMMA-*ran*-PHEMA and P2VP-*ran*-PHEMA, by measuring the contact angle for water and diiodomethane. The results corresponded to 47 and 53 $mJ\,m^{-2}$ for PMMA-*ran*-PHEMA and P2VP-*ran*-PHEMA respectively, and this indicated that PMMA-*ran*-PHEMA is relatively more hydrophobic.

Next, we examined a BCP thin film in which the composition ratio of random copolymers of the base materials was changed as necessary. The results revealed that with respect to PMMA-*ran*-PHEMA, when the composition ratio of PHEMA increased from 5 to 24 mol%, the ratio of the lamellar structure in

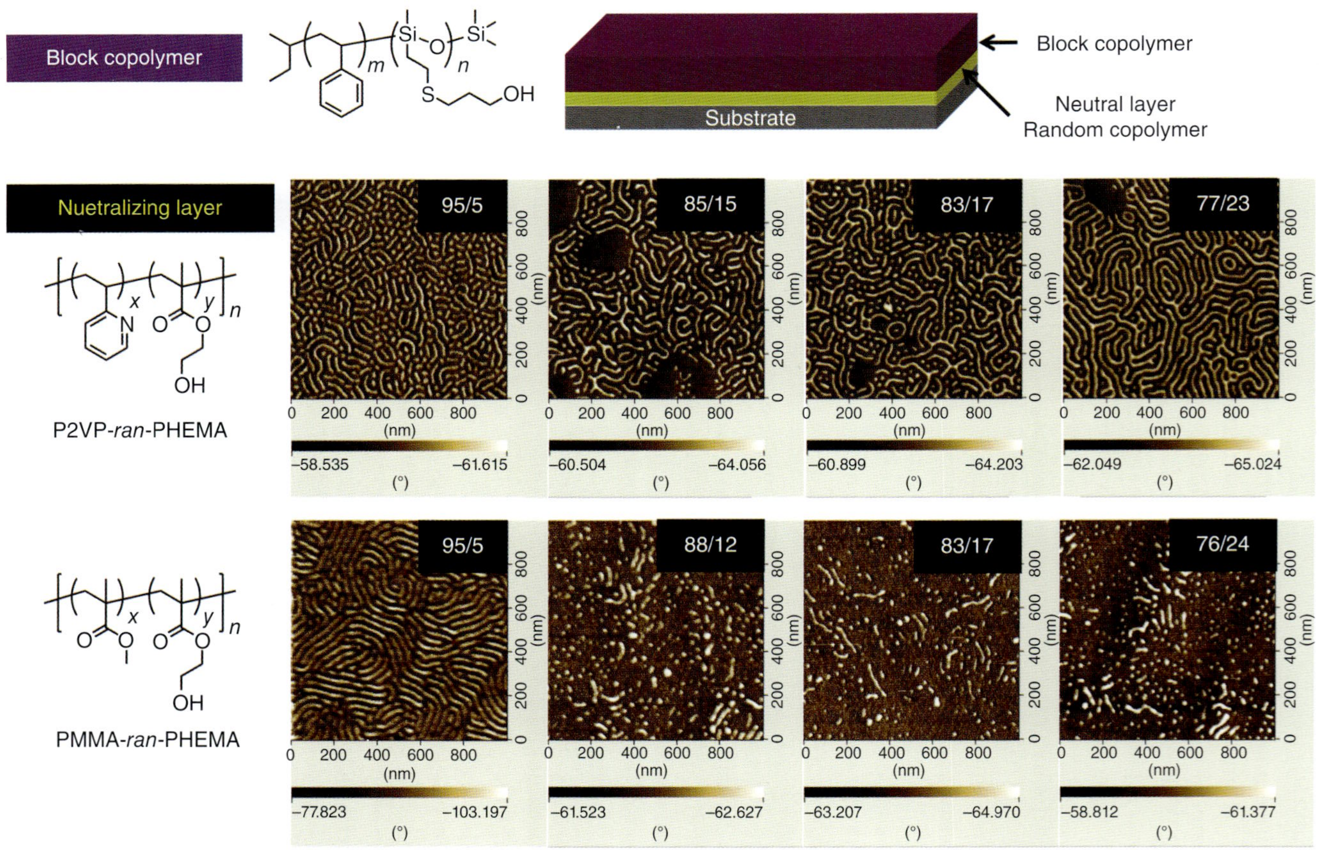

Figure 4.7 Atomic force microscopy images of microphase-separated nanostructures for PS-*b*-PMHxOHS thin films with a different kind of neutralizing layers of poly(2-vinylpyrridine-*random*-hexylmethylacrylate) (P2VP-*ran*-PHEMA) and poly(methyl methacrylate-*random*-hexylmethylacrylate) (PMMA-*ran*-PHEMA).

the thin film dramatically decreased. This behavior depends on the phenomenon in which as the ratio of highly hydrophilic PHEMA increases, the PMHxOHS polymer with the hydroxyl group with high polarity becomes selectively wet at the interface with BCP. Therefore, the PMHxOHS contact area is high near the base material interface and forms a BCP film with the random orientation of the microphase-separated structure. Conversely, the ratio of point-like structure in P2VP-*ran*-PHEMA decreased when the ratio of PHEMA increased from 5 to 23 mol%. Although the details remain unclear, this is potentially due to the balance of interface free energy at the interface of P2VP-*ran*-PHEMA and BCP in a manner similar to PMMA-*ran*-PHEMA.

We used the previously examined PMMA derivative (PMMA-*ran*-PHEMA, 95 : 5 mol%) as the base film (neutralizing film) and created a BCP(PS_{100}-*b*-$PMHxOHS_{25}$) film on the silicon substrate with the base film [22]. This was followed by performing thermal annealing at 130 °C for one minute and then by performing oxygen plasma etching treatment. In the AFM observation prior to the etching treatment, a fingerprint-like pattern was observed in the entire film. In the AFM and SEM observation after the etching, concave–convex patterns derived from the perpendicular orientation of lamellar microphase-separated domains were clearly observed (Figure 4.8). The width of the linear pattern was approximately 8.5 nm, thereby indicating that this corresponded to a BCP that satisfies the requirement of sub-10 nm. The introduction of a hydroxyl group with high polarity in the side change of polysiloxane adjusted the surface tension between both blocks of PS and PMHxOHS such that it was similar. Thus, perpendicular orientation was promoted.

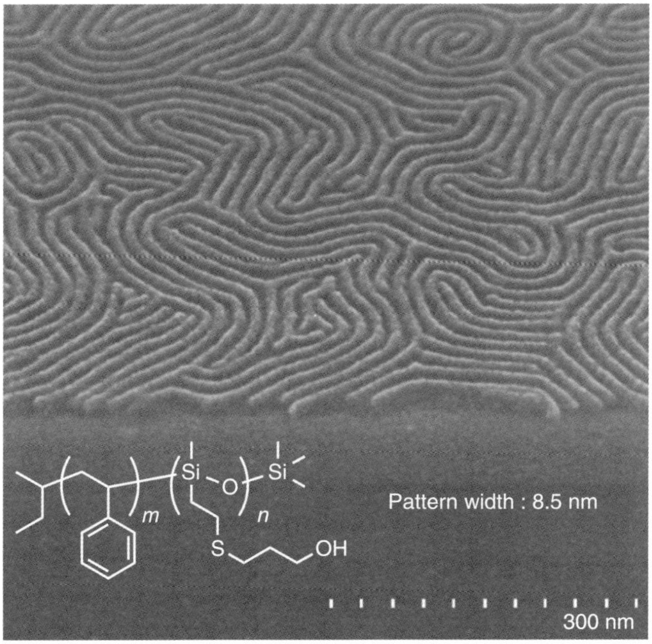

Figure 4.8 Scanning electron micrograph of PS-*b*-PMHxOHS thin film after etching with oxygen plasma for perpendicularly oriented lamellar nanostructures.

4.3 Conclusions

In this study, we introduced perpendicular orientation control of microphase-separated domains. This poses a challenge in BCP lithography with high-χBCP with the aim of forming sub-10 nm patterns. In conjunction with BCP, improvements in individual accuracies of surrounding materials, such as base film, and DSA technology that realizes ordered structural control over long distance and high area are also necessary to create patterns without structural defect in high-resolution imaging that suit the manufacture of semiconductor circuits based on self-assembly technology of polymers. Given these goals, it is essential to include polymer synthesis based on precision polymerization, control of the polymer structure and physical characteristics (especially an understanding of structural control at the surface interface), and use of these findings and technology. Furthermore, the application of perpendicular orientation technology of polymers to a wide range of materials is expected to contribute to the development of devices that are based on diverse optical, electronic, and biological materials.

Acknowledgments

This work was funded by the Japan Science and Technology Agency (JST), the Precursory Research for Embryonic Science and Technology (PRESTO) on the Molecular Technology and Creation of New Functions, and a Grant-in-Aid for Scientific Research (B) (JSPS KAKENHI Grant Number 17H03113.

References

1 International Technology Roadmap for Semiconductors. www.itrs.net.
2 Gronheid, R. and Nealey, P. (2015). *Directed Self-assembly of Block Copolymers for Nano-Manufacturing*, 1e. Cambridge: Woodhead Publishing.
3 Ruiz, R., Kang, H., Detcheverry, F.A. et al. (2008). Density multiplication and improved lithography by directed block copolymer assembly. *Science* 321: 936–939.
4 Kim, B.H., Kim, J.Y., and Kim, S.O. (2013). Directed self-assembly of block copolymers for universal nanopatterning. *Soft Matter* 9: 2780–2786.
5 Griffiths, R.A., Williams, A., Oakland, C. et al. (2013). Directed self-assembly of block copolymers for use in bit patterned media fabrication. *J. Phys. D: Appl. Phys.* 46: 1–29.
6 Tavakkoli, K.G.A., Gotrik, K.W., Hannon, A.F. et al. (2012). Templating three-dimensional self-assembled structures in bilayer block copolymer films. *Science* 336: 1294–1298.
7 Kim, B.H., Park, S.J., Jin, H.M. et al. (2015). Anomalous rapid defect annihilation in self-assembled nanopatterns by defect melting. *Nano Lett.* 15: 1190–1196.
8 Jin, H.M., Lee, S.H., Kim, J.Y. et al. (2016). Laser writing block copolymer self-assembly on graphene light absorbing layer. *ACS Nano* 10: 3435–3442.

9 Bates, C.M., Maher, M.J., Janes, D.W. et al. (2014). Block copolymer lithography. *Macromolecules* 47: 2–12.
10 Jeong, S.-J., Kim, J.Y., Kim, B.H. et al. (2013). Directed self-assembly of block copolymers for next generation nanolithography. *Mater. Today* 16: 468–476.
11 Kim, S.O., Solak, H.H., Stoykovich, M.P. et al. (2003). Epitaxial self-assembly of block copolymers on lithographically defined nanopatterned substrates. *Nature* 424: 411–414.
12 Segalman, R.A., Yokoyama, H., and Kramer, E.J. (2001). Graphoepitaxy of spherical domain block copolymer films. *Adv. Mater.* 13: 1152–1155.
13 Hirai, T., Leolukman, M., Hayakawa, T. et al. (2008). *Macromolecules* 41: 4558–4560.
14 Hirai, T., Leolukman, M., Liu, C.C. et al. (2009). One-step direct-patterning template utilizing self-assembly of POSS-containing block copolymers. *Adv. Mater.* 21: 4334–4338.
15 Tada, Y., Yoshida, H., Ishida, Y. et al. (2012). Directed self-assembly of POSS containing block copolymer on lithographically defined chemical template with morphology control by solvent vapor. *Macromolecules* 45: 292–304.
16 Maeda, R., Higuchi, T., Okuhara, K. et al. (2016). *Polym. J.* 48: 533–538.
17 Kato, F., Chandra, A., Horiuchi, S., and Hayakawa, T. (2016). Morphological dependence on the addition of a soft middle block segment to rigid POSS-containing triblock copolymers for forming cylindrical nanostructures. *RSC Adv.* 6: 62172–62180.
18 Takano, H., Wang, L., Tanaka, Y. et al. (2015). Perpendicular oriented lamellar formation of fluorine- and silicon-containing block copolymers without neutral layers. *J. Photopolym. Sci. Technol.* 28: 649–652.
19 Nakatani, R., Takano, H., Wang, L. et al. (2016). Precise synthesis of fluorine-containing block copolymers via RAFT. *J. Photopolym. Sci. Technol.* 29: 705–708.
20 Nakatani, R., Takano, H., Chandra, A. et al. (2017). *ACS Appl. Mater. Interfaces.*
21 Seshimo, T., Utsumi, Y., Dazai, T. et al. (2016). Perpendicular orientation control in thin films of POSS-containing block copolymer domains with a top-coat surface treatment. *Polym. J.* 48: 407–411.
22 Seshimo, T., Maeda, R., Odashima, R. et al. (2016). Perpendicularly oriented sub-10-nm block copolymer lamellae by atmospheric thermal annealing for one minute. *Sci. Rep.* 6: 19481–19488.
23 Seshimo, T., Odashima, R., Maeda, R. et al. (2016). Perpendicularly oriented cylinders of Si-containing block copolymers by atmospheric thermal annealing. *J. Photopolym. Sci. Technol.* 29 (5): 689–693.
24 Kim, S.H., Misner, M.J., Xu, T. et al. (2004). Highly oriented and ordered arrays from block copolymers via solvent evaporation. *Adv. Mater.* 16: 226–231.
25 Son, J.G., Gotrik, K.W., and Ross, C.A. (2012). High-aspect-ratio perpendicular orientation of PS-*b*-PDMS thin films under solvent annealing. *ACS Macro Lett.* 1: 1279–1284.
26 Bates, C.M., Seshimo, T., Maher, M.J. et al. (2012). Polarity-switching top coats enable orientation of sub-10 nm block copolymer domains. *Science* 338: 775–779.

5

Molecular Simulations of Deformation and Fracture Processes of Crystalline Polymers

Yuji Higuchi

Tohoku University, Institute for Materials Research, Materials Design Division, 2-1-1 Katahira, Sendai, Aoba-ku 980-8577, Japan

5.1 Introduction

Crystalline polymers such as polyethylene, polypropylene, polyvinyl alcohol, and polyoxymethylene, and their composites are used in many industrial materials, such as elastomers, plastics, and fibers. The mechanical properties of these polymers are important for their safe use, and fracturing is a particularly severe problem. Thus, there are numerous studies [1–6] of the fracture process for improving the mechanical properties of these materials. The deformation and fracture processes during stretching are intriguing because they are complicated and follow nonequilibrium paths. In addition, the hierarchical structures of crystalline polymers are attracting interest. Crystalline polymers generally consist of crystalline and amorphous layers, which are called semicrystalline polymers. Figure 5.1 shows a schematic of the hierarchical structures in polyethylene. In a folded polymer chain, the straight parts of the chain form the crystalline region and the bent parts form the amorphous region. The lamellar structure, consisting of crystalline and amorphous layers, is on the scale of 10 nm. The spherulite structure is observed on a larger scale of 50–500 μm. This complicated hierarchical structure makes the fracture process complex.

It is essential to understand fracture processes on the molecular scale to develop tough materials. Based on experimental results, theoretical molecular-scale models have been suggested. Block structures, consisting of crystalline layers in polyethylene, were observed using electron microscopy after the lamellar structure was broken by stretching [7]. The crystalline regions were solid and acted as a block. Nitta et al. suggested the lamellar cluster model for the deformation and fracture processes of semicrystalline polymers [8–11]. The lamellar clusters retain the block structure and their positions are rearranged during stretching. Polymer chains connecting the lamellar clusters play an important role during stretching, and are known as tie chains [3, 4, 6]. Strobl and coworkers [12, 13] suggested that the lamellar blocks slip first, then move together, fibrils form during further stretching, and then disentanglement starts and finally the structure changes. In the final step, memory loss occurs and the

Molecular Technology: Materials Innovation, Volume 3, First Edition.
Edited by Hisashi Yamamoto and Takashi Kato.

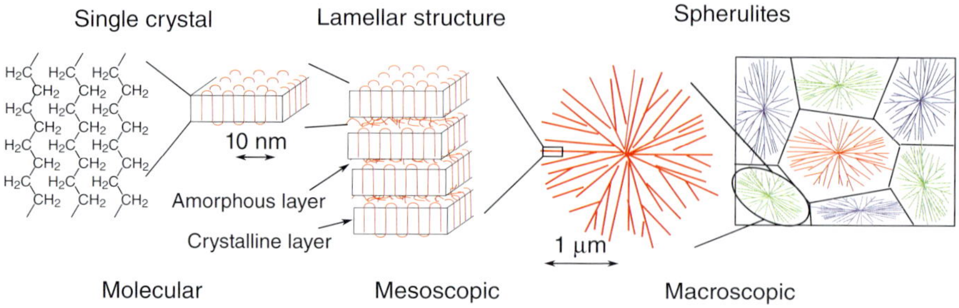

Figure 5.1 Schematic of the semicrystalline polymer structure of polyethylene. The polymer chains align in the same direction on the atomic scale. The lamellar structure consists of amorphous and crystalline layers, where polymer chains fold. Spherulites composed of lamellar bundles form the macroscopic structure.

path becomes irreversible. There are many parameters that affect the mechanical properties. Mandelkern revealed that crystallinity and crystal thickness are important factors in the mechanical properties [14]. In addition to these parameters, molecular mass is also an important factor [15, 16]. There are still many parameters to be clarified; thus, a simple model was prepared. To simplify the problem, the deformation and fracture processes of lamellar structures were investigated. The lamellar structures were oriented roughly in the same direction and were studied. Zhou and Wilkes revealed the mechanical properties depending on the stretching direction and suggested that the morphology changes on the molecular scale [17]. Fukuoka et al. also prepared a highly oriented structure and investigated the effect of entanglement in the amorphous layers [18]. However, the dynamics of polymer chains cannot be observed directly by experiments. Furthermore, there are many parameters, such as crystallinity, crystal thickness, and molecular weight, which make the dynamics complicated. Therefore, the deformation and fracture processes of crystalline polymers on the molecular scale are still controversial, and molecular technology is required.

Computational simulations are useful for elucidating polymer chain dynamics and can be applied to the deformation and fracture processes of polymers. The processes for polymer glasses and elastomers have been well studied, whereas those of crystalline polymers have not been for the following reasons. (i) In addition to the necessity of atomic-scale information, long-time and large-scale simulation is required. (ii) Crystalline polymers have a hierarchical structure; thus, the structure is not homogeneous, making the problem complicated. In this chapter, we review computational simulations of the deformation and fracture processes of crystalline polymers. We focus on how computational simulations can treat the processes. First, coarse-grained models are introduced, and the crystallization, deformation, and fracture processes of homogeneous structures are described. We explain the difficulties in investigating crystalline polymers. Second, we discuss the deformation and fracture processes of crystalline polymers. The results of these studies are then compared and discussed. Better understanding of these similarities and differences should contribute to the development of the molecular simulations and understanding of the deformation and fracture processes of semicrystalline polymers on a molecular scale, leading to the design of tough crystalline polymers.

5.2 Coarse-Grained Molecular Simulations

In this section, coarse-grained models are explained. Then, the applications of coarse-grained models to the deformation and fracture processes of glass polymers and elastomers are described. Coarse-grained models, which do not include atomic information, reveal the mechanism on a molecular scale well. However, during crystallization, atomic information is required, and we introduce the challenges of simulating the crystallization process of polymers.

The dynamics of polymers are long, large-scale phenomena for computational simulations. Therefore, the properties of polymers are simulated by coarse-grained models instead of atomic models. Here, we explain coarse-grained models using polyethylene as an example. One unit of polyethylene, CH_2, is represented by one bead, and the beads are connected by springs in the bead–spring model. At one more level of coarse-graining, the $(CH_2)_2$ group is represented by one bead. Coarse-graining gradually removes information about atomic properties. There are two approaches to coarse-grained models: one is based on atomic information, focused on quantitative properties and polymer composition, and the other is based on polymer chain dynamics, focused on qualitative properties and the universal behavior of polymers, rather than on the detailed polymer composition. In both approaches, the chain length must be long enough to represent the properties of polymers. Kremer and Grest [19] revealed that polymer chain length should be several times as long as the entanglement length; therefore, short chain lengths do not represent the polymer's properties. This is also highlighted in the fracture process of polymer glass.

5.2.1 Deformation and Fracture Processes of Glass Polymers and Elastomers

The deformation and fracture processes of polymers have been studied with coarse-grained models. The fracture process of polymer glass has been studied extensively in bulk materials [20–22] and reviewed [23] and also in thin films [24]. The deformation and fracture processes of rubbery and glassy polymers were compared [25]. For adhesion and separation, deformation and fracture processes were observed during unloading, and these processes were simulated in elastomers [26, 27] and polymer glass [28]. Even though a coarse-grained model can treat polymers at a larger scale, it still cannot simulate a full experimental system. Rottler et al. calculated the crack propagation of polymer glasses by using a coarse-grained molecular dynamics simulation [20, 21]. Figure 5.2 shows the model and snapshots of the fracture process. The deformation at low strain in Figure 5.2a corresponds to the growth of the craze network at the boundary between dense and crazed polymer glasses, as shown in the schematic model in Figure 5.2. Polymer chains are entangled and fibrils are observed showing crazing (Figure 5.2b). Finally, failure at the crack tip is observed at high strain (Figure 5.2c). Flexible and semiflexible chains change the entanglement length. In the stress–strain curve (Figure 5.3a), a peak is observed at low strain and the stress is constant. The stress decreases for $N/N_e = 2$ and $N/N_e = 4$, whereas the stress increases for $N/N_e = 8$ and $N/N_e = 12$, where N is the number of beads in a polymer chain and N_e is the entanglement length. S and S_{max}

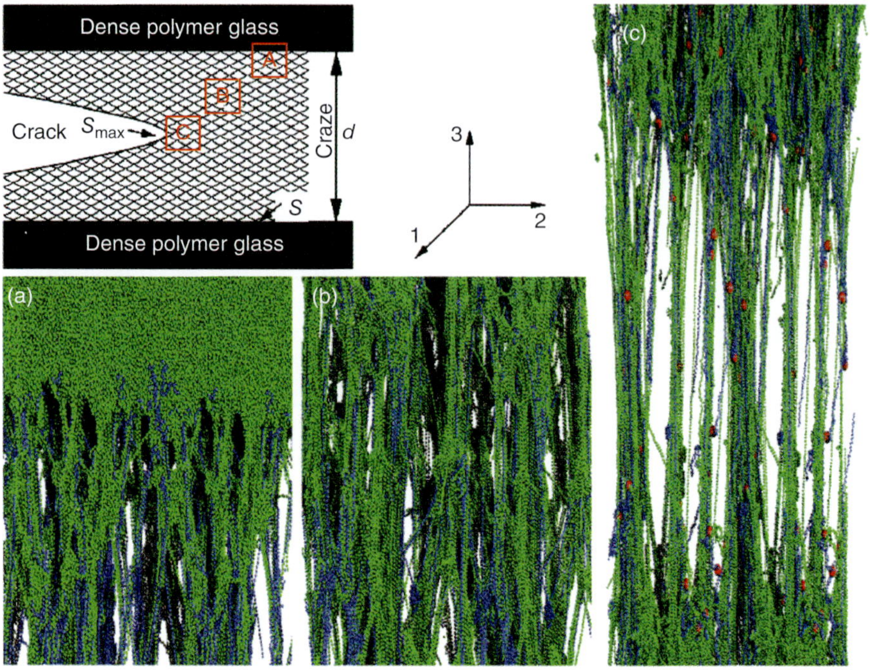

Figure 5.2 Schematic model and snapshots of the fracture process of polymer glass. A crack propagates in crazing sandwiched by dense polymer glass. As polymer glass stretches, (a) craze formation at the boundary between dense and crazed polymer glasses, (b) deformation in the craze, and (c) failure at the tip of the crack are observed. The maximum stress, S_{max}, and constant plateau stress, S, correspond to the stress at (c) and (a), respectively. Source: Reproduced with permission from Rottler et al. [20]. Copyright 2002, American Physical Society.

are the constant plateau of stress and the maximum stress after the plateau, respectively. S corresponds to the stress required to draw fibrils from dense regions (Figure 5.2a). The stress does not correspond to the experimental stress directly. However, the physical meanings of the stress–strain curve can be used to estimate the mechanical properties, such as fracture energy. S_{max} indicates the stress at the tip of the crack at A in Figure 5.2. Figure 5.3b shows the calculated value of S/S_{max}. In flexible and semiflexible cases, the value of S/S_{max} becomes saturated when the chain is eight times longer than the entanglement length. Thus, the chain length should be at least eight times longer than the entanglement length to reveal the deformation and fracture processes in molecular simulations. Based on the stress in Figure 5.3 and the corresponding model in Figure 5.2, the fracture energy and the value of d/D_0 were estimated, where d and D_0 are the craze width and the fibril spacing, respectively. The trends in these values are consistent with the experimental results. Therefore, the simulation results represent the deformation and fracture processes of polymer glasses. These results indicate that the chain length should be long enough to represent the mechanical properties of polymers and that coarse-grained models are valuable for observing deformation and fracture processes on a molecular scale.

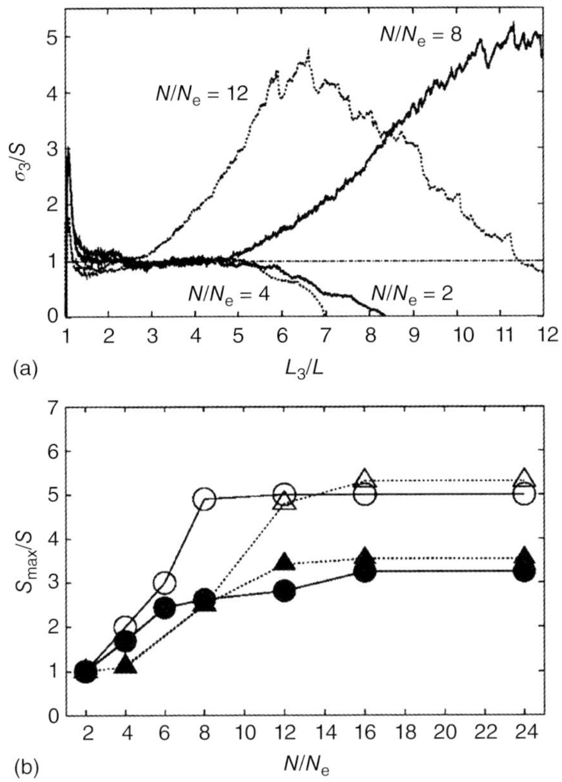

Figure 5.3 Stresses of polymer glasses during the stretching process. (a) Normalized stress, σ^3/S, vs. stretching ratio, L_3/L, where σ is the length unit, S is the constant plateau stress, and L_3 and L are the cell length and initial length in the stretching direction, respectively. As the chain length, N, increases relative to the entanglement length, N_e, the maximum stress, S_{max}, increases. (b) S_{max}/S increases and becomes saturated as the chain length, N, increases relative to the entanglement length, N_e, for flexible (○) and semiflexible (△) chains at higher (open symbols) and lower (solid symbols) temperatures. Source: Reproduced with permission from Rottler et al. [20]. Copyright 2002, American Physical Society.

Molecular simulations are effective for investigating homogeneous structures, such as polymer glasses and elastomers. However, inhomogeneous structures, such as semicrystalline polymers, consist of amorphous and crystalline regions and phase separation structures. For example, block copolymers exhibit phase separation and inhomogeneous structures. Makke et al. [29] studied a lamellar structure consisting of glassy and rubbery layers, but not containing crystalline regions (Figure 5.4). The lamellar structure is stretched perpendicular to the lamellar direction (z-direction) and voids are generated in the rubbery layers. The stress increased with an increase in the number of tie chains. Although the lamellar structure did not contain crystalline layers, this result clearly shows the importance of tie chains in the deformation and fracture processes of the lamellar structure. This importance of tie chains is common to semicrystalline polymers [3, 4, 6]. The effect of tie chains on the mechanical properties can be revealed by molecular simulations. Makke et al. [30, 31] also studied the

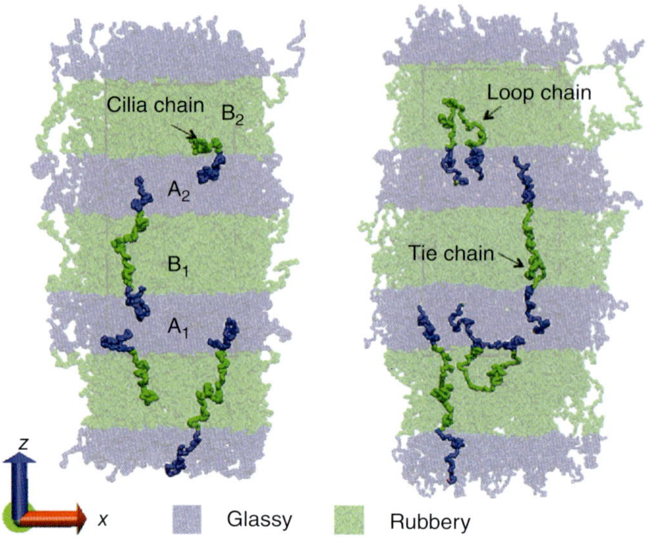

Figure 5.4 Schematic of lamellar structures consisting of noncrystalline glassy and rubbery layers. Tie chains connect solid glassy layers, increasing stress as the strain increased. Source: Reproduced with permission from Makke et al. [29]. Copyright 2012, American Chemical Society.

mechanical properties of the lamellar structure. The simulation size parallel to the lamellar direction increased and voids were generated in the rubbery layers. Furthermore, buckling instability following void generation was observed. Buckling always occurred when samples were large enough, whereas it was not observed in a small simulation [29]. Therefore, the buckling was cyclic and occurred at a characteristic size. These studies used noncrystalline polymers; however, comparing this behavior with the behavior in semicrystalline polymers is important to determine their similarities and differences. Observations of the block structure of polyethylene by electron microscopy during the fracture process [7] indicated that there is also a characteristic size for buckling in semicrystalline polymers.

Coarse-grained models can be used to reveal the deformation and fracture processes of homogeneous structures, such as glassy polymers and elastomers, on the molecular scale. Furthermore, the mechanical properties in a lamellar structure, consisting of glassy and rubbery layers, have been well studied on the molecular scale with a coarse-grained model. These simulations do not contain atomic information and focus on the properties of polymer chains. However, the results reveal the dynamics at the molecular scale.

5.2.2 Molecular Simulation of Polymer Crystallization

In the previous section, we showed that coarse-grained models can reveal the deformation and fracture processes of glassy and rubbery polymers. Atomic-scale information can be ignored and only the polymer chain dynamics are important. However, during crystallization, atomic-scale properties are important. Furthermore, a long simulation is needed because the process is too long to simulate by an all-atomic model. Thus, the coarse-grained model should include atomic information. We introduce molecular simulations of polymer crystallization.

Polymer crystallization has been studied extensively, and coarse-grained models have been used to reveal the process at the molecular scale. The coarse-grained model of polyethylene is widely used, where the CH_2 group is represented by one bead. This model includes the torsional potential and has a zigzag structure. At one more level of coarse-graining, the torsional potential is included in the bending potential to accelerate the calculation. The polymer chain is flexible and is likely to be a straight conformation. A local minimum exists at $\theta = 90°$ in the bending potential, representing the torsional effect in polyethylene, where θ is the bond angle. In both models, the bond length between beads is short compared with the radius of the bead. This leads to a highly attractive interaction between polymer chains and increases crystallization. These models represent crystalline polymers by including the atomic information. A molecular dynamics method is generally used in crystallization simulations. Initially, crystallization simulations of n-alkanes were performed [32–34]. Nucleation of polyethylene was observed in a molecular dynamics simulation of a polymer chain [35], and the nucleation pathway of polyethylene in a single chain was also studied [36]. Polyvinyl alcohol has also been investigated [37–39]. To consider the effect of solvent and viscosity on crystallization, Brownian or Langevin dynamics have been used [40, 41]. These studies revealed the details of the nucleation process at the molecular scale, which cannot be determined by experimental methods alone, and these findings improve our fundamental understanding of polymer crystallization.

As computers have developed, the simulation size and chain length have increased. Nucleation and growth have been observed [42], and the critical nuclei of polyethylene were estimated. Some polymer chains assemble in the nucleus [43]. Crystallization processes under flow or large deformation were also studied [44–46], and fiber or early stage shish-kebab crystallization was observed. The simulation size was limited to 10^4 to 10^5 beads, although the simulation size increased by 10^6 beads, where a bead represented one unit of CH_2 for example. The crystal growth pathway was observed in a large-scale simulation on the order of 10^6 beads [47], although the attractive interaction was ignored. In contrast, in a large-scale simulation on the order of 10^6 beads using poly(vinylidene fluoride) and polyethylene models, the simulation time was limited and the molecular dynamics simulations showed spinodal-assisted crystallization [48]. Therefore, polycrystal structures were formed and the lamellar structure was not observed.

Polymer crystallization has been studied extensively, and nucleation and crystal growth at the molecular scale have been observed and their processes gradually revealed. Molecular simulations have contributed to understanding the crystallization process of polymers, allowing the development of molecular simulations of the deformation and fracture processes of semicrystalline polymers. However, it is still difficult to construct lamellar structures. Therefore, to reveal the deformation and fracture processes of semicrystalline polymers, a quicker method for constructing semicrystalline structures is required.

5.3 Deformation and Fracture Processes of Semicrystalline Polymers on the Molecular Scale

5.3.1 Deformation and Fracture Process

Coarse-grained molecular dynamics simulations of fracture and deformation processes of semicrystalline polymers are discussed in this section. Few simulations of the deformation and fracture of crystalline polymers have been reported. Rutledge and coworkers are pioneers of molecular simulations in

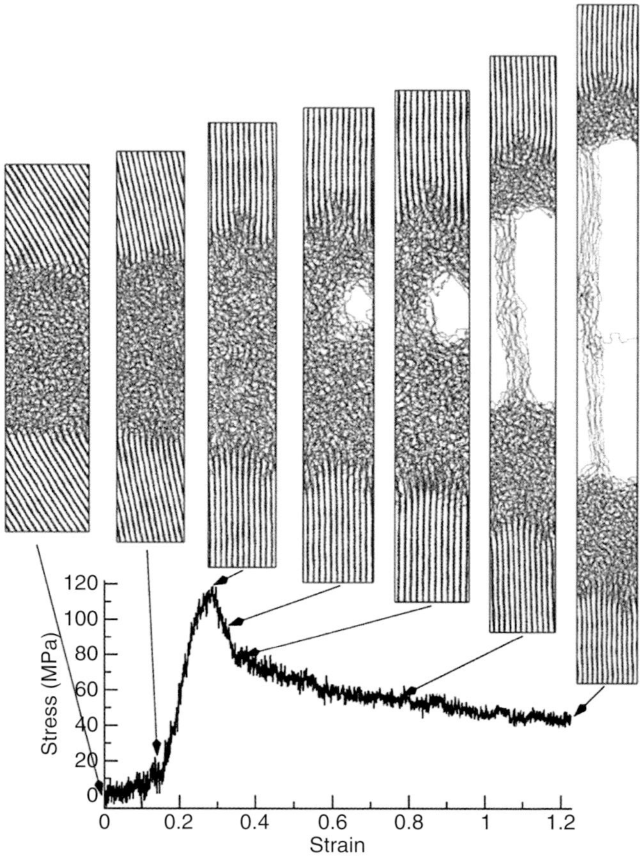

Figure 5.5 Deformation and fracture process of a semicrystalline polymer. An amorphous layer generated by the Monte Carlo simulation is sandwiched by an ideal crystalline layer. The stress increases with the strain until a peak is observed corresponding to the generation of a void in the amorphous layer. Then, the stress decreases with the growth of the void. Source: Reproduced with permission from Yeh et al. [54]. Copyright 2015, American Chemical Society.

the deformation and fracture simulations of crystalline polymers. They developed a modeling method for a lamellar structure by using the Monte Carlo method [49–51]. The lamellar structures of polyethylene were stretched by using the molecular dynamics method [52–55], where an amorphous layer is sandwiched by a crystalline layer (Figure 5.5). They focused on the behavior at the interface between an amorphous and a crystalline layer during the deformation and fracture processes. The amorphous layer initially fractured during the stretching before it grew, and the importance of entanglement in the amorphous layer was revealed [52]. Next, in addition to the stretching, the compression and shear processes were investigated [53], and slip between the amorphous and crystalline layers was observed at the interface. The simulation size was enlarged [54] to reveal the molecular structure, including tie chains and entanglement. In these simulations, the amorphous regions melted, and then crystallized. The crystalline regions have ideal structures and the amorphous region is constructed by the Monte Carlo method. Therefore, the effect of the molecular structure, such as tie chains, loops, and entanglement, on the mechanical properties can be revealed at the molecular scale. These findings are difficult to observe experimentally. Therefore, molecular simulations are valuable in understanding the fundamental mechanisms in the deformation and fracture processes.

Yamamoto examined the crystallization of semicrystalline polymers [56] and the deformation of polyethylene fiber [57]. The initial structures are usually prepared by Monte Carlo simulation or other methods in the molecular dynamics simulation of deformation and fracture processes, because the crystallization process is too long for the molecular simulations to focus on the mechanical properties. The semicrystalline structure was constructed based on a previous study of the crystallization process [56]. Yamamoto performed a crystallization simulation of polyethylene under flow, and then stretched the crystallized structure, which comprised three amorphous and crystalline layers. In the stretching parallel to the crystal direction (Figure 5.6), crystallization of the amorphous layers and void generation were observed. In the stretching perpendicular to the crystal direction, crystal breaking and recrystallization along the stretching direction were observed. Thus, the crystallization process under flow and the large deformation of polyethylene were observed at the molecular scale.

Higuchi and Kubo [58] investigated highly oriented semicrystalline polymers containing randomly located crystalline domains with a more coarse-grained model. The structure corresponded to the block structure around the crack observed by electron microscopy [7]. The mechanical properties of the melt and semicrystalline states were compared. The relaxation of the folded polymer chains in semicrystalline polymers was much slower than that of the random coiled state in the melt. This led to bond dissociation in the semicrystalline polymers. The structure was not realistic but the study contributed to the fundamental understanding of the relaxation process in semicrystalline polymers at the molecular scale.

Next, we discuss the deformation process simulated with an all-atomic model. Monasse et al. studied the mechanical properties of polyethylene using an all-atomic model [59]. Two amorphous layers were constructed between ideal crystalline layers, and voids were generated in the amorphous layers. The effect

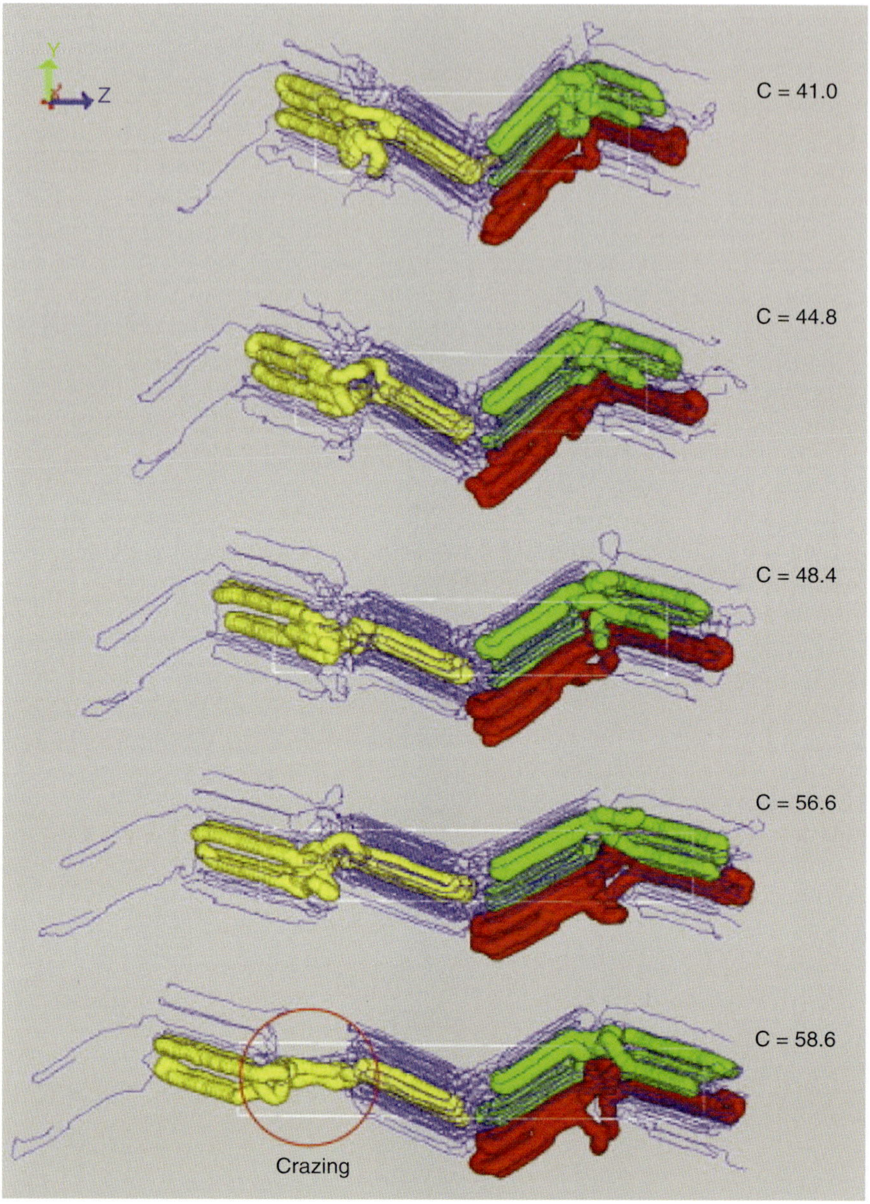

Figure 5.6 Snapshots of the deformation and fracture process of a semicrystalline polymer. The structure is constructed after crystallization under flow. During stretching, crystallization of the amorphous layer and void generation in the amorphous layer are observed. Source: Reproduced with permission from Yamamoto [57]. Copyright 2013, Elsevier.

of tie chains was also investigated. An increase in the number of tie chains increased the stress as the strain increased. After the elongation of tie chains in the amorphous layer, the polymer chains in the crystalline layer were pulled out. The model and results are similar to the results published by Rutledge and coworkers [52–55], demonstrating the accuracy of the coarse-grained model. O'Connor and Robbins investigated the yield stress of polyethylene fibers by a molecular dynamics simulation using an all-atomic model [60]. Polymer chains were aligned completely in the same direction, corresponding to a model of ultrahigh molecular weight polyethylene. The slip of polyethylene from the free chain ends was observed, indicating that the chain ends acted as defects; chain ends are the only defects in the ideal crystal structure. As the chain length increased, the yield stress became saturated when the number of CH_2 units was on the order of 10^3. Thus, a chain length on the order of at least 10^3 is required to determine the mechanical properties in molecular simulations. The saturated yield stress calculated by the simulation agreed with experimental results. The infinite chain model was also calculated and chain scission was observed, and the yield stress was consistent with that estimated by the carbon–carbon bond strength. These quantitative calculations are suitable for all-atomic simulations.

These simulations revealed the deformation and fracture processes at the molecular scale, such as the crystallization of amorphous layers, the generation of nanovoids, and the slip of chain ends. Larger scale phenomena, such as the fragmentation of crystalline layers and the stress of lamellar layers, should be investigated. Jabbari-Faouji et al. [61, 62] performed a large-scale molecular dynamics simulation consisting of 4.3×10^6 beads using a coarse-grained model of polyvinyl alcohol. To construct a semicrystalline structure, small crystal grains were prepared and then copied to enlarge the simulation size. A polycrystalline structure was used. Figure 5.7 shows the fragmentation of large crystalline domains and reorientation along the stretching direction. The polycrystalline structure is not consistent with the classical structure observed experimentally. However, fragmentation could not be observed in small-scale simulations on the order of 10^4 beads. Therefore, large-scale simulations are useful for investigating the deformation and fracture processes of semicrystalline polymers at the

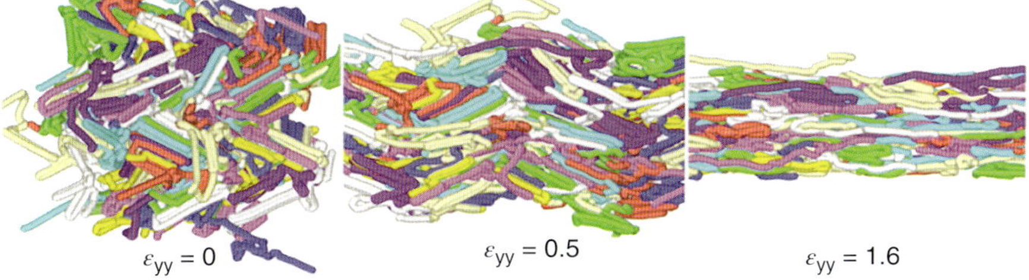

Figure 5.7 Snapshots of the large-scale simulation of the deformation process of a semicrystalline polymer on the order of 10^6 beads. The polycrystalline structure is stretched, and fragmentation of the crystalline regions and reorientation are observed. Source: Reproduced with permission from Jabbari-Farouji et al. [61]. Copyright 2015, American Chemical Society.

molecular scale. These simulation results demonstrate that the deformation and fracture processes of lamellar layers are a major challenge for molecular simulations, particularly the construction of the lamellar layers and the large simulation size. Higuchi and Kubo [63] studied the deformation and fracture processes of polyethylene by using a coarse-grained molecular dynamics simulation. They constructed a highly oriented lamellar structure by applying cyclic forces to promote crystallization. A large-scale simulation consisting of 3.0×10^6 beads was performed. The lamellar thickness and the crystallinity were set as typical experimental values [14, 15]. The lamellar structure, consisting of four amorphous and crystalline layers, was stretched parallel and perpendicular to the crystal direction. The vacuum space was located in one direction and the other directions were under periodic conditions, which enabled partial deformation. The trends of the stress–strain curve in both stretching directions agreed with a previous experiment [17], where a highly oriented lamellar structure was stretched. Figure 5.8 shows the deformation and fracture processes for stretching perpendicular to the lamellar direction. Crystallization of the amorphous layers and void generation in the amorphous layers were observed. Fragmentation of the crystalline layers was also observed because the simulation scale was on the order of 10^6 beads. Crystallization and fragmentation were validated quantitatively by the change in the crystallinity along the stretching direction (Figure 5.9). The four minima corresponded to the amorphous layers, and the crystallinity increased around the amorphous layers, indicating crystallization. Interestingly, the crystallinity of the crystalline layers, which is high at a strain of 0.00, decreased with stretching. This indicates the fragmentation of crystalline layers. During stretching parallel to the lamellar direction, polymer chains gradually aligned in the stretching direction. Furthermore, the effect of molecular structures was analyzed during the deformation and fracture processes. During stretching perpendicular to the lamellar direction, the polymer chain ends moved from the amorphous to the crystalline layers. The polymer chain ends

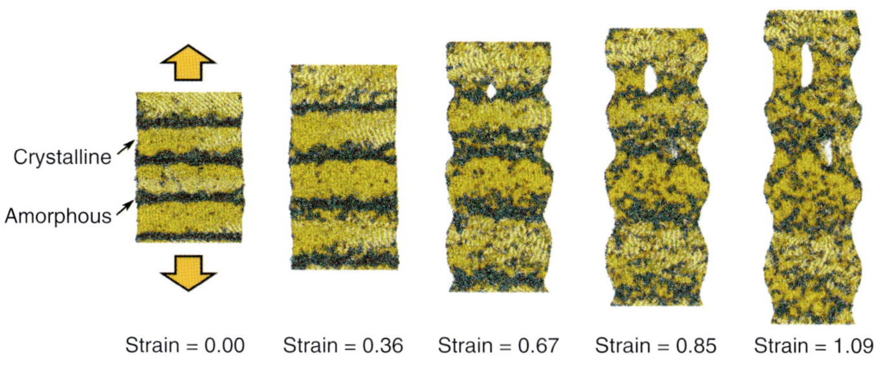

Figure 5.8 Snapshots at different strains of the large-scale simulation of the deformation and fracture process of semicrystalline polymer on the order of 10^6 beads. A lamellar structure consisting of four amorphous and crystalline layers is constructed. During stretching, void generation in the amorphous layer, void growth, and crystallization of amorphous layers are observed.

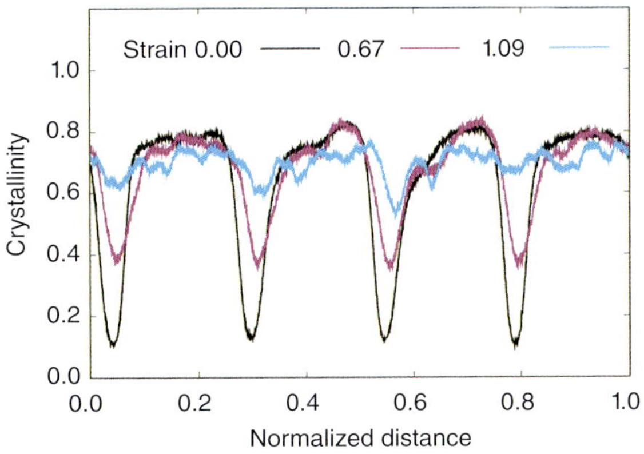

Figure 5.9 Variation of the averaged orientational order, η, vs. the normalized distance of the cell length in the stretching direction during stretching parallel to the crystal direction in a large-scale simulation of the deformation and fracture process of a semicrystalline polymer on the order of 10^6 beads. Orientational order indicates the degree of crystallinity.

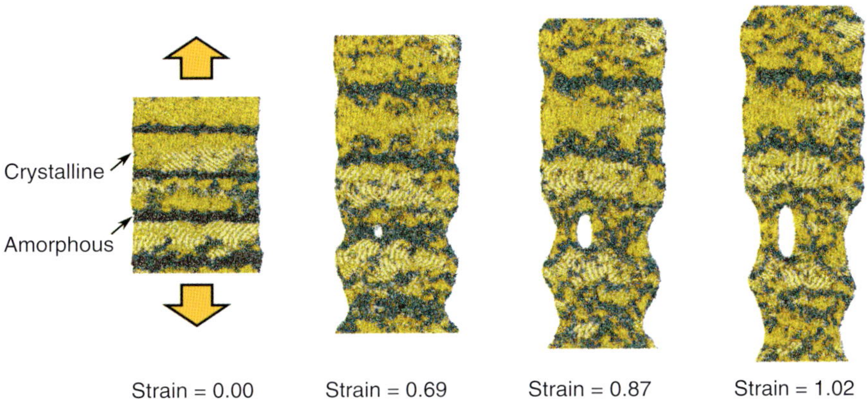

Figure 5.10 Snapshots at different strains of the large-scale simulation of the deformation and fracture process of a semicrystalline polymer on the order of 10^6 beads. The lamellar structure consists of four amorphous and crystalline layers. During stretching, void generation and growth in the amorphous layer, where polymer chains are concentrated, are observed.

were concentrated in the lower parts of the sample (Figure 5.10), leading to void generation. The maximum stress decreased, which resulted in easy fracturing due to the concentration of polymer chain ends. Thus, the deformation and fracture processes of the lamellar structure were elucidated on the molecular scale.

5.3.2 Discussion

These findings in coarse-grained and all-atomic simulations may reveal the deformation and fracture processes of semicrystalline polymers at the molecular scale, which are difficult to obtain experimentally. Furthermore, large-scale simulations on the order of 10^6 beads elucidate the dynamics of polymer chains and molecular assembly. Here, we discuss the similarities and differences in the phenomena and structures in these simulations.

5.3.2.1 Comparison of Simulation Results

Crystallization of amorphous layers and increase in crystallinity were observed in the coarse-grained model by Rutledge and coworkers [52–55], Yamamoto [57], Jabbari-Faouji et al. [61, 62], and Higuchi and Kubo [63]. Crystallization has also been observed experimentally by X-ray scattering [17]. The simulation results are validated by the crystallization of amorphous layers. Void generation in amorphous layers was observed in a coarse-grained model by Rutledge and coworkers [52–55], in a large-size simulation by Higuchi and Kubo [63], and in an all-atomic model by Monasse et al. [59]. The crystalline regions were assembled by strong attractive interactions compared with the amorphous regions and were solid. The lamellar cluster model [8–11], which was based on the experimental results, also indicated that the crystalline regions are solid. Therefore, void generation in amorphous regions is likely to be a universal phenomenon in the fracture process of semicrystalline polymers. In a lamellar structure consisting of rubbery and glassy polymers, void generation was observed in the rubbery layers by a coarse-grained molecular dynamics simulation [30, 31]; thus, void generation in soft layers is also a general phenomenon. Void generation in polyethylene was investigated experimentally for different crystallite thicknesses and stretching rates by Humbert et al. [64], and they referred to the generation and growth of voids as cavitation. They found that homogeneous cavitation changed to localized cavitation, and then to no cavitation as the number of tie chains and amount of entanglement increased in the amorphous layers. The minimum void size measured was 40 nm, which was larger than that in the simulation results. Therefore, void generation and growth in the molecular simulations still do not correspond to the experimental observations directly. However, these processes are difficult to observe experimentally on a molecular scale and molecular simulations help clarify the mechanisms. Related to the cavitation, whitening is a difficult problem that currently can only be investigated experimentally. Farge et al. [65] discussed the explanation for whitening during stretching of semicrystalline polymers. They focused on cavitation and summarized the stretching process on a molecular scale. They suggested that whitening is not related to the cavitation, although previous studies have proposed that it is. To resolve this problem, further development of simulations and theoretical approaches is desirable, possibly through increasing simulation sizes.

Many factors affect the mechanical properties of semicrystalline polymers. At a molecular scale, tie chains [3, 4, 6] and entanglement in the amorphous layers are important factors [18]. Humbert et al. [64, 66] also suggested that stress transmitters, such as tie chains and entanglements in amorphous layers that connect to solid crystalline layers, are important. However, the effects of the molecular structure are interrelated and difficult to determine directly by experiments. Molecular simulations have been used to solve this problem. Rutledge and coworkers prepared an ideal crystalline layer and an amorphous layer was constructed with a Monte Carlo simulation [49–51]. Molecular structures such as tie chains and entanglements have also been analyzed [52–55] under different initial conditions and in different numbers. The results confirmed the importance of entanglement and tie chains on the mechanical properties. In a similar model, where an amorphous layer was sandwiched by ideal crystalline layers, the number of tie chains was changed [59]. The increase in the stress was confirmed by an all-atomic model. Similarly, Makke et al. [29] also observed that the stress increased with an increase in the number of tie chains in a model of a lamellar structure consisting of noncrystalline rubbery and glassy layers. Thus, the importance of tie chains and entanglements in amorphous layers was confirmed by a molecular simulation. The simulation model is ideal, where the interface of the crystalline and amorphous layers is discontinuous. However, there are too many important factors to solve the problems on a molecular scale. Therefore, a simple model should be used initially. In addition to tie chains, chain ends are also important. In a completely crystalline model, the only defects are chain ends. This was confirmed by all-atomic molecular dynamics simulation with an ideal model, where polyethylene was completely aligned in the same direction [60]. It was shown experimentally that as the molecular mass of polyethylene decreases, the yield stress decreases [16]. It is predicted that chain ends act as defects in a lamellar structure. This effect was confirmed by Higuchi and Kubo [63] by using a lamellar structure model consisting of four amorphous and crystalline layers. The movement of chain ends from the amorphous to the crystalline layers led to void generation in the amorphous layers. As the number of chain ends in the amorphous layers increased, the number of tie chains in the amorphous layer decreased. Therefore, the effect of chain ends is consistent with the effect of tie chains. This behavior is also difficult to observe experimentally.

Slip between polymer chains, slip at the interfaces between amorphous and crystalline layers, and slip at the interfaces between lamellar blocks are different scale phenomena. In an ideal crystalline structure, the chain slip from chain ends was observed in an all-atomic molecular dynamics simulation [60]. In the lamellar structure, the movement of chain ends was observed by a coarse-grained molecular dynamics simulation [63]. The slip at the interface between an amorphous and a crystalline layer was observed by a coarse-grained molecular dynamics simulation [53]. Although fragmentation of the crystalline regions was observed in a large-scale coarse-grained molecular dynamics simulation [61–63], the slip of lamellar blocks was not observed. An all-atomic model is required to reveal the chain slip. Large-scale simulations are essential for modeling the slip at the interface and of lamellar blocks; therefore, a coarse-grained model is useful. The slips are related beyond the scales. Therefore, molecular simulations can help tackle this unsolved hierarchical problem.

5.3.2.2 Degradation and Mechanical Properties of Polymers

Next, we discuss aging and degradation. Chemical degradation, such as scission and cross-linking, has been studied extensively. However, the effect of degradation on the mechanical properties on a molecular scale is still controversial. Fayolle et al. [67] reviewed the relationship between chain scission and embrittlement of semicrystalline polymers. Chain scission leads to a decrease in molar mass, which decreases the number of tie chains in amorphous layers and in interlamellar space layers through the crystallization of amorphous layers. Although the molecular structure is changed by chain scission, the details are still unknown. These structural changes at the molecular scale should be investigated by molecular simulations. We propose that chain scission should be modeled and relaxation calculations for the crystallization of the amorphous layers performed. Then, to reveal the effect of dissociated tie chains and crystallization, stretching simulations should be performed under various initial conditions. The results should then be compared with experimental results. The physical aging of polymers is also an important problem [23, 68]. However, in semicrystalline polymers, the mechanism is still not understood well. For example, recycled polypropylene becomes brittle although its molecular mass does not change [69]. Structural changes on the molecular scale are thought to play a role in this process. To solve this problem, different models consisting of different molecular structures should be examined by molecular simulations. For example, the model where polymer chain ends are concentrated fractures easily [63]. This is a possible cause of physical aging, although the diffusion of polymer chain ends is long range. Rutledge and coworkers [54, 55] performed stretching simulations in different structures. The number of tie chains, loops, and entanglements was changed, and the stresses were compared. These investigations contribute to the understanding of the ductile structure.

5.3.2.3 Future Work

Finally, future work and the application of molecular simulations are discussed. As computers develop, the value of molecular simulations also increases because simulation times and sizes can be increased. Molecular simulations are valuable in revealing mechanisms in complicated soft matter that depend on many parameters and factors. Simulations model one aspect of an experimental system and can simplify a problem. Inhomogeneous structures are complicated compared with homogeneous structures and are difficult to understand. For example, double-network (DN) gels [70] consist of highly and slightly cross-linked networks, known as the first and second networks, respectively. Although the essential factors and mechanisms have been suggested [71], it is desirable to understand them at a molecular level. To increase the toughness, it is essential to design the network structure appropriately. Ducrot et al. made elastomers that showed large Young's moduli and ductility consisting of the same network structure as DN gels [72]. In hydrogels composed of polymers, single gels are well studied, whereas inhomogeneous structures, such as DN gels, have not been studied on the molecular scale. Although Jang et al. studied the mechanical properties of DN gels with an all-atomic molecular dynamics simulation [73], the mechanism of the network structure is still not revealed. In addition to the complicated problem of the inhomogeneous structure, atomic information is also important. Although coarse-grained simulations simplify and enlarge the simulation, including atomic information is still difficult. In this review, simple models of linear crystalline polymers, such as polyethylene and polyvinyl alcohol, were reviewed. In the future work, more complicated crystalline polymers, such as polypropylene and polyoxymethylene, should be clarified. These polymers have a helical structure and are difficult to simulate. Molecular simulations of the crystallization of these polymers are still being studied [74–77], and their mechanical properties are not understood and should be investigated. It is of particular interest to clarify the behavior of the helical structure during stretching at the molecular scale. Molecular simulations can reveal the polymer chain dynamics on a molecular scale, and it is important to simplify complicated problems by modeling and coarse-graining. However, atomic information is crucial for modeling the chemical properties of polymers; thus, atomic information should be included in models.

5.4 Conclusions

In this review, we summarized molecular simulations of the deformation and fracture processes of semicrystalline polymers. First, molecular simulations of the deformation and fracture processes of noncrystalline polymers were introduced. The results indicated that the simulations model one aspect of the experiments and can reveal the dynamics on a molecular scale. Then, the lamellar structure consisting of noncrystalline glassy and rubbery layers was described. The simulations became complicated for inhomogeneous structures such as lamellae, compared with homogeneous structures such as glassy or rubbery states. Crystalline polymers were more difficult to simulate because atomic information is required to model their structures. Crystallization simulations have been developed. The initial simulation focused on the folding of one polymer chain, and a subsequent simulation revealed the growth of a lamellar structure and a polycrystal consisting of polymer chains. Based on the model for crystalline polymers, molecular simulations have been used to study the deformation and fracture processes of semicrystalline polymers. Initially, the deformation and fracture process of an amorphous and a crystalline layer was studied. The crystallization of the amorphous layer, the generation of voids in the amorphous layer, and the slip at the interface between the amorphous and crystalline layer were observed. Furthermore, the importance of tie chains and entanglement in the amorphous layer was revealed. These findings would be difficult to obtain experimentally. Therefore, molecular simulations are useful for revealing phenomena at a molecular level. The fragmentation of the lamellar region and void generation caused by the movement of chain ends were observed in large-scale simulations. Understanding the deformation and fracture processes on a molecular scale is essential for increasing toughness against stretching and compression, and durability against degradation by chemical reactions and structural changes during prolonged use. Therefore, molecular technology based on molecular simulations to understand the mechanisms of the deformation and fracture processes on the molecular scale will contribute to the development of tough polymers to improve safety and save resources.

References

1 Strobl, G. (2007). *The Physics of Polymers: Concepts for Understanding Their Structures and Behavior*. Springer.

2 Peterlin, A. (1971). Molecular model of drawing polyethylene and polypropylene. *J. Mater. Sci.* 6 (6): 490–508.

3 Séguéla, R. (2005). Critical review of the molecular topology of semicrystalline polymers: the origin and assessment of intercrystalline tie molecules and chain entanglements. *J. Polym. Sci., Part B: Polym. Phys.* 43 (14): 1729–1748.

4 Séguéla, R. (2007). On the natural draw ratio of semi-crystalline polymers: review of the mechanical, physical and molecular aspects. *Macromol. Mater. Eng.* 292 (3): 235–244.

5 Galeski, A. (2003). Strength and toughness of crystalline polymer systems. *Prog. Polym. Sci.* 28 (12): 1643–1699.

6 Patlazhan, S. and Remond, Y. (2012). Structural mechanics of semicrystalline polymers prior to the yield point: a review. *J. Mater. Sci.* 47 (19): 6749–6767.

7 Adams, W., Yang, D., and Thomas, E. (1986). Direct visualization of microstructural deformation processes in polyethylene. *J. Mater. Sci.* 21 (7): 2239–2253.

8 Nitta, K. and Takayanagi, M. (2006). Application of catastrophe theory to the neck-initiation of semicrystalline polymers induced by the intercluster links. *Polym. J.* 38 (8): 757–766.

9 Kuriyagawa, M. and Nitta, K. (2011). Structural explanation on natural draw ratio of metallocene-catalyzed high density polyethylene. *Polymer* 52 (15): 3469–3477.

10 Nitta, K. and Kuriyagawa, M. (2012). Application of catastrophe theory to neck initiation of metallocene-catalyzed high-density polyethylene. *Polym. J.* 44 (3): 245–251.

11 Mizushima, M., Kawamura, T., Takahashi, K., and Nitta, K. (2014). In situ near-infrared spectroscopic studies of the structural changes in polyethylene during tensile deformation. *Polym. Test.* 38: 81–86.

12 Men, Y., Rieger, J., and Strobl, G. (2003). Role of the entangled amorphous network in tensile deformation of semicrystalline polymers. *Phys. Rev. Lett.* 91 (9): 095502.

13 Hong, K., Rastogi, A., and Strobl, G. (2004). A model treating tensile deformation of semicrystalline polymers: quasi-static stress-strain relationship and viscous stress determined for a sample of polyethylene. *Macromolecules* 37 (26): 10165–10173.

14 Mandelkern, L. (1985). The relation between structure and properties of crystalline polymers. *Polym. J.* 17 (1): 337–350.

15 Kennedy, M.A., Peacock, A.J., and Mandelkern, L. (1994). Tensile properties of crystalline polymers: linear polyethylene. *Macromolecules* 27 (19): 5297–5310.

16 Hosoda, S. and Uemura, A. (1992). Effect of the structural distribution on the mechanical properties of linear low-density polyethylenes. *Polym. J.* 24 (9): 939–949.

17 Zhou, H. and Wilkes, G.L. (1998). Orientation-dependent mechanical properties and deformation morphologies for uniaxially melt-extruded high-density polyethylene films having an initial stacked lamellar texture. *J. Mater. Sci.* 33 (2): 287–303.

18 Fukuoka, M., Aya, T., Saito, H. et al. (2006). Role of amorphous region on the deformation behavior of crystalline polymers. *Polym. J* 38 (6): 542–547.

19 Kremer, K. and Grest, G.S. (1990). Dynamics of entangled linear polymer melts: a molecular–dynamics simulation. *J. Chem. Phys.* 92 (8): 5057–5086.

20 Rottler, J., Barsky, S., and Robbins, M.O. (2002). Cracks and crazes: on calculating the macroscopic fracture energy of glassy polymers from molecular simulations. *Phys. Rev. Lett.* 89 (14): 148304.

21 Rottler, J. and Robbins, M.O. (2003). Growth, microstructure, and failure of crazes in glassy polymers. *Phys. Rev. E* 68 (1): 011801.

22 Papakonstantopoulos, G.J., Riggleman, R.A., Barrat, J.L., and de Pablo, J.J. (2008). Molecular plasticity of polymeric glasses in the elastic regime. *Phys. Rev. E* 77 (4): 041502.

23 Rottler, J. (2009). Fracture in glassy polymers: a molecular modeling perspective. *J. Phys. Condens. Matter* 21 (46): 463101.

24 Yoshimoto, K., Jain, T.S., Nealey, P.F., and de Pablo, J.J. (2005). Local dynamic mechanical properties in model free-standing polymer thin films. *J. Chem. Phys.* 122 (14): 144712.

25 Makke, A., Perez, M., Lame, O., and Barrat, J.L. (2009). Mechanical testing of glassy and rubbery polymers in numerical simulations: role of boundary conditions in tensile stress experiments. *J. Chem. Phys.* 131 (1): 014904.

26 Morita, H., Yamada, M., Yamaguchi, T., and Doi, M. (2005). Molecular dynamics study of the adhesion between end-grafted polymer films. *Polym. J.* 37 (10): 782–788.

27 Morita, H., Miura, H., Yamada, M. et al. (2007). Molecular dynamics study of the adhesion between end-grafted polymer films II - effect of grafting density. *Polym. J.* 39 (1): 73–80.

28 Sides, S.W., Grest, G.S., Stevens, M.J., and Plimpton, S.J. (2004). Effect of end-tethered polymers on surface adhesion of glassy polymers. *J. Polym. Sci., Part B: Polym. Phys.* 42 (2): 199–208.

29 Makke, A., Lame, O., Perez, M., and Barrat, J.L. (2012). Influence of tie and loop molecules on the mechanical properties of lamellar block copolymers. *Macromolecules* 45 (20): 8445–8452.

30 Makke, A., Perez, M., Lame, O., and Barrat, J.L. (2012). Nanoscale buckling deformation in layered copolymer materials. *Proc. Natl. Acad. Sci. U.S.A.* 109 (3): 680–685. https://doi.org/10.1073/pnas.1111367109.

31 Makke, A., Lame, O., Perez, M., and Barrat, J.L. (2013). Nanoscale buckling in lamellar block copolymers: a molecular dynamics simulation approach. *Macromolecules* 46 (19): 7853–7864.

32 Rigby, D. and Roe, R.J. (1988). Molecular dynamics simulation of polymer liquid and glass. II. Short-range order and orientation correlation. *J. Chem. Phys.* 89 (8): 5280–5290.

33 Esselink, K., Hilbers, P.A.J., and van Beest, B.W.H. (1994). Molecular dynamics study of nucleation and melting of n-alkanes. *J. Chem. Phys.* 101 (10): 9033–9041.

34 Takeuchi, H. (1998). Structure formation during the crystallization induction period of a short chain-molecule system: a molecular dynamics study. *J. Chem. Phys.* 109 (13): 5614–5621. https://doi.org/10.1063/1.477179.
35 Kavassalis, T.A. and Sundararajan, P.R. (1993). A molecular-dynamics study of polyethylene crystallization. *Macromolecules* 26 (16): 4144–4150.
36 Fujiwara, S. and Sato, T. (1997). Molecular dynamics simulations of structural formation of a single polymer chain: bond-orientational order and conformational defects. *J. Chem. Phys.* 107 (2): 613–622.
37 Meyer, H. and Müller-Plathe, F. (2001). Formation of chain-folded structures in supercooled polymer melts. *J. Chem. Phys.* 115 (17): 7807–7810.
38 Meyer, H. and Müller-Plathe, F. (2002). Formation of chain-folded structures in supercooled polymer melts examined by MD simulations. *Macromolecules* 35 (4): 1241–1252.
39 Vettorel, T. and Meyer, H. (2006). Coarse graining of short polyethylene chains for studying polymer crystallization. *J. Chem. Theory Comput.* 2 (3): 616–629.
40 Liu, C. and Muthukumar, M. (1998). Langevin dynamics simulations of early-stage polymer nucleation and crystallization. *J. Chem. Phys.* 109 (6): 2536–2542.
41 Welch, P. and Muthukumar, M. (2001). Molecular mechanisms of polymer crystallization from solution. *Phys. Rev. Lett.* 87 (21): 218302.
42 Muthukumar, M. and Welch, P. (2000). Modeling polymer crystallization from solutions. *Polymer* 41 (25): 8833–8837.
43 Yi, P., Locker, C.R., and Rutledge, G.C. (2013). Molecular dynamics simulation of homogeneous crystal nucleation in polyethylene. *Macromolecules* 46 (11): 4723–4733.
44 Koyama, A., Yamamoto, T., Fukao, K., and Miyamoto, Y. (2002). Molecular dynamics simulation of polymer crystallization from an oriented amorphous state. *Phys. Rev. E* 65 (5): 050801.
45 Dukovski, I. and Muthukumar, M. (2003). Langevin dynamics simulations of early stage shish-kebab crystallization of polymers in extensional flow. *J. Chem. Phys.* 118 (14): 6648–6655.
46 Jabbarzadeh, A. and Tanner, R. (2009). Crystallization of alkanes under quiescent and shearing conditions. *J. Non-Newtonian Fluid Mech.* 160 (1): 11–21. Complex flows of complex fluids.
47 Luo, C. and Sommer, J.U. (2011). Growth pathway and precursor states in single lamellar crystallization: MD simulations. *Macromolecules* 44 (6): 1523–1529.
48 Gee, R.H., Lacevic, N., and Fried, L.E. (2009). Atomistic simulations of spinodal phase separation preceding polymer crystallization. *Nat. Mater.* 8 (2): 159–159.
49 Balijepalli, S. and Rutledge, G.C. (1998). Molecular simulation of the intercrystalline phase of chain molecules. *J. Chem. Phys.* 109 (16): 6523–6526.
50 in 't Veld, P.J. and Rutledge, G.C. (2003). Temperature-dependent elasticity of a semicrystalline interphase composed of freely rotating chains. *Macromolecules* 36 (19): 7358–7365.

51 in 't Veld, P.J., Hütter, M., and Rutledge, G.C. (2006). Temperature-dependent thermal and elastic properties of the interlamellar phase of semicrystalline polyethylene by molecular simulation. *Macromolecules* 39 (1): 439–447.

52 Lee, S. and Rutledge, G.C. (2011). Plastic deformation of semicrystalline polyethylene by molecular simulation. *Macromolecules* 44 (8): 3096–3108.

53 Kim, J.M., Locker, R., and Rutledge, G.C. (2014). Plastic deformation of semicrystalline polyethylene under extension, compression, and shear using molecular dynamics simulation. *Macromolecules* 47 (7): 2515–2528.

54 Yeh, I.C., Andzelm, J.W., and Rutledge, G.C. (2015). Mechanical and structural characterization of semicrystalline polyethylene under tensile deformation by molecular dynamics simulations. *Macromolecules* 48 (12): 4228–4239.

55 Yeh, I.C., Lenhart, J.L., Rutledge, G.C., and Andzelm, J.W. (2017). Molecular dynamics simulation of the effects of layer thickness and chain tilt on tensile deformation mechanisms of semicrystalline polyethylene. *Macromolecules* 50 (4): 1700–1712.

56 Yamamoto, T. (2009). Computer modeling of polymer crystallization – toward computer-assisted materials' design. *Polymer* 50 (9): 1975–1985.

57 Yamamoto, T. (2013). Molecular dynamics in fiber formation of polyethylene and large deformation of the fiber. *Polymer* 54 (12): 3086–3097.

58 Higuchi, Y. and Kubo, M. (2016). Coarse-grained molecular dynamics simulation of the void growth process in the block structure of semicrystalline polymers. *Modell. Simul. Mater. Sci. Eng.* 24 (5): 055006.

59 Monasse, B., Queyroy, S., and Lhost, O. (2008). Molecular dynamics prediction of elastic and plastic deformation of semi-crystalline polyethylene. *Int. J. Mater Form-Springer/ESAFORM* 1 (1): 1111–1114.

60 O'Connor, T.C. and Robbins, M.O. (2016). Chain ends and the ultimate strength of polyethylene fibers. *ACS Macro Lett.* 5 (3): 263–267.

61 Jabbari-Farouji, S., Rottler, J., Lame, O. et al. (2015). Plastic deformation mechanisms of semicrystalline and amorphous polymers. *ACS Macro Lett.* 4 (2): 147–150.

62 Jabbari-Farouji, S., Rottler, J., Lame, O. et al. (2015). Correlation of structure and mechanical response in solid-like polymers. *J. Phys. Condens. Matter* 27 (19): 194131.

63 Higuchi, Y. and Kubo, M. (2017). Deformation and fracture processes of a lamellar structure in polyethylene at the molecular level by a coarse-grained molecular dynamics simulation. *Macromolecules* 50 (9): 3690–3702.

64 Humbert, S., Lame, O., Chenal, J.M. et al. (2010). New insight on initiation of cavitation in semicrystalline polymers: in-situ SAXS measurements. *Macromolecules* 43 (17): 7212–7221.

65 Farge, L., André, S., Meneau, F. et al. (2013). A common multiscale feature of the deformation mechanisms of a semicrystalline polymer. *Macromolecules* 46 (24): 9659–9668.

66 Humbert, S., Lame, O., and Vigier, G. (2009). Polyethylene yielding behaviour: what is behind the correlation between yield stress and crystallinity? *Polymer* 50 (15): 3755–3761.

67 Fayolle, B., Richaud, E., Colin, X., and Verdu, J. (2008). Review: degradation-induced embrittlement in semi-crystalline polymers having their amorphous phase in rubbery state. *J. Mater. Sci.* 43 (22): 6999–7012.
68 Hutchinson, J.M. (1995). Physical aging of polymers. *Prog. Polym. Sci.* 20 (4): 703–760.
69 Yao, S., Tominaga, A., Fujikawa, Y. et al. (2013). Inner structure and mechanical properties of recycled polypropylene. *Nihon Reoroji Gakkaishi* 41 (3): 173–178.
70 Gong, J., Katsuyama, Y., Kurokawa, T., and Osada, Y. (2003). Double-network hydrogels with extremely high mechanical strength. *Adv. Mater.* 15 (14): 1155–1158.
71 Gong, J.P. (2010). Why are double network hydrogels so tough? *Soft Matter* 6 (12): 2583–2590.
72 Ducrot, E., Chen, Y., Bulters, M. et al. (2014). Toughening elastomers with sacrificial bonds and watching them break. *Science* 344 (6180): 186–189.
73 Jang, S.S., Goddard, W.A., and Kalani, M.Y.S. (2007). Mechanical and transport properties of the poly(ethylene oxide)–poly(acrylic acid) double network hydrogel from molecular dynamic simulations. *J. Phys. Chem. B* 111 (7): 1729–1737.
74 Yamamoto, T. and Sawada, K. (2005). Molecular-dynamics simulation of crystallization in helical polymers. *J. Chem. Phys.* 123 (23): 234906.
75 Yamamoto, T. (2006). Crystallization of helical oligomers with chirality selection. I. A molecular dynamics simulation for bare helix. *J. Chem. Phys.* 125 (6): 064902.
76 Yamamoto, T. (2014). Molecular dynamics of crystallization in a helical polymer isotactic polypropylene from the oriented amorphous state. *Macromolecules* 47 (9): 3192–3202.
77 Romanos, N.A. and Theodorou, D.N. (2010). Crystallization and melting simulations of oligomeric α1 isotactic polypropylene. *Macromolecules* 43 (12): 5455–5469.

6

A Tale of Chirality Transfer, Multistep Chirality Transfer from Molecules to Molecular Assemblies, Organic to Inorganic Materials, Then to Functional Materials

Reiko Oda[1], *Emilie Pouget*[1], *Thierry Buffeteau*[2], *Sylvain Nlate*[1], *Hirotaka Ihara*[3], *Yutaka Okazaki*[1,3], and *Naoya Ryu*[1,4]

[1] *CNRS – Universite Bordeaux, Institute of Chemistry and Biology of Membranes and Nanoobjects (UMR5248 CBMN), Chemistry Department, Bordeaux INP, 2 rue Robert Escarpit Pessac, 33607, France*
[2] *CNRS, University of Bordeaux, Institute of Molecular Science (UMR5255 ISM), Chemistry Department, 351, Cours de la Libération, 33405 Talence, France*
[3] *Kumamoto University, Department of Applied Chemistry and Biochemistry, 2-39-1 Kurokami, Chuo-ku, Kumamoto 860-8555, Japan*
[4] *Kumamoto Industrial Research Institute, Materials Development Department, 3-11-38 Higashimachi, Higashi-ku, Kumamoto 862-0901, Japan*

6.1 Introduction

Living organisms are based on complex dynamic supramolecular organization, are structured hierarchically through the macroscopic scale, and adopt at each instance the optimized functional form. Intramolecular and intermolecular transfer of information is crucial for the good functioning of the systems, allowing conformational or configurational information to propagate from one structure to another. One particular aspect controlling molecular structure is chirality. Most of the biomolecules are chiral and this very aspect is often propagated and amplified through dynamic recognition processes to create macroscopic objects. In this chapter, we will focus on how chirality transfer is observed in various systems with different levels of complexity, particularly for nanometric molecular self-assemblies. We will then present our own studies in which the concept of chirality transfer is used at different levels, starting from a chiral molecule to functional materials with new chiral properties.

6.2 Chirality Induction and Chirality Transfer

Chirality is a geometric property generally described for molecules, supramolecules, macromolecules, or even larger objects which are non-superposable to their mirror images. This concept is of great importance because most biomolecules and pharmaceuticals are chiral. More than 80% of the active compounds used by pharmaceutical companies are chiral, and this fraction is increasing, as the development of active compounds continues to be

Molecular Technology: Materials Innovation, Volume 3, First Edition.
Edited by Hisashi Yamamoto and Takashi Kato.

improved. This is also an important question in the agrochemical, flavor, and fragrance industries. For example, the two enantiomers of limonene smell of oranges and lemons respectively. If the chirality is well known at the molecular level, chiral structures can be detected at different dimensions and are often closely related to their functions: from a nanometric to submicron scale (DNA, peptides, viruses, etc.) as well as macroscopic levels as observed in the objects such as hands, tree trunks, snail shells, and even spiral galaxies. For example, shell-coiling is important as snail's sexual organs are usually twisted and it is difficult for snails of opposite handedness to reproduce [1].

6.2.1 Notion and Examples of Chirality Induction and Chirality Transfer

A concept that is of particular importance is the transmission of chirality information, generally named chirality transfer or chirality induction. This transfer of chirality can be observed at different levels: from a part of a molecule to another part of the same molecule (intramolecular), from a chiral molecule to another molecule (intermolecular), from a molecule to the nanometric or micrometric self-assembly (chirality amplification) and inversely from a chiral self-assembly or a patterned surface to a non-chiral molecule, and also from organic structures to inorganic ones. A number of biological phenomena result from this chirality transfer principles, and a large panel of applications are being developed such as the control of light, chiral recognition, enantioselective catalysis, or sensing.

6.2.2 From Molecule to Molecule

The concept of chirality transfer from a molecule to another molecule or several other molecules has been extensively studied in the context of host–guest chemistry, for example, between chiral molecules and achiral or prochiral helically folded polymers or foldamers, cages, capsules, ligands, or cavities (or vice versa, i.e. a- or prochiral guest with chiral host) [2–6]. Beyond the fundamental aspect with a number of elegant molecular designs, which showed a remarkable development in the last decades, the potential application of such systems includes chiral switching, sensing, and chiroptical devices [7–11]. Another field that shows an important development is asymmetric catalysis, involving the transformation of prochiral starting material into enantioenriched products with the help of chiral catalysts. Today, a large number of chiral catalysts are able to catalyze a variety of transformation with enantiomeric excess higher than 98% [12, 13]. However, the major remaining challenges are to overcome substrate specificity, the high cost of many chiral ligands and metal precursors, and/or in many, their availability in large quantities. A class of very important coordination complexes with crystalline structures, metal-organic frameworks (MOFs), which have infinite network structures built with multioptic organic ligands and metal ions, have been thoroughly studied as potential asymmetric catalysts [8]. Also, a promising family of enantiopure dendritic polyoxometalate (POM) hybrids were built by assembling chiral dendritic amines and achiral POM moieties, giving significant induced optical activity in the POM cluster [14].

6.2.3 From Molecule to Self-assembled Systems

There has been great interest in mimicking biomolecular structures such as protein, DNA, or viruses, which adopt hierarchical organization from precisely defined and functional structures in order to better understand the crucial link between their structures and functions. Generally, the construction of these superstructures, which are most of the time chiral, are based on nanoscale bottom-up self-assembly where various molecules can be held together hierarchically through noncovalent interactions such as π–π stacking, hydrogen bonds, hydrophobic interactions, and electrostatic interactions. The transfer of chirality and the resulting cooperative chirality amplification from a chiral building block to molecular assemblies is widely found, providing complex and controllable superstructures as the chirality adds a symmetry dimension to the multi-block constructions [15, 16]. The asymmetry information of each building block or each molecule is transmitted to its neighbor to propagate chirality but also the asymmetric interactions between different units can allow the recognition and amplification of the chirality in order to create chiral superstructures creating 2D or 3D structures. It is important to note that not all the chiral molecules express their chirality when assembled (i.e. no chirality amplification). Extremely delicate balance of inter–intramolecular interaction is required. It is therefore crucial to design suitable molecules or building blocks capable of expressing the chirality at higher dimension. An important class of structures showing chiral morphology based on molecular assembly is the fibrillary structures, in other words, helices, which are extensively covered in a number of literature reports since the very first examples reported in the 1940s–1960s [17–19]. Then, the mechanisms of formation were more thoroughly investigated in the 1980s [20–24]. Since then, the number of the reports on these chiral fibrillary systems showed exponential increase and resulted in a number of comprehensive reviews [25–29]. These chiral self-assembled structures are also very promising candidates for various applications. In biology, for example, it has been determined that the handedness of the self-assembled nanotwists has a profound effect on cell adhesion in the supramolecular hydrogels [30] whereas others have demonstrated the importance of the chirality of a fibrillary peptidic system for the drug delivery [31]. In photonics, chiral nanoarchitectures provided by the helical arrangements of luminophores make them excellent candidates for emitting circular polarized light (CPL) as proposed by Shen et al. [32] who synthesized supramolecular gels with strong CPL activity. Several interesting examples of the use of chiral self-assemblies for chiral recognition instead of the classical molecular receptors can also be found [33, 34].

6.2.4 From Molecular Assemblies to Molecules

If the chirality transfer induced from chiral molecules gives chiral self-assembled nanostructures, these supramolecular assemblies in bulk or at a surface can also influence in turn the chirality at the molecular level. In this case, the self-assembly is used as host to induce chirality to non-chiral molecules. In order to produce the induced chirality, it is necessary for the achiral molecule to have a strong interaction with the chiral host through a noncovalent bond. For example, chirally patterned surfaces or self-assembled monolayers can transfer the chirality to achiral or prochiral molecules adsorbed on their surfaces [35–39]. In another example, achiral fluorophores are incorporated in the chirally arranged alkyl chains forming chiral fibers, where the chirality is transferred to the dye molecule during the co-assembly, resulting in CPL [40, 41]. In view of the rapidly increasing demand for novel CPL materials, these works develop new strategies to achieve low-cost CPL flexible materials. Such chirality transfer from the molecular assembly to the molecule is also observed in heterogeneous asymmetric catalysis. For example, polymeric supramolecular chiral catalysts constructed on the basis of self-assembly show excellent asymmetric induction and reusability [42, 43].

6.2.5 Inorganic Chiral Structures

Inorganic chiral structures are less common than the organic ones, particularly in the artificial compounds that are not directed by genetics. While at the microscale, lithographical methods are powerful tools capable of creating complex patterns on flat substrates, they are in general not adequate for creating small (<100 nm) and hierarchically organized structures throughout the macroscopic scale [44–46]. Indeed, at the nanoscale synthetic methodologies and theories are still very limited, particularly regarding chiral structures [47–52]. One of the most efficient methods to form chiral inorganic nano-objects is to use organic nanostructures as templates by adsorbing or depositing the inorganic materials at their surfaces, thus transferring the chirality of the templates to the inorganic structures. Indeed, highly developed molecular design of chiral nanostructures as mentioned above has triggered the approach of using them as templates to transpose their finely tuned structures to inorganic materials. In order to assure efficient chiral transfer from the organic template to the inorganic material or precursors, the inorganic material should not disrupt the organic structure while having specific affinities with it. The first strategy is to graft the prefabricated nanoparticles on the organic self-assembly in order to create helical arrays of inorganic particles. The most studied materials are the grafting of gold nanoparticles (GNPs) or nanorods. When grafted on superstructures with chiral configurations, plasmon–plasmon interactions of the plasmonic building blocks appear and lead to strong chiroptical activity [53]. Reports on self-assembly of plasmonic building blocks (NPs) onto preexisting chiral nanostructures include the use of fibers [54–56], helices [57–65], and films [66]. Regarding silver-based hybrids, it has been shown that silver ions can complex with nucleic acids and that this affinity can be used to grow chiral chains of silver nanoparticles exhibiting chiroptical properties [67]. Finally, motivated by the development of chiral emitting probes, some works on the effect of the optical properties of chiral molecules in close interaction with nanocrystals such as quantum dots are reported. For example, the chirality of the stabilizers of CdTe nanocrystals has an effect both on the growth kinetics and the optical properties of CdTe [68, 69]. Similar to the systems with GNPs, self-assembled nanofibers (helical or straight) have been used for the nucleation of cadmium sulfide (CdS) nanocrystals. It has been shown that the chiral nature of the template influences the shape of the CdS nanocrystals and their photoluminescence activity [70].

The second strategy is to permanently replicate the chiral organic structure in inorganic materials through a sol–gel polycondensation. This templating route, which is prevalent in biomineralization, has been developed for chiral objects after the pioneering work of Shinkai's group followed by research by many others to produce inorganic replicas of diverse chiral organogels, leading to the formation of nanotubes, ribbons, or fibers [71–77]. Shinkai's group reported the first example of the creation of both "right- and left-handed helical silica structures" by a sol–gel transcription method [78]. Also, hierarchical helical constructions transcribed from organic templates give interesting chiral mesoporous silica structures (for a review, see Qiu et al., [48]). If silica is the most common metal oxide for the inorganic replication, other minerals can be envisaged, particularly titanium dioxide because of the interest it has created in photovoltaics and photocatalysis. Because of the high reactivity of the titania precursors, direct organic templating is difficult to achieve. Some examples have been published on the use of amino-derivative structures [73, 79, 80], peptidic lipids [81], or even biological structures such as M13 viruses [82] and amyloid fibrils [83] as templates. Among other materials, Stupp et al. reported the use of CdS to replicate self-assembled nanohelices [57, 84]. In this case, nanocrystals of CdS, which were adsorbed on the surface of the organic helices, were observed to coalesce to form continuous CdS helices.

6.2.6 Characterization Methods of Chiral Assemblies

Depending on the dimension of the chirality expressed in the systems, different characterization techniques are used. Nuclear magnetic resonance (NMR) and X-ray diffraction (XRD) structural analysis are extremely powerful techniques in determining the absolute configuration and conformation of the chiral molecules and biomolecules at the atomic level. However, XRD requires crystalline samples, whereas many of these self-assembled systems with chirality expressed at mesoscopic levels are not crystalline and thus cannot be investigated directly with this technique. Nevertheless, when crystallized, their molecular configurations and packing information obtained from XRD analysis can help understand the self-assembly process. Solid-state NMR, on the other hand, allows the configuration/conformation analysis of noncrystalline systems, but requires labeling of atoms. Chiroptical techniques such as electronic circular dichroism (ECD) and vibrational circular dichroism (VCD) on the other hand allow direct investigation of these chiral nanostructures at a molecular and supramolecular level in the bulk in solution or suspension and are well adapted to investigate chiral structures (determination of the configuration of chiral molecules, determination of the conformation of chiral objects such as handedness in helical structures) and to reveal the transfer of chirality in achiral materials. The transfer of chirality can be easily followed by VCD and ECD since these two techniques are very sensitive to the chiral arrangement of molecules (or subunits) in chiral materials, due to the excitonic coupling of the elementary entities. In addition, ECD and VCD are complementary chiroptical techniques since ECD will be more suitable for the investigation of chromophore groups (dyes for example) or for revealing chiral effects in the UV–vis spectral range (plasmonic chirality for example) whereas VCD will be more relevant to show transfer of chirality in organic compounds due to the large number of the vibrational modes in the mid-infrared spectral range (3N-6 vibrations), allowing generally a clear differentiation between the chiral contributions of the chiral inducer and achiral derivative.

Finally, at the mesoscopic morphological level, the rapid development since several decades of imaging techniques such as atomic force microscopy (AFM), scanning tunneling microscopy (STM), scanning electron microscopy (SEM), and transmission electron microscopy (TEM) allowed detailed characterization and analysis of the typical chiral morphologies such as nanometric helical or twisted ribbons. Unfortunately, these imaging techniques require thermal or chemical fixation and/or drying of the sample and enhancement of the contrast. When the chirality is expressed at more than a few hundred nanometer level, optical microscope can be used to visualize them. The advantage of this technique is that it allows *in situ* observation of the solvated objects, following the kinetics or dynamics of the structure modification.

6.2.7 Aim of This Chapter

In all the examples cited in this introduction, there is a chemical or physical transfer of chirality at different levels: (i) breaking of the symmetry in achiral or racemic systems to express chirality; (ii) transcribing the chirality of one object to others to propagate chirality; and (iii) organizing the asymmetric interactions between different units to recognize or amplify chirality. In this chapter, we will present "a story of chirality transfer" starting from a simple molecule to molecular assemblies, from organic to inorganic systems through amplification of the chirality, then finally to bring toward functional nanostructures.

6.3 Molecular and Supramolecular Chirality from Gemini-Tartrate Templates

6.3.1 Gemini-Tartrate Amphiphiles

6.3.1.1 Formation of Gels with Chiral Nanoribbon Structures

We have reported that achiral cationic gemini surfactants having the formula C_2H_4-1,2-($(CH_3)_2N^+C_{16}H_{33})_2$, denoted as 16-2-16, self-assemble in aqueous solution in the presence of chiral tartrate counterions to form nanometrical twisted or helical ribbons as well as tubules as observed by TEM (Figure 6.1) [85, 86]. The handedness of the ribbons depends on the D- or L-chiral configuration of the tartrate (right-handed ribbons are formed in the presence of L-tartrate and left-handed ribbons are formed with D-tartrate). Their twist pitch as well as their width decrease continuously upon increasing the enantiomeric excess (ee), from flat multilayer membranes for the racemate (infinite pitch) to ribbons of the pure enantiomer (pitch is about 100 nm) as shown in Figure 6.1e,f [87].

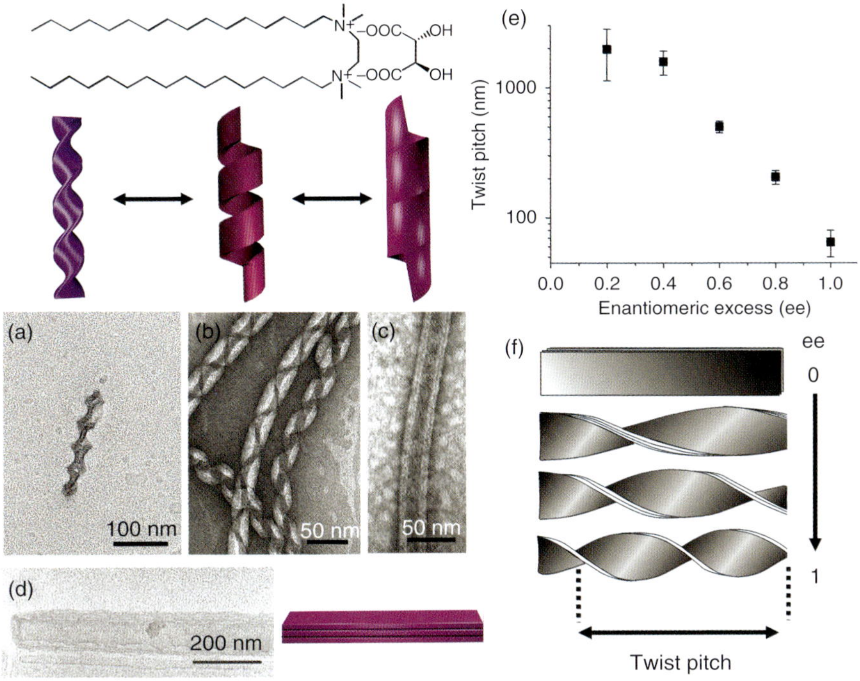

Figure 6.1 Chemical structure of 16-2-16 L-tartrate. Transmission electron microscopy (TEM) images and schematic representation of twisted ribbons (a), helical ribbons (b), and tubules (c) of 16-2-16 L-tartrate; TEM image and schematic representation of flat ribbons of racemic 16-2-16 tartrate (d). The morphologies of the multilayer ribbons depend on the ee of the tartrate counterion. For a pure enantiomer, twisted ribbons form first and transform into helical ribbons and tubules with time [4]. This transformation into helical ribbons or tubules subsists at an ee as low as 0.8, but below ee = 0.6, only twisted ribbons are observed. The pitch of the twist increases with decreasing ee (e and f), up to an infinite value for ee = 0 (flat ribbon).

The morphologies and dimensions of these chiral nanostructures can be easily controlled and the interconversion between these structures can be induced by modifying several parameters. For example, (i) varying the hydrophobic chain length or adding gemini having bromide counterions (1%) or the opposite enantiomer (10%) leads to an increase in the diameter of membrane tubules from 33 to 48.5 nm; (ii) further addition of gemini bromide (1.5%) or a slight increase in temperature induces a transition from tubules to twisted ribbons; (iii) the twist pitch of the ribbons can be continuously tuned by varying enantiomeric excess; and (iv) it was also observed that the morphologies of these ribbons much evolve with time. Such unprecedented observations over easy tuning of the chiral supramolecular structures are very intriguing and are clearly related to the original feature that the induction of chirality is solely due to the counterions, which are much more mobile than the amphiphiles.

In order to shed light on the mechanism of chirality transfer from counterions to gemini amphiphiles, and from molecular level to supramolecular level, we have performed various experiments by ^{1}H NMR, ECD and VCD, as well as by XRD techniques.

6.3.1.2 Specific Recognition Between Dication Amphiphiles and Tartrates

The *n*-2-*n* dimeric amphiphiles consist of two charged quaternary centers separated by an ethylene spacer. Both steric and electrostatic effects strongly stabilize the anti conformation of the spacer with respect to the gauche conformations. Rotations about the bonds linking the nitrogens to the spacer carbons give rise to seven conformers [88]. Enantiomeric conformers can be distinguished by the conformations of the space 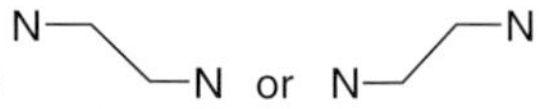, which are chiral in the two-dimensional spacer plane. The ^{1}H NMR spectra of 12-2-12 L-tartrate in the mixture of deuterated chloroform and methanol (e.g. 9 : 1 $CDCl_3/CD_3OD$) is compared with those of 12-2-12 Br. We observed a splitting of the signals of the cationic headgroup into a diastereotopic pattern (Figure 6.2), which is due to the

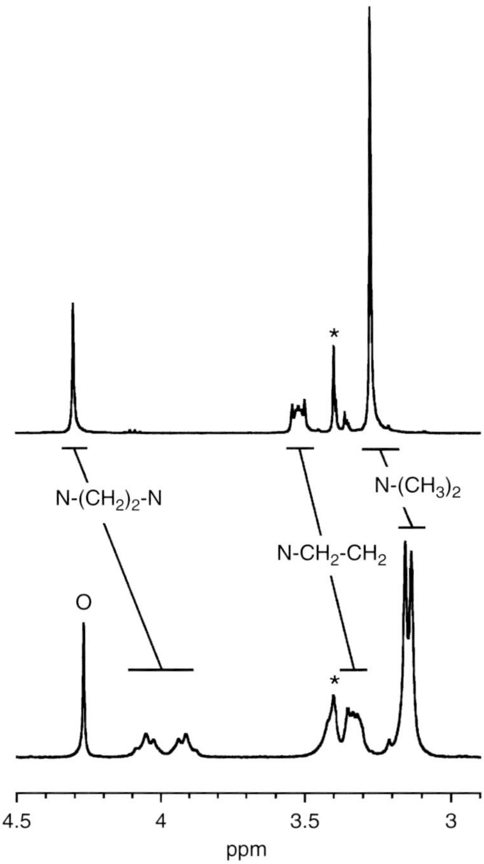

Figure 6.2 ^{1}H NMR spectra of 12-2-12 surfactant in 9 : 1 $CDCl_3/CD_3OD$ (10 mM), with bromide (top) and tartrate (bottom) counterions. The asterisks indicate methanol residual solvent peak (CHD_2OD). The circle indicates the CHOH signal of the tartrate counterion. The signals of the N^+CH_3 appear as two singlets ($\Delta\delta = 8$ Hz), and those of the ethylene spacer appear as an AA′BB′ motif ($\Delta\delta = 68$ Hz). We also note a significant upfield shift (up to 0.3 ppm) of the signals of the protons belonging to the headgroup in the presence of tartrate ions compared to bromide ions. Source: Reprinted with permission from Berthier et al. [88]. Copyright 2002, American Chemical Society.

nonequivalence of some protons of the cation in the presence of a chiral anion. This demonstrates the presence of chiral conformers of the cation in solution as characterized in the solid state. Indeed, protons which are enantiotopic in a given chiral conformer of the cation become diastereotopic under the influence of a chiral anion. It also implies a close proximity and a tight recognition between dianionic tartrate and *n*-2-*n* dications.

6.3.1.3 Conformation of Tartrate Ions in Solution, Micellar Aggregates, and Twisted Ribbons

The conformation of tartrate salts has been evaluated using ECD in the absorption region of the carboxylate chromophores, at 190–250 nm, as shown in Figure 6.3 [88, 89]. In aqueous solution, the ECD spectrum of sodium L-tartrate shows two negative bands at 194 and 211 nm (Figure 6.3a). The absence of exciton coupling between the carboxylates is consistent with the expected trans conformation, in which the transition moments of the two chromophores are coplanar and remote from each other. Similar ECD spectra were observed when L-tartrate is complexed with cetyl-trimethylammonium (CTA) or dimeric surfactant 12-2-12, two cationic amphiphiles forming micellar aggregates.

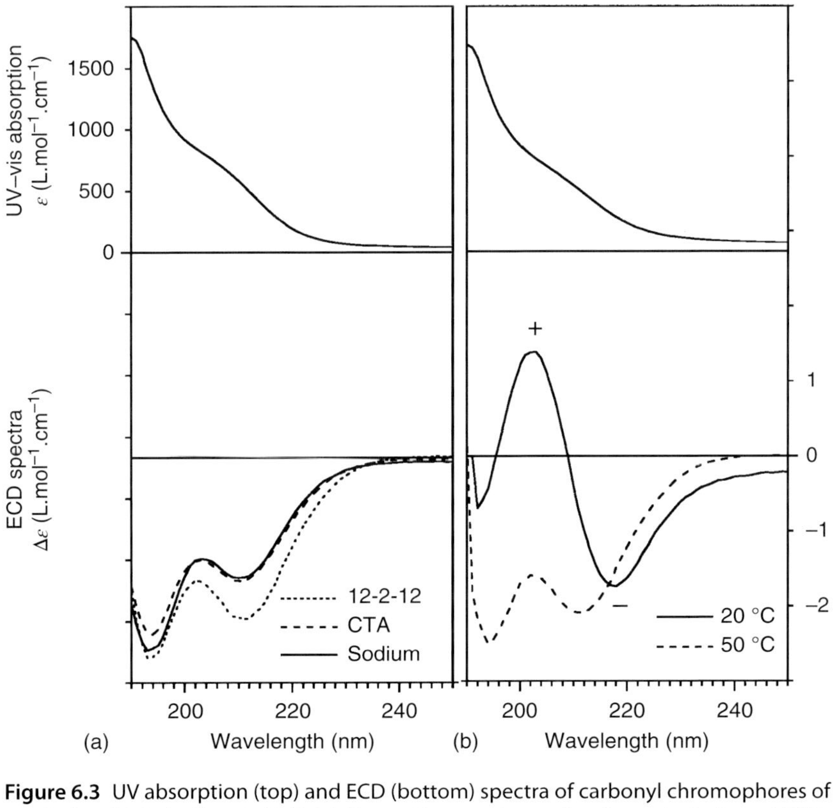

Figure 6.3 UV absorption (top) and ECD (bottom) spectra of carbonyl chromophores of L-tartaric acid derivatives in H_2O. (a) Absorption spectrum of sodium L-tartrate and ECD spectra of sodium L-tartrate, CTA L-tartrate, and 12-2-12 L-tartrate. (b) Absorption spectrum of 16-2-16 L-tartrate and its ECD spectra at 20 and 50 °C. Source: Reprinted with permission from Berthier et al. [88]. Copyright 2002, American Chemical Society.

The ECD spectrum of L-tartrate is completely different when they are incorporated into chiral bilayers in the presence of 16-2-16 surfactants (Figure 6.3b). A strong positive–negative exciton coupling between the carboxylates is then observed at 20 °C, with a positive band at 202 nm and a negative band at 219 nm. Such negative exciton coupling suggests, in a first qualitative analysis, a counterclockwise arrangement of the electric transition moments of carboxylate groups. This counterclockwise arrangement may be at the intramolecular level (*G*-conformation of the tartrate molecules) or at the intermolecular level (the COO^- groups of the neighboring tartrate molecules are organized in a counterclockwise chiral arrangement; the molecules of tartrate can then adopt *G*- or trans conformation) to form submicrometric chiral aggregates.

Complementary information can be drawn from VCD experiments [89]. The VCD spectrum of sodium L-tartrate solution or L-tartrate in micellar aggregates exhibits a negative couplet (positive at higher wavenumbers and negative at lower wavenumbers) for the $\nu_a CO_2^-$ mode at 1602 cm^{-1} whereas no bands are observed for the $\nu_s CO_2^-$ mode at 1392 cm^{-1} (Figure 6.4a,b, or c). On the basis of the degenerate coupled-oscillator model, this non observation of VCD bands is consistent with the trans conformation in which the transition moments of the two $\nu_s CO_2^-$ modes are coplanar both in tartrate solution and in micellar aggregates.

In contrast, the VCD spectra of tartrates in twisted or helical ribbons (Figure 6.4d) reveal strong modifications. The VCD spectrum of the tartrate in chiral membranes also features a bisignate band for the $\nu_a CO_2^-$ mode, but the band is much more intense (by a factor 6) and its sign is inverted. The spectrum also shows a weaker negative band for the $\nu_s CO_2^-$ mode, not seen in the spectrum of micellar or molecular solution of tartrates. What would be the origin of these spectral modifications? Would it be due to a conformational change of tartrate ions upon formation of twisted membranes as mentioned above for ECD experiments, or due to the intermolecular organization? A complementary information was obtained by the VCD signals of C*–O stretching vibration region of 16-2-16 L-tartrate in $CHCl_3$ solution (Figure 6.4e), which shows a negative couplet (positive band for the symmetric combinations of C*-O stretching motions at 1114 cm^{-1} and negative band for the antisymmetric combinations of C*-O stretching motions at 1063 cm^{-1}) not observable in D_2O. Considering the exciton coupling theory, such a negative couplet is expected for the trans conformation of L tartrates whereas the *G*_conformation should lead to a positive couplet. The ensemble of the ECD and VCD results indicate that these self-assembled systems are governed by the chiral intermolecular arrangement of the tartrate ions (trans conformation) in the membranes.

6.3.1.4 Induction of Chiral Conformation in the Cationic Amphiphile

Direct observation of the chiral induction from chiral dianion, tartrate, to non-chiral dicationic surfactant can be achieved by VCD measurements, since in the infrared region, vibrators belonging to tartrate dianions and the amphiphilic cations can be easily distinguished. As shown in Figure 6.4d, the VCD spectrum of membranes of 16-2-16 L-tartrate in D_2O features a negative and a weak positive band for the two split components of the bending mode of methylene groups (δCH_2) located at 1470 and 1466 cm^{-1}, respectively. Two bands exhibiting opposite signs are found with the D-tartrate derivative. The VCD spectra of the same samples in the 3000–2800 cm^{-1} region associated with the antisymmetric ($\nu_a CH_2$) and symmetric ($\nu_s CH_2$) stretching vibrations

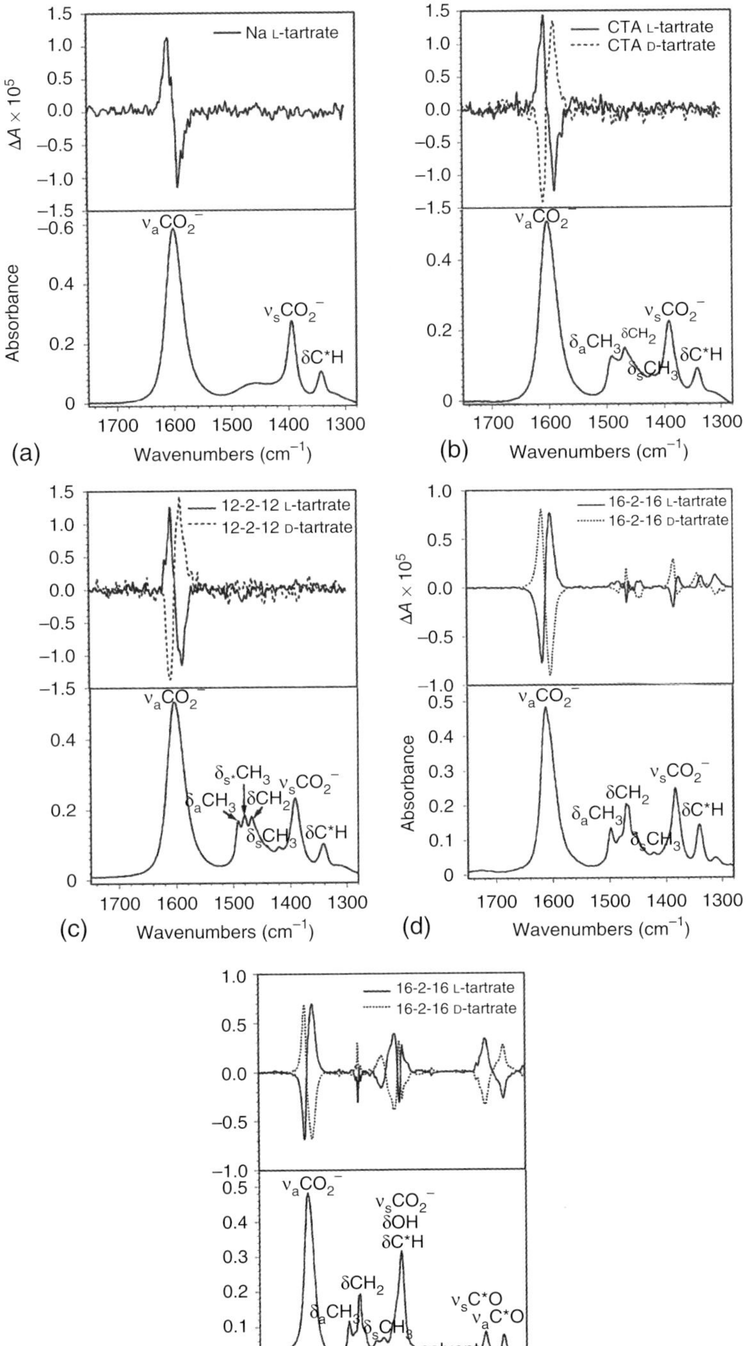

Figure 6.4 VCD (top) and IR (bottom) spectra of (a) sodium L-tartrate, (b) CTA L- and D-tartrate, (c) 12-2-12 L- and D-tartrate, and (d) 16-2-16 L- and D-tartrate, recorded at 100 mM in D_2O at room temperature with a path length of 50 μm. (e) VCD (top) and IR (bottom) spectra of 16-2-16 L- and D-tartrate recorded at 20 mM in $CHCl_3$ with a path length of 250 μm. Source: From Brizard et al. 2009 [89].

of the methylene groups confirmed such observations. As shown in Figure 6.5a, the VCD spectra of 16-2-16 L- or D-tartrate counterions reveal very intense induced bands both for the $\nu_a CH_2$ mode at 2918 cm^{-1} and for the $\nu_s CH_2$ mode at 2850 cm^{-1}.

The frequencies obtained for the $\nu_a CH_2$ and $\nu_s CH_2$ modes, along with the δCH_2 mode, which shows a splitting, indicate that the alkyl chains are well organized in an orthogonal packing with mainly all trans conformation. For micellar solutions of CTA or 12-2-12 L-tartrate (Figure 6.5b), the stretching vibrations of methylene groups are observed at higher frequencies (2925 and 2854 cm^{-1} for the $\nu_a CH_2$ and $\nu_s CH_2$ modes, respectively) and the splitting of the δCH_2 band disappears, indicating the presence of many gauche conformations and disordered chains. The VCD spectrum of micelles does not reveal induced bands in the CH_2 stretching region. No chiral induction from the anion to the cation takes place in the dynamic micellar systems in which cations and anions are poorly associated. Thus, the chiral induction only takes place in membranes when cations and anions cooperatively organize in a multilayer structure.

6.3.1.5 Effect of Enantiomeric Excess Studied by ECD, VCD, and XRD

Pure 16-2-16 L-tartrate forms right-handed helical ribbons, whereas the D-enantiomer forms ribbons of opposite handedness. Various ribbons were prepared by mixing the two pure enantiomers to follow the dependence of the ECD and VCD intensities on ee and compared with the TEM images [87, 89].

The ECD spectra of various mixtures of the two enantiomers of 16-2-16 tartrate are reported in Figure 6.6a. While a decrease of the enantiomeric excess from 1 to 0 causes a decrease in the intensity of the positive exciton coupling between the carboxylates, a positive exciton coupling is observed regardless of the

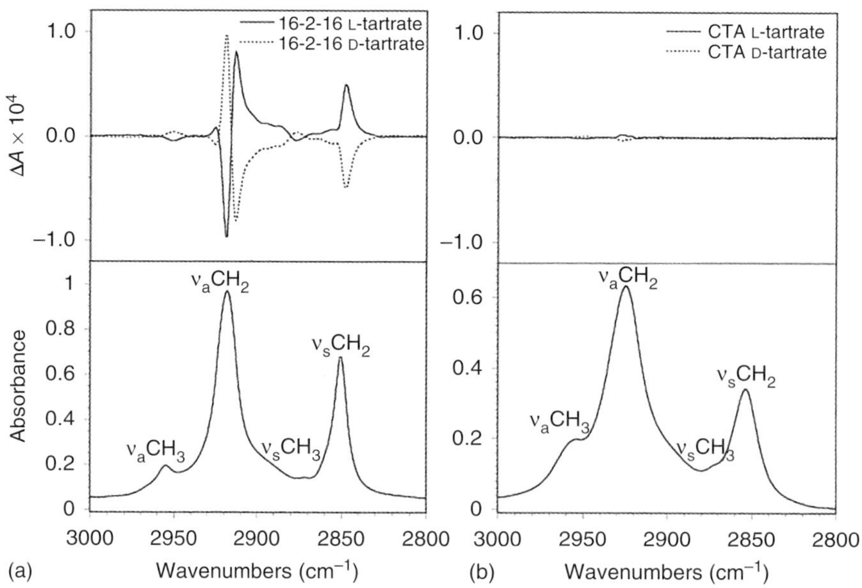

Figure 6.5 VCD (top) and IR (bottom) spectra of (a) 16-2-16 L- and D-tartrate and (b) CTA L- and D-tartrate, recorded at 100 mM in D_2O at room temperature with a path length of 50 μm. Source: Reprinted with permission from Brizard et al. [89]. Copyright 2009, Wiley.

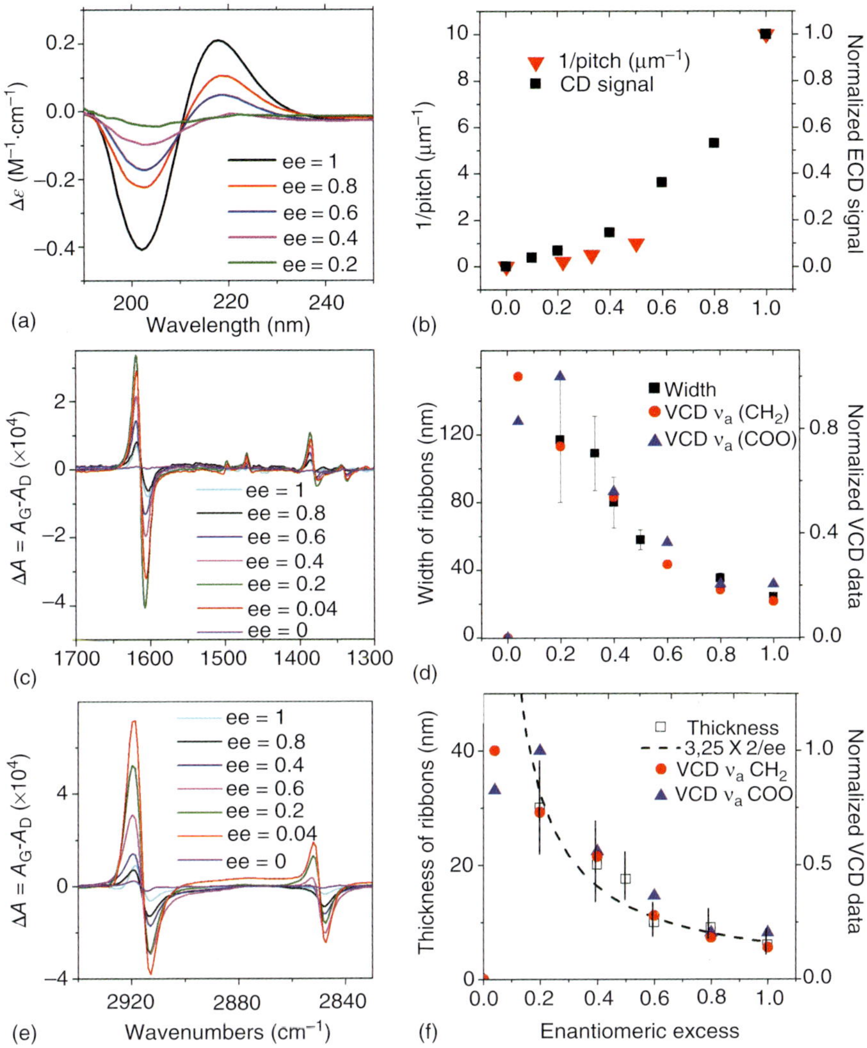

Figure 6.6 (a) ECD spectra of various mixtures of the two enantiomers of 16-2-16 tartrate recorded at room temperature at 10 mM in H_2O with a path length of 1 mm. (b) Variation of the inverse of the pitch period (right) and the normalized electronic circular dichroism (CD) data (left) as a function of enantiomeric excess. VCD spectra of various mixtures of the two enantiomers of 16-2-16 tartrate recorded at room temperature at 100 mM in D_2O with a path length of 50 μm (c) in the 1700–1300 cm^{-1} region and (e) in the 2970–2820 cm^{-1} region. Variation of the (d) width and the (f) thickness of the ribbons and the normalized VCD data as a function of enantiomeric excess. Using the bilayer thickness obtained from X-ray diffraction (i.e. 3.25 nm), ribbons thickness can be calculated as 3.25 × 2/ee nm. Source: From Brizard et al. 2009 [89].

value of ee, indicating a similar chiral environment of the tartrate dianions. Linear dependence could be expected for a complete phase separation of the two enantiomers into twisted ribbons of opposite handedness or if the two enantiomers form racemic flat ribbons first, and the excess 16-2-16 D-tartrate enantiomer gives rise to the ECD signal. In the present case, TEM images show continuous evolution of the ribbon morphology, continuous variation of twist pitch, width, and thickness (Figure 6.1e–f) [87], with a handedness of the ribbons given by the enantiomer in excess. As is shown in Figure 6.6b, a nonlinear dependence of the ECD intensity with ee is observed. It is noteworthy that the ECD intensity shows very good correlation with the inverse of the pitch period as observed with TEM images, a parameter which is directly related to the supramolecular chirality. This result suggests again that the ECD intensities reproduce not only the variation of the molecular chirality but also the emergence of a supramolecular chirality.

The dependence of VCD spectra as a function of ee is quite different (Figure 6.6c,e). Unlike the dependence of the ECD intensity on ee, VCD intensities increase when the enantiomeric excess decreases and reaches a maximum for ee at around 0.1–0.2, before rapidly going down to 0 when ee is strictly equal to 0. The lower the molecular chirality is, the higher the VCD intensities. This unexpected behavior cannot be simply explained by supramolecular chirality (the pitch of the chiral ribbons). Other than the pitch period of the twisted ribbons, the width and the thickness (or number of bilayers) of the ribbons formed by 16-2-16 tartrate are also strongly dependent on the ee. We have observed that the mean width of ribbons decreases from 200 (ee = 0) to 25 (ee = 1) nm with an increase of the enantiomeric excess from 0 to 1 and thickness variation from 30 nm (ee 0.2) to 6 nm (ee 1). Thus, the increase of the VCD intensity when the ee decreases could be related to the increase in the size of the chiral ribbons and better molecular arrangement of the cationic amphiphiles. Indeed, the evolution of the normalized intensities of the antisymmetric stretching vibrations of the carboxylate and methylene groups is correlated to the width (Figure 6.6d) and the thickness (Figure 6.6f) of the twisted ribbons.

Small-angle X-ray scattering (SAXS) and wide-angle X-ray scattering (WAXS) associated with molecular modeling revealed (Figure 6.7) that the ribbons formed with 16-2-16 tartrate follow well-defined molecular organization [90]. They are made of stacked heterochiral bilayers comprised of homochiral monolayers, and this organization is conserved through large morphology transitions from racemic (i.e. flat multilayered ribbons) to helical ribbons or tubules for the pure enantiomer. The chirality of mono/bilayers is defined not only by the molecular chirality of tartrate (D- or L-) and through the propagation of hydrogen bonds between tartrates and crystalline water molecules, but also by the dications that are intrinsically achiral but adopt two mirror-image conformations of their headgroups, Z or Σ (i.e.  "Z" or  "Σ") through the formation of strong ion pair with dianion tartrate. The contact domain between adjacent bilayers is homochiral: both gemini headgroups and tartrate of the symmetrical leaflet have the same chirality.

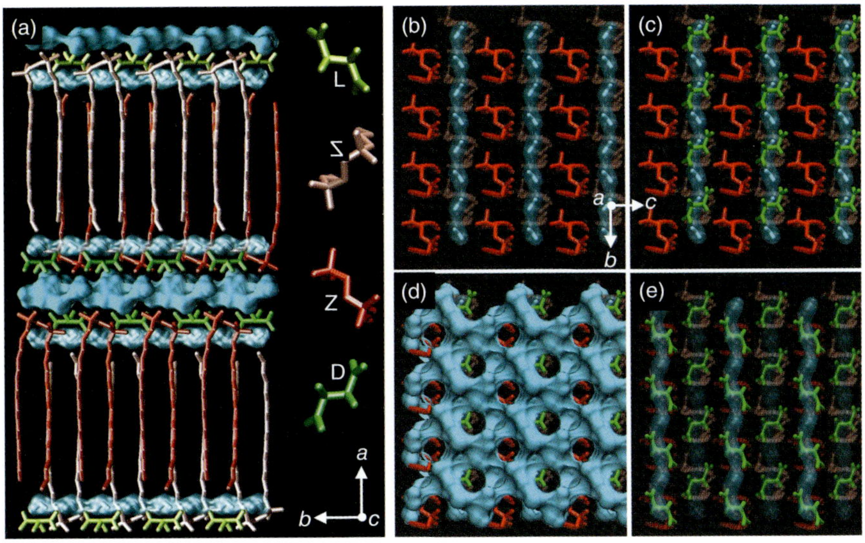

Figure 6.7 Molecular packing of racemic 16-2-16 tartrate giving the best fit between experimental and simulated powder diffraction patterns. (a) View from the side, along the *c* axis, and color code used to represent D- and L-tartrate as well as the two enantiomeric conformers of the gemini headgroup. (b–e) Views from the top (down the *a* axis) of the surface of a bilayer. In (b), only the gemini headgroups and water columns between them are shown. In (c) and (d), tartrate counterions and interstitial water have been successively added on top of the view in (b). In (e), the tartrate ions belonging to the opposite bilayer have been added, but interstitial water was removed for clarity. Source: Reprinted with permission from Oda et al. [90]. Copyright 2008, American Chemical Society.

In the racemate, the infinite alternation of domains between bilayers having DZ DZ and LƧ LƧ arrays of headgroups and tartrate ions is possible without chirality frustration (Figure 6.8). In contrast, following this rule, only two bilayers can stack for pure enantiomer, separated by a domain with an LƧ·LƧ (for instance) array of headgroups and tartrate ions. In the external leaflets of these double-bilayer ribbons, the heterochiral bilayer packing requires that the gemini conformation is opposite that in the inner leaf, and obviously, the only available tartrate (L) cannot be accommodated within the chiral cavity where D-tartrates are expected. From this rule, the number of bilayers (nbs) of the ribbons can be expected to follow the relationship nb = 2/ee. The ribbon thickness estimated from TEM images and XRD confirms this model (Figure 6.6f). The ribbon thickness is very close to (2/ee) × 3.25 nm as obtained from X-ray powder diffraction.

This stacking of bilayers model suggests that while the molecular organization is crystal-like at the inner surfaces of the bilayers, those on the outer surfaces and at the ribbons edges are much more loosely organized, reminiscent of micellar surfaces. The increase in the ratio between well-organized and disorganized molecules along with the increase in the size of the chiral ribbons and better molecular arrangement may explain in part the enhancement of the VCD signal when ee decreases.

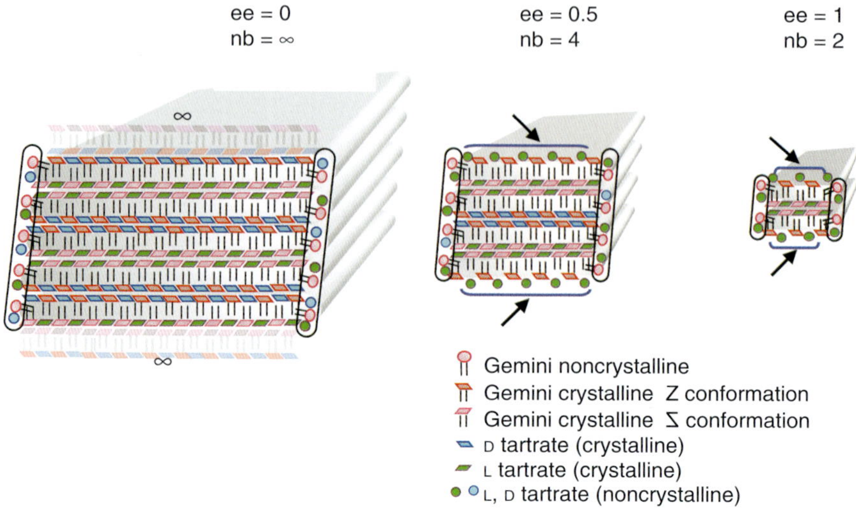

Figure 6.8 Schematic representation of the effect of enantiomeric excess on the width and thickness (number of bilayers) of ribbons for the case of an excess in L-tartrate. Source: From Brizard et al. 2009 [89].

Thus, since all chiral molecules, regardless of their D- or L- absolute configuration, are involved in ribbons that overall possess the same macroscopic handedness (dictated by the enantiomer in excess), a larger number of molecules are at the origin of the enhancement of the VCD intensity at lower ee. At very low enantiomeric excess (ee < 0.1), VCD intensity decreases and tends toward a zero value as expected. This behavior clearly shows that VCD intensities in self-organized systems are dependent on the size and the molecular organization of the investigated objects.

The ensemble of the analysis performed by ECD, VCD, and XRD allowed us to have a global view on this simple but beautiful system. Under certain conditions, chiral counterions and non-chiral gemini surfactants form crystalline bilayer organization with well-defined chiral organization as demonstrated by the XRD with a long range hydrogen-bonded network, leading to the formation of chiral ribbons. The chirality is transferred from tartrate to gemini molecules and leads to chiral assemblies expressing chirality at the mesoscopic level. The chirality evidenced by ECD is directly related to the expressed chirality and thus the intermolecular organization, i.e. the smaller the helical pitches and larger the angle between the adjacent molecules, the higher the ECD signals, whereas the VCD signal is related to the overall chirality of the nano-objets.

6.3.2 Organic–Inorganic Nanohelices

The structural diversity of the nanometric chiral nanoribbons based on self-assembled gemini-tartrate as described above inspired us to exploit them as templates for inorganic chiral nanostructures fabrication with controlled pitch and handedness by sol–gel reaction of tetraethoxysilane (TEOS) [75, 76]

(Figure 6.9A). The morphologies of these nanocomposite structures are sensitive to the molecular composition of the self-assemblies but they also vary with time (aging process) as well as the parameters of the polycondensation, which gives us access to various nanostructures from twisted ribbons (with Gaussian curvature, Figure 6.9B(a,b)), helical ribbons (with cylindrical curvature, Figure 6.9C(a,b)), as well as tubes. Calcination was performed by heating the silica–lipid hybrid nanoribbons after transcription at 600 °C for two hours under air. After the calcination process, both twisted and helical morphologies were maintained (Figure 6.9B,C(c)) but they shrank in all three dimensions (thickness of silica wall, pitch and width of the ribbons as defined in Figure 6.9D) by about 16–17% [91].

6.3.2.1 Chirality of Silica Nanohelices Evidenced by VCD

If the chiral shape of the helical and twisted silica nanoribbons is evident from the TEM observations, the chirality at a molecular level inside the silica walls is not obvious. Infrared (IR) measurements of these silica nanoribbons exhibit specific vibrational signals of silica in the 900–1250 cm^{-1} spectral range (Figure 6.10a,b, bottom) [91]. The band located at 1090 cm^{-1} and its shoulder at 1240 cm^{-1} are associated with the transverse optic (TO) and longitudinal optic (LO) modes of the Si–O–Si asymmetric stretching vibration (ν_aSi–O–Si), respectively. The band at 960 cm^{-1} is assigned to the Si–OH stretching vibration of free silanol groups. Surprisingly, VCD measurements clearly demonstrated Cotton effect both for the LO and TO modes. The VCD spectra of L-silica–lipid hybrid twisted nanoribbons (L-tw) presented a strong negative–positive (from long to short wavenumbers) Cotton effect for the TO mode at 1090 cm^{-1} and a much weaker negative Cotton effect for the LO mode (Figure 6.10a, top, dotted line). Opposite spectra were obtained for D-silica–lipid hybrid twisted nanoribbons (D-tw). The two components of the 1090 cm^{-1} band are probably due to the coupling of the different ν_aSi–O–Si vibrators that are distributed in a chiral arrangement within the silica network. Similar phenomenon was observed with the $\nu_a CH_2$ vibration of 16-2-16 alkyl chains in Figure 6.5a. These results unambiguously indicate that chiral information on organic template was transcribed to the silica nanoribbons. Although a few VCD spectra of siloxane-based materials have been reported in the literature for ladder-like silsesquioxane polymers [92] and for micrometric chiral mesoporous silica particles [93], no such strong VCD signals of silica chiral nanostructures with clear opposite signals for two enantiomers have been reported before, which enable quantitative evaluation of the chiral information on silica nanoribbons. After the calcination at 600 °C, the intensity of the Si–OH stretching band at 960 cm^{-1} reduced (Figure 6.10a,b, bottom, solid line with respect to dotted line) in agreement with the Si—O—Si bond formation (silanol condensation reaction) during calcination. Strikingly, an important increase in VCD signal for the bands at 1100 and 1240 cm^{-1} was observed upon calcination (Figure 6.10a, top, solid line). To compare the chiral information on these silica nanoribbons, the dimensionless anisotropic ratio $g = \Delta A/A$ was calculated (ΔA being the differential absorbance of left and right circularly polarized light). The g-factor of calcinated silica twisted or helical nanoribbons ($g_{\text{tw-calc}}$ or $g_{\text{hel-calc}}$) is about 1.7 or 1.9 times larger than that of uncalcinated silica–lipid hybrid twisted or helical nanoribbons (g_{tw} or g_{hel}),

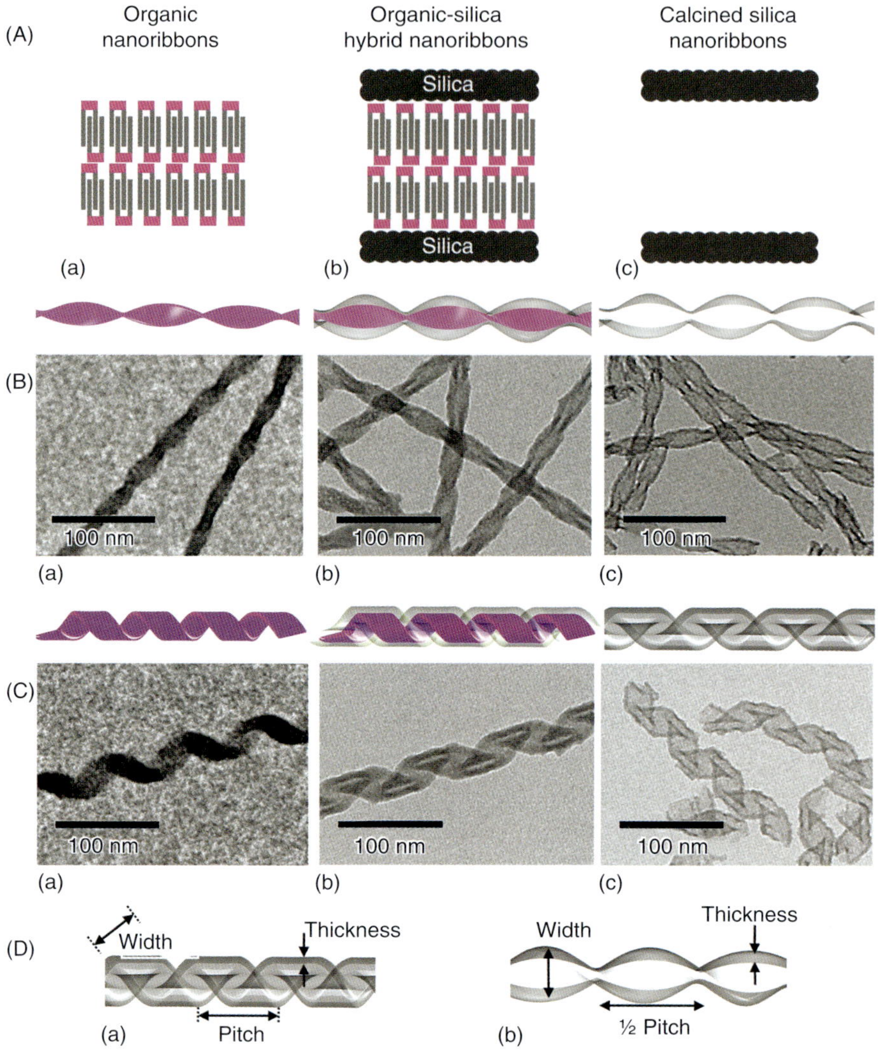

Figure 6.9 (A) Schematic illustrations of cross-sectional surfaces of lipidic nanoribbons (a), silica–lipid hybrid nanoribbons before (b) and after (c) calcination. (B) TEM images of lipidic twisted nanoribbons prepared by one hour aging of 16-2-16 L-tartrate (1 mM) aqueous solution at 20 °C (a), silica–lipid hybrid twisted nanoribbons prepared by sol–gel reaction of TEOS using lipidic twisted nanoribbons as template (b) and silica twisted nanoribbons calcinated at 600 °C for two hours (c). (C) TEM images of lipidic helical nanoribbons prepared by three days aging of 16-2-16 L-tartrate (1 mM) aqueous solution at 20 °C (a), silica–lipid hybrid helical nanoribbons prepared by sol–gel reaction of TEOS using lipidic helical nanoribbons as template (b) and silica helical nanoribbons calcinated at 600 °C for two hours (c). (D) Schematic illustration of a silica (a) helical and (b) twisted nanoribbon for quantitative morphological evaluation. Source: From Okazaki et al. 2016 [91].

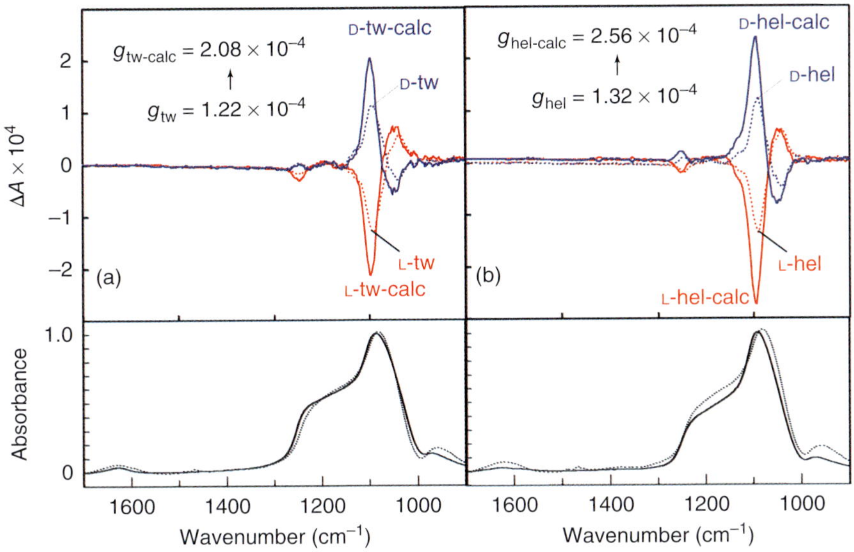

Figure 6.10 IR and VCD spectra of silica nanoribbons. (a) IR and VCD spectra of silica–lipid hybrid twisted nanoribbons before (dotted line) and after (solid line) calcination. (b) IR and VCD spectra of silica–lipid hybrid helical nanoribbons before (dotted line) and after (solid line) calcination. All VCD spectra were normalized for an absorbance intensity of the 1090 cm^{-1} band equals to one. Source: From Okazaki et al. 2016 [91].

respectively. This is likely related to the global shrinkage of nanoribbons and stronger silica network constraint after calcination as described above.

Therefore, chirality probably originates from the siloxane network distortion and the resulting chirality enhancement through shrinkage of the network and decrease in pitch. The VCD spectra of helical ribbons are slightly stronger than those of twisted ribbons (approximately 10% before calcination and 25% after calcination). This study demonstrated, for the first time, strong VCD signals of siloxane network forming right-handed vs. left-handed twisted and helical ribbon-like nanostructures. It was demonstrated that the small structural changes due to dehydration upon calcination induce the large enhancement of VCD signals, which reveals that the observed silica chirality originates from a chirally arranged siloxane network expressed from chiral organization of Si—O—Si bonds.

6.3.3 Silica Nanohelices as Platform to Organize Non-chiral Objects

These inorganic or hybrid nano-objects with controlled chirality can be used as the template to induce chirality in achiral functional molecules or nanoparticles.

6.3.3.1 Induction of CD on Non-chiral Dye

As we have shown above, the chiral arrangement of gemini surfactant molecules induces the formation of chiral silica network around them. We then demonstrated that such gemini surfactants confined in chiral silica matrices can be

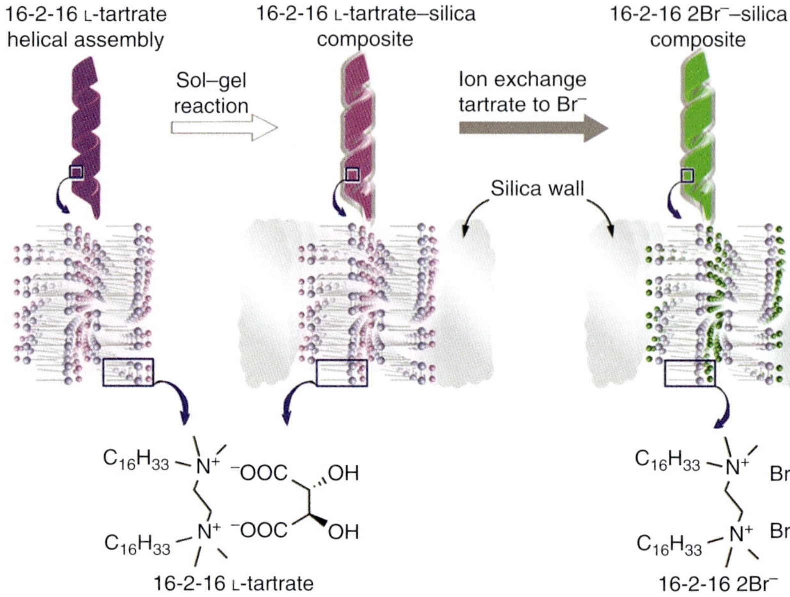

Figure 6.11 Schematic illustration of the gemini surfactant–silica composite prepared from 16-2-16 L-tartrate and the organized structure of the gemini surfactant in the composite. Source: From Ryu et al. 2016 [94].

maintained even after the removal of the chiral source, namely, the counterion tartrate (Figure 6.11), using an achiral dye (methyl orange, MO) as a probe [94].

For this work, helical and twisted ribbons of gemini (16-2-16 tartrate) surfactant–silica composites were prepared, and then tartrate anions in the composites replaced by bromide anions. When these 16-2-16 $2Br^-$–silica composite dispersions with increasing concentration were added to a MO aqueous solution, the absorption band at 505 nm (free MO in solution) gradually decreased and a new absorption band appeared at around 373 nm (Figure 6.12b,c, top), along with change of the color of the solution from original orange to yellow at a composition concentration of around 0.06 mg ml^{-1} (Figure 6.12a). A split Cotton effect appeared in the region of the new absorption band as assessed by ECD (Figure 6.12b,c, bottom), which is attributed to R-chiral (right-handed) H-aggregates of unprotonated MO molecules according to exciton chirality theory.

This result indicates that surprisingly, non-chiral cationic molecules confined in the chiral nanospace in silica matrices adopted chiral arrangement despite the absence of any chiral molecules, which then interacted with non-chiral anionic MO molecules through electrostatic interaction, inducing H-aggregates of MO molecules to which the chirality is transferred.

It is important to note that in the absence of the silica matrix, the assemblies of gemini surfactant molecules lose the chiral arrangement, and chiral ribbon structures are destroyed when enantiomeric tartrate counterions are replaced by bromide anions.

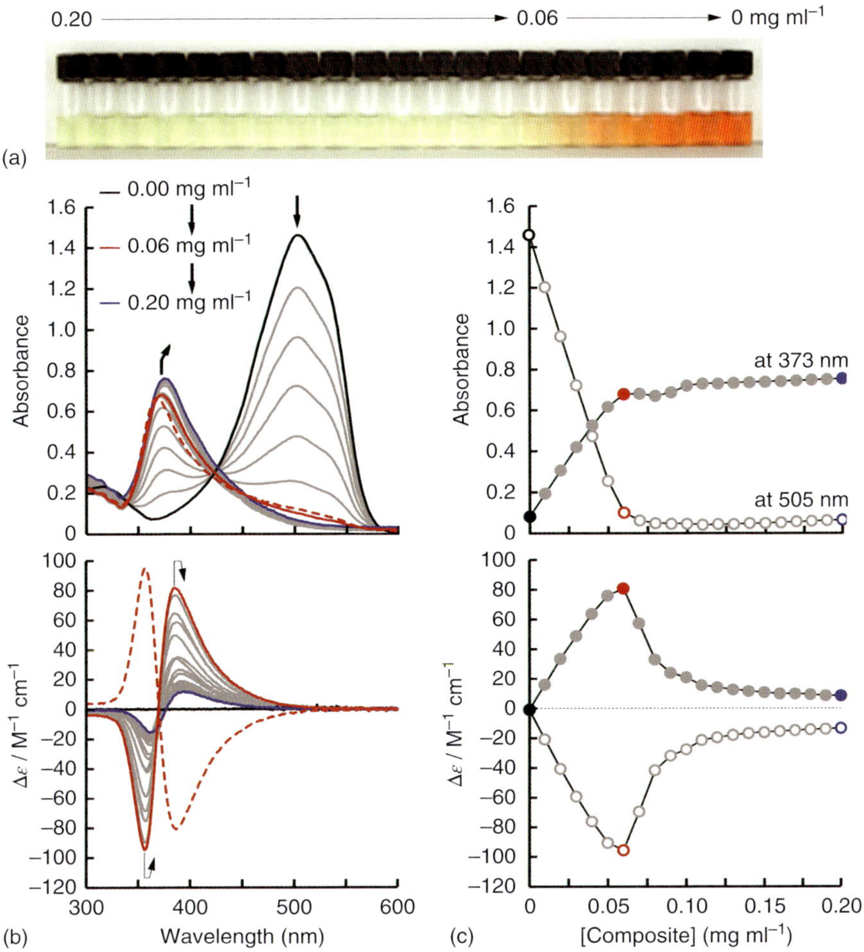

Figure 6.12 (a) Photo of MO (0.04 mM) in the presence of various concentrations of the 16-2-16 2Br$^-$–silica composite prepared from 16-2-16 L-tartrate in water at pH 3.0 (room temperature, [composite] = 0–0.20 mg ml^{-1} [right to left]). (b) UV–vis absorption (top) and ECD spectra (bottom) of MO (0.04 mM) in the presence of various concentrations of the 16-2-16 2Br$^-$–silica composite prepared from 16-2-16 L- (0–0.20 mg ml^{-1}, solid line) and D-tartrate (0.06 mg ml^{-1}, dashed line) in water at pH 3.0 (temperature = 20 °C, path length = 1.0 cm). The spectra are normalized to 0.04 mM MO. (c) Effect of concentration of composite prepared from 16-2-16 L-tartrate on the absorbance (top) and molar circular dichroism (bottom) of MO (0.04 mM) in water at pH 3.0 (temperature = 20 °C, path length = 1.0 cm). Source: From Ryu et al. 2016 [94].

6.3.3.2 Gold Nanoparticles Forming 3D Helical Superstructures with Controlled Morphology and Strong Chiroptical Property

These nanometric silica helices are also used as chiral templates in order to organize GNPs to prepare a collection of helical GNPs superstructures (called Goldhelices hereafter) [95]. These nano-objects exhibit well-defined arrangement of GNPs (with diameters varying between 4 and 10 nm) following

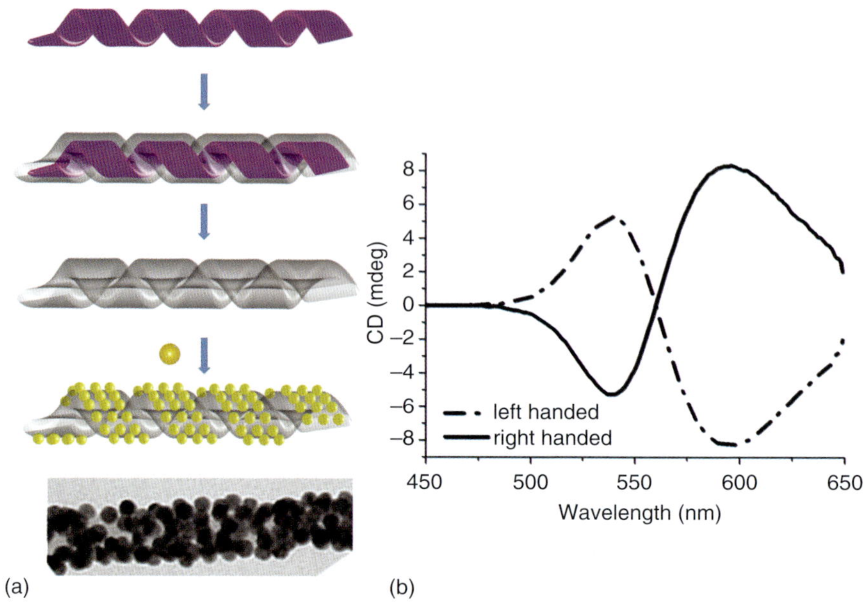

Figure 6.13 Silica nanohelices synthesized from sol–gel condensation of TEOS on self-assembled organic nanostructures are used as a matrix to prepare helical GNPs superstructures (goldhelix) with tunable and well-defined handedness. Goldhelix made using GNPs with various sizes. (a) representative TEM images of 5–10 nm GNPs@ SiO_2 helix. (b) corresponding electronic ECD spectra of the Goldhelices. Dashed lines represent left-handed Goldhelix and solid lines represent right-handed goldhelix. Source: From Cheng et al. 2017 [96].

the helicity of the silica template. Strong chiroptical activity is evidenced by ECD spectroscopy at the wavelength of the surface plasmon resonance (SPR) of the GNPs with an anisotropy factor (g-factor) of the order of 10^{-4}, i.e. 10-fold larger than what is typically reported in the literature [96]. Our study clearly showed that these optical activities depend both on the size and the organization of the GNPs on the surface of the helical silica template, which are closely correlated with the surface chemistry of both the GNPs and the silica helices (Figure 6.13).

6.4 Conclusion

In the system presented in this chapter, we observe striking multistep chirality transfer. First, the chirality of a dianionic molecule (tartrate), which is in close contact with nonchiral dicationic n-2-n amphiphile, self-organizes together into right- or left-handed chiral nanostructures. The conformations of both tartrate and gemini surfactants in multilayered chiral assemblies were investigated using ECD, VCD, NMR, and X-ray powder diffraction, which demonstrated the chiral arrangement of the tartrate molecules at the surface of the bilayers. Moreover,

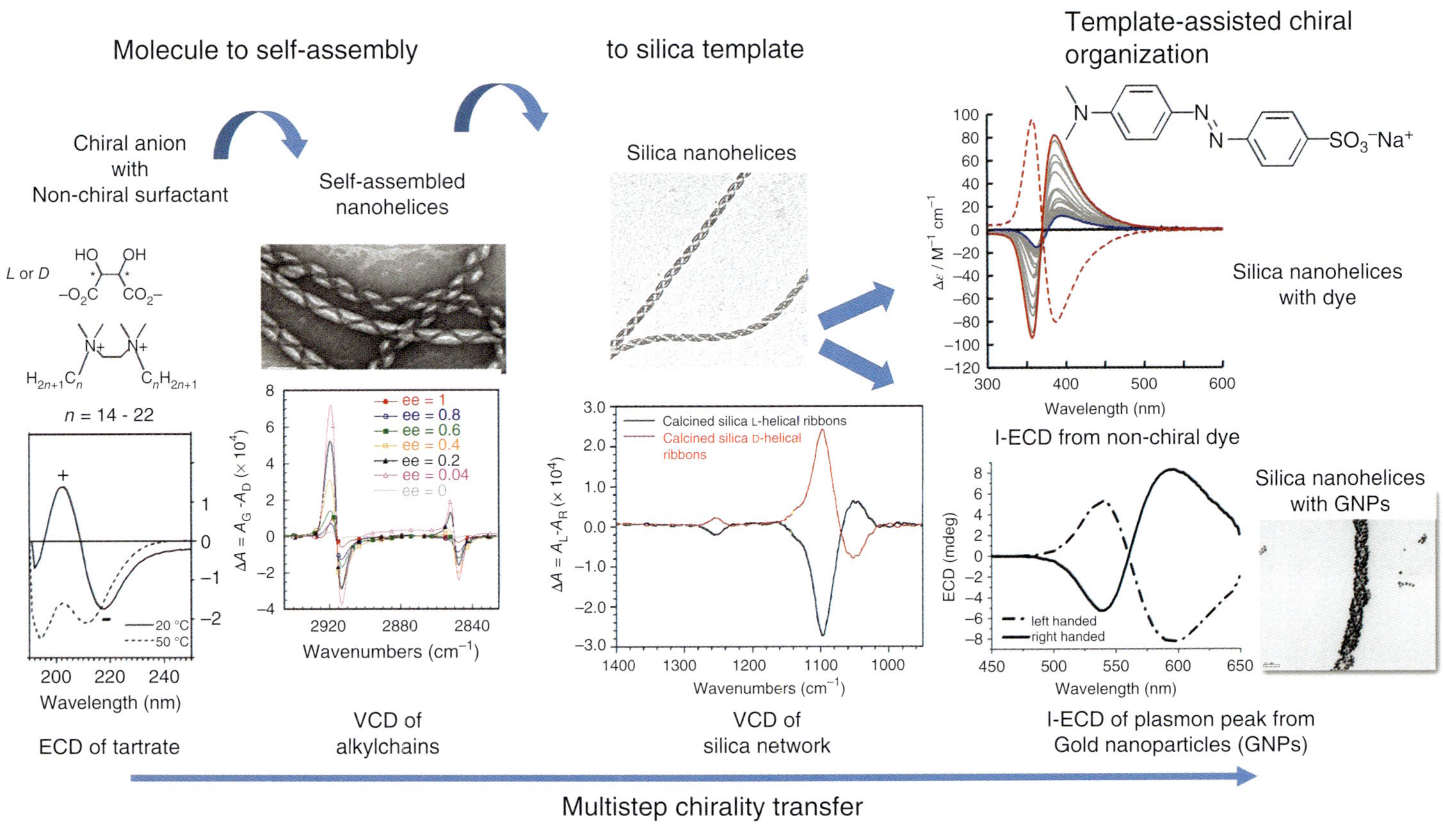

Figure 6.14 Multistep chirality transfer from tartrate anions to surfactant, to silica nanohelices, then to dye molecules or to GNPs. The only chiral component is the original tartrate, the chirality of which is transferred step by step to these various non-chiral molecules or materials.

chirality transfer from counterions to achiral amphiphiles has been clearly evidenced by VCD. The ECD and VCD intensities do not vary linearly with the enantiomeric excess of the anion, and different behaviors have been observed from the two spectroscopic methods: ECD intensities are correlated with the pitch of the ribbons, whereas the VCD intensities are correlated with the dimension (size) of the chiral ribbons.

These chiral molecular assemblies with controllable chirality were used to synthesize chiral inorganic or hybrid nanohelices, and for the first time, chirality on noncrystalline silica nanoribbons was directly demonstrated.

These inorganic or hybrid nano-objects with controlled chirality were then used as the template to induce chirality in achiral functional molecules or nanoparticles. Two examples were presented in this chapter. In the first one, the interaction between the non-chiral organic surfactant molecules and the non-chiral dye, methyl orange, confined in the chiral silica matrices led to chiral aggregates of the latter, leading to a strong induced ECD signal. In the second example, the surfaces of these silica nanohelices were functionalized for the grafting of GNPs, leading to a chiral organization of the GNPs on the surface of nanohelices evidenced by ECD signal of their plasmon bands. These Goldhelices based on inorganic templates are much more robust than previously reported organic-based chiroptical nanostructures making them good candidates for complex hierarchical organization, providing a new promising approach for light management and new benefits in applications such as circular polarizers, chiral metamaterials, or chiral sensing in the visible range.

Such an ensemble of unprecedented multistep chirality transfer from molecule to molecular assemblies, from organic to inorganic templates, then to other non-chiral molecules or particles (steps summarized in the scheme Figure 6.14) is a very interesting example of how to achieve finely controllable chirality of various functional materials. Indeed, with the robustness and fine tunability of these inorganic platforms, it makes such systems very attractive chiral platforms for chiroptical materials, chiral recognition, or asymmetric catalysis.

References

1 Kuroda, R., Endo, B., Abe, M., and Shimizu, M. (2009). Chiral blastomere arrangement dictates zygotic left–right asymmetry pathway in snails. *Nature* 462 (7274): 790–794.

2 Ferrand, Y., Chandramouli, N., Kendhale, A.M. et al. (2012). Long-range effects on the capture and release of a chiral guest by a helical molecular capsule. *J. Am. Chem. Soc.* 134 (27): 11282–11288.

3 Ajami, D. and Rebek, J. (2009). Compressed alkanes in reversible encapsulation complexes. *Nat. Chem.* 1 (1): 87–90.

4 Luo, X., Cao, Y., Wang, T. et al. (2016). Host–guest chirality interplay: a mutually induced formation of a chiral ZMOF and its double-helix polymer guests. *J. Am. Chem. Soc.* 138 (3): 786–789.

5 Hou, J.-L., Shao, X.-B., Chen, G.-J. et al. (2004). Hydrogen bonded oligohydrazide foldamers and their recognition for saccharides. *J. Am. Chem. Soc.* 126 (39): 12386–12394.
6 Nonaka, K., Yamaguchi, M., Yasui, M. et al. (2014). Guest-induced supramolecular chirality in a ditopic azoprobe–cyclodextrin complex in water. *Chem. Commun.* 50 (70): 10059–10061.
7 Akine, S., Sairenji, S., Taniguchi, T., and Nabeshima, T. (2013). Stepwise helicity inversions by multisequential metal exchange. *J. Am. Chem. Soc.* 135 (35): 12948–12951.
8 Crassous, J. (2012). Transfer of chirality from ligands to metal centers: recent examples. *Chem. Commun.* 48 (78): 9687–9695.
9 Miyake, H. and Tsukube, H. (2012). Coordination chemistry strategies for dynamic helicates: time-programmable chirality switching with labile and inert metal helicates. *Chem. Soc. Rev.* 41 (21): 6977–6991.
10 Dhamija, A., Ikbal, S.A., and Rath, S.P. (2016). Induction and rationalization of supramolecular chirality in the tweezer–diamine complexes: insights from experimental and DFT studies. *Inorg. Chem.* 55 (24): 13014–13026.
11 Lin, R., Zhang, H., Li, S. et al. (2011). pH-Switchable inversion of the metal-centered chirality of metallabenzenes: opposite stereodynamics in reactions of ruthenabenzene with L-and D-Cysteine. *Chem. Eur. J.* 17 (8): 2420–2427.
12 Jacobsen, E.N., Pfaltz, A., and Yamamoto, H. (2003). *Comprehensive Asymmetric Catalysis: Supplement 1*, vol. 1. Springer Science & Business Media.
13 Ojima, I. (2004). *Catalytic Asymmetric Synthesis*. Wiley.
14 Jahier, C. and Nlate, S. (2012). L-proline-derived dendritic tetrakis (diperoxotungsto) phosphate: synthesis and enantioselective oxidation catalysis. *Eur. J. Inorg. Chem.* 2012 (5): 833–840.
15 Grason, G.M. (2015). Chirality transfer in block copolymer melts: emerging concepts. *ACS Macro Lett.* 4 (5): 526–532.
16 Green, M.M., Park, J.W., Sato, T. et al. (1999). The macromolecular route to chiral amplification. *Angew. Chem. Int. Ed.* 38 (21): 3138–3154.
17 Weitkamp, A.W. (1945). The acidic constituents of degras. A new method of structure elucidation. *J. Am. Chem. Soc.* 67 (3): 447–454.
18 Hotten, B. and Birdsall, D. (1952). Fine structure and rheological properties of lithium soap-oil dispersions. *J. Colloid Sci.* 7 (3): 284–294.
19 Tachibana, T. and Kambara, H. (1965). Enantiomorphism in the helical aggregate of lithium 12-hydroxystearate. *J. Am. Chem. Soc.* 87 (13): 3015–3016.
20 Nakashima, N., Asakuma, S., Kim, J.-M., and Kunitake, T. (1984). Helical superstructures are formed from chiral ammonium bilayers. *Chem. Lett.* 13 (10): 1709–1712.
21 Yager, P. and Schoen, P.E. (1984). Formation of tubules by a polymerizable surfactant. *Mol. Cryst. Liq. Cryst.* 106 (3–4): 371–381.
22 Yamada, K., Ihara, H., Ide, T. et al. (1984). Formation of helical super structure from single-walled bilayers by amphiphiles with oligo-L-glutamic acid-head group. *Chem. Lett.* 13 (10): 1713–1716.

23 Singh, A., Burke, T.G., Calvert, J.M. et al. (1988). Lateral phase separation based on chirality in a polymerizable lipid and its influence on formation of tubular microstructures. *Chem. Phys. Lipids* 47 (2): 135–148.
24 Georger, J.H., Singh, A., Price, R.R. et al. (1987). Helical and tubular microstructures formed by polymerizable phosphatidylcholines. *J. Am. Chem. Soc.* 109 (20): 6169–6175.
25 Zhang, L., Wang, T., Shen, Z., and Liu, M. (2016). Chiral nanoarchitectonics: towards the design, self-assembly, and function of nanoscale chiral twists and helices. *Adv. Mater.* 28 (6): 1044–1059.
26 Wang, Y., Xu, J., Wang, Y., and Chen, H. (2013). Emerging chirality in nanoscience. *Chem. Soc. Rev.* 42: 2930–2962. (Copyright (C) 2014 American Chemical Society (ACS). All Rights Reserved.).
27 Li, C., Deng, K., Tang, Z., and Jiang, L. (2010). Twisted metal–amino acid nanobelts: chirality transcription from molecules to frameworks. *J. Am. Chem. Soc.* 132 (23): 8202–8209.
28 Liu, M., Zhang, L., and Wang, T. (2015). Supramolecular chirality in self-assembled systems. *Chem. Rev.* 115 (15): 7304–7397.
29 Yashima, E., Ousaka, N., Taura, D. et al. (2016). Supramolecular helical systems: helical assemblies of small molecules, foldamers, and polymers with chiral amplification and their functions. 116: 13752–13990.
30 Liu, G.F., Zhang, D., and Feng, C.L. (2014). Control of three-dimensional cell adhesion by the chirality of nanofibers in hydrogels. *Angew. Chem. Int. Ed.* 53 (30): 7789–7793.
31 Liu, J., Liu, J., Chu, L. et al. (2014). Self-assembling peptide of D-amino acids boosts selectivity and antitumor efficacy of 10-hydroxycamptothecin. *ACS Appl. Mater. Interfaces* 6 (8): 5558–5565.
32 Shen, Z., Wang, T., Shi, L. et al. (2015). Strong circularly polarized luminescence from the supramolecular gels of an achiral gelator: tunable intensity and handedness. *Chem. Sci.* 6 (7): 4267–4272.
33 So, S.M., Moozeh, K., Lough, A.J., and Chin, J. (2014). Highly stereoselective recognition and deracemization of amino acids by supramolecular self-assembly. *Angew. Chem. Int. Ed.* 53 (3): 829–832.
34 Kodama, K., Kobayashi, Y., and Saigo, K. (2007). Two-component supramolecular helical architectures: creation of tunable dissymmetric cavities for the inclusion and chiral recognition of the third components. *Chem. Eur. J.* 13 (7): 2144–2152.
35 Sun, X., Silly, F., Maurel, F., and Dong, C. (2016). Supramolecular chiral host–guest nanoarchitecture induced by the selective assembly of barbituric acid derivative enantiomers. *Nanotechnology* 27 (42): 42LT01.
36 Zhang, H., Gong, Z., Sun, K. et al. (2016). Two-dimensional chirality transfer via on-surface reaction. *J. Am. Chem. Soc.* 138 (36): 11743–11748.
37 De Feyter, S., Grim, P., Rücker, M. et al. (1998). Expression of chirality by achiral coadsorbed molecules in chiral monolayers observed by STM. *Angew. Chem. Int. Ed.* 37 (9): 1223–1226.
38 Schöck, M., Otero, R., Stojkovic, S. et al. (2006). Chiral close-packing of achiral star-shaped molecules on solid surfaces. *J. Phys. Chem. B* 110 (26): 12835–12838.

39 Zhang, H.M., Xie, Z.X., Mao, B.W., and Xu, X. (2004). Self-assembly of normal alkanes on the Au (111) surfaces. *Chem. Eur. J.* 10 (6): 1415–1422.
40 Okazaki, Y., Goto, T., Sakaguchi, R. et al. (2016). Facile and versatile approach for generating circularly polarized luminescence by non-chiral, low-molecular dye-on-nanotemplate composite system. *Chem. Lett.* 45 (4): 448–450.
41 Goto, T., Okazaki, Y., Ueki, M. et al. (2017). Induction of strong and tunable circularly polarized luminescence of nonchiral, nonmetal, low-molecular-weight fluorophores using chiral nanotemplates. *Angew. Chem. Int. Ed.* 56 (11): 2989–2993.
42 Shi, L., Wang, X., Sandoval, C.A. et al. (2006). Engineering a polymeric chiral catalyst by using hydrogen bonding and coordination interactions. *Angew. Chem. Int. Ed.* 45 (25): 4108–4112.
43 Yu, L., Wang, Z., Wu, J. et al. (2010). Directed orthogonal self-assembly of homochiral coordination polymers for heterogeneous enantioselective hydrogenation. *Angew. Chem. Int. Ed.* 49 (21): 3627–3630.
44 Yang, S.M., Jang, S.G., Choi, D.G. et al. (2006). Nanomachining by colloidal lithography. *Small* 2 (4): 458–475.
45 Guo, L.J. (2007). Nanoimprint lithography: methods and material requirements. *Adv. Mater.* 19 (4): 495–513.
46 Cavallini, M., Albonetti, C., and Biscarini, F. (2009). Nanopatterning soluble multifunctional materials by unconventional wet lithography. *Adv. Mater.* 21 (10–11): 1043–1053.
47 Luo, Z. and Zhang, S. (2012). Designer nanomaterials using chiral self-assembling peptide systems and their emerging benefit for society. *Chem. Soc. Rev.* 41 (13): 4736–4754.
48 Qiu, H. and Che, S. (2011). Chiral mesoporous silica: chiral construction and imprinting via cooperative self-assembly of amphiphiles and silica precursors. *Chem. Soc. Rev.* 40 (3): 1259–1268.
49 Yu, J. and Xu, R. (2010). Rational approaches toward the design and synthesis of zeolitic inorganic open-framework materials. *Acc. Chem. Res.* 43 (9): 1195–1204.
50 Zhang, Y. and Zheng, L. (2010). Towards chirality-pure carbon nanotubes. *Nanoscale* 2 (10): 1919–1929.
51 Gautier, C. and Bürgi, T. (2009). Chiral gold nanoparticles. *ChemPhysChem* 10 (3): 483–492.
52 Yang, M. and Kotov, N.A. (2011). Nanoscale helices from inorganic materials. *J. Mater. Chem.* 21 (19): 6775–6792.
53 Lan, X. and Wang, Q. (2016). Self-assembly of chiral plasmonic nanostructures. *Adv. Mater.* 28 (47): 10499–10507.
54 Fu, X., Wang, Y., Huang, L. et al. (2003). Assemblies of metal nanoparticles and self-assembled peptide fibrils – formation of double helical and single-chain arrays of metal nanoparticles. *Adv. Mater.* 15 (11): 902–906.
55 Zhang, S., Ni, W., Kou, X. et al. (2007). Formation of gold and silver nanoparticle arrays and thin shells on mesostructured silica nanofibers. *Adv. Funct. Mater.* 17 (16): 3258–3266.

56 Guerrero-Martínez, A., Auguié, B., Alonso-Gómez, J.L. et al. (2011). Intense optical activity from three-dimensional chiral ordering of plasmonic nanoantennas. *Angew. Chem. Int. Ed.* 50 (24): 5499–5503.
57 Sone, E.D., Zubarev, E.R., and Stupp, S.I. (2002). Semiconductor nanohelices templated by supramolecular ribbons. *Angew. Chem. Int. Ed.* 41 (10): 1705–1709.
58 Hwang, L., Chen, C.-L., and Rosi, N.L. (2010). Preparation of 1-D nanoparticle superstructures with tailorable thicknesses using gold-binding peptide conjugates. *Chem. Commun.* 47 (1): 185–187.
59 Chen, C.-L., Zhang, P., and Rosi, N.L. (2008). A new peptide-based method for the design and synthesis of nanoparticle superstructures: construction of highly ordered gold nanoparticle double helices. *J. Am. Chem. Soc.* 130 (41): 13555–13557.
60 Kuzyk, A., Schreiber, R., Fan, Z. et al. (2012). DNA-based self-assembly of chiral plasmonic nanostructures with tailored optical response. *Nature* 483 (7389): 311–314.
61 Sharma, J., Chhabra, R., Cheng, A. et al. (2009). Control of self-assembly of DNA tubules through integration of gold nanoparticles. *Science* 323 (5910): 112–116.
62 Guha, S., Drew, M.G., and Banerjee, A. (2008). Construction of helical nanofibers from self-assembling pseudopeptide building blocks: modulating the handedness and breaking the helicity. *Small* 4 (11): 1993–2005.
63 Govorov, A.O., Gun'ko, Y.K., Slocik, J.M. et al. (2011). Chiral nanoparticle assemblies: circular dichroism, plasmonic interactions, and exciton effects. *J. Mater. Chem.* 21 (42): 16806–16818.
64 Song, C., Blaber, M.G., Zhao, G. et al. (2013). Tailorable plasmonic circular dichroism properties of helical nanoparticle superstructures. *Nano Lett.* 13 (7): 3256–3261.
65 Wang, Y., Wang, Q., Sun, H. et al. (2011). Chiral transformation: from single nanowire to double helix. *J. Am. Chem. Soc.* 133 (50): 20060–20063.
66 Qi, H., Shopsowitz, K.E., Hamad, W.Y., and MacLachlan, M.J. (2011). Chiral nematic assemblies of silver nanoparticles in mesoporous silica thin films. *J. Am. Chem. Soc.* 133 (11): 3728–3731.
67 Shemer, G., Krichevski, O., Markovich, G. et al. (2006). Chirality of silver nanoparticles synthesized on DNA. *J. Am. Chem. Soc.* 128 (34): 11006–11007.
68 Zhou, Y., Yang, M., Sun, K. et al. (2010). Similar topological origin of chiral centers in organic and nanoscale inorganic structures: effect of stabilizer chirality on optical isomerism and growth of CdTe nanocrystals. *J. Am. Chem. Soc.* 132 (17): 6006–6013.
69 Nakashima, T., Kobayashi, Y., and Kawai, T. (2009). Optical activity and chiral memory of thiol-capped CdTe nanocrystals. *J. Am. Chem. Soc.* 131 (30): 10342–10343.
70 Elliott, S.D., Moloney, M.P., and Gun'ko, Y.K. (2008). Chiral shells and achiral cores in CdS quantum dots. *Nano Lett.* 8 (8): 2452–2457.
71 Ono, Y., Nakashima, K., Sano, M. et al. (1998). Organic gels are useful as a template for the preparation of hollow fiber silica. *Chem. Commun.* (14): 1477–1478.

72 Van Bommel, K.J., Friggeri, A., and Shinkai, S. (2003). Organic templates for the generation of inorganic materials. *Angew. Chem. Int. Ed.* 42 (9): 980–999.

73 Kobayashi, S., Hamasaki, N., Suzuki, M. et al. (2002). Preparation of helical transition-metal oxide tubes using organogelators as structure-directing agents. *J. Am. Chem. Soc.* 124 (23): 6550–6551.

74 Seddon, A.M., Patel, H.M., Burkett, S.L., and Mann, S. (2002). Chiral templating of silica–lipid lamellar mesophase with helical tubular architecture. *Angew. Chem. Int. Ed.* 41 (16): 2988–2991.

75 Sugiyasu, K., Tamaru, S.-i., Takeuchi, M. et al. (2002). Double helical silica fibrils by sol–gel transcription of chiral aggregates of gemini surfactants. *Chem. Commun.* (11): 1212–1213.

76 Delclos, T., Aimé, C., Pouget, E. et al. (2008). Individualized silica nanohelices and nanotubes: tuning inorganic nanostructures using lipidic self-assemblies. *Nano Lett.* 8 (7): 1929–1935.

77 Pouget, E., Dujardin, E., Cavalier, A. et al. (2007). Hierarchical architectures by synergy between dynamical template self-assembly and biomineralization. *Nat. Mater.* 6 (6): 434–439.

78 Jung, J.H., Ono, Y., Hanabusa, K., and Shinkai, S. (2000). Creation of both right-handed and left-handed silica structures by sol–gel transcription of organogel fibers comprised of chiral diaminocyclohexane derivatives. *J. Am. Chem. Soc.* 122 (20): 5008–5009.

79 Kobayashi, S., Hanabusa, K., Hamasaki, N. et al. (2000). Preparation of TiO_2 hollow-fibers using supramolecular assemblies. *Chem. Mater.* 12 (6): 1523–1525.

80 Liu, S., Han, L., Duan, Y. et al. (2012). Synthesis of chiral TiO_2 nanofibre with electron transition-based optical activity. *Nat. Commun.* 3: 1215.

81 Ji, Q. and Shimizu, T. (2005). Chemical synthesis of transition metal oxide nanotubes in water using an iced lipid nanotube as a template. *Chem. Commun.* (35): 4411–4413.

82 Dang, X., Yi, H., Ham, M.-H. et al. (2011). Virus-templated self-assembled single-walled carbon nanotubes for highly efficient electron collection in photovoltaic devices. *Nat. Nanotechnol.* 6 (6): 377–384.

83 Bolisetty, S., Adamcik, J., Heier, J., and Mezzenga, R. (2012). Amyloid directed synthesis of titanium dioxide nanowires and their applications in hybrid photovoltaic devices. *Adv. Funct. Mater.* 22 (16): 3424–3428.

84 Sone, E.D., Zubarev, E.R., and Stupp, S.I. (2005). Supramolecular templating of single and double nanohelices of cadmium sulfide. *Small* 1 (7): 694–697.

85 Oda, R., Huc, I., and Candau, S.J. (1998). Gemini surfactants as new, low molecular weight gelators of organic solvents and water. *Angew. Chem. Int. Ed.* 37 (19): 2689–2691.

86 Oda, R., Huc, I., Schmutz, M. et al. (1999). Tuning bilayer twist using chiral counterions. *Nature* 399 (6736): 566–569.

87 Brizard, A., Ahmad, R.K., and Oda, R. (2007). Control of nano-micrometric twist and helical ribbon formation with gemini-oligoalanine via interpeptidic beta-sheet structure formation. *Chem. Commun.* (22): 2275–2277.

88 Berthier, D., Buffeteau, T., Leger, J.M. et al. (2002). From chiral counterions to twisted membranes. *J. Am. Chem. Soc.* 124 (45): 13486–13494.

89 Brizard, A., Berthier, D., Aimé, C. et al. (2009). Molecular and supramolecular chirality in gemini-tartrate amphiphiles studied by electronic and vibrational circular dichroisms. *Chirality* 21 (1E): E153–E162.

90 Oda, R., Artzner, F., Laguerre, M., and Huc, I. (2008). Molecular structure of self-assembled chiral nanoribbons and nanotubules revealed in the hydrated state. *J. Am. Chem. Soc.* 130 (44): 14705–14712.

91 Okazaki, Y., Buffeteau, T., Siurdyban, E. et al. (2016). Direct observation of siloxane chirality on twisted and helical nanometric amorphous silica. *Nano Lett.* 16 (10): 6411–6415.

92 Kaneko, Y. and Iyi, N. (2009). Sol-gel synthesis of ladder polysilsesquioxanes forming chiral conformations and hexagonal stacking structures. *J. Mater. Chem.* 19 (38): 7106–7111.

93 Guo, Z., Du, Y., Chen, Y. et al. (2010). Understanding the mechanism of chirality transfer in the formation of a chiral MCM-41 mesoporous silica. *J. Phys. Chem. C* 114 (34): 14353–14361.

94 Ryu, N., Okazaki, Y., Hirai, K. et al. (2016). Memorized chiral arrangement of gemini surfactant assemblies in nanometric hybrid organic–silica helices. *Chem. Commun.* 52 (34): 5800–5803.

95 Tamoto, R., Lecomte, S., Si, S. et al. (2012). Gold nanoparticle deposition on silica nanohelices: a new controllable 3D substrate in aqueous suspension for optical sensing. *J. Phys. Chem. C* 116 (43): 23143–23152.

96 Cheng, J., Le Saux, G., Gao, J. et al. (2017). GoldHelix: gold nanoparticles forming 3D helical superstructures with controlled morphology and strong chiroptical property. *ACS Nano* https://doi.org/10.1021/acsnano.6b08723.

7

Solution Plasma Reactions and Materials Synthesis

Gasidit Panomsuwan[1], Tomonaga Ueno[2], Hiroharu Yui[3], Jun Nakamura[4], and Nagahiro Saito[2]

[1] *Kasetsart University, Faculty of Engineering, Department of Materials Engineering, Chatuchak, Bangkok 10900, Thailand*
[2] *Nagoya University, Graduated School of Engineering, Department of Chemical Systems Engineering, Chikusa, Nagoya 464-8603, Japan*
[3] *Tokyo University of Science, Faculty of Science Division I, Department of Chemistry, Shinjuku, Tokyo 162-8601, Japan*
[4] *The University of Electro-Communications, Department of Engineering Science, Chofu, Tokyo 182-8585, Japan*

7.1 General Introduction

Plasma is known as the fourth state of matter beyond the three common states: solid, liquid, and gas. Plasma is an ionized gas that contains an equal number of positive ions and negative electrons ($n_i \approx n_e$). In addition, neutral, metastable, or excited atoms or molecules, reactive radicals, and ultraviolet light are also generated from plasma. Historically, the term plasma was coined by Irving Langmuir in 1928, who studied gas discharge in a vacuum tube [1]. The discovery of plasma is the initiating point of several important breakthroughs in the last century, such as radio-frequency communication, nuclear fusion, plasma astrophysics, and nanotechnology. Today, plasma technology has already been extended to a viable industrial manufacturing process across many current high-tech products in our daily lives, such as solar cells, chips, optics, sensors, and energy devices [2, 3]. Gas-phase plasma operated at low pressure is the most common type of plasma that is usually used in the processing and fabrication of materials, which involves thin-film coating, surface treatment (e.g. chemical modification, activation, cleaning), and etching of various material substrates [4, 5]. Without plasma technology, we will have no high-tech devices to make our life simpler and more convenient.

Liquid-phase plasma, another branch of plasma technology, has also been studied and developed in parallel with gas-phase plasma a long time ago [6]. In the past, the main applications of liquid-phase plasma were in decontamination of organic pollutants and water sterilization [7–9], while no significant attention was paid to its application in material synthesis. In fact, the synthesis of materials by liquid-phase plasma was first proposed by Gubkin in 1887 before the term

Molecular Technology: Materials Innovation, Volume 3, First Edition.
Edited by Hisashi Yamamoto and Takashi Kato.

plasma originated [10]. They used plasma to reduce Ag^+ ions into Ag using $AgNO_3$ solution as precursor. However, there were no continuing efforts to make further progress in the synthesis of other materials by utilizing plasma in liquid after this inspiring work. This may likely be due to lack of interest from researchers and unavailable equipment to probe the materials in nanoscale at that time. Over the last decade, liquid-phase plasma has started to awaken from its long slumber and become a powerful tool for synthesis of numerous materials, which is reflected in the fruitful achievement of published works in this emerging area.

7.2 Solution Plasma

Liquid-phase plasma can be operated in an open reactor system at atmospheric pressure unlike gas-phase plasma, which requires a complicated and expensive vacuum chamber. Many phenomena of plasma occurring under liquid environment are more complex and difficult to understand than gas-phase plasma at low pressure or even at atmospheric condition owing to its discontinuous gas–liquid boundaries and the higher molecular density of liquid. In general, plasma can exhibit a broad spectrum of states ranging from nonequilibrium (nonthermal plasma) to almost complete thermal equilibrium (thermal plasma), which can be classified by electron and ion temperatures generated in plasma. For thermal plasma, electrons and ions have almost the same temperatures ($T_e \approx T_i$, T_n) or thermal equilibrium. The temperature is relatively homogeneous in the system throughout the atoms, molecules, ions, and electrons. On the other hand, for nonthermal plasma, the plasma in under nonequilibrium state where the electron temperature is several orders of magnitude higher than ion and neutral molecule temperatures ($T_e \gg T_i$, T_n). Importantly, the occurrence of electron collision reactions is dominant in nonequilibrium plasma, resulting in the dissociation of molecules into reactive and charged species.

It is well known that the type of power supply used to generate plasma is crucial in determining the plasma state. High-voltage pulse with short pulse duration (micro or nanoscale second) is an ideal type of power supply to realize nonthermal plasma in liquid as also to prevent its transition to arc discharge at longer discharge time. Even though a liquid environment under nonthermal plasma can be maintained at low temperature, the temperature at the center of plasma is still relatively high, about 3000–4000 K. At present, the specific term used to define the liquid-phase plasma in nonequilibrium state is *"solution plasma (SP),"* which was developed by Saito's group [11–13]. In fact, various types of electrode configuration and design reactors have been extensively used to generate plasma in liquid, which varies depending on each research group [14]. Figure 7.1a depicts a typical setup of SP system with a symmetric wire-to-wire electrode type connected with a bipolar pulsed DC power supply [11, 12, 15, 16]. The size of plasma generated between two wire electrodes under liquid is typically in millimeter and submicrometer scales. In brief, when the voltage is applied, bubbles containing water vapor are formed at the electrode surface by Joule heating effect. Once the voltage increases up to the breakdown point, an electrical discharge occurs

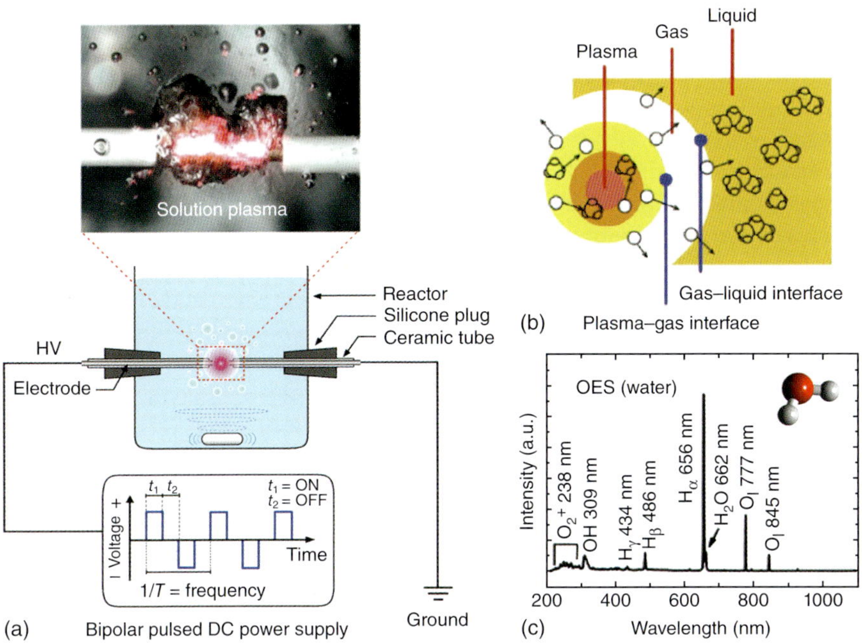

Figure 7.1 (a) Schematic illustration of typical experimental setup for SP with wire-to-wire electrode (Saito laboratory) [11, 12, 15, 16]. (b) A model of SP. [17] (c) Optical emission spectrum of SP generated in water [18].

along the plasma–liquid surface or across the bubbles, thus generating plasma at the gap between electrodes. A model of SP is revealed in Figure 7.1b. Plasma is generated at the center and surrounded by a liquid phase, resulting in an extremely large temperature gradient at the plasma–liquid interface. Various reactive species produced in gas-phase plasma can diffuse to the gas–liquid interface, and finally enter the liquid phase. On the other hand, the molecular species in liquid can diffuse to the gas-phase plasma at the same time; therefore, a dissociation reaction can proceed continuously. A combination of high vapor density and nonequilibrium plasma state can drive a high rate of chemical reaction, which in turn creates unique reactive species on rapid time scales. It has been considered that the interaction at plasma–liquid interface where various physical and chemical reactions take place plays an important role in material synthesis [13, 19–21]. In addition, the electrical discharge can be produced in various types of liquids, such as water, aqueous solution, organic solvent, mixed solution thereof, or even liquid nitrogen. The flexible use of multiple liquids is the major merit of SP to create various reactive species and novel reaction fields, which open up a broader possibility for materials synthesis and processing.

7.3 Materials Synthesis by Solution Plasma

Recently, SP has become one of the powerful tools for synthesis and molecular technology across a wide range of nanomaterials. Its unique advantages can meet the requirements of an ideal synthesis method, such as low energy consumption, simple equipment, one-step synthesis, operation at low temperature and atmospheric pressure, short processing time, minimized use of chemicals, and flexible use of liquid medium. These fascinating features of SP have attracted the interest of researchers significantly, leading to a major boost and rapid growth in the number of publications on materials synthesis by SP over the past decade. The synthesis of materials by SP is mostly based on bottom-up or top-down approach or even a combination of these two approaches, depending on the actual requirements of a particular nanofabrication process.

Bottom-up approach: As we know, a large number of reactive species are produced *in situ* by SP. Most experiments are normally carried out in aqueous water solution or nonaqueous organic solution. In case of aqueous solution, the dominant species generated from SP comprise of both reductive species (e.g. H radical, free electron, H_2, H_2O_2) and oxidative species (e.g. OH radical and atomic O) [11, 15, 18, 22–24]. These reductive species play a key role in the reduction reactions of metal ions to metal nanoparticles, while oxidative species are important for the oxidation of transition metals as well as degradation of organic compounds and biopolymers. Some chemicals or gases may be added into the solutions in order to promote more reduction or oxidation reaction in the synthesis. In addition, the dissolution of N_2 in aqueous solution under air exposure may cause acidification through the formation of NO_2 and HNO_3 owing to the reaction of plasma with the nitrogen species dissolved in solution. In contrast, the reactive species produced by SP in organic solvents can result in the generation of C_2, CH, and H radicals, which sometimes lead to the formation of lower/higher molecular weight compounds and carbonaceous materials as by-products. For this synthesis approach, the chemical process plays a dominant role in the formation of materials rather than the physical process. Therefore, the electrode materials must be carefully selected to completely avoid impurities as much as possible in the bottom-up synthesis approach. Normally, tungsten is used as the electrode due to its highest melting point among metallic materials (3695 K).

Top-down approach: In addition to the bottom-up approach via chemical process, an opposite route for material synthesis relies on a top-down approach via physical process (i.e. sputtering and erosion). In general, the electrical discharge is performed through a symmetric or asymmetric pair of metal electrodes. There are two possible conceptual pathways for the formation of metal nanoparticles: (i) Disintegration of electrodes by high-energy atomic or ionic bombardments (sputtering). Metal nanoparticles can be formed through subsequent diffusion, growth, and finally condensation in liquid medium. (ii) Erosion of electrode by electrical discharge. The concentration of current at the small gap between the electrodes results in a localized high-temperature region on the electrode surface from where the molten particles are released and subsequently quenched in the solution. In this case, several materials can be produced directly from the corresponding electrode materials, which can avoid the use of chemicals in the synthesis process. Furthermore, it may sometimes combine with chemical process by the interaction of sputtered or evaporated atoms with reactive species – for example, metal oxides and metal carbides.

Thanks to the significant research efforts from researchers who have worked in this area, many different types of materials have been achieved through the above two synthesis approaches. Herein, we will give the recent advancement and overall progress on the utilization of SP reaction in materials synthesis with emphasis on nanoparticles (NPs). Mostly, controllable synthesis of NPs with the desired size, shape, and structure as well as understanding of synthetic mechanisms is the main focus of SP synthesis. Comprehensive and perceptive discussion on these interesting topics is also provided in detail. Finally, the future challenge for SP in the area of material research is proposed as a useful guideline for researchers for further development and study.

7.3.1 Noble Metal Nanoparticles

Noble metal NPs (i.e. Au, Ag, Pd, and Pt) can be synthesized by a chemical reduction method in aqueous or nonaqueous solutions using a reducing agent. Although the particle size can be controlled with small size and narrow distribution by this conventional method, it inevitably requires a reducing agent, multistep process, and long processing time. Therefore, a simple and fast method with minimal use of chemicals is greatly needed for the synthesis of these noble metal NPs for scientific research and practical application. Recently, SP has shown great potential to synthesize various noble metal NPs by fast reduction of metal salt precursors (e.g. $HAuCl_4$, $AgNO_3$, H_2PtCl_6, $PdCl_2$) without the addition of a reducing agent. Based on recent studies, most of the research efforts have been devoted to synthesis of Au NPs and Ag NPs since their formation could be visualized easily by change in solution color, the so-called localized surface plasmon resonance (LSPR), whereas there is no LSPR for Pd and Pt. It is well known that the shape and size of these metal NPs affect strongly their optical and catalytic properties. Therefore, intense studies have focused on how to control the shape, size, and monodispersity by rationally adjusting the surrounding chemistry in solution, such as solution composition, temperature, pH, conductivity, and surfactant.

Saito et al. studied the formation mechanisms of Au NPs at different experimental conditions, including the concentration of Au ions, applied voltage, and discharge time [17, 25]. The use of higher initial concentration operated at higher applied voltage could induce the faster reduction rate of Au ions and formation rate of Au NPs. Interestingly, the dendrite shape with size larger than 100 nm (loose assembly of small Au nanoclusters) was initially observed at short discharge times; however, the particle size decreased with the appearance of anisotropic shape (i.e. triangular, pentagonal, and hexagonal) at longer discharge times (45 minutes) (Figure 7.2). The reduction of particle size at longer discharge time indicated that the dissolution of Au NPs into the solution occurred because the pH was lower than the critical value. Subsequently, anisotropic shape NPs were formed from the dissolved Au ions as a precursor, while Au NPs that were not completely dissolved acted as seeds for nucleation. Hieda et al. proposed a strategy to control the size of Au NPs by changing the volume of water droplets in a reverse micelle solution (a mixture of sodium bis(2-ethylhexyl) sulfosuccinate (AOT), dodecane, and water) [26]. Reduction

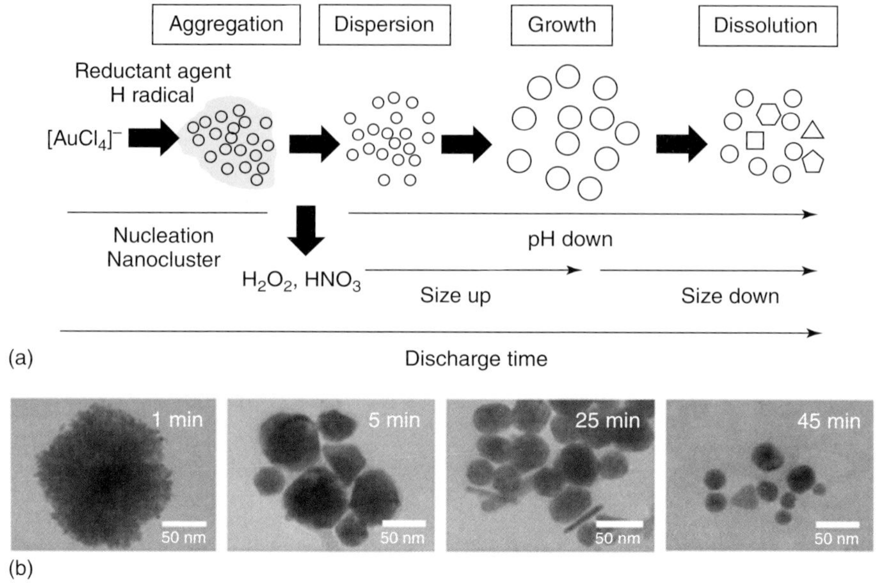

Figure 7.2 (a) Formation mechanism of Au NPs synthesized from SP in aqueous solution [17]. (b) TEM images of Au NPs at different discharge times.

of Au ions and formation of Au NPs proceeded only inside the water droplets formed by reverse micelles, which acted as nanoreactors. The size of Au NPs could be reduced from 15 to 4 and also exhibited better uniformity as the reverse micelle size decreased. The effect of pH on the size of Au NPs was also investigated by Bratescu et al. [27] The size of Au NPs was found to vary from 1 to 10 nm as the pH increased from 3 to 12. The reduction mechanisms of Au ions at acidic (HCl) and basic conditions (NaOH) proceeded via a distinct pathway by H radicals owing to the difference in the surrounding adsorbed molecules (Cl^- for acid and OH^- for basic). The relationship between the size of Au NPs and pH could be understood on the basis of the redox standard potential and electrostatic repulsion force, and surfactant protective layer. In addition, the solution temperature during synthesis also played a significant impact on the formation of Au NPs. Heo et al. found that the size of Au NPs became smaller and had a narrower size distribution when the solution was kept at low temperature [28]. Saito et al. proposed that the H radicals generated from water dissociation could act as a reducing agent in the reduction of Au ions to neutral Au NPs. The role of H radicals in the reduction process in SP has been proved based on the reliable measurement of electron spin resonance (ESR) [17]. Furthermore, Shirai et al. studied the reduction mechanism of both Au NPs and Ag NPs by a DC microplasma liquid system with an H-shaped glass reactor from $AgNO_3$ and $HAuCl_4$, respectively [29]. They obviously found that the Au NPs and Ag NPs were preferentially formed near the anode and cathode, respectively. This finding indicated that the reduction mechanism of Ag NPs should be different from that of Au NPs. Considering the nature of the Au and Ag ions in the solution, Ag ions in the solution take the form of Ag^+, while Au ions mainly take the form of

$[AuCl_x(OH)_{4-x}]^-$ ($x \leqslant 2$ at low pH, $x < 2$ at high pH). Therefore, the Au and Ag ions will move in the solution in different directions under applied electric field. They conclude that electrons and H radicals are candidates to reduce Ag and Au ions, respectively. This finding can be used to explain the reduction mechanisms of Pt NPs (from H_2PtCl_6 solution by H radicals) [30, 31] and Pd NPs (from $PdCl_2$ solution by electrons) [32] as well. The reactive species involved in the reduction process of Au NPs was not limited to only H radicals, but other radicals may participate when the other solute was added in water.

Very recently, Sudare et al. investigated the influence of ethanol mole fraction ($X_{ethanol}$) in the ethanol–water mixture on the reduction rate of Au ions to Au NPs under SP. [33, 34] Interestingly, the reduction rate reached a maximum value at the significant fraction ($X = 0.089$), which was 35 times faster than in pure water. At this concentration, the microscopic structure of ethanol–water mixture changed drastically. They also proved that the fastest reduction at this point was mainly induced by hydroxyethyl radicals ($C_2H_5OH \rightarrow CH_3CHOH\bullet + H^+ + e^-$), not only by H radicals. A similar phenomenon was also observed for other short-chain alcohols (i.e. methanol and 2-proponol).

The microstructural characteristic of Au NPs synthesized by SP was elucidated by Cho et al. [35] It was found that Au NPs mainly exhibited thermodynamically stable forms of the multiply twinned particles (MTPs) of icosahedral (Ih) and incomplete MTP (94%), which was higher than those synthesized from the conventional chemical reduction method. The favorable formation of MTP is most likely due to the unique characteristic and high energy of the SP system, in which the reactive species produced result in a rapid chemical reduction process. Furthermore, a seed-mediated approach of gold ions was studied by Ashkarran. In this work, although the discharge process stopped, the formation of Au NPs still continued until eventually reaching saturation after several hours or days. This result indicated that the Au seeds primarily formed served as nucleation sites for further growth of reduced Au ions that remained in solution without discharging [36]. In all studies, although a unique point regarding the SP is the non-requirement of reducing agent, various organic surfactants, such as polyvinylpyrrolidone (PVP) [37], sodium dodecyl sulfate (SDS) [17, 25], cetyltrimethylammonium bromide (CTAB) [38], and cetyltrimethylammonium chloride (CTAC) [27], were necessarily added in the solution to inhibit the aggregation and to maintain the dispersity of NPs. Negative charges on the NPs surface during and after SP would generate repulsive forces among the NPs, thus preventing them from aggregation. Nevertheless, the surface layer covering the NPs may inhibit the active sites for catalytic reaction, which gives a negative effect for catalytic applications. Therefore, the synthesis of noble metal NPs without surfactant is still needed.

Another synthesis route of noble metal NPs by SP has been achieved by a physical top-down approach through an electrical discharge between metallic electrodes in liquid medium. Compared with the former bottom-up approach, the major merits of this top-down approach is non-requirement of chemicals (i.e. metal salt precursor and surfactant), clean particle, and very short synthesis time (within a few minutes). Lung et al. studied the synthesis of colloidal Au

NPs by electrical discharge in water without the stabilizer and the surfactant [39]. The resulting Au NPs were well crystalized and their size varied between 7 and 40 with a broad size distribution. In their further study, the authors found that Au NPs synthesized in ethanol could provide a smaller size and narrower size distribution (2–20 nm) than in water [40, 41]. The zeta potentials of Au NPs synthesized in water or ethanol were about −30 to −50 mV, indicating that the Au NPs were stable in a colloid form. The high dispersity of Au NPs in water/ethanol can be explained by the fact that active oxygen molecules produced by plasma were adsorbed on the surface of Au NPs, creating the hydrogen bonding in water/ethanol environment and finally forming a negatively charged micelle on the surface [39]. The repulsive electrostatic forces between Au NPs protect them against agglomeration and sedimentation in the suspension. The same research group also prepared the Ag NPs by changing the electrode to Ag. The average particle size was larger than 10 nm with a broad size distribution (5–50 nm) [42]. The presence of large-size particles and broad distribution may be caused by the combined effects of sputtering and erosion processes, in which erosion seems to be dominant owing to the high-temperature reaction. In a continuing effort with the aim to reducing the size and size distribution, the synthesis should be conducted at low input energy and low temperature. Hu et al. reported the synthesis of ultrafine Au NPs with the size less than 2 nm and narrow size distribution by discharging in liquid nitrogen (−196 °C) [43]. However, these ultrafine Au NPs could not be obtained when discharged in water or ethanol at 0 °C (∼3–5 nm) [44]. This result clearly indicated that the solution temperature had a strong influence in inhibiting the growth of particles, resulting in size reduction. Moreover, the effect of HCl (acid), NaCl (neutral), and NaOH (base) solutions on the Au NPs was also studied by Mizutani et al. [45] All solutions yielded an almost similar particle size of about 2–3 nm, but Au NPs were dissolved into Au ions after synthesis in HCl solution. The Au NPs had good dispersion in HCl and NaCl solutions due to the presence of Cl^- ions, whereas aggregation of Au NPs was observed in NaOH. This synthesis route did not limit only Au NPs [36, 39–41, 43–45] and Ag NPs [42, 46, 47], but other noble metal NPs (i.e. Pt [48–50] and Pd [51, 52]) could also be synthesized simply by changing the type of electrodes, as reported elsewhere. The noble metal NPs synthesized by SP can be mixed with various kinds of support materials (carbon, metal oxide, mesporous silica, biopolymer, etc.) for a broad range of applications, which will be discussed below.

7.3.2 Non-noble Metal Nanoparticles

Various kinds of non-noble metal NPs (e.g. Cu [53–55], Sn [56–59], Fe [60], Co [61], Ni [62]) can also be synthesized by SP through two approaches as in the case of noble metal NPs as described above. Compared with the noble metal NPs, the formation of pure non-noble metal NPs is more difficult and challenging due to the ease of oxidation under exposure to strong oxidizing species under discharge or after discharge. To address this critical issue, suitable capping agents, and surfactants were necessarily added into the solution to immediately protect the nascent particles against oxidation during synthesis. It has been demonstrated that pure Cu NPs could be obtained by discharging in CTAB/ascorbic acid [53], gelatin/ascorbic acid [55], and citrate buffer solutions [63], either by consuming Cu electrode or by reducing Cu ions. Without a suitable type and amount of capping agent, oxidized Cu phases (i.e. Cu_2O and CuO) were inevitably formed [53, 55, 63]. Moreover, Lo et al. studied the effect of ethylene glycol/water mixture ratio on the oxidation of Cu particles [54]. It was observed that pure Cu-phase nanoparticles were only formed in pure ethylene glycol, while other mixture ratios resulted in the formation of Cu_2O and CuO.

For the synthesis of Sn NPs, Saito et al. found that a sufficiently large amount of CTAB in solution was required to cover the whole surface of Sn NPs synthesized by Sn electrode, resulting in the complete prevention of oxidation [58]. A similar effect of CTAB on Sn NPs was also observed with the use of $SnCl_2$ as precursor [56]. Further studies demonstrated that Sn NPs could be obtained in water without the addition of any capping agent and surfactant, but optimal electrolyte concentration (KCl) and applied voltage conditions were essential for Sn NP formation [57]. The selection of suitable capping agent and surfactant does not only prevent the oxidation of metal particles but also results in small-size particles (<10 nm) with good dispersity and narrow size distribution. Another interesting result is the synthesis of pure α–Fe NPs (98%) in microemulsion solution obtained by Kelgenbaeva et al. [60] In this work, toluene:water with 95 : 5 (%) ratio was found to be the most appropriate medium for the complete formation of Fe NPs. The Fe NPs were formed at the water:toluene interface and stabilization occurred through the generation of protective layers (oxide or polymer) immediately after particle formation. In other words, both water (oxygen source) and toluene (carbon source) were unable to react directly with the Fe particles formed.

7.3.3 Bimetallic and Alloy Nanoparticles

SP is not only applicable in the synthesis of pure metal NPs, but can also be used for the synthesis of various bimetallic or alloy NPs. Terashima et al. first used SP to synthesize the PtAu alloy through Au–Pt pair electrodes in water without a reducing agent and stabilizer [64]. However, the PtAu NPs were partially obtained with the mixture of monometallic Au NPs and Pt NPs in the final products. Further study using a similar synthetic approach by Hu et al. demonstrated that complete PtAu NPs were successfully synthesized with absence of monometallic Au and Pt phases (Figure 7.3a) [65]. They pointed out that a small gap between the Au and Pt electrodes (~0.3 mm) was optimal to complete the formation of PtAu NPs. At wider electrode gap, pure Au and Pt clusters began to be observed, while PtAu alloy phase could not be formed owing to un-overlapped plasma region. The PtAu NPs with size of about 2 nm were highly dispersed in water or even ethanol with/without stabilizer. Moreover, they also applied this approach for the synthesis of PdAu NPs from the Pd–Au pair electrodes [67]. Such PtAu NPs could not only be produced successfully in aqueous solution, but were also possible in the presence of an organic solvent (i.e. benzene), as shown by Kang and Saito [68]. All studies described above used a bipolar pulsed power supply in the synthesis in order to obtain an equally symmetric sputtering from both metallic electrodes.

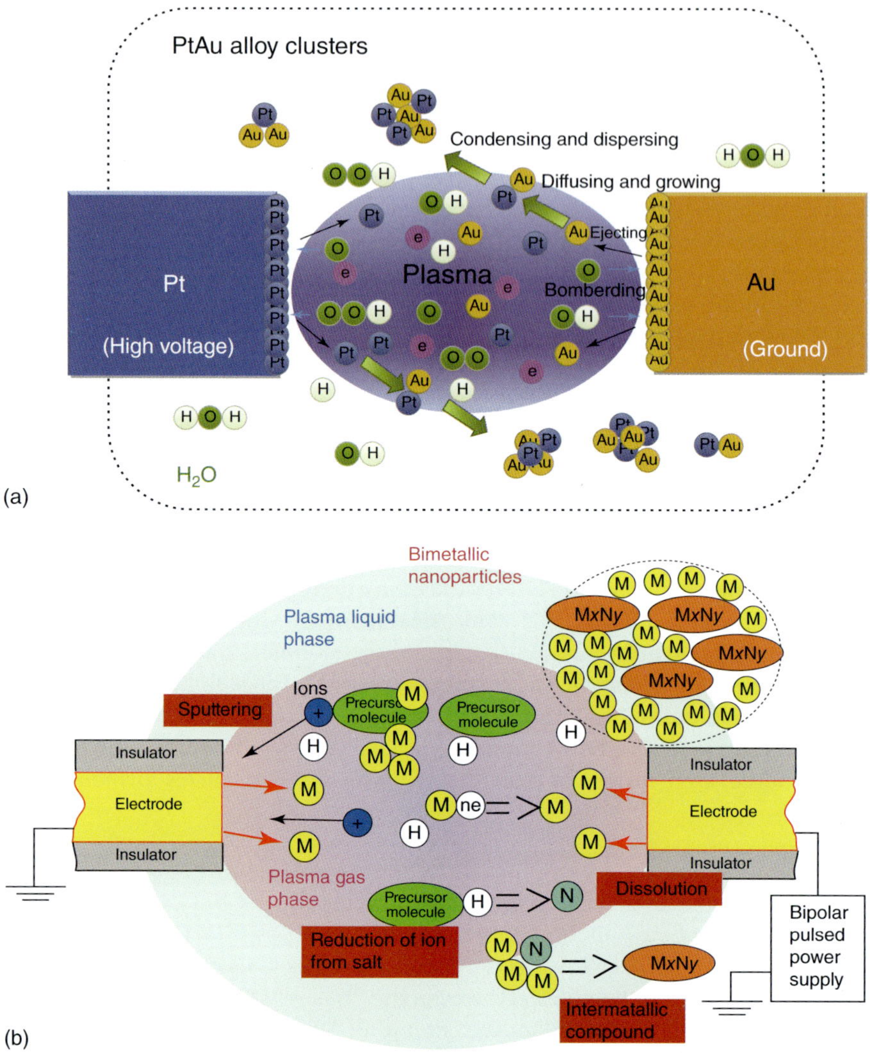

Figure 7.3 (a) Formation mechanism model of PtAu NPs in water by SP through sputtering [65]. (b) Formation mechanism of Au bimetallic NPs in SP through a simultaneous combination of the reduction of metal ions (metal nitrate therein) and sputtering of Au electrodes. In this diagram, the Au atom is represent by the circle with M, and other metals are marked by the circle with N (i.e. Fe, Co, Ni, Cu, Zn, Ga, Pd, Ag, Cd, In). The intermetallic compound is marked as M_xN_y [66].

It is important to note that the type of power supply used had a strong influence on the formation of bimetallic or alloy NPs. For instance, Pootawang et al. tried to synthesize PtAg NPs using a unipolar pulsed voltage [69]. The Ag (cathode) and Pt (anode) were used as opposite electrodes in this system. The result showed that the synthesized NPs were in the form of a bimetallic composite, not alloy, in which small amounts of Pt (~5%) dispersed in the Ag matrix (~95%). A comparative study of the synthesis of alloy NPs using two types of power sources (i.e. a direct current and bipolar power supplies) was investigated by Kim et al. [70] Obviously, the composition of PtPd NPs was changed by the type of power supply and electrode configuration. In case of direct current, the metal (Pd or Pt) at the cathode side was found to be the dominant phase in the product. On the other hand, PtPd NPs synthesized using bipolar pulsed voltage were composed of pure Pt, pure Pd, and PtPd alloy phase. A continuing effort by Kim et al. found that the synthesis of PtAg NPs could be achieved by using a bipolar-pulse power supply and their compositions were controlled by varying the pulse duration [71]. These results from the literature suggest that the type of power supply and experimental conditions (electrode gap, pulse duration, etc.) are very crucial factors in determining the formation of pure alloy NPs by SP through a physical top-down process.

Recently, Bratescu et al. demonstrated the synthesis of bimetallic Au NPs by SP through a simultaneous combination of the reduction of metal ions (metal nitrate therein) and sputtering of Au electrodes [66, 72]. Bimetallic Au NPs were synthesized by alloying with various types of metals, including sp metal (Zn, Cd), trivalent sp metals (Ga and In), 3d metals (Fe, Co, Ni, Cu), and 4d metal (Ag, Pd). The synthesized bimetallic Au NPs tended to aggregate and their size was investigated in the range of 5–20 nm. High percentage of Au (~80–90%) was detected for Au alloyed with 3d metals. Moreover, Au NPs alloyed with sp trivalent and 4d metal had good solubility in water, whereas those with divalent sp metal and 3d metal were insoluble. The composition of the various bimetallic Au NPs depends on electrochemical factors, charge factors, charge transfer between gold and other metals, and initial concentration of metal ions in solution. The mechanism of the formation of bimetallic Au NPs is shown in Figure 7.3b. Besides the possible synthetic routes that we discussed above, direct use of alloy materials with specific composition as electrodes and co-reduction of metal salts by SP is an alternative route for further effective synthesis of various alloy NPs [73].

7.3.4 Metal Oxide

Another recent accomplishment in SP has been the synthesis of metal oxide NPs. Compared with non-noble metal NPs, metal oxides are relatively easier to synthesize due to the ease of oxidation under exposure to strong oxidizing species during discharge or after discharge. The *in situ* production of superoxide (O_2^-) and hydroxyl (OH^-) created by the reaction between plasma and water molecules is very effective for the oxidization process. The original route to synthesize metal oxide by SP is the sputtering or erosion from metallic electrodes and the subsequent oxidation under plasma. Ashkaran's group demonstrated the synthesis of various metal oxides from metallic electrodes in water (i.e. ZnO [74, 75], ZrO_2 [76], TiO_{2-x} [77], and WO_3 [78]). They observed that the size of oxide particles became larger at higher current of plasma. The synthesized metal oxide particles normally exhibited a spherical shape due to the effect of surface tension in minimizing their surface area (Figure 7.4a). However, only ZnO preferentially formed a rod-like shape and transformed to a spherical-like structure at high current where the reaction had high temperature (Figure 7.4b) [75]. This morphological feature can be explained by the faster cooling rate and condensation

Figure 7.4 SEM and TEM images of various metal oxides with different shapes synthesized from SP through physical (a–e) and chemical processes (f–k): (a) TiO_{2-x} spheres [77], (b) ZnO nanorods [75], (c) ZnO nanoflowers [79], (d) CuO nanofibers [80], (e) SnO nanoplates [81], (f) FeO(OH) nanorices [82], (g) ZnO nanospheres [83], (h) FeOF nanococoons [84], (i) TiO_2 nanoleaves [85], (j) NiO nanoflowers [86], and (k) MnO_2 nanosheets [87].

of the product, leading to freezing of their shape and inadequate time to form the rod-like structure. For metal oxide synthesized by a top-down approach in water, an undesirable effect is the formation of mixed crystalline phases of oxidized and pure metallic NPs due to incomplete oxidation. For example, with the use of Ti electrode, almost same results were found by several works – that the products exhibited a nano/microspherical shape with the mixture of several crystalline phases, such as Ti, TiO, Ti_2O_3, and TiO_2 (rutile and anatase) [77, 88–90]. A similar phenomenon was also observed for the synthesis of other oxides (i.e. Zn/ZnO [74, 75, 91, 92], Sn/SnO [59], etc). To suppress the formation of oxygen-deficient and pure metallic phases, various strategies have been used to promote the oxidation reaction, such as bubbling water with O_2 gas [91], addition of oxidizing agent (H_2O_2) [77], post-annealing in air at high temperature [59, 75, 77, 78], and aging in the solution after synthesis [91]. Although oxidation level could be improved significantly, pure crystalline phase and complete oxidation of these metal oxides are still difficult to achieve in some cases. Addition of some chemicals and operation under specific conditions were considered as effective strategies to obtain a pure crystalline phase of the metal oxide. For example, ZrO_2 normally exhibited a mixture of monoclinic and tetragonal phases by direct discharge in water or even in the mixture of H_2O_2 [76]. Chen et al. successfully synthesized pure tetragonal phase ZrO_2 NPs with smaller size in alkaline solution (high pH value) in the presence of ammonia [93]. Kelgenbaeva et al. found that the purity of magnetite Fe_3O_4 phase could be enhanced by increasing the amount of cetylpyridinium bromide (CPyB) in water [94]. Without the addition of CPyB, the synthesized products comprised a mixed phase of Fe, FeO, and Fe_3O_4.

Over the past few years, various strategies have been proposed by many researchers to precisely control metal oxide NPs with the desirable size and shape through the optimization of physical (e.g. low voltage, low temperature) or chemical processes (pH, surfactant) under SP. Several studies have shown that the addition of organic surfactants (e.g. SDS, CTAB, PEG) into the solution could prevent aggregation, giving rise to a significant reduction in size. Moreover, physical ultrasonic wave applied into the system during synthesis and synthesis in cooling systems were also effective ways to prevent aggregation and retain small-sized particles. Moreover, it is possible to alter their common spherical shape into structures with specific shape, such as nanoflower, square plate, and nanofiber by synthesizing under specific controlled conditions and environments. Akiyama's group studied the effects of several parameters, such as input voltage, concentration of electrolyte (K_2CO_3), solution temperature, and agitation on the formation of ZnO nanoflowers by SP. [79, 95] Specifically, ZnO nanoflowers were preferentially formed at a high temperature in a static system (no agitation). The use of agitation led to both amorphous ZnO and metallic Zn nanospheres. Additionally, high enough voltage and appropriate solution conductivity also played a crucial role in the complete formation of ZnO nanoflowers. The formation mechanism of ZnO nanoflowers occurred through different pathways unlike the spherical shape. Several nanorods grew on the surface of the formed ZnO spherical particles along the [0001] direction by precipitation from $Zn(OH)_4^{2-}$ ions, as shown in Figure 7.4c [79]. The formation of CuO nanoflowers in K_2CO_3 solution could reasonably be explained in the same manner by precipitation of $Cu(OH)_4^{2-}$ ions, but these nanorods grew along the [010] direction [63]. In contrast, spindle-shaped or needle-like CuO particles were more easily formed in pure water without involvement of any surfactant or electrolyte (Figure 7.4d) [53, 80]. The same research group also reported that the type of surfactant and temperature strongly affected the shape and size of SnO particles [81]. Single-crystalline SnO plates were only formed at high temperature (>95 °C) with the use of SDS as surfactant. The plate-like crystal was formed from precipitation of $Sn(OH)_3^-$ along the (001) plane face (Figure 7.4e).

In addition to the direct synthesis from metallic electrode, metal oxides are also synthesized by the reduction of dissolved metal ions followed by subsequent oxidation in solution – a bottom-up approach. First, metal ions were reduced into metal NPs by H radicals. Second, these metal NPs were then oxidized by oxidative species, forming metal oxide (i.e. OH radical and atomic O). Therefore, competition between the reduction and oxidation reactions in SP is also important for the formation of metal oxide. Bratescu et al. studied the redox reaction rate of $FeCl_3$ solution with the addition of urea and PEG by SP on the formation of FeO and FeO(OH) (Figure 7.4f) [82]. They proposed that the oxidation rate was found to be higher than the reduction rate in this system, as evidenced by the OES (Optical emission spectroscopy) results. Tong et al. found that uniform spherical ZnO particles could be formed by discharging $ZnCl_2$ solution with urea; however, no ZnO was obtained without urea (Figure 7.4g) [83]. Furthermore, Fu et al. prepared mesoporous amorphous FeOF nanococoons using NaF and $FeCl_3$ solution as precursor (Figure 7.4h) [84]. They noticed that the cocoon-like structure was only formed by the addition of an appropriate amount of urea, while addition of other organic amines (e.g. pyridine, amine, amine oxide) could not yield the nanococoon structure. This finding indicated that CO_2 released from the hydrolysis of urea acted as the guiding agent in the formation of small nanorods along a specific direction. The TiO_2 nanoleaves were successfully synthesized from titanium bis(ammonium lactate) dihydroxide by He et al. [85] The 2D willow leaf-like structure was composed of many small particles that were attached to each other in an orderly manner and also shared the same crystallographic orientation (Figure 7.4i). In this work, the presence of an appropriate amount of nonionic surfactant Brij-30 is essential for the formation of the leaf-like nanostructure. The above findings indicate that the addition of appropriate chemicals or surfactants into solution plays a significant role in controlling the formation and growth of metal oxide with specific morphological features. In addition to the metal oxide described above, CoO [96], NiO (Figure 7.4j) [86], SnO_2 [97], and MnO_2 (Figure 7.4k) [87] were successfully synthesized via the reduction and oxidation of metal ions. Recently, these metal oxides synthesized by SP have been extensively utilized in several applications, including photocatalysts, gas sensors, and anode materials in batteries. Owing to its flexibility in the use of chemical and plasma conditions, there is plenty of room for more studies and design of advanced functional metal oxides in future.

7.3.5 Metal Carbide, Boride, and Sulfide

In addition to metal oxide, metal carbide, boride, and sulfide materials have also been studied concurrently, but only limited work has been done to date as compared with metal oxide. Mashimo's group has synthesized various metal carbide NPs coated with graphitic layers (e.g. ZrC [98], CoC_x [99], WC_{1-x} [100], TiC [100]), as illustrated by TEM images in Figure 7.5b–e. The discharge was performed between two metal electrodes that were submerged in ethanol. The conceptual synthesis route of all metal carbides can be explained in the same manner as with the cases of metal and metal oxide NPs, but suitable carbon source solution is needed in the reaction. The proposed formation mechanism of metal carbide is depicted in Figure 7.5a. The synthesis involved both bottom-up and top-down processes, in which metal atoms were vaporized from electrodes and combined with C and/or C_2 atoms dissociated from ethanol molecules, then forming metal carbide particles. Subsequently, encapsulation of metal carbide NPs in graphitic carbon shells will occur during the cooling process in solution. Such graphitic carbon shells covering metal carbide particles could protect them from oxidation and corrosion and also result in high-temperature stability. Another work of this group showed that only magnetic Fe NPs and Ni NPs coated with graphitic carbon shells were obtained rather than Fe and Ni carbides by using the same experiment and electrode configuration [104, 105]. However, there is no explanation why Fe and Ni carbides could not be obtained in their studies. The carbon encapsulated magnetic NPs showed a potential possibility for use in targeted drug delivery or cancer therapy.

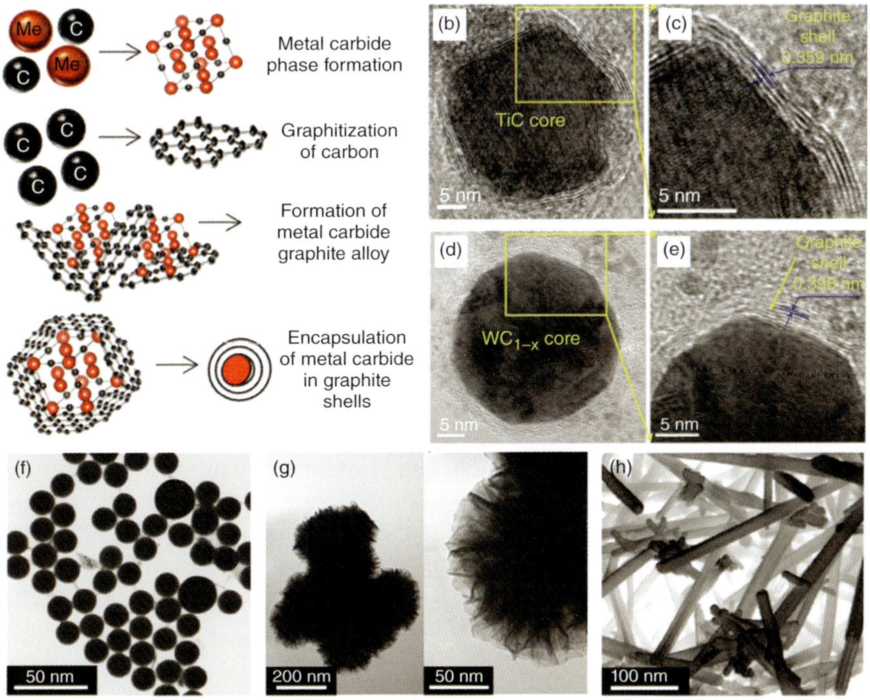

Figure 7.5 (a) Proposed formation mechanism of metal carbide encapsulated in graphitic shells by SP. [100] (b,c) TEM image of TiC. (d,e) TEM images of WC_{1-x}. The Co–B particles with different shapes: (f) monodispersed Co–B nanospheres [101], (g) honeycomb-like Co–B [102], and (h) Co–B–N–H nanowires [103].

The pioneering work on preparation of metal boride NPs by SP was carried out by Tong et al. [101–103, 106] They proved that SP can be used to prepare noncrystalline metal boride with a well-defined nanostructure. For example, mono-dispersed Co–B nanospheres were synthesized by the direct reduction of Co^{2+} and BH_4^- in the ammonia solution containing CTAB [101]. By adding appropriate amount of triethanolamine (TEA) instead of CTAB, honeycomb-like Co–B could be completely formed at long enough discharge time (Figure 7.5g) [102]. A further study of their group successfully prepared N- and H-co-doped mesoporous Co–B nanowires (Figure 7.5h) [103]. In this study, polyethylene glycol hexadecyl ether (Brij-58) was introduced in the solution with the aim to decrease the surface energy in some directions on Co–B–N–H particles, leading to nanowire formation. These results clearly suggest that the type of surfactant plays an important role in controlling the growth of particles in a particular direction, which results in a unique shape and morphology. The prepared Co–B particles showed excellent catalytic activity in the hydrolysis of hydrous hydrazine with high selectivity of hydrogen. More interestingly, they found that SP could also be used to reactivate the catalysts due to surface oxidation after catalytic test [102, 103].

In the other direction, studies on metal sulfide synthesized by SP are relatively rare compared with metal carbide and boride. There are a few reports available regarding metal sulfide synthesis by SP. Alexandrou et al. reported that MoS_2 could be produced by electrical discharge in water [107]. Small MoS_2 core–shell NPs were produced from the MoS_2 powders packed in the anode electrode. Furthermore, wurtzite ZnMgS NPs were synthesized by discharging between ZnMg electrodes submerged in molten sulfur. Annealing was required to obtain pure ZnMgS NPs [108].

7.3.6 Carbon Materials

Synthesis of carbon materials by SP has been gaining considerable interest since it has the ability to unlock many constraints of conventional methods (chemical vapor deposition, arc discharge, pyrolysis, combustion, etc.), such as its simplest equipment without high-cost vacuum system and operation at room temperature and atmospheric pressure. Carbon sources that can be used for the synthesis of carbon materials by SP include either graphite electrode (top-down) or liquid organic compounds (bottom-up). In the early 2000s, arc discharge between graphite electrodes submerged in water was proposed by Sano et al. as a simple and economical method for producing spherical carbon nano-onions in large quantities at low temperature without the use of vacuum equipment [109, 110]. High temperature occurring in the region between graphite electrodes resulted in the generation of C atomic vapors, and the subsequent interaction of the C atoms led to the formation of carbon nano-onions. By using this method, the formation of carbon nanotubes (CNTs) was also achieved, but metal ionic salt solutions were used instead of pure water [111, 112]. Recently, Lee et al. reported that controlling the input energy of SP could tailor the formation of carbon materials in water [113]. At low input energy, graphene flakes were fabricated by exfoliating from the surface of graphite electrodes, whereas use of high input energy resulted in carbon nano-onions (Figure 7.6). However, they

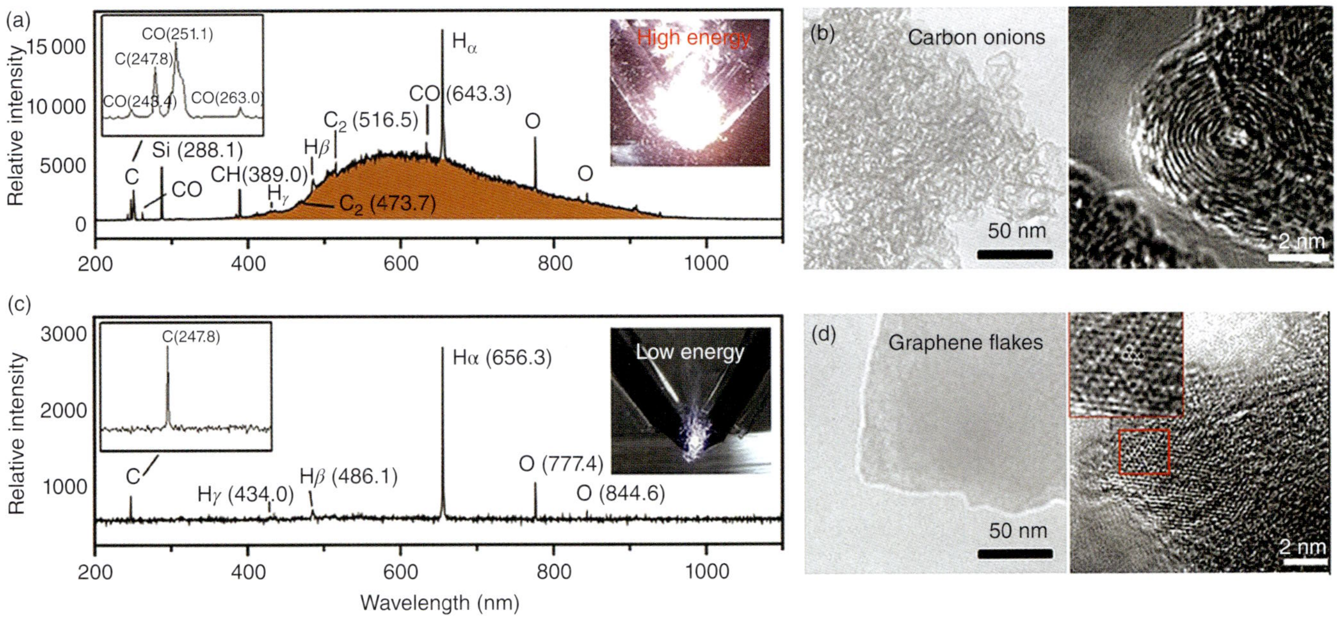

Figure 7.6 (a) Optical emission spectra and images of plasma generated during the synthesis of (a) carbon onion at high energy and (c) graphene flakes at low energy. TEM and HRTEM images of (b) carbon onions and (d) graphene flakes [113].

could not precisely control the lateral size and thickness of the graphene flakes. From the report by Omurzak et al., the number of carbon shell layers of carbon nano-onions can be controlled by adjusting the pulse duration as well as selecting the liquid medium [114]. By using longer pulse duration, carbon nano-onions had larger hollow size and smaller number of carbon shells. With regard to the liquid medium, carbons produced in toluene grew more carbon shells than those in ethanol, which is due to the higher number of C atoms in toluene.

For the bottom-up approach, various organic hydrocarbons have recently been employed as the precursors for carbon synthesis, such as benzene [115–123], hexane [124], dodecane [125], alcohol [126, 127], and acetone [125]. The most recent studies on SP have mostly used liquid benzene as the precursor for synthesis because it provides the fastest production rate ($\sim$10–20 mg min^{-1}) among other liquid organic precursors: derivative aromatic hydrocarbon ($<$10 mg min^{-1}) and aliphatic hydrocarbon ($<$1 mg min^{-1}). Figure 7.7a displays the formation of CNPs by SP under submerged in benzene. Kang et al. reported that the carbons derived from benzene exhibited a quasi-spherical shape with the size ranging between 20 and 40 nm and they aggregated to form 3D interconnected pore channels in different directions in the meso- and macrometer length scales [116, 117]. However, their crystal structures revealed the dominant amorphous phase. On being applied as electrode materials in Li–air battery, the carbon nanoparticles (CNPs) from benzene proved to have a better performance in discharge capacity than the commercial Ketjen black. It is important to note that although CNPs could be derived from benzene at room temperature and atmospheric pressure, subsequent thermal treatment was still required to further enhance the conductivity and surface area. After thermal treatment (700–900 °C) in inert atmosphere (N_2 or Ar), the conductivity was improved by about two orders of magnitude (from $\sim 10^2$ to 1–10 Ω cm) and the surface area was also enhanced from 150 to 300 m^2 g^{-1}. Yun et al. reported that the surface area of CNPs from benzene could reach almost 1000 m2 g^{-1} after heating at 450 °C in air atmosphere [119]. The hierarchical porous structure of CNPs from benzene is beneficial to use as anode materials in Li ion batteries and metal–air batteries.

With continuing effort to improve the degree of graphitization and conductivity of CNPs, adjustment of experimental parameters and solution composition have been investigated. Kang et al. [117] and Hyun et al. [128] reported that the degree of graphitization and electrical conductivity of CNPs derived from benzene were improved significantly as the pulse frequency increased ($>$60 kHz, normally about 15–30 kHz) (Figure 7.7c–e) . This result was attributed to the higher input energy applied during synthesis at high frequencies. However, an increase in frequency caused adverse effects, such as larger particle size (80–100 nm) and a substantial decrease of surface area. Lee et al. found that electrode gap had a strong influence on the crystalline structure and amount of hydrogen atoms in CNPs. By adjusting the electrode gap at 0.75 mm, the conductivity improved, which related with the decrease in hydrogen atoms. However, a wider electrode gap (1 mm) could lead to an unstable plasma condition [123]. Another strategy has been proposed by Li et al. who reported the enhancement in conductivity of CNPs by the addition of carbon tetrachloride (CCl_4) into benzene [129]. It is proposed that Cl radicals promoted the elimination of hydrogen via the formation of HCl. This resulted in the desorption of hydrogen atoms (sp^3) from the carbon structure, thereby enhancing electrical conductivity. Besides the CCl_4 addition, the addition of solid naphthalene and anthracence in benzene could also enhance the conductivity of CNPs compared with the use of only pure benzene. Furthermore, Yun et al. found that the addition of appropriate amount of water in benzene could slightly enhance the crystallinity and pore size of the CNPs synthesized [120–122].

The possible mechanism of CNPs from benzene by SP can be explained as follows [50, 115, 118, 130]. The benzene molecules are vaporized by an intense electric field between electrodes. The benzene vapor converts to plasma after breakdown, resulting in the complete decomposition of benzene molecules into small fragments (i.e. C_2, CH, and H radicals) at the plasma center, as evidenced by OES in Figure 7.7b. The C_2 molecules further interact with each under high temperature (3000–4000 K), forming the nanocrystalline graphite and finally carbon particles. On the one hand, polymerization reaction induced by radicals dominantly takes place at the plasma–liquid interface rather than decomposition due to its sufficiently low temperatures. Therefore, CNPs formation occurs through a combination of the above two reaction pathways under SP. Considering which is the dominant pathway in the formation of CNPs from benzene it is noted that benzene containing unsaturated π bonds (C=C) favors the formation of stable free radical monomers and initiates the polymerization reaction. Polymerization seems to play a dominant role over carbonization in case of benzene because the plasma–liquid interface volume is much larger than the plasma center, resulting

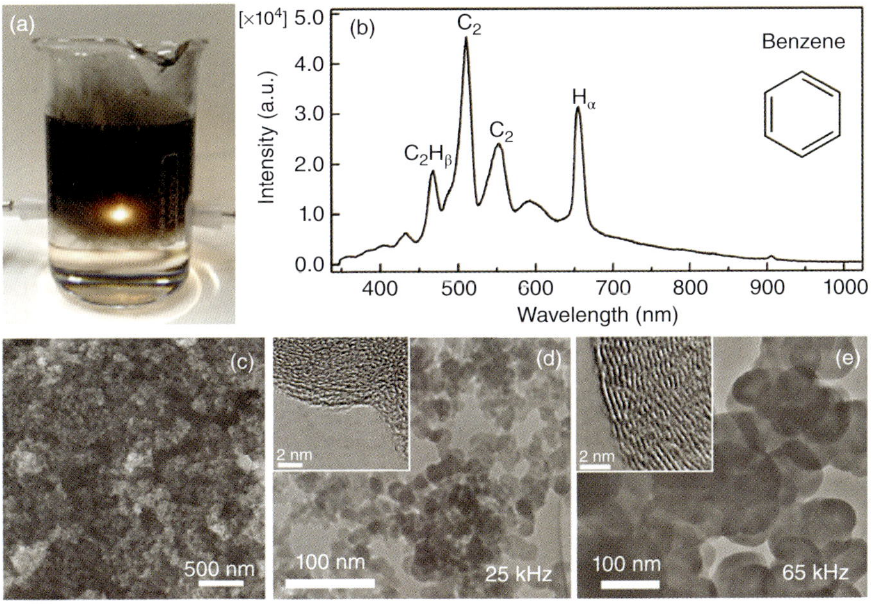

Figure 7.7 (a) Synthesis of CNPs from benzene by SP. (b) The OES spectrum of plasma generated in benzene [130]. (c) SEM image of CNPs synthesized from benzene at 20 kHz showing hierarchical meso/macro pore structure. TEM image of CNPs synthesized from benzene at (d) 25 kHz and (e) 65 kHz [117].

in high production rate of CNPs. In contrast, in case of aliphatic hydrocarbons, the formation mechanism possibly proceeds through carbonization at the plasma center, while polymerization is not favored mainly due to the presence of the saturated σ bond (C—C). This is the reason why production rate of CNPs from aliphatic hydrocarbon is very low compared with benzene. However, an attempt to fully understand the formation mechanisms of CNPs is still under progress.

While considerable progress has occurred on CNPs from benzene, synthesis of heteroatom-doped carbons by SP is currently gaining momentum at a significant pace. It is known that heteroatom doping can tune the physical and chemical properties as well as electronic structure of carbon materials, which is highly essential for a new platform in electrocatalysis research. Based on most recent literature, the main application of heteroatom-doped carbons by SP was use as cathode catalyst for oxygen reduction reaction (ORR) in fuel cell and metal–air battery. To date, various heteroatom-doped carbons (i.e. N, B, P, F, and their mixture) have been successfully synthesized by SP with the use of heteroatom-containing organic solvents as precursors. By using SP, homogeneous heteroatom doping in carbon structure could be achieved via *in situ* doping throughout the surface and bulk of CNPs during growth and formation. Among various heteroatoms studied, N-doped carbon nanoparticles (NCNPs) have received considerable attention. Panomsuwan et al. showed that the N content in synthesized NCNPs could be tuned easily by adjusting the amount of pyrazine (N source) in liquid benzene (30–70 vol%) [131]. A similar trend was also investigated by Kim et al. where phthalocyanine was used as N source mixed with benzene and dodecane [132]. However, the main problem of mixed solid–liquid precursors is the solubility limit of solid compounds in liquid solvent at high concentrations, which thus leads to difficulty in enhancing N doping content. It is therefore worth using the N-containing organic solvents, based on either aliphatic or aromatic molecules as a single-source precursor (that serves as both C and N sources). In the last few years, various types of N-containing organic solvents have been used for synthesis, such as acrylonitrile [133], pyrrole [134], pyridine [135, 136], and aniline [128]. Panomsuwan et al. prepared the NCNPs from a series of cyano-aromatic molecules [137, 138]. They found that the N doping contents depended on the nitrogen/carbon (N/C) mole ratio of organic molecules, which is in the following order: cyanopyrazine (0.60) > cyanopyridine (0.33) > benzonitrile (0.14). The results also revealed that higher N/C mole ratio resulted in higher N content in as-synthesized NCNPs; however, production rate became lower due to lack of carbon atoms. Importantly, precise control of the proportion of N states that existed in the structure of NCNPs was difficult and depended on the source precursors. Under SP synthesis, the surface of NCNPs seemed to be enriched by N-containing functional groups, such as nitrile, amine, or amino, while graphitic N and pyridinic N were found as the minor states. For these reasons, subsequent heat treatment (700–900 °C) was required to tailor the bonding state of nitrogen as well as maximizing the proportions of graphitic N and pyridinic N because they were considered to be catalytically active sites for ORR [132, 137, 139]. Although as-synthesized NCNPs possessed high N contents (4–6 at%), their content was reduced significantly after heat treatment owing to their desorption and decomposition at high temperature. The way to enhance N content with high proportion of graphitic

N and pyridinic N without heat treatment still remains a big challenge in SP synthesis. Considering the formation mechanism of NCNPs, in addition to C_2 radicals, the CN radicals dissociated from organic molecule play an important role in the *in situ* doping during growth and formation of NCNPs [131, 133].

Very recently, Hyun et al. successfully synthesized N-doped carbon with sheet-like structure from *N*-methyl pyrrolidone (NMP) by SP. [140, 141] This result presented an interesting finding in SP for the synthesis of carbon materials with sheet-like structure, which has never been found in previous works. Specifically, the sheet-like morphology could be only obtained from NMP at a high frequency of 200 kHz, as revealed in Figure 7.8. By operating at the same conditions with the use of aniline and benzene as precursors, unlike NMP, only carbon particles with larger size were observed without the observation of nanosheet structure. This evidence clearly indicated that specific composition and molecular structure of organic precursor as well as appropriate plasma conditions are very important factors to create an unusual formation of sheet-like structure. However, the formation mechanism of sheet-like structure from NMP is still unclear at present. More understanding and in-depth investigation are needed for further development in the right direction.

The synthesis of heteroatom-doped carbons by SP is not limited to only N doping, but other heteroatom dopings beyond nitrogen can also be processed by choosing an appropriate organic precursor containing the desired heteroatoms. For example, Ishizaki et al. reported that the addition of 1,4 dioxane into benzene could result in the alteration of oxygen-containing functional groups on the surface of carbon products, of which carboxylic and hydroxyl groups were increased with the addition of 1,4 dioxane [142]. Panomsuwan et al. synthesized F-doped carbon nanoparticles (FCNPs) from the mixture of toluene and trifluorotoluene at different mixing ratios [143]. The F atoms preferred to exist in the carbon structure in the form of ionic bonding at low doping content and became dominant by semi-ionic bonds at higher doping content. The ionic and semi-ionic bonds were considered to be the important sites in promoting ORR activity. The synthesis of B-doped carbon nanoparticles (BNCPs) was also reported in the works by Panomsuwan et al. [144] and Kim et al. [136], in which triphenylborane and boric acid were mixed with benzene as boron source, respectively. Furthermore, Kim et al. studied the binary co-doping of B and P with N into the carbon structure [136]. The pyridine was mixed with boric acid (B source) or phosphoric acid (P source) and then used as the precursor

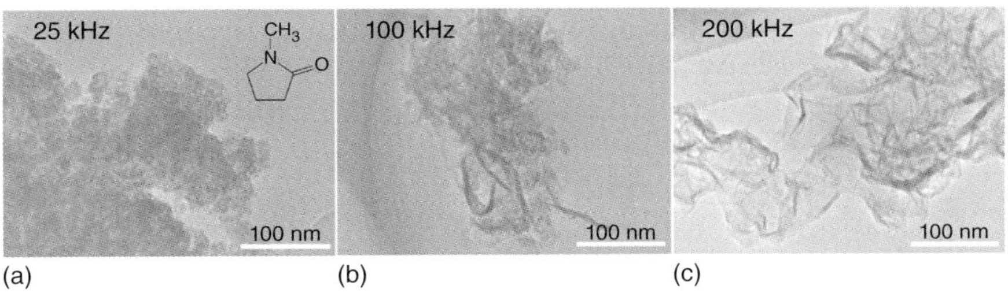

Figure 7.8 TEM images of N-doped carbon nanosheets synthesized from NMP at different frequencies: (a) 25 kHz, (b) 100 kHz, and (c) 200 kHz [140].

for B/N- or P/N-doped carbons, respectively. The results showed that the binary co-doping of B or P with N into carbon induced a remarkable ORR enhancement compared to N-doped carbons. This result is evidence that SP approach can potentially be applied to binary or ternary co-doping of heteroatoms into carbons. The studies on another doping element beyond N and multiple doping are currently quite limited in the field of SP and call for future research. Considering carbon source precursors, liquid biomass or waste from petrochemical industry seems to be the promising choice in future. Moreover, progress in the chemistry of ionic liquid may also provide new efficient precursors.

7.3.7 Mesoporous Silica

According to the report of Pootawang et al., the organic surfactant template inside mesoporous silica prepared by sol–gel method could be removed by SP within 15 minutes, which is much faster than conventional calcination methods [145–147]. The organic surfactant template was decomposed into small organic compounds through oxidation reaction by the generated OH radical and atomic O species under SP (Figure 7.9a). These reactive species could penetrate inside the pores of mesoporous silica and decompose the organic template. The liquid medium could serve as an eluant for transporting the decomposed products out of pore channels. The ability for template removal showed better performance in either acid or basic conditions as compared with that in neutral condition. This result was due to the fact that the number of generated active species in acid and base solutions is relatively higher than those generated in neutral solution. The pore feature of mesoporous silica transformed from a disordered wormlike structures (Figure 7.9b) to an ordered 2D hexagonal structure (Figure 7.9c) at lower acidity of sol-gel solution. After SP treatment, pore feature of mesoporous silica remained unchanged. However, higher surface area was found for mesoporous silica treated by SP at pH 3 and 11 as compared to neutral condition and conventional calcination process. In their further studies, they reported that the SP was not only utilized for template removal, but also accelerated the gelation time in the preparation of mesoporous silica [148, 149]. The solidification of polysilicate network via hydrolysis and condensation during sol–gel reaction was accelerated by highly reactive species and could be rapidly achieved within three minutes, which was much faster than by a conventional sol–gel method (>12 hours). Interestingly, the morphology of mesoporous silica accelerated gelation by SP exhibited a rope-like shape with a 2D hexagonal mesopore arrangement, which was distinct from that obtained by conventional sol–gel method, which showed the aggregate of irregular particle shape (Figure 7.9d,e). It suggested that the rapid solidification of polysilicate under unusual SP reaction resulted in the thermodynamically favorable formation of rope-like morphology with higher surface area and larger pore size.

7.3.8 Low Molecular Weight Biopolymer

Another area of application for SP is the depolymerization of biolpolymer from high molecular weight materials to lower MW (Microwave) for practical

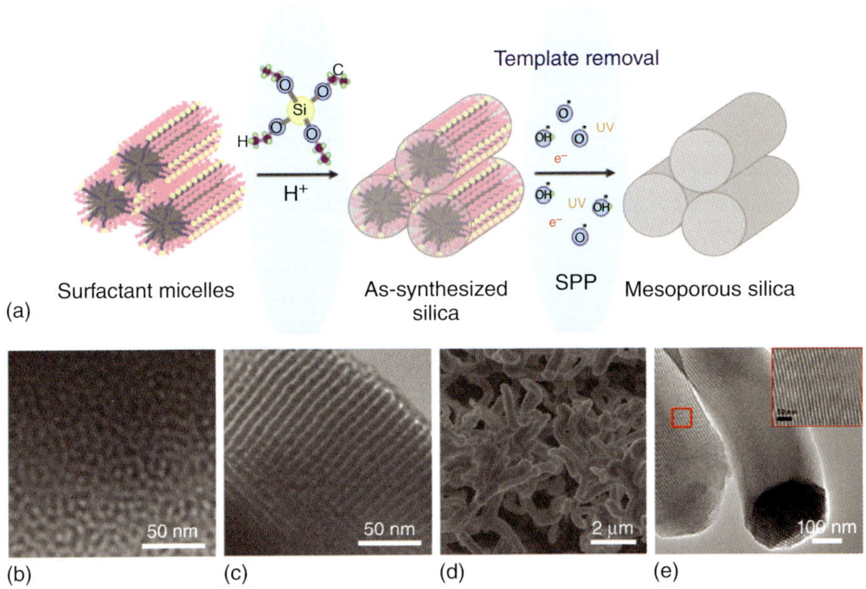

Figure 7.9 (a) Schematic representation of synthesis of mesoporous silica by SP using ternary surfactant system as template and TEOS (Tetraethyl orthosilicate) as silica precursor [145]. (b) Disordered wormlike structures and (d) ordered 2D hexagonal structures of mesoporous silica synthesized by SP at pH 3 solutions [145]. (d) SEM and (e) TEM images of rope-like mesoporous silica synthesized by accelerating gelation time by SP. [148]

application in biological fields. It has been proved that OH radicals play a key role in the decomposition and dissociation of various organic compounds [11, 145]. By using SP, low molecular weight chitosan ($\sim 10^3$–10^4 Da) with high water solubility could be realized as reported in the literature [150–153]. It offers several significant advantages over conventional chemical and enzymatic degradation methods, such as simplicity, no chemical use, short processing time, and single-step process. Prasertsung et al. studied the degradation of two-types of high molecular weight chitosans: α-chitosan and β-chitosan by SP. [151, 152] The molecular weight of α-chitosan and β-chitosan solutions decreased significantly after treatment. The degradation rate of β-chitosan was found to be higher than that of α-chitosan, which was due to the more rigid crystalline structure of α-chitosan. The production of reactive species, especially OH radicals and atomic O, is believed to be the key factor in the degradation of chitosan. The most recent trends in this direction have focused on the enhancement of degradation efficiency of chitosan solution through the addition of metal ions. For instance, Pornsunthorntawee et al. studied the degradation of chitosan–metal complexes by the SP. [150] Four different types of metal ions, including Ag^+, Zn^{2+}, Cu^{2+}, Fe^{3+}, were used to form the complexations with chitosan via the interaction with $-OH$ and $-NH_2$ group. The addition of Cu^{2+} and Fe^{3+} could effectively promote the degradation of chitosan, whereas the addition of Ag^+ and Zn^+ had negligible influence in the degradation of chitosan. Another study on the effect of metal ions on degradation of chitosan was demonstrated by Tangtipapol et al. [153]

They found a similar tendency that degradation rate of chitosan was significantly improved by using Fe or Cu as electrodes in comparison with tungsten electrode. The Fe and Cu ions produced by the sputtering from electrodes during discharge are the important factors that contribute to the degradation of chitosan. The promoting effect on degradation of chitosan by addition of metal ions can reasonably be explained as follows: [1] the coordination of high-affinity metal ions (Cu^{2+} and Fe^{3+}) weakens the covalent bonds near the coordinating sites, which thus create the weak points on the chitosan chain for depolymerization. [2] The Cu^{2+} and Fe^{3+} could enhance the amount of OH radicals in the reaction by decomposition of H_2O_2 generated in the SP system (Fenton reaction). The results obtained in all above reports have shown similar trends that SP had a significant effect on decrease in molecular weight, viscosity, and crystalline structure of chitosan, but not on its chemical structure, regardless of the addition of metal ions or change of electrodes. The low molecular weight chitosan or water soluble chitosan obtained from SP can be further used in antitumor, antifungal, and antibacterial applications. In addition to chitosan, Watthanaphanit and Saito used SP to depolymerize alginate polymer at different concentrations [154]. As expected, alginate polymers could be deopolymerized gradually with increasing treatment time, reflected by lower molecular weight and lower viscosity. The depolymerization mechanisms can be explained in the same manner as with the case of chitosan. Although the SP showed high efficiency for biopolymer depolymerization, the process is still random and involves nonselective bond cleavage, sometimes resulting in a wide range of molecular weight in final products. Therefore, the way to control high selectivity of bond cleavage is an ongoing challenge.

7.3.9 Composite Materials

Rational integration of two or three distinct materials into a composite system has resulted in the creation of advanced novel materials inheriting the advantages of each material or some specific functions for the desired applications. The potential utilization of SP to synthesize various types of materials can also bring us immense benefit in the fabrication of composite materials. To date, there are a number of published works on the involvement of SP in the preparation of several composite systems. Most of them have emphasized on the noble metal NPs supported on various support materials (carbon, metal oxide, etc.), while other composite systems have received less attention. To give a clear overview of the composite systems studied, we have divided the composite systems by the type of constituent materials as in the following.

7.3.9.1 Noble Metal Nanoparticle/Carbon Composite

The research trend today is mainly toward the preparation of this composite type for use as electrodes in electrochemical devices, especially fuel cells. Noble metals (e.g. Pt, Au, Pd) and alloys (e.g., PtAu, PtPd, PtAg) were synthesized and supported on various carbon materials (e.g. carbon nanotube, CNP, commercial carbon black, and graphene). Based on the literature studies, these composites were prepared by several synthetic routes, which can be categorized as follows:

(i) Noble metal NPs were synthesized separately by SP through either chemical or physical processes, and then the composites were obtained by direct mixing of carbon materials and the noble metal NPs prepared in solution [52, 65, 70, 71].
(ii) Noble metal NPs were synthesized by SP through either chemical or physical process in the presence of carbon materials that are well dispersed in solution, forming noble metal NP/carbon composites [30, 49, 51, 65]. Figure 7.10a shows the typical TEM image of Pt/MWCNT composite prepared by this route [155].
(iii) Carbon materials were synthesized by discharge between graphite electrodes while noble metal NPs were generated simultaneously through reduction of metal ions [31, 156]. This could lead to the formation of noble metal NP/carbon composites in a one-step process.
(iv) In contrast to route (iii), carbon materials were synthesized from an organic precursor (i.e. benzene), while noble metal NPs were generated simultaneously by the sputtering process [50, 68]. In this case, noble metal NP/carbon composites could also be formed through the one-step process route (iii). However, the unique feature of this synthetic route is that the noble metal NPs seemed to be both deposited on the surface and embedded inside the carbon materials (Figure 7.10b), which is unlike routes (i)–(iii) where NPs are only deposited on the surface of the carbon supports.

7.3.9.2 Metal Oxide/Carbon Composite

Jung's group revealed the preparation of various types of metal oxide/carbon composites for use as electrode in supercapacitors [157–159]. From their works, the small-sized Fe_2O_3, RuO, and MnO_2 particles were formed via reduction and oxidation reactions under SP, and then deposited on carbon black to form a composite (Figure 7.10c).

7.3.9.3 Metal Nanoparticle/Metal Oxide Composite

Noble metal NPs supported on metal oxide were synthesized by either reduction or sputtering process in the solution suspended by metal oxides. Recently, studies have been attempted on Au/TiO_2 [160, 161], Au/$SrTiO_3$ (Figure 7.10d) [162], and Ag/Ga_2O_3 [163], which were mainly proposed to be used in photocatalytic applications. The enhanced photocatalytic activity was observed after depositing Au NPs or Ag NPs, which can possibly be explained by the improvement in charge separation and adsorption in the visible light region due to the LSPR effect.

7.3.9.4 Metal Nanoparticle/Mesoporous Silica Composite

Pootawang et al. studied the incorporation of Ag NPs in mesoporous silica [147–149]. As can be seen in Figure 7.10e, Ag NPs were observed both inside and outside pore channels. This finding indicated that the reduction of Ag ions occurred both inside and outside pore channels because mesoporous silica was suspended in $AgNO_3$ solution. The mesoporous silica incorporated with Ag NPs can be used as catalysts for 4-nitrophenol reduction and oleic acid hydrogenation.

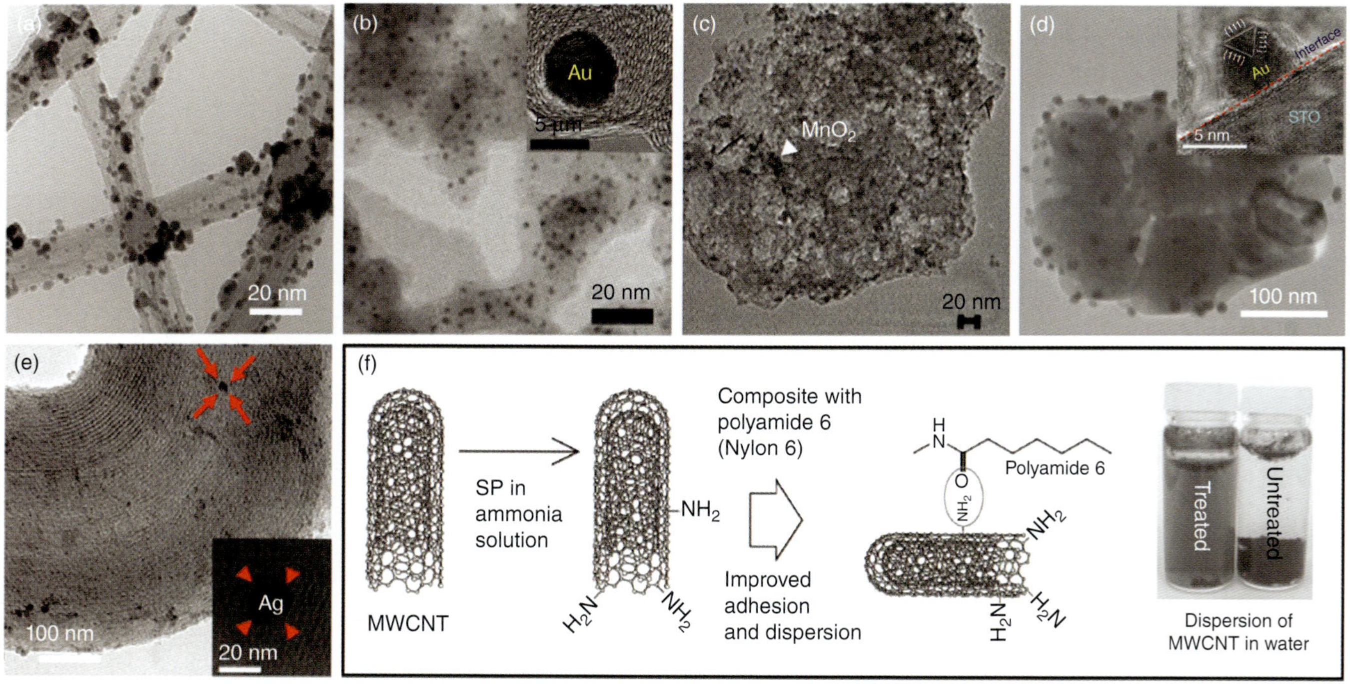

Figure 7.10 Typical TEM images of several composite systems synthesized by SP: (a) Pt/MWCNT [155], (b) Au/CNP [50], (c) MnO_2/carbon [158], (d) Au/$SrTiO_3$ [162], (e) Ag/mesoporous silica [148]. (f) Schematic illustration of functionalization of MWCNT (Multi-wall carbon nanotube) with amino group by SP under ammonia solution [164]. Hydrogen bonds between NH_2-functionalized MWCNT and oxygen in polyamide 6 resulted in the improved adhesion and dispersion of MWCNT in polyamide 6 matrix. The right panel shows good dispersion of treated MWCNT in water compared with untreated MWCNT.

7.3.9.5 Metal Nanoparticle/Biopolymer Composite

It was demonstrated that SP is the most effective and safest method to synthesize the composites of metal NPs and biopolymer composites for biomedical applications. Lee's group published several works on the incorporation of Ag NPs into biopolymers (i.e. gelatin [165–167], cellulose [168], chitosan [169], and alginate [170]) by SP for antibacterial applications. From their works, these bio-composites exhibited a broad spectrum of antimicrobial activity against various pathogens with a minimal inhibition concentration. Moreover, Watthanaphanit et al. synthesized Au NPs in alginate polymer solution with the expectation to apply the same in cancer therapy [171, 172]. The cancer cell treatment test of Au NPs in alginate solution is still under progress.

7.3.9.6 Polymer/Carbon Composite

The CNT has become a major important filler material in high strength and low weight composites; however, aggregation and hydrophobicity are the major hurdles for composite processing. Surface modification and size reduction of CNTs are needed to obtain better homogeneous dispersion and adhesion of CNTs in the polymer matrix. Recently, Tong et al. successfully cut CNTs into shorter length by using SP through OH radical attacks [173]. Additionally, oxygen functional groups (i.e. OH and COOH) were also introduced onto the sidewall of CNTs, resulting in more hydrophilicity [173, 174]. Further effort to modify the surface of CNTs by SP was taken by Shirafuji et al. [164] The CNTs were treated by SP under ammonia solution (Figure 7.10f). Both nitrogen and oxygen groups were formed on the surface of CNTs. Improved hydrophilicity was evidenced by good dispersion in aqueous solution. The composite of polymide 6 and treated CNTs showed better mechanical properties than untreated CNTs because of its good dispersion and homogeneity. Surface modification by attaching atoms and molecular groups to the surface of CNTs or other materials can offer new functionalities and applications.

7.4 Summary and Future Challenge

Ten years ago, the term "solution plasma (SP)" was unknown to the material research community. During the past decade, researchers have made significant contributions to gain knowledge and understanding of SP in the aspects of materials synthesis, which finally bridges the gap between SP and material synthesis. Today, SP has become one of the most discussed topics and is more familiar to the researchers as a powerful tool for the synthesis of a broad range of materials in various areas of applications, such as electrocatalysis, photocatalysis, sensors, and antibacterials. Owing to its fast reaction and operation at low temperature and atmospheric pressure, SP is thus expected to be more practical and to expand to a broader area in the near future. Although significant progress has been made on SP, it is still at a young technological stage and many issues are not fully resolved yet. To realize full potential applications of SP, future studies should be focused on the following four issues.

7.4.1 Highly Controllable Synthesis of Materials

The major challenge in materials synthesis by SP is how to precisely control the NPs with the desirable size, shape, structure, and function through the adjustment of both physical and chemical parameters. Considering noble metal NPs (e.g. Au, Ag, Pt. Pd), their particle size seems to be easy to control simply by increasing the input energy or by adding an organic surfactant. On the other hand, the shape of noble metal NPs synthesized by SP is usually round or spherical, while other shapes are difficult to form regardless of changing experimental parameters. For example, rod-shaped Au NPs cannot be successfully synthesized by SP under usual conditions, but specific chemicals may need to control growth of Au NPs into a specific direction. Recently, Tsukeda et al. tried to synthesize Au nanorod using SP by a combination of reduction of metal salt and sputtering from electrode [175]. However, they could not obtain homogeneous Au nanorods, but only mixed with spherical shape. In case of metal oxide, it has been shown that various metal oxides with uncommon shapes, such as SnO nanoplate [81], ZnO nanoflower [79, 95], FeOF nanococoon [84], and TiO_2 nanoleaf [85], could be obtained via suitable selection of surfactants, which is essential for the controllable growth of particles along a specific direction. Therefore, varying the surfactant or addition of chemical in SP can possibly lead to the formation of metal oxide with uncommon and unique shapes. Moreover, the synthesis of heteroatom-doped metal oxides (N-doped TiO_2, N-doped ZnO, etc.) would be expected by discharging in the presence of a solution containing the desired heteroatoms. For carbon synthesis, as described above, the formation of nanosheet structure was surprisingly found by Hyun et al. [140, 141] This work suggested that proper selection of organic precursor (i.e. NMP) and high input energy are both required for the formation of unusual nanosheet structure. However, the mechanisms of nanosheet formation in SP are still unclear and further in-depth investigation is needed. Therefore, there is still plenty of room for studies to understand and link the underlying mechanisms to nucleation, growth, surface termination, and orientation of NPs under SP in order to attain the materials with desired size, shape, structure, and function.

7.4.2 High-Precision Tools and Measurements

The development of reliable power supply with precisely controllable current–voltage waveforms can help attain high selectivity and certainty of reaction fields in SP, which is very useful for materials synthesis. Moreover, power supply that provides ultrahigh frequency and ultrashort pulse duration can also lead to the discovery of new materials with unusual physicochemical properties. More importantly, future research on SP should be focused on an in-depth study and understanding of the underlying physicochemical mechanisms created by SP, which are related with plasma physics, plasma chemistry, thermodynamics, and interactions at gas–plasma–electrode interfaces. The characteristic nature of most reactive species and reactions generated by SP has already been unraveled on the basis of the OES and ESR analyses [15, 17, 18, 23]. However, at present, a conclusive understanding alongside reliable measurements of some radicals (e.g. benzene radical cation, H_3O, etc) is still unclear and there is lack of solid evidence due to the limitation of instrument performance. Therefore, the development of a novel *in situ* spectroanalysis method to identify and detect these reactive species is greatly needed, such as time-resolved OES 2D mapping with nanosecond time resolution [176, 177] and coherent anti-Stokes Raman spectroscopy (CARS) [178, 179]. Therefore, groundbreaking discoveries in SP reaction may be expected from these *in situ* spectroanalysis methods.

7.4.3 Computational Simulation

Based on the recent literature, most research efforts on SP have been mainly focused only on the experimental aspects of material synthesis. In contrast, to the best of our knowledge, there are no intensive studies related to theoretical simulations on SP, as reflected by the lack of research works in this field. Perhaps it is time to start theoretical studies on SP because we have sufficient knowledge and background data on SP, which can be used as a useful database for computational simulation. The complex formation of reactive species and reactions created by SP sometimes cause more difficulty in understanding based on only experimental results. Therefore, a combination of experimental results and computational simulation may be helpful for us to provide new insights and address the unexplained phenomena in future. Moreover, computational simulation is very useful for the rational design of shape, size, and configuration of electrode and reactor [180].

7.4.4 Large-Scale Synthesis

Currently, SP is carried out in only a small-cell size or a laboratory scale, which limits the yield of synthesized materials. Its potential extension to real industrialization is still far from being realized. Therefore, our next challenge is to design and scale up SP to a larger scale and ultimately extend to practical manufacturing with high productivity and continuous process. Figure 7.11a reveals the prototype of an SP reactor with multi-electrode pairs for large-scale production from Saito's group. Another possible way is to increase the plasma–liquid interface region by generating a large number of tiny plasma (micro plasma) within the reactor through a porous dielectric material as illustrated in Figure 7.11b. This kind of plasma system is named "3D integrated micro-solution plasma (3D IMSP)" by Shirafuji et al. [180–182] Although the distribution of the 3D IMSP was

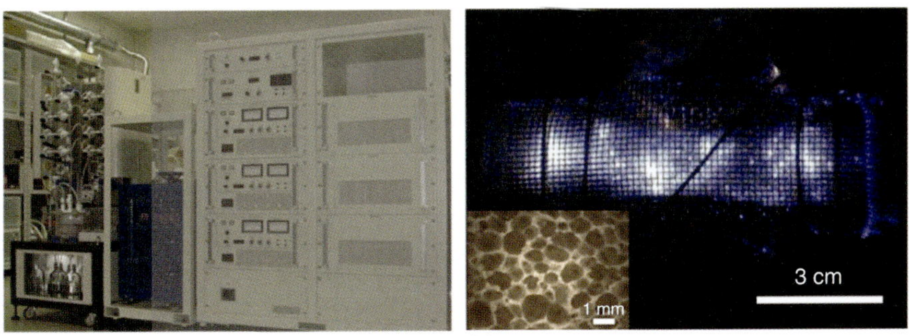

(a) (b)

Figure 7.11 (a) A prototype of large-scale SP system with multiple electrode pairs developed by Saito's group [20]. (b) 3D integrated micro-solution plasma (3D IMSP) system developed by Shirafuji's group [180–182].

not yet uniform, it provided better performance than conventional SP with single wire-to-wire electrode, as evidenced by faster decomposition of the organic dye [181]. More importantly, maintaining uniformity of material properties over a large area is one of most important concerns in the large-scale synthesis that needs to be addressed. Furthermore, combination of SP reactor with other techniques as an all-in-one synthesis system will be beneficial to the flexible design and fabrication of various functional materials in the future.

In summary, we have given the recent advancements in SP in the areas of materials synthesis from the past to the present and also outlined its future prospects in the years ahead. SP is a fertile field for brighter materials research society, in which we can find many exciting opportunities to explore a new class of materials and groundbreaking results may be expected. We believe that SP will possess great potential to further extend across the whole synthesis of materials in the near future. Lastly, we hope that this chapter will serve as one of the most valuable references and a useful guideline for researchers who are interested in SP or related fields.

References

1 Mott-Smith, H.M. (1971). *Nature* 233: 219.
2 Sankaran, R.M. (2012). *Plasma Processing of Nanomaterials*. USA: CRC Press; Taylor & Francis Group.
3 Charitidis, C.A., Georgiou, P., Koklioti, M.A. et al. (2014). *Manuf. Rev.* 1: 11.
4 Hossain, M.M. (2008). *Plasma Technology for Deposition and Surface Modification*. Berlin, Germany: Logos Verlag Berlin GmbH.
5 Ostrikov, K., Cvelbar, U., and Murphy, A.B. (2011). *J. Phys. Appl. Phys.* 44: 174001.
6 Sunka, P. (2001). *Phys. Plasmas* 8: 2587–2594.
7 Locke, B.R., Sato, M., Sunka, P. et al. (2006). *Ind. Eng. Chem. Res.* 45: 882–905.
8 Yang, Y., Cho, Y.I., and Fridman, A. (2012). *Plasma Discharge in Liquid: Water Treatment and Applications*. USA: CRC Press; Taylor & Francis Group.
9 Jiang, B., Zheng, J., Qiu, S. et al. (2014). *Chem. Eng. J.* 236: 348–368.
10 Gubkin, J. (1887). *Annalen der Physik (Leipzig)* 268: 114.
11 Baroch, P., Abnita, V., Saito, N., and Takai, O. (2008). *J. Electrostrat.* 66: 294–299.
12 Baroch, P. and Saito, N. (2008). *IEEE Trans. Plasma Sci.* 36: 1156–1157.
13 Takai, O. (2008). *Pure Appl. Chem.* 80: 2003–2011.
14 Saito, G. and Akiyama, T. (2015). *J. Nanomater.* 2015: 123696.
15 Potocky, S., Saito, N., and Takai, O. (2009). *Thin Solid Films* 518: 918–923.
16 Baroch, P., Potocky, S., and Saito, N. (2011). *Plasma Sources Sci. Technol.* 20: 034017.
17 Saito, N., Hieda, J., and Takai, O. (2009). *Thin Solid Films* 518: 912–917.
18 Miron, C., Bratescu, M.A., Saito, N., and Takai, O. (2011). *J. Appl. Phys.* 109: 123301.

19 Chen, Q., Li, J., and Li, Y. (2015). *J. Phys. D: Appl. Phys.* 48: 424005.
20 Takai, O. (2014). *J. Photopolym. Sci. Technol.* 27: 379–384.
21 Graham, W.G. and Stalder, K.R. (2011). *J. Phys. D: Appl. Phys.* 44: 174037.
22 Sunka, P., Babicky, V., Clupek, M. et al. (1999). *Plasma Sources Sci. Technol.* 8: 258–265.
23 Miron, C., Bratescu, M.A., Saito, N., and Takai, O. (2010). *Plasma Chem. Plasma Proc.* 30: 619–631.
24 Miron, C., Bratescu, M.A., Saito, N., and Takai, O. (2011). *Curr. Appl. Phys.* 11: S154–S158.
25 Hieda, J., Saito, N., and Takai, O. (2008). *J. Vac. Sci. Technol., A* 26: 854–856.
26 Hieda, J., Saito, N., and Takai, O. (2008). *Surf. Coat. Technol.* 202: 5343–5346.
27 Bratescu, M.A., Cho, S.P., Takai, O., and Saito, N. (2011). *J. Phys. Chem. C* 115: 24569–24576.
28 Heo, Y.K., Bratescu, M.A., Ueno, T., and Saito, N. (2014). *J. Appl. Phys.* 116: 024302.
29 Shirai, N., Uchida, S., and Tochikubo, F. (2014). *Jpn. J. Appl. Phys.* 53: 046202.
30 Ichino, Y., Mitamura, K., Saito, N., and Takai, O. (2009). *J. Vac. Sci. Technol. A* 27: 826–830.
31 Lee, Y.J., Kim, S.M., Kim, J.W., and Lee, S.Y. (2014). *Mater. Lett.* 123: 184–186.
32 Bera, D., Kuiry, S.C., McCutchen, M., and Seal, S. (2004). *J. Appl. Phys.* 96: 5152–5157.
33 Sudare, T., Ueno, T., Watthanaphanit, A., and Saito, N. (2015). *J. Phys. Chem. A* 119: 11668–11673.
34 Sudare, T., Ueno, T., Watthanaphanit, A., and Saito, N. (2015). *Phys. Chem. Chem. Phys.* 17: 30255–30259.
35 Cho, S.P., Bratescu, M.A., Saito, N., and Takai, O. (2011). *Nanotechnology* 22: 455701.
36 Ashkarran, A.A. (2012). *J. Theor. Appl. Phys.* 6: 14.
37 Kim, S.M., Kim, G.S., and Lee, S.Y. (2008). *Mater. Lett.* 62: 4354–4356.
38 Lee, H., Park, S.H., Jung, S.C. et al. (2013). *J. Mater. Res.* 28: 105–110.
39 Lung, J.K., Huang, J.C., Tien, D.C. et al. (2007). *J. Alloys Compd.* 434–435: 655–658.
40 Tseng, K.H., Liao, C.Y., Huang, J.C. et al. (2008). *Mater. Lett.* 62: 3341–3344.
41 Tseng, K.H., Huang, J.C., Liao, C.Y. et al. (2009). *J. Alloys Compd.* 472: 446–450.
42 Lo, C.H., Tsung, T.T., and Lin, H.M. (2007). *J. Alloys Compd.* 434–435: 659–662.
43 Hu, X., Cho, S.P., Takai, O., and Saito, N. (2012). *Cryst. Growth Des.* 12: 119–123.
44 Hu, X., Takai, O., and Saito, N. (2013). *J. Phys. Conf. Ser.* 417: 012030.
45 Mizutani, T., Murai, T., Nameki, H. et al. (2014). *Jpn. J. Appl. Phys.* 53, 11RA03.
46 Tien, D.C., Tseng, K.H., Liao, C.Y., and Tsung, T.T. (2009). *J. Alloys Compd.* 473: 298–302.

47 Ashkarran, A.A. (2010). *Curr. Appl. Phys.* 10: 1442–1447.
48 Hu, X., Takai, O., and Saito, N. (2013). *Jpn. J. Appl. Phys.* 52, 01AN05.
49 Yaowarat, W., Li, O.L., and Saito, N. (2016). *Jpn. J. Appl. Phys.* 55, 01AE23.
50 Kang, J., Li, O.L., and Saito, N. (2013). *Nanoscale* 5: 6874–6882.
51 Shi, J., Hu, X., Zhang, J. et al. (2014). *Prog. Nat. Sci. Mater. Int.* 24: 593–598.
52 Jo, Y.G., Kim, S.M., and Lee, S.Y. (2016). *Jpn. J. Appl. Phys.* 55, 01AE01.
53 Xie, S.Y., Ma, Z.J., Wang, C.F. et al. (2004). *J. Solid State Chem.* 177: 3743–3747.
54 Lo, C.H., Tsung, T.T., and Chen, L.C. (2005). *J. Cryst. Growth* 277: 636–642.
55 Pootawang, P., Saito, N., and Takai, O. (2013). *Nanotechnology* 24: 055604.
56 Lee, H., Park, S.H., Kim, S.J. et al. (2014). *Microelectron. Eng.* 126: 153–157.
57 Saito, G., Azman, W.O.S.B.W.M., Nakasugi, Y., and Akiyama, T. (2014). *Adv. Powder Technol.* 25: 1038–1042.
58 Saito, G., Zhu, C., and Akiyama, T. (2014). *Adv. Powder Technol.* 25: 728–732.
59 Kelgenbaeva, Z., Omurzak, E., Ihara, H. et al. (2015). *Phys. Status Solidi A* 12: 2951–2957.
60 Kelgenbaeva, Z., Omurzak, E., Takebe, S. et al. (2014). *J. Nanopart. Res.* 16: 2603.
61 Kim, H.G., Lee, H., Kim, B.-H. et al. (2013). *Jpn. J. Appl. Phys.* 52, 01AN03.
62 Saito, G., Hosokai, S., Akiyama, T. et al. (2010). *J. Phys. Soc. Jpn.* 79: 083501.
63 Saito, G., Hosokai, S., Tsubota, M., and Akiyama, T. (2011). *J. Appl. Phys.* 110: 023302.
64 Terashima, C., Iwai, Y., Cho, S.P. et al. (2013). *Int. J. Electrochem. Sci.* 8: 5407–5420.
65 Hu, X., Shen, X., Takai, O., and Saito, N. (2013). *J. Alloys Compd.* 552: 351–355.
66 Bratescu, M.A., Takai, O., and Saito, N. (2013). *J. Alloys Compd.* 562: 1072–1080.
67 Hu, X., Shi, J., Zhang, J. et al. (2015). *J. Alloys Compd.* 619: 452–457.
68 Kang, J. and Saito, N. (2015). *RSC Adv.* 5: 29131–29134.
69 Pootawang, P., Saito, N., Takai, O., and Lee, S.Y. (2012). *Nanotechnology* 23: 395602.
70 Kim, S.M., Lee, Y.J., Kim, J.W., and Lee, S.Y. (2014). *Thin Solid Films* 572: 260–265.
71 Kim, S.M., Jo, Y.G., and Lee, S.Y. (2015). *Electrochim. Acta* 174: 1244–1252.
72 Bratescu, M.A. and Saito, N. (2013). *J. Phys. Chem. C* 117: 26804–26810.
73 Saito, G., Nakasugi, Y., Yamashita, T., and Akiyama, T. (2014). *Nanotechnology* 25: 135603.
74 Ashkarran, A.A., Iraji Zad, A., Mahdavi, S.M., and Ahadian, M.M. (2009). *Mater. Chem. Phys.* 118: 6–8.
75 Ashkarran, A.A., Iraji Zad, A., Mahdavi, S.M., and Ahadian, M.M. (2010). *Appl. Phys. A* 100.
76 Ashkarran, A.A., Afshar, S.A.A., Aghigh, S.M., and Kavianipour, M. (2010). *Polyhedron* 29: 1370–1374.
77 Ashkarran, A.A., Kavianipour, M., Aghigh, S.M. et al. (2010). *J. Cluster Sci.* 21: 753–766.

78 Ashkarran, A.A. (2011). *J. Cluster Sci.* 22: 233–266.
79 Saito, G., Hosokai, S., and Akiyama, T. (2011). *Mater. Chem. Phys.* 130: 79–83.
80 Hu, X., Zhang, X., Shen, X. et al. (2014). *Plasma Chem. Plasma Process.* 34: 1129–1139.
81 Saito, G., Hosokai, S., Tsubota, M., and Akiyama, T. (2012). *Cryst. Growth Des.* 12: 2455–2459.
82 Bratescu, M.A., Saito, N., and Takai, O. (2011). *Curr. Appl. Phys.* 11: S30–S34.
83 Tong, D.G., Wu, P., Su, P.K. et al. (2012). *Mater. Lett.* 70: 94–97.
84 Fu, S.Y., Li, Y.Z., Chu, W. et al. (2015). *J. Mater. Chem. A* 3: 16716–16727.
85 He, Y.J., Peng, J.F., Chu, W. et al. (2014). *J. Mater. Chem. A* 2: 1721–1731.
86 Wu, P., Sun, J.H., Huang, Y.Y. et al. (2012). *Mater. Lett.* 82: 191–194.
87 Kim, H., Watthanaphanit, A., and Saito, N. (2016). *RSC Adv.* 6: 2826–2834.
88 Zhang, Z.K., Bai, M.L., Guo, D.Z. et al. (2011). *Chem. Commun.* 47: 8439–8441.
89 Nakasugi, Y., Saito, G., Yamashita, T., and Akiyama, T. (2014). *J. Appl. Phys.* 115: 123303.
90 Panomsuwan, G., Watthanaphanit, A., Ishizaki, T., and Saito, N. (2015). *Phys. Chem. Chem. Phys.* 17: 13794–13799.
91 Burakov, V.S., Nevar, A.A., Nedelko, M.I., and Tarasenko, N.V. (2008). *Tech. Phys. Lett.* 34: 679–681.
92 Tarasenko, N.V., Nevar, A.A., and Nedelko, M.I. (2010). *Phys. Status Solidi A* 207: 2319–2322.
93 Chen, L.C., Mashimo, T., Omurzak, E. et al. (2011). *J. Phys. Chem. C* 115: 9370–9375.
94 Kelgenbaeva, Z., Omurzak, E., Takebe, S. et al. (2013). *Jpn. J. Appl. Phys.* 52, 11NJ02.
95 Saito, G., Nakasugi, Y., Yamashita, T., and Akiyama, T. (2014). *Appl. Surf. Sci.* 290: 419–424.
96 Chen, Q., Kaneko, T., and Hatakeyama, R. (2012). *Appl. Phys. Exp.* 5: 096201.
97 Zhang, J., Hu, X., Shi, J. et al. (2016). *Jpn. J. Appl. Phys.* 55, 01AE17.
98 Chen, L., Iwamoto, C., Omurzak, E. et al. (2011). *RSC Adv.* 1: 1083–1088.
99 Chen, L., Mashimo, T., Iwamoto, C. et al. (2013). *Nanotechnology* 24: 045602.
100 Abdullaeva, Z., Omurzak, E., Iwamoto, C. et al. (2013). *RSC Adv.* 3: 513–519.
101 Tong, D.G., Zeng, X.L., Chu, W. et al. (2010). *Mater. Res. Bull.* 45: 442–447.
102 Tong, D.G., Chu, W., Wu, P., and Zhang, L. (2012). *RSC Adv.* 2: 2369–2376.
103 Yang, F., Li, Y.Z., Li, C., and Tong, D.G. (2014). *Catal. Sci. Technol.* 4: 3168.
104 Abdullaeva, Z., Omurzak, E., Iwamoto, C. et al. (2013). *Jpn. J. Appl. Phys.* 52, 01AJ01.
105 Abdullaeva, Z., Omurzak, E., Iwamoto, C. et al. (2012). *Carbon* 50: 1776–1785.
106 Yang, F., Fu, S.Y., Li, C., and Tong, D.G. (2014). *RSC Adv.* 4: 45838–45843.
107 Alexadrou, I., Sano, N., Burrows, A. et al. (2003). *Nanotechnology* 14: 913–917.

108 Omurzak, E., Shimokawa, W., Taniguchi, K. et al. (2011). *Jpn. J. Appl. Phys.* 50, 01AB09.
109 Sano, N., Wang, H., Chhowalla, M. et al. (2001). *Nature* 414: 506–507.
110 Sano, N., Wang, H., Alexadrou, I. et al. (2002). *J. Appl. Phys.* 92: 2783–2788.
111 Zhu, H.W., Li, X.S., Jiang, B. et al. (2002). *Chem. Phys. Lett.* 366: 664–669.
112 Lange, H., Sioda, M., Huczko, A. et al. (2003). *Adv. Mater.* 13: 1617.
113 Lee, H., Bratescu, M.A., Ueno, T., and Saito, N. (2014). *RSC Adv.* 4: 51758–51765.
114 Omurzak, E., Abdullaeva, Z., Iwamoto, C. et al. (2015). *J. Nanosci. Nanotechnol.* 15: 3703–3709.
115 Li, O.L., Kang, J., Urashima, K., and Saito, N. (2012). *Int. J. Plasma Environ. Sci. Technol.* 7: 31–36.
116 Kang, J., Li, O.L., and Saito, N. (2013). *J. Power Sources* 261: 156–161.
117 Kang, J., Li, O.L., and Saito, N. (2013). *Carbon* 60: 292–298.
118 Li, O.L., Kang, J., Urashima, K., and Saito, N. (2013). *J. Inst. Electrostat. Jpn.* 37: 22–27.
119 Yun, K.S., Kim, B.R., Kang, W.S. et al. (2013). *J. Nanosci. Nanotechnol.* 13: 7381–7385.
120 Yun, K.S., Kim, B.R., Kim, S.C. et al. (2014). *Res. Chem. Intermed.* 40: 2559–2564.
121 Yun, K.S., Kim, B.R., Noh, E. et al. (2014). *Sci. Adv. Mater.* 6: 1594–1598.
122 Yun, K.S., Kim, B.R., Noh, E. et al. (2014). *Int. J. Precis. Eng. Manuf.* 15: 1689–1693.
123 Lee, H., Ueno, T., and Saito, N. (2015). *J. Miner. Met. Mater. Soc.* 67: 2550–2556.
124 Kroushawi, F., Latifi, H., Hosseini, S.H. et al. (2012). *Plasma Chem. Plasma Process.* 32: 959–968.
125 Fisher, K.B. and Tahagard, S.M. (2012). *Plasma Chem. Plasma Process.* 32: 919–931.
126 Hagino, T., Kondo, H., Ishiwara, K. et al. (2012). *Appl. Phys. Exp.* 5: 035101.
127 Matsushima, M., Noda, M., Yoshida, T. et al. (2013). *J. Appl. Phys.* 113: 114304.
128 Hyun, K., Ueno, T., and Saito, N. (2016). *Jpn. J. Appl. Phys.* 55, 01AE18.
129 Li, O.L., Hayashi, H., Ishizaki, T., and Saito, N. (2016). *RSC Adv.* 6: 51864–51870.
130 Morishita, T., Ueno, T., Panomsuwan, G. et al. (2016). *Sci. Rep.* 6: 36880.
131 Panomsuwan, G., Chiba, S., Kaneko, Y. et al. (2014). *J. Mater. Chem. A* 2: 18677–18686.
132 Kim, D.W., Li, O.L., and Saito, N. (2014). *Phys. Chem. Chem. Phys.* 16: 14905–14911.
133 Panomsuwan, G., Saito, N., and Ishizaki, T. (2016). *ACS Appl. Mater. Interfaces* 8: 6962–6971.
134 Kim, D.W., Li, O.L., Pootawang, P., and Saito, N. (2014). *RSC Adv.* 4: 16813–16819.
135 Thongwichit, N., Li, O.L., Yaowarat, W. et al. (2016). *Jpn. J. Appl. Phys.* 55, 01AE10.

136 Kim, D.W., Li, O.L., and Saito, N. (2015). *Phys. Chem. Chem. Phys.* 17: 407–413.
137 Panomsuwan, G., Saito, N., and Ishizaki, T. (2016). *Carbon* 98: 411–420.
138 Panomsuwan, G., Saito, N., and Ishizaki, T. (2016). *Mater. Today Proc.* 2: 4302–4308.
139 Panomsuwan, G., Saito, N., and Ishizaki, T. (2015). *Phys. Chem. Chem. Phys.* 17: 6227–6323.
140 Hyun, K., Ueno, T., Li, O.L., and Saito, N. (2016). *RSC Adv.* 6: 6690–6996.
141 Hyun, K., Ueno, T., Panomsuwan, G. et al. (2016). *Phys. Chem. Chem. Phys.* 18: 10856–10863.
142 Ishizaki, T., Chiba, S., Kaneko, Y., and Panomsuwan, G. (2014). *J. Mater. Chem. A* 2: 10589–10598.
143 Panomsuwan, G., Saito, N., and Ishizaki, T. (2015). *J. Mater. Chem. A* 3: 9972–9981.
144 Panomsuwan, G., Saito, N., and Ishizaki, T. (2015). *Electrochem. Commun.* 59: 81–85.
145 Pootawang, P., Saito, N., and Takai, O. (2010). *Jpn. J. Appl. Phys.* 49: 12602.
146 Pootawang, P., Saito, N., and Takai, O. (2011). *Thin Solid Films* 519: 7030–7035.
147 Pootawang, P., Saito, N., and Takai, O. (2011). *Mater. Lett.* 65: 1037–1040.
148 Pootawang, P., Saito, N., and Takai, O. (2012). *Mater. Lett.* 80: 1–4.
149 Pootawang, P., Saito, N., and Takai, O. (2012). *Mater. Res. Bull.* 47: 2726–2729.
150 Pornsunthorntawee, O., Katepetch, C., Vanichvattanadecha, C. et al. (2014). *Carbohydr. Polym.* 102: 504–512.
151 Prasertsung, I., Damrongsakkul, S., and Saito, N. (2013). *Polym. Degrad. Stab.* 98: 2089–2093.
152 Prasertsung, I., Damrongsakkul, S., Tereshima, C. et al. (2012). *Carbohydr. Polym.* 87: 2745–2749.
153 Tantipapol, T., Singsawat, Y., Narongsil, N. et al. (2015). *Innovative Food Sci. Emerg. Technol.* 32: 116–120.
154 Watthanaphanit, A. and Saito, N. (2013). *Polym. Degrad. Stab.* 98: 1072–1080.
155 Matsuda, N., Nakashima, T., Kato, T., and Shiroishi, H. (2014). *Electrochim. Acta* 146: 73–78.
156 Guo, J., Wang, X., and Xu, B. (2009). *Mater. Chem. Phys.* 113: 179–182.
157 Lee, H., Park, S.H., Kim, S.-J. et al. (2014). *J. Nanomater.* 2014: 132032.
158 Lee, H., Park, S.H., Kim, S.-J. et al. (2015). *Int. J. Hydrogen Energy* 40: 754–759.
159 Lee, H., Kim, S.-J., Kim, J.-S. et al. (2016). *Adv. Mater. Lett.* 7: 98–103.
160 Mizutani, T., Murai, T., Nomoto, T. et al. (2014). *Surf. Interface Anal.* 46: 1125–1128.
161 Nakasugi, Y., Saito, G., Yamashita, T. et al. (2015). *Thin Solid Films* 583: 135–141.
162 Panomsuwan, G., Zettsu, N., and Saito, N. (2013). *MRS Proc.* 1509, mrsf12-1509-cc1509-1514.

163 Yamamoto, M., Yoshida, T., Yamamoto, N. et al. (2015). *Nucl. Instrum. Methods Phys. Res. B* 359: 64–68.
164 Shirafuji, T., Noguchi, Y., Yamamoto, T. et al. (2013). *Jpn. J. Appl. Phys.* 52: 125101.
165 Kim, S.C., Kim, J.W., Yoon, G.J., and Nam, S.W. (2013). *Curr. Appl. Phys.* 13: S48–S53.
166 Pootawang, P., Kim, S.C., Kim, J.W., and Lee, S.Y. (2013). *J. Nanosci. Nanotechnol.* 13: 589–592.
167 Kim, S.-C., Kim, S.M., Yoon, G.J. et al. (2014). *Curr. Appl. Phys.* 14: S172–S179.
168 Davoodbasha, M., Lee, S.Y., Kim, S.C., and Kim, J.W. (2015). *RSC Adv.* 5: 35052–35060.
169 Davoodbasha, M., Kim, S.C., Lee, S.Y., and Kim, J.S. (2016). *Arch. Biochem. Biophys.* 605: 49–58.
170 Nam, S.W., Davoodbasha, M., and Kim, J.W. (2016). *J. Nanomater.* 2016: 4712813.
171 Watthanaphanit, A., Heo, Y.K., and Saito, N. (2014). *J. Taiwan Inst. chem. Eng.* 45: 3099–3103.
172 Watthanaphanit, A., Panomsuwan, G., and Saito, N. (2014). *RSC Adv.* 4: 1622–1629.
173 Tong, D.G., Luo, Y.Y., Chu, W. et al. (2010). *Plasma Chem. Plasma Process.* 30: 897–905.
174 Tong, D.G., Chu, W., Wu, P. et al. (2013). *J. Mater. Chem. A* 1: 358.
175 Tsukeda, C., Mizutani, T., Ogawa, S. et al. (2016). *Surf. Interface Anal.* .
176 Banno, M., Kanno, K., Someya, Y., and Yui, H. (2015). *Jpn. J. Appl. Phys.* 54: 066101.
177 Banno, M., Kanno, K., and Yui, H. (2016). *RSC Adv.* 6: 16030–16036.
178 Bratescu, M.A., Hieda, J., Umemura, T. et al. (2011). *J. Vac. Sci. Technol. A* 29: 031302.
179 Bratescu, M.A. and Saito, N. (2016). *J. Instrum.* 11: C01009.
180 Shirafuji, T., Nomura, A., and Himeno, Y. (2014). *Plasma Chem. Plasma Process.* 34: 523–534.
181 Shirafuji, T. and Himeno, Y. (2013). *Jpn. J. Appl. Phys.* 52, 11NE03.
182 Shirafuji, T., Ueda, J., Nakamura, A. et al. (2013). *Jpn. J. Appl. Phys.* 52: 126202.

8

Global Reaction Route Mapping Strategy: A Tool for Finding New Chemistry in Computers

Satoshi Maeda, Yu Harabuchi, and Kenichiro Saita

Hokkaido University, Department of Chemistry, Faculty of Science, Kita 10, Nishi 8, Kita-ku, Sapporo, 060-0810, Japan

8.1 Introduction

Quantum chemical calculation has become a powerful tool in studies of reaction mechanisms [1–6]. The calculation can reveal structures of reaction intermediates (IMs) and transition states (TSs). The intrinsic reaction coordinate (IRC) is calculated starting from TSs as an approximate path of chemical reactions [7, 8]. These calculations provide deep insights into reaction mechanisms. Quantum chemical calculation has contributed to the design of compounds, catalysts, reaction conditions, and so forth, in the development of "molecular technology."

Structures of IMs and TSs have been explored by geometry optimization [9, 10]. Geometry optimization can give a single IM or TS starting from a guess structure, and thus is very useful to study reaction profiles for assumed mechanisms. When there are two or more plausible mechanisms, all IMs and TSs along all the assumed mechanisms are determined by geometry optimization, and the most probable mechanism is identified by comparing barriers at the rate-determining steps. This task, however, can be difficult if there are too many reaction steps along the assumed mechanism and/or if there are too many possible mechanisms.

To solve this problem, automated reaction path search methods combined with quantum chemical calculation have been developed [11–33]. We have developed automated reaction path search methods for practical applications under the title of global reaction route mapping (GRRM) strategy. In this chapter, we focus on the artificial force induced reaction (AFIR) method [34–38]. Despite its extremely simple idea, the AFIR method has been applicable to a variety of chemical events such as conformational rearrangement, hydrogen bond rearrangement, chemical bond rearrangement, coordination bond rearrangement, metal–metal bond rearrangement, nonadiabatic transition, and so forth. To date, the AFIR method is the one that has been applied to organic reactions and photoreactions most extensively [39–57]. In the next section, its theoretical details are described. Then, its applications to a variety of chemical reactions such as organic reaction, organometallic catalysis, metal cluster catalysis, and photoreaction are showcased with brief explanations.

Molecular Technology: Materials Innovation, Volume 3, First Edition.
Edited by Hisashi Yamamoto and Takashi Kato.

8.2 Methodology

8.2.1 Artificial Force Induced Reaction Method

In the AFIR method [34], various molecular deformations are induced by pushing fragments A and B together, or by pulling them apart, using the following AFIR function:

$$F^{\mathrm{AFIR}}(\mathbf{Q}) = E(\mathbf{Q}) + \rho\alpha \frac{\sum_{i\in A}\sum_{j\in B} \omega_{ij} r_{ij}}{\sum_{i\in A}\sum_{j\in B} \omega_{ij}} \tag{8.1}$$

In Eq. (8.1), $E(\mathbf{Q})$ is an adiabatic potential energy surface (PES) of geometrical parameters $\mathbf{Q}$. The second term applies the force and is given as a weighted sum of the interatomic distance r_{ij} between atoms i and j. The weight ω_{ij} is defined as

$$\omega_{ij} = \left[\frac{R_i + R_j}{r_{ij}}\right]^6 \tag{8.2}$$

In Eq. (8.2), R_i is the covalent radius for atom i. The constant parameter α in Eq. (8.1) defines the strength of the force, and ρ is either 1.0 or −1.0 to push fragments together or to pull them apart, respectively. The force parameter α is obtained using a model equation given as

$$\alpha = \frac{\gamma}{\left[2^{-\frac{1}{6}} - \left(1 + \sqrt{1 + \frac{\gamma}{\varepsilon}}\right)^{-\frac{1}{6}}\right] R_0} \tag{8.3}$$

In Eq. (8.3), R_0 and ε are parameters of the Lennard–Jones potential and set to the standard Ar–Ar values, i.e. 3.8164 Å and $\varepsilon = 1.0061\ \mathrm{kJ\,mol^{-1}}$, respectively. The model collision energy parameter γ is a user-defined parameter. The α value determined with this equation corresponds to the mean force that the Ar–Ar pair feels from the minimum point to the turning point on the Lennard–Jones potential in their direct collision. The γ value determines an approximate upper limit of the barrier height that the system can overcome by minimizing the AFIR function. In this study, minimization of the AFIR function has been made by the rational function optimization (RFO) method [58].

The path that the system follows during the minimization of the AFIR function is called the AFIR path. The AFIR path is expected to pass a low-energy region near the corresponding TS. It was shown that the maximum energy point along the AFIR path was close to the corresponding TS [35]. Approximate minima and TSs can therefore be obtained by scanning the energy variation along the AFIR path. Geometry optimization from the approximate TS may fail, especially when γ is much larger than the barrier for the corresponding elementary step. To reduce such cases, relaxation of the AFIR path has been made before optimization of TSs. There are several methods to relax a path [59–63], and any of these methods can be used for this purpose. In our recent implementation, the locally updated planes (LUP) [60] method is employed. In the LUP method, path points are evenly distributed along a given path, and moved to lower energy points in the hyperplane perpendicular to the path tangent. After the relaxation of an AFIR path, actual TSs are optimized starting from the maximum energy points on the relaxed AFIR path. IRC paths that start from all the TSs obtained are also calculated. There is an option that aims at only searching for minima (denoted by "minimum-only search") [38]. In the minimum-only search, path relaxation and TS optimization are skipped and energy minima on the AFIR path are further optimized.

The flow of the calculation of a single path is described as below:

1. Minimization of the AFIR function for a fragment pair A and B to obtain the AFIR path
2. Relaxation of the AFIR path by the LUP method (skipped in the minimum-only search)
3. Optimization of actual TSs starting from maximum energy points along the relaxed path (skipped in the minimum-only search)
4. Calculation of IRC paths starting from all obtained TSs (skipped in the minimum-only search)
5. Optimization of actual local minima starting from minimum energy points along the relaxed path (or starting from those along the AFIR path in the minimum-only search).

By calculating many paths with this five-step procedure, many actual local minima and TSs can be explored. Two algorithms that automatically obtain many paths are described in Sections 8.2.3 and 8.2.4. In addition, an extension to nonadiabatic paths is described in Section 8.2.4.

8.2.2 Multicomponent Algorithm (MC-AFIR)

The AFIR method has originally been developed to investigate the reaction path in which two or more molecules react together, i.e. reactions of type $A + B \rightarrow X\ (+Y)$ [34–36]. An algorithm developed for this purpose has been termed MC-AFIR. In MC-AFIR, fragments A and B in Eq. (8.1) are reactant molecules or a part of them. When a third reactant C exists, two additional terms to apply forces between A and C and between B and C are added.

MC-AFIR searches for many paths with the following procedures:

I. Initial structure generation by randomly defining the mutual orientation and position of the reactant molecules
II. Minimization of the AFIR function starting from the initial structure generated in Step I

By repeating Steps I and II, many AFIR paths are obtained automatically. This cycle is terminated when no new product is found in the last n AFIR paths. Then, Step 2–5 described in the last section, i.e. path relaxation by LUP, TS optimization, IRC calculation, and local minimum optimization, are applied successively to all (or selected) AFIR paths automatically, to obtain local minima, TSs, and IRC paths. In MC-AFIR, the AFIR path calculations can be executed in parallel.

8.2.3 Single-Component Algorithm (SC-AFIR)

An automated algorism that systematically generates fragment pairs in an entire system (molecule or complex) was introduced [37]. In SC-AFIR, AFIR paths for the systematically generated fragment pairs are calculated one after another, to obtain many AFIR paths and related local minima, TSs, and IRC paths automatically.

Fragments A and B are defined for two atoms k and l. At first, the atoms k and l are chosen from the entire system, and ρ in Eq. (8.1) is set to either 1 or -1. Then, the distance between the atoms k and l is either decreased or increased. The distance between k and l is decreased by x% when $\rho = 1$, while the distance is increased by $(x + 30)$% when $\rho = -1$, where $x = 20$ in our latest default setup. In this procedure, the atoms k and l are moved with the same amount along the line connecting k and l. The positions of other atoms are then optimized by a least squares fitting method so as to reproduce the distance matrix of the original structure as much as possible.

In this perturbed structure, fragments A and B are defined around the atoms k and l. The atoms bonded to the atom k and the atoms bonded to the atoms bonded to the atom k are selected as members of the fragment A, where atoms i and j are regarded to be bonded when the normalized distance $r_{ij}/(R_i + R_j)$ is smaller than 1.25. Similarly, the fragment B is defined around the atom l. Atoms of fragment B whose normalized distance to the atom k is less than the normalized distance between the atoms k and l are excluded from the fragment B. Some atoms in the fragment A are also excluded with this condition. Atoms i in the fragment A and j in the fragment B are further excluded if the normalized distance between them is shorter than the normalized distance between the atoms k and l.

Starting from the perturbed structure, the AFIR function with the fragments A and B is minimized to obtain the AFIR path. If the fragment pair is not reactive at a given γ, a behavior in which the structure returns to the original structure occurs. Therefore, the path is ignored if the last m successive minimization steps increased the value of the second term in Eq. (8.1), where m is set to 10 in default. Once the AFIR path reaches a local minimum on the AFIR function, then local minima, TSs, and IRCs are identified through Step 2–5 in Section 8.2.1.

By applying these procedures to various combinations of k, l, and ρ in a starting structure, many paths connected to the starting structure can be obtained. In default, if $\rho = 1$ and the normalized distance between the atoms k and l is longer than 8.0, or if $\rho = -1$ and the normalized distance is longer than 1.25, the corresponding combination of k, l, and ρ is omitted.

Each AFIR path is distinguished by four indices, i.e. the structure number m, atom k, atom l, and ρ. The structure number is given to all obtained local minima in the order of their discovery. In default, all the AFIR paths are calculated to find all accessible local minima, TSs, and IRC paths with a given γ, where each AFIR path is not calculated more than once. There are several options to reduce the total computational costs. If such an option is applied, AFIR paths associated with the lowest lying local minima, AFIR paths associated with local minima whose bond connectivity matrix is identical to the one of the initial structure, AFIR paths associated with local minima that have a specified bond connectivity pattern, and/or AFIR paths associated with the initial structure are only calculated. There is an option to limit the atoms that are selected as the atoms k and l for fragments generation. By using these limited-search options, SC-AFIR can be applied to large systems including 50–150 atoms with density functional theory (DFT) calculations. The calculations of different AFIR paths can be done in parallel.

8.2.4 Search for Potential Crossing Points

MC- and SC-AFIR have been also used in automated exploration of minimum energy seam of crossing (MESX) and minimum energy conical intersection (MECI) structures between two different adiabatic PESs. MESX corresponds to the minimum energy point in the geometrical hyperspace in which the two adiabatic PESs that have different spin multiplicity or different space symmetry degenerate. MECI is the minimum energy point in the hyperspace in which the two adiabatic PESs that have the same spin multiplicity and the same space symmetry degenerate. Nonadiabatic transition between two states occurs frequently in the geometrical hyperspace in which the two states degenerate. MESXs and MECIs have been calculated as energetically most favorable points within the degenerate hyperspace.

MECIs and MESXs can be explored systematically by applying an automated reaction path search method to the following model function termed seam model function (SMF) [64, 65].

$$F^{SMF}(\mathbf{Q}) = \frac{1}{2}(E^{\mathrm{X}}(\mathbf{Q}) + E^{\mathrm{Y}}(\mathbf{Q})) + \frac{(E^{\mathrm{X}}(\mathbf{Q}) - E^{\mathrm{Y}}(\mathbf{Q}))^2}{\beta} \tag{8.4}$$

In Eq. (8.4), E^{X} and E^{Y} are the adiabatic PESs of the two target electronic states X and Y, respectively, and β is a constant parameter, which is usually set to the standard value $\beta = 30\,\mathrm{kJ\,mol^{-1}}$ [20, 65]. At local minima on the SMF, both the mean energy of the two states and their energy gap are small. Thus, local minima on the SMF can be good guesses for further optimizations of actual MESX or MECI geometries. Any automated reaction path search method can be combined with the SMF method. In this study, the MC- and SC-AFIR methods were used to explore local minima on the model function.

Such guess MESX/MECI geometry obtained as a local minimum on the model function is further optimized to the actual MECI or MESX geometry, by a standard geometry optimization method [66–74]. The gradient projection (GP) method, which is one of MECI/MESX optimization methods [68, 73], has been employed in this study. In the GP method, a composed gradient vector $\mathbf{g}^{\mathrm{GP}}$ given in Eq. (8.5) is used to optimize an MECI or MESX geometry between the two states X and Y.

$$\mathbf{g}^{\mathrm{GP}}(\mathbf{Q}) = 2(E^{\mathrm{X}}(\mathbf{Q}) - E^{\mathrm{Y}}(\mathbf{Q}))\mathbf{v}^{\mathrm{DGV}} + \mathbf{P}\frac{1}{2}\left(\frac{\mathrm{d}E^{\mathrm{X}}(\mathbf{Q})}{\mathrm{d}\mathbf{Q}} + \frac{\mathrm{d}E^{\mathrm{Y}}(\mathbf{Q})}{\mathrm{d}\mathbf{Q}}\right) \tag{8.5}$$

The first term in Eq. (8.5) corresponds to the gradient of the energy gap squared between the two states, where $\mathbf{v}^{\mathrm{DGV}}$ is the difference gradient vector (DGV). The projection matrix $\mathbf{P}$ used in MESX optimization is defined as below:

$$\mathbf{P} = \mathbf{1} - \mathbf{s}\mathbf{s}^{\mathrm{T}} \tag{8.6}$$

P in this equation eliminates the projection of the vector **s** from the mean energy gradient, where **s** is the unit vector along $\mathbf{v}^{DGV}$, i.e. $\mathbf{s} = \mathbf{v}^{DGV}/|\mathbf{v}^{DGV}|$. On the other hand, the projection matrix **P** for MECI optimization is defined as below:

$$\mathbf{P} = \mathbf{1} - \mathbf{s}\mathbf{s}^{T} - \mathbf{t}\mathbf{t}^{T} \tag{8.7}$$

P in this equation eliminates the projection of the plane formed by **s** and **t**, called the branching plane (BP). BP is the plane between the DGV and the derivative coupling vector (DCV). DGV can be easily obtained from gradient vectors of the two adiabatic PESs. To obtain the DCV, some additional calculations are required. The BP update (BPU) method was therefore proposed to avoid the calculation of DCV [74]. In the BPU, **t** is estimated as the unit vector perpendicular to **s** on the BP, using DGVs calculated in the previous optimization steps. The GP method combined with the BPU can optimize an MECI geometry without the calculation of DCV.

The automated search of MESX/MECI geometries is done in the following two steps. First, possible local minima on the SMF are explored by any automated local minima search method, where MC- or SC-AFIR method is employed in this study. Second, actual MESX or MECI geometries are optimized starting from local minima on the SMF by any standard MESX/MECI optimization method, where combined GP and BPU methods are applied in this study. This approach is called GP/SMF/AFIR approach.

Very recently an alternative approach termed GP/AFIR was proposed [57]. In the GP/AFIR approach, the AFIR functions of the two target states are substituted to E^{X} and E^{Y} of Eq. (8.5), respectively, and local minima on Eq. (8.5) are searched directly by MC- or SC-AFIR. The computational costs of GP/AFIR and GP/SMF/AFIR approaches were compared in the search for MECIs between the lowest two singlet states of benzene and naphthalene, and GP/AFIR was found to be more efficient. The results shown below are obtained by the GP/SMF/AFIR approach.

8.3 Results and Discussion

8.3.1 Aldol Reaction

As a case study, MC-AFIR was applied to a simple bimolecular reaction, i.e. aldol reaction between formaldehyde and vinyl alcohol [38]. In the search, both cis- and trans- isomers of vinyl alcohol were considered, where one of them was randomly chosen in the initial random structure generation (Step I of the MC-AFIR algorithm). The MC-AFIR was terminated when no new product was obtained in the last 100 AFIR paths. The search was done at the spin-restricted B3LYP/6-31G level. Two different searches were conducted with different γ values.

In the first test, the initial γ was randomly given between 106.9 and 1069 kJ mol^{-1}, and the γ value was gradually increased up to 1069 kJ mol^{-1}. As a result, 30 independent IRC paths were obtained. Figure 8.1 shows the best 10 IRC paths in the ascending order of relative Gibbs free energy value of the TS. These energy values are relative to the sum of energies of separately

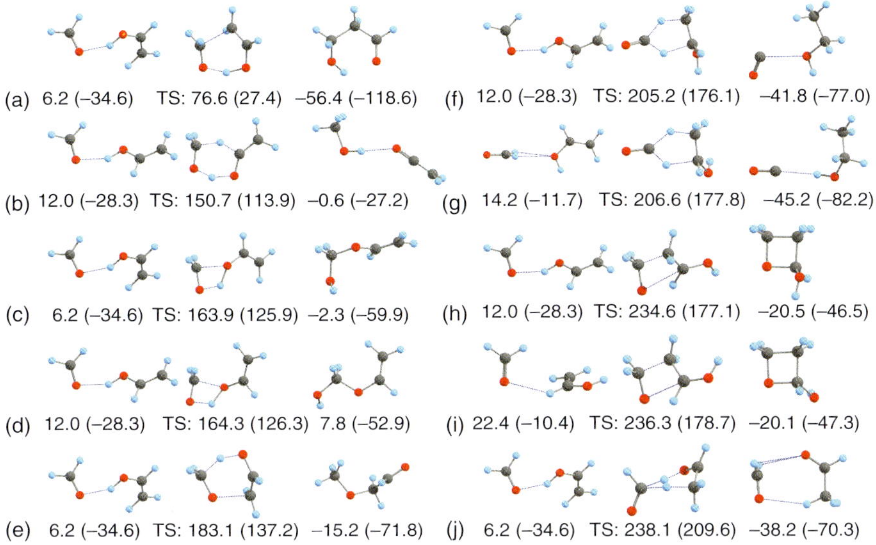

Figure 8.1 Reaction pathways obtained by MC-AFIR for the aldol reaction of formaldehyde and vinyl alcohol, in ascending order of the Gibbs free energy barrier ($T = 298.15$ K). Gibbs free energy and electronic energy in parentheses relative to separately optimized reactants are shown in kJ mol^{-1} [38].

optimized formaldehyde and cis-isomer of vinyl alcohol. The lowest barrier path corresponds to the path of the aldol reaction. Many high-energy pathways that generate radical pairs such as HCO + CH_3—CH—OH without TS were also found. In this search, 270 287 gradient and 11 919 Hessian calculations were needed. Among many paths obtained in this search, only the aldol path is kinetically accessible in mild experimental conditions.

In the second test, γ was set to 106.9 kJ mol^{-1}. This search found only one path, which was the path of aldol reaction. This test calculation was done 10 times with different random number seeds. The path of aldol reaction was found in all the 10 test calculations. In these 10 test calculations, 6716 times gradient calculations and 233 times Hessian calculations were required on average before completion of the search. This result demonstrates that the computational cost can be dramatically reduced by choosing a reasonable γ value when only low barrier paths are required. In a study on a reaction that is expected to have a barrier of less than δ kJ mol^{-1}, we usually set the γ value to 1.2δ–1.5δ.

8.3.2 Passerini Reaction

Multicomponent reactions, in which three or more molecules react in one pot, have been developed extensively to achieve highly atom-, step-, and energy-economic organic synthesis [75]. The Passerini reaction, which formally is a three-component reaction, gives an α-(acyloxy)-carboxamide from a carboxylic acid, an aldehyde (or ketone), and an isocyanide [76]. The MC-AFIR method was applied to the simplest case of the Passerini reaction with HCOOH, HCHO, and

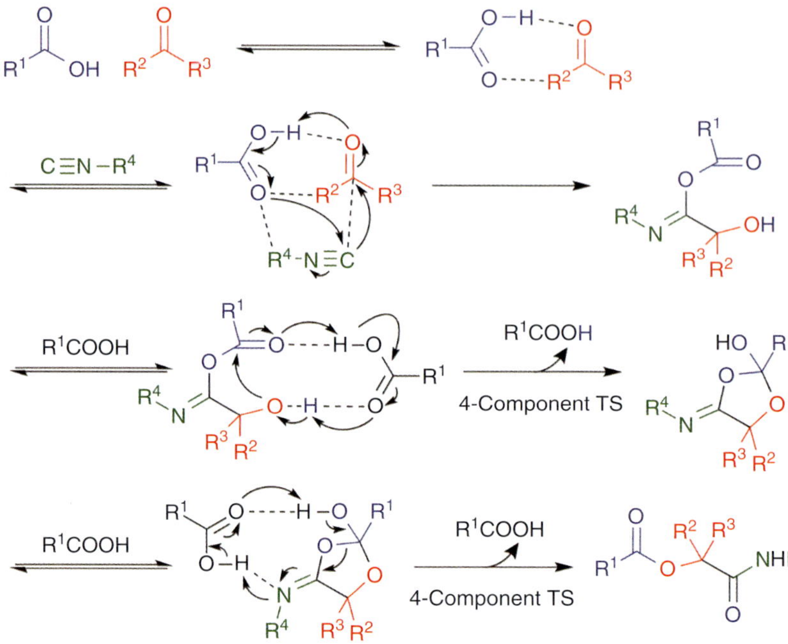

Figure 8.2 Detailed mechanism of the Passerini reaction proposed by the MC-AFIR method [39].

CH_3NC molecules [39]. In the MC-AFIR search at the M06/6-31+G(d,p) level of theory with $\gamma = 150\,\text{kJ}\,\text{mol}^{-1}$, two-component reactions for three reactant pairs, i.e. HCOOH–HCHO, HCOOH–CH_3NC, and HCHO–CH_3NC, and also a three-component reaction among HCOOH, HCHO, and CH_3NC, were considered. All of the reaction paths located for the two- and three-component reactions have relatively high-energy TSs. When an extra HCOOH molecule was introduced, a pathway with a low-energy four-component TS was obtained. At the TS, the fourth component (HCOOH) acts as both a proton donor and a proton acceptor. Figure 8.2 summarizes the mechanisms of the reaction from this MC-AFIR study. It is important to note that the extra HCOOH molecule in the system is involved in proton exchange in two steps. In summary, the Passerini reaction is a four-component reaction in which an extra HCOOH is a catalyst.

8.3.3 Claisen Rearrangement

As a case study, the SC-AFIR method was applied to Claisen rearrangement of allyl-vinyl-ether (Scheme 8.1) [38]. The search was initiated from a single conformer of allyl-vinyl-ether. The SC-AFIR method was applied to all conformers of allyl-vinyl-ether obtained during the search. In other words, the SC-AFIR was not applied to isomers that had different bond connectivity from allyl-vinyl-ether. This reaction occurs at high temperature, ~470.15 K [77]. Thus, a relatively large $\gamma = 171.5\,\text{kJ}\,\text{mol}^{-1}$ was employed. The search was done at the spin-restricted B3LYP/6-31G level.

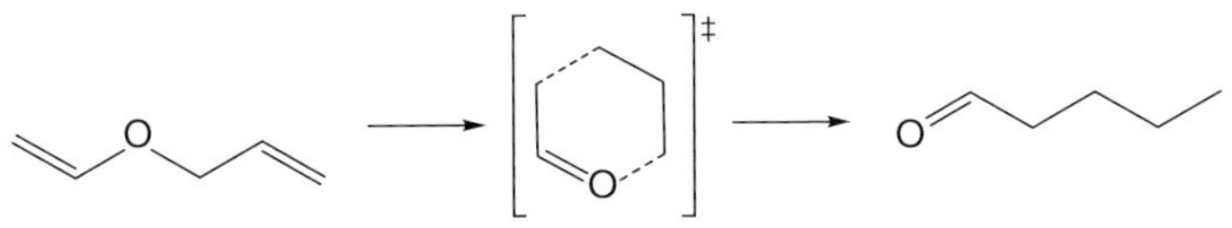

Scheme 8.1 Claisen rearrangement of allyl-vinyl-ether.

All the local minima and TSs obtained are listed in Figure 8.3 with both relative Gibbs free energy and electronic energy in parentheses. TS$x/y$ is a TS along the IRC path connecting x and y. There are two bond rearranging TSs, i.e. TS2/11 and TS6/14, and these two correspond to the chair- and boat-type TSs of Claisen rearrangement, respectively [78]. It should be emphasized that no kinetically unimportant product was found in the search with $\gamma = 171.5\,\text{kJ mol}^{-1}$. In addition, many conformational rearrangement paths of allyl-vinyl-ether were found. The conformational rearrangement path network was found to be important in the accurate prediction of the overall rate constant, taking account of the conformational entropy in the reactant's basin [79, 80].

8.3.4 Co-catalyzed Hydroformylation

The full catalytic cycle of a simple organometallic catalysis, hydroformylation [81], was examined by the combined use of the MC- and SC-AFIR methods [38]. The mechanism of this reaction has been studied well both experimentally and theoretically [82–86]. There are three reactants, i.e. H_2, CO, and ethylene (C_2H_4), and a cobalt hydride carbonyl complex $HCo(CO)_3$ **1** in Figure 8.4 as the catalyst. In the calculation, it was assumed that each one molecule was involved in the reaction. The search was done in a step-by-step manner, by manually choosing the (kinetically and/or thermodynamically) best intermediate among those obtained by MC- or SC-AFIR calculations, where γ was set to $153.0\,\text{kJ mol}^{-1}$, throughout. In SC-AFIR, the search was not applied to local minima that have the different bond connectivity from the initial structure in each calculation. These calculations were done at the B3LYP/6-31G level.

Figure 8.4 summarizes the best paths obtained in each MC- or SC-AFIR calculation. At first, the coordination pathways of H_2, CO, or C_2H_4 to $HCo(CO)_3$ were searched by three different MC-AFIR calculations. All the obtained coordination paths were barrierless, and their kinetic importance was equivalent. Thermodynamically, **2** was most preferable. An application of SC-AFIR to **2** did not generate any important path. Bimolecular paths were further searched by two different MC-AFIR calculations between **2** and H_2 or C_2H_4, although these calculations also did not give any preferable path. Thus, **2** was found to be just a resting state. Then, SC-AFIR was applied to the second most preferable coordination complex **3**. This calculation gave a preferable intermediate **5**. Further application of SC-AFIR to **5** did not give any important intermediate. Therefore, paths between **5** and H_2 or CO were searched by two different MC-AFIR calculations. Among the paths obtained, the one giving **6** was found to be kinetically most preferable. Although SC-AFIR of **6** gave **8** as a kinetically accessible intermediate, **8** was somewhat less stable than **6**. Thus, paths between

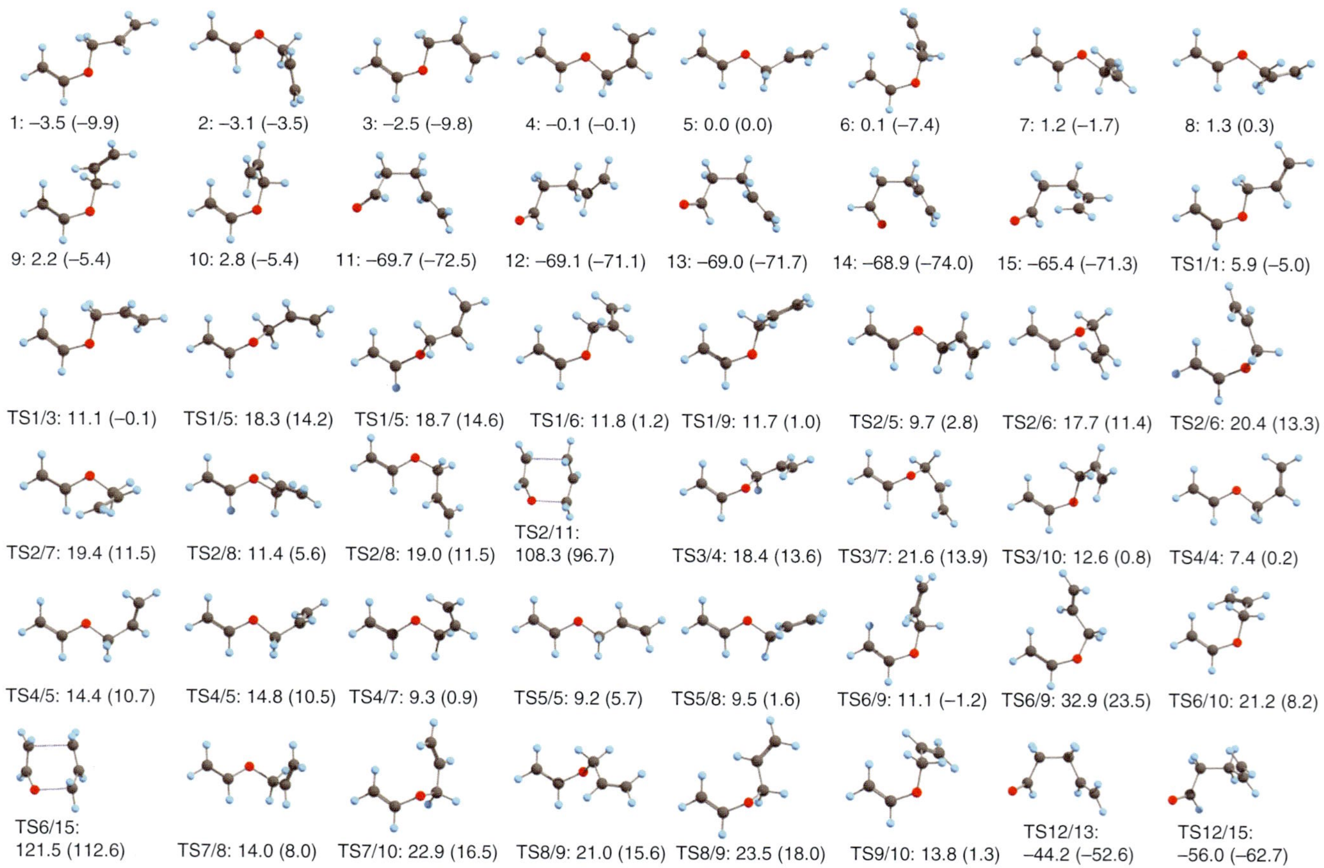

Figure 8.3 Local minima and TSs obtained by the SC-AFIR method starting from a single conformer of allyl-vinyl-ether **5**. TS$x/y$ is a TS connecting x and y. The Gibbs free energy ($T = 470.15$ K) and the electronic energy in parentheses, relative to **5,** are shown in kJ mol^{-1} [38].

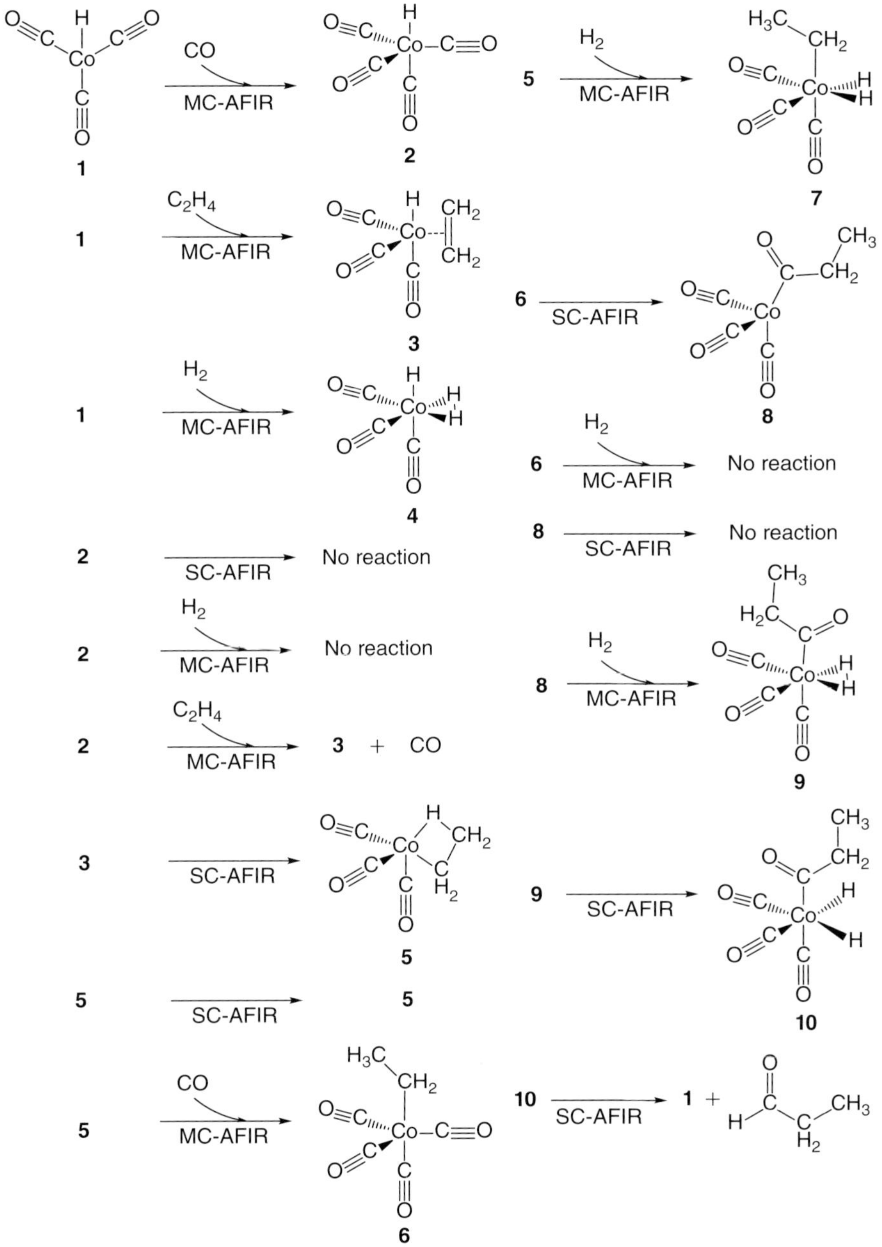

Figure 8.4 Best paths obtained in each MC- or SC-AFIR calculation. "No reaction" indicates the case where any kinetically accessible path was not obtained [38].

6 and H_2 were searched by MC-AFIR, although this calculation did not give any preferable intermediate. SC-AFIR was then applied to the second most preferable intermediate **8**, although this calculation did not give any important intermediate. Therefore, paths between **8** and H_2 were further searched by MC-AFIR, and a H_2-coordinated complex **9** was obtained. SC-AFIR of **9** gave **10** as the most preferable intermediate, and further application of SC-AFIR to **10** finally afforded propionaldehyde and reproduced **1**. In this example, the full catalytic cycle was obtained systematically without using any guess on the mechanism.

8.3.5 Lanthanide-Ion-Catalyzed Mukaiyama Aldol Reaction

Theoretical prediction of stereoselectivity is one of the most important topics. The AFIR method was applied to the Mukaiyama aldol reaction catalyzed by lanthanide Lewis acid [87, 88], using the model shown in Scheme 8.2 [41]. This reaction gives *syn-* and *anti-* products. The *syn*/*anti* ratio can be discussed by comparing the energies of *syn*-TS and *anti*-TS. However, in the case of large and flexible molecules, there could be a large number of TSs because of structural fluctuation. The MC-AFIR search with $\gamma = 50\,\text{kcal mol}^{-1}$ at the B3LYP/RECP,6-31G level and subsequent geometry re-optimization at the B3LYP-D3/RECP,6-31+G(d) level gave unique 165 TSs (91 *syn-* and 74 *anti*-TSs) for this C—C bond formation step. Most TSs are able to be categorized into six groups (Figure 8.5), A, B, and C for *syn*-structures and D, E, and F for *anti*-structures based on the C–C–C–O dihedral angle ϕ, since they are localized only around 60°, 180°, and 300°, respectively. And, in the same group, energy distribution is caused by the structure fluctuation of water molecules around Eu^{3+}. The *syn*/*anti* ratio was calculated as 75/25 by taking Boltzmann distribution of all the TSs obtained, whereas the calculated ratio considering only the most stable *syn-* and *anti*-TSs was 66/34. The former reproduces the experimental ratio of 73/27 [88] quantitatively. Because the existing probabilities of TSs whose relative energies are more than $2\,\text{kcal mol}^{-1}$ are less than 1%, the *syn*/*anti* ratio could be determined by the lowest 11 *syn-* and 6 *anti*-TSs, at room temperature.

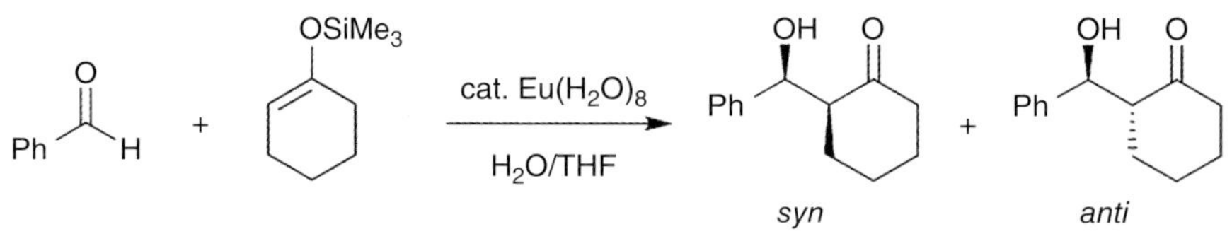

Scheme 8.2 Lanthanide catalyzed Mukaiyama aldol reaction.

8.3.6 Base-Mediated Borylation with a Silylborane

Organoboronate esters are important building blocks in organic synthesis. Recently, Ito and coworkers reported the BBS (base-mediated borylation with a silylborane) method [89], in which an arylboronic acid pinacol ester is

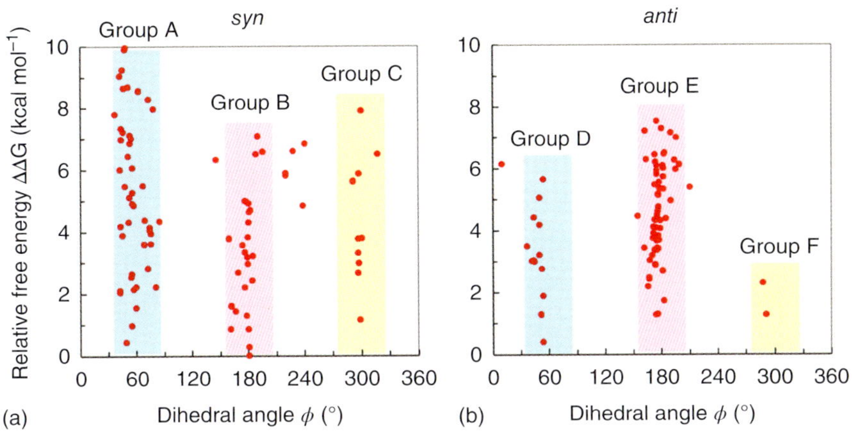

Figure 8.5 The relative Gibbs free energy $\Delta\Delta G$ (in kcal mol^{-1}) and dihedral angle ϕ (in degrees) of TSs for C—C bond-forming step. TSs producing *syn*- and *anti*-products are shown in (a) and (b), respectively. Those included in blue, pink, and yellow bars are categorized as groups A, B, and C, for *syn*-, and D, E, and F for *anti*-products, respectively. The Gibbs free energy barrier of the lowest *syn*-TS is 5.5 kcal mol^{-1} which is used as a reference for the relative free energies of other TSs [41].

efficiently produced from organic halides, a silylborane, and alkoxy bases, under transition-metal-free conditions. Their discovery was completely unexpected because silylboranes usually react as silylating agents. Therefore, systematic exploration using the MC-AFIR method was performed, without any estimated reaction paths [45].

In this study, a (2-bromoethyl)benzene, a (dimethylphenylsilyl)boronic acid pinacol ester [$PhMe_2Si$–B(pin)], and a potassium methoxide were used as reactants. The AFIR ($\gamma = 200$ kJ mol^{-1}) paths obtained by the B3LYP level calculation with small basis sets (STO-3G for Me and Ph groups, and 6-31G for the others) were subsequently optimized using the LUP method at the M06L/6-311+G(d) level, and important pathways were further optimized at the M06L/6-311+G(2d,p) level.

A silyl anion $[PhMe_2Si]^-$, which is generated due to heterolytic cleavage of the Si—B bond of $[PhMe_2Si\text{-}B(pin)OMe]^-K^+$, attacks the Br atom of $PhCH_2CH_2Br$ so that the negative charge transfers from the silyl anion to the phenethyl group. Then, a highly reactive carbanion species ($PhCH_2CH_2^-$) is produced. This carbanion intermediate undergoes borylation through a TS ($\Delta\Delta G = 5.2$ kJ mol^{-1}), giving a precursor of the borylation product.

Figure 8.6 shows the whole reaction profile. As described above, CP4A exclusively gives the borylation product through a tiny barrier. On the other hand, CP4A has a conformer CP4B, and CP4B undergoes both borylation through TS(CP4B/CP5) and silylation via TS(CP4B/CP_{Si}). Using these barriers, the borylation/silylation ratio (B/Si) can be estimated. The calculated B/Si ratio is 97.2/2.8, which qualitatively reproduces the observed B/Si = 91/9 [89]. These results show that the AFIR method is successful in construction of a reasonable reaction profile starting from no estimation.

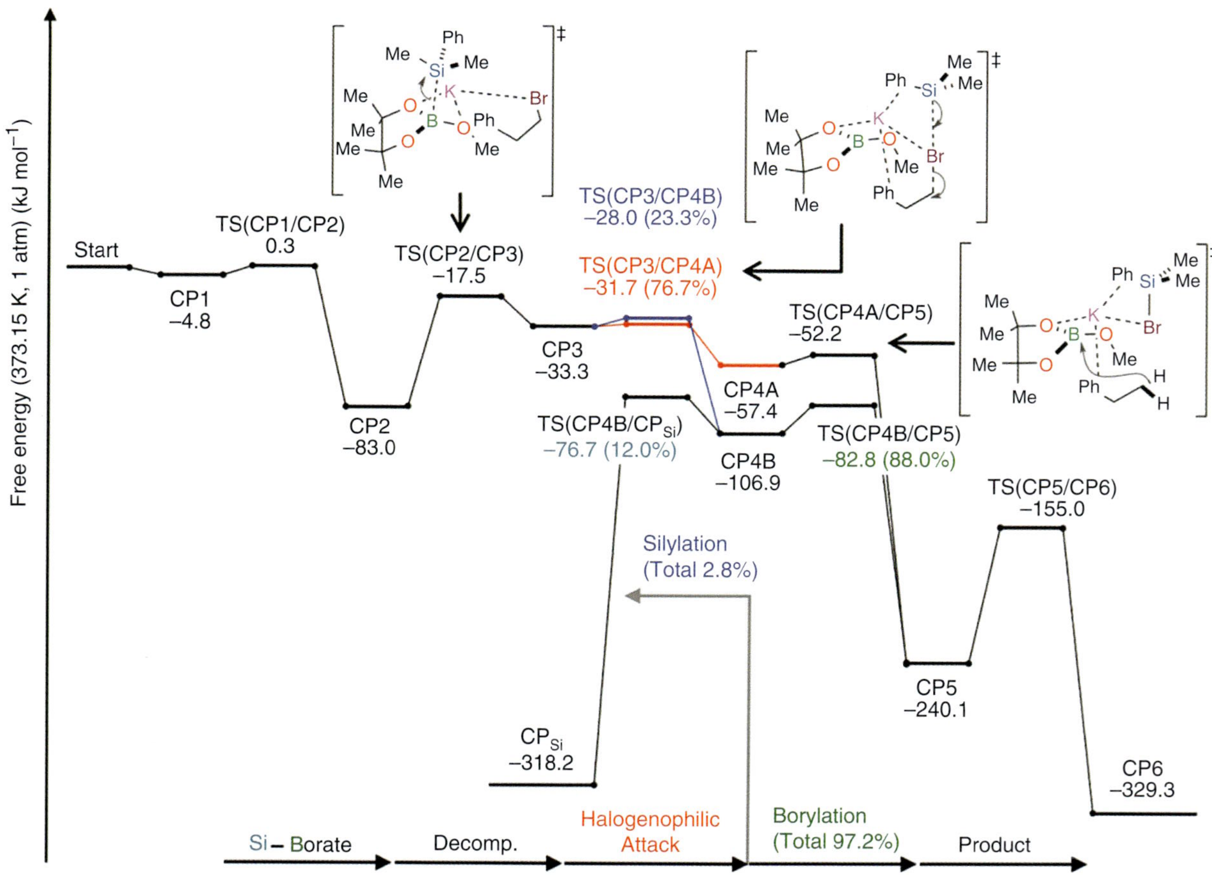

Figure 8.6 The reaction profile for base-mediated borylation with a silylborane [45].

8.3.7 Search for Cluster Structures by AFIR

The AFIR method can be also used for sampling of cluster structures. As a case study of the stochastic SC-AFIR algorithm, a search for the structure of Buckminsterfullerene C_{60} was performed [38]. Starting from a random C_{60} vapor, the SC-AFIR search with $\gamma = 1238.9\,\text{kJ mol}^{-1}$ found a C_{60} fullerene (Figure 8.7), at the self-consistent-charge density functional tight-binding (SCC-DFTB [DFTBA in Gaussian 09]) level using the minimum-only algorithm. In addition, about 1300 independent local minima were obtained during the searches. This result demonstrates that SC-AFIR can be used for the prediction of unknown structures.

Another application to a small gold cluster, which would catalyze the H–H activation, was performed at the PBE/DZP level of theory using the SIESTA program package [90]. By an MC-AFIR search with $\gamma = 100\,\text{kJ mol}^{-1}$, 20 TS structures for H_2 dissociation on Au_8 cluster were automatically found. The 20 IRC paths obtained are shown in Figure 8.8. The i-th lowest-energy Au_8 isomer (without H_2) is labeled as $C_{8,i}$, and $A_{8,i,j}$, $TS_{8,i,j}$, and $IM_{8,i,j}$ mean the j-th lowest adsorption state, transition state, and intermediate state between $C_{8,i}$ and H_2, respectively. $P_{8,i,j}$ stands for the IRC path between $A_{8,i,j}$ and $IM_{8,i,j}$. While the most stable Au_8 isomer and the H_2 adsorption state are $C_{8,0}$ and $A_{8,0,0}$, the most low-energy TS

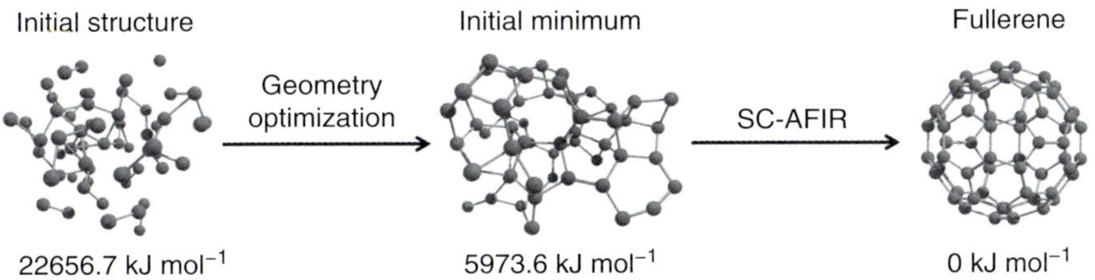

Figure 8.7 An initial random structure, the initial local minimum structure obtained by optimization of the random structure, and the final structure seen in a SC-AFIR search [38].

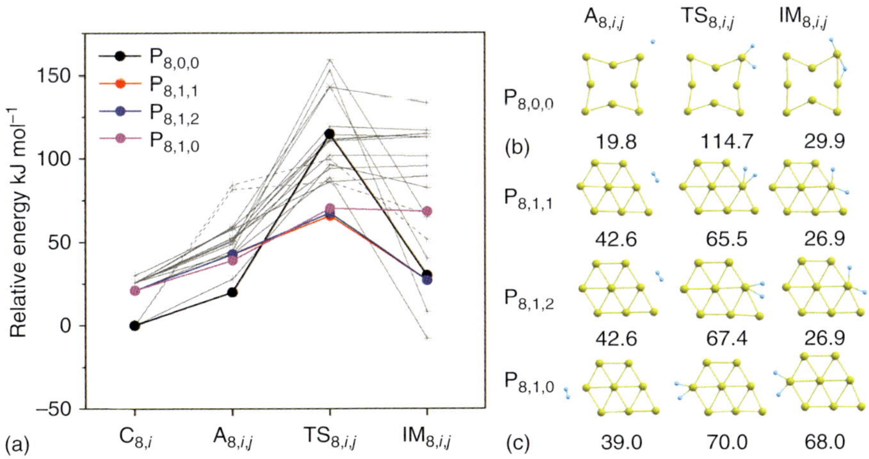

Figure 8.8 Free energy profiles (298.15 K, 1 atm) along 20 IRC paths for the H_2 dissociation step on the Au_8 clusters. Free energy values are calculated relatively to $C_{8,0} + H_2$ [90].

is $TS_{8,1,1}$. A number of theoretical investigations have been reported; however, most of those studies have considered only the most stable cluster structures as an effective catalyst; it was thought in conventional discussions that an Au_8 cluster does not catalyze the H–H activation well since $TS_{8,0,0}$ is located at high energy (114.7 kJ mol^{-1}). This AFIR study demonstrates that the most stable structures of the gold clusters are not always highly reactive and several isomeric structures must be taken into account for adequate description of the reaction rates at finite temperatures.

8.3.8 The Paternò–Büchi Reaction

The Paternò–Büchi (PB) reaction is a photochemical [2+2] cycloaddition that forms four-membered oxetane-rings from a carbonyl and an alkene [91, 92]. Its mechanism has been extensively studied both experimentally and theoretically [93–101], and it is known that the singlet first excited (S_1) and lowest triplet (T_1) electronic states are involved as well as the singlet ground (S_0) state. Thus, the PB reaction is ideal for benchmarking the applicability of the GP/SMF/MC-AFIR approach.

In this study [49], the S_0, S_1, and T_1 states for the simplest $H_2CO + C_2H_4$ system were considered. The initial SMF/MC-AFIR (γ = 100 kJ mol^{-1}) search was performed at the CAS(6,5)-PT2/6-31G level (CAS, complete active space), and the structures obtained were re-optimized by the GP method with the CAS(6,5)-PT2/aug-cc-pVDZ level of theory.

The reaction pathways found by the AFIR search are summarized in Figure 8.9. At first, a H_2CO is excited to the S_1 (n–π*) state and generates a •C–O• biradical.

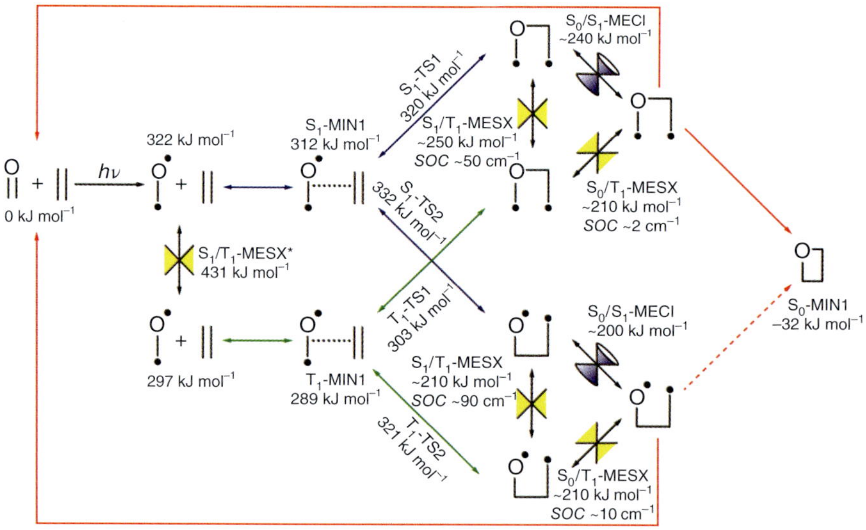

Figure 8.9 An overview of reaction pathways for the Paternò–Büchi reaction obtained by the automated reaction path search [49]. Energy values by CAS(6,5)-PT2/aug-cc-pVDZ level are shown in each structure label. Spin–orbit coupling (SOC) values are also presented for MESX structures. For simplicity, energies and SOCs are averaged over similar MESX and MECI structures. S_1/T_1-MESX* is the lowest MESX between S_1 and T_1 PESs for isolated H_2CO taken from Ref. 20.

It may either stay on the S_1, or undergo intersystem crossing (ISC) to the T_1 state. Then, the •C–O• biradical reacts with a C_2H_4, and two types of biradical intermediates, •C–O–C–C• type and •O–C–C–C• type, are produced. The •C–O–C–C• type biradical passes through the S_0/S_1-MECI between the S_0 and S_1, or the S_0/T_1-MESX between the S_0 and T_1 states, to the oxetane product without barriers in the S_0 state. On the other hand, there is a small barrier (5 kJ mol^{-1}) on the pathway to the product from the •O–C–C–C• type biradical in the S_0 state, after the IC or ISC. Therefore, the final fate of the •O–C–C–C• biradical in the S_0 should be controlled by the stability of the barrier for the final ring closure with the O—C bond formation.

Figure 8.9 is a comprehensive view of the PB reaction. The pathways have been reported in previous studies; however, they have been located in different studies and discussed separately. It should be noteworthy that the AFIR search found all previously suggested pathways without any guesses. In addition, the comprehensive view gave new insights into the mechanism; the ISC pathways between the T_1 and S_1 excited biradicals were found to play crucial roles in conventional triplet and singlet routes.

8.3.9 Minimum Energy Conical Intersection Structures of 1,3-Butadiene and Benzene

As a case study, S_0/S_1-MECI structures of 1,3-butadiene and benzene were explored. Although the GP/SMF/AFIR approach can be combined with any electronic structure calculation method, either the time-dependent density functional theory (TDDFT) or the spin-flip-time-dependent density functional theory (SF-TDDFT) has been used most frequently in our studies.

The GP/SMF/SC-AFIR approach with SF-TDDFT was applied to *cis*-butadiene, *trans*-butadiene, and benzene. In these applications, only AFIR paths starting from a given initial structure were calculated to find MECIs near the Franck–Condon (FC) structure. The searches located all low-lying MECIs previously reported for 1,3-butadiene [72, 73, 102–117] and benzene [118, 119]. The MECIs obtained for 1,3-butadiene and benzene were further optimized at the CAS(4,4)PT2/cc-pVDZ and CAS(6,6)PT2/cc-pVDZ levels, respectively. In Figure 8.10, MECI structures of the SF-TDDFT method (black) are compared with those by the CASPT2 method (red). The geometries of the SF-TDDFT method well reproduced those by the CASPT2 method.

8.3.10 Application of SMF/SC-AFIR for Medium-Sized Molecules

The GP/SMF/SC-AFIR search for S_0/S_1-MECI structures has been applied to six medium-sized organic molecules, i.e. thymine, coumarin, naphthalene, phenanthrene, anthracene, and pyrene, and 12, 18, 12, 26, 14, and 17 S_0/S_1-MECIs were found, respectively [50, 51]. In these applications, only AFIR paths starting from the FC structure were calculated. In Figure 8.11, the five lowest S_0/S_1-MECI geometries among those easily accessible from the FC structure are depicted for each molecule. There were several studies that reported S_0/S_1-MECI geometries for thymine [120–130] and coumarin [131]. All of the reported geometries were obtained by the automated search, and many unknown S_0/S_1-MECI geometries were also discovered.

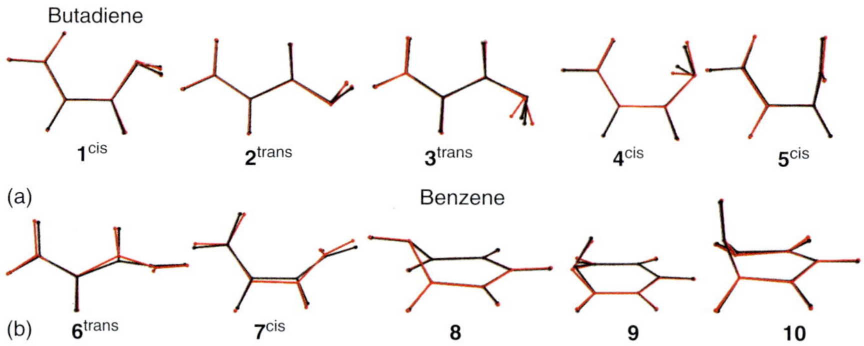

Figure 8.10 S_0/S_1-MECI structures of 1,3-butadiene and benzene obtained by the GP/SMF/AFIR method starting from the FC geometry. The lowest seven and the lowest three S_0/S_1-MECIs among those accessible from the FC structure are shown for 1,3-butadiene and benzene, respectively, in the ascending order of energy. In the labels of MECIs for 1,3-butadiene, those obtained by the search starting from cis- or trans-isomer are indicated as $\mathbf{x}^{cis}$ or $\mathbf{x}^{trans}$, respectively. Structures by the SF-TDDFT method and those by the CASPT2 method are shown in black and red, respectively [50, 51].

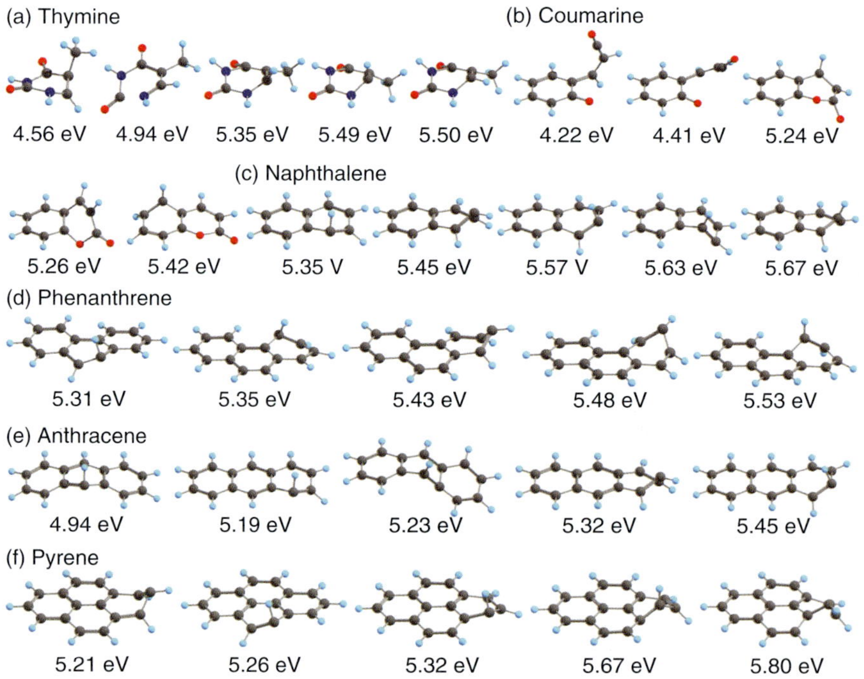

Figure 8.11 S_0/S_1-MECI structures of (a) thymine, (b) coumarin, (c) naphthalene, (d) phenanthrene, (e) anthracene, and (f) pyrene, obtained by GP/SMF/SC-AFIR starting from the corresponding FC geometry. The lowest five S_0/S_1-MECIs among those easily accessible from the FC structure are shown in the ascending order of relative energy. Energies relative to the ground state geometry are indicated in electronvolts [50, 51].

Nonradiative decay can be a major process when there is a MECI easily accessible from the FC structure. The search automatically identified the lowest S_0/S_1-MECI near the FC structure. In thymine and coumarin, the lowest S_0/S_1-MECI is low enough to be accessed from the FC structure in a short timescale. This is consistent with the fact that these molecules show ultrafast nonradiative decay. On the other hand, we discussed a relationship between the height of the lowest S_0/S_1-MECI relative to the S_1 minimum and the experimental fluorescence quantum yield, and obtained a clear correlation between these two quantities [51]. This suggests that fluorescence quantum yield of a given molecule can be discussed theoretically by identifying the lowest S_0/S_1-MECI with an exhaustive search. It would also be noteworthy that GP/SMF/SC-AFIR with SF-TDDFT has been applied to molecules including up to 40–50 atoms with laboratory-sized computer clusters. Use of GP/SMF/SC-AFIR in the design of fluorescent molecules would be an interesting future subject.

8.3.11 Ultrafast Nonradiative Decay in Organometallic Complex

Internal conversion and intersystem crossing pathways starting from a high-lying electronic state can also be searched systematically by the GP/SMF/SC-AFIR method. In this case, the search is applied to all possible PES pairs below the excitation photon energy. This procedure was applied to the rhenium(I) carbonyl bipyridine halide complexes [Re(X)$(CO)_3$(bpy)] (X = Cl, Br, I; bpy = 2,2′-bipyridine), and the ultrafast decay in these complexes observed after 400 nm photoexcitation was discussed. In this chapter, we only show the case of X = Br.

Previously, absorption/emission wavelengths of the Re complexes have been investigated experimentally and theoretically [132]. Decay times were measured by time-resolved emission spectroscopy [133], where three luminescence peaks that decay in different timescales were observed after 400 nm photoexcitation to the bright state S_2 [133], at 530 nm (τ = 128 fs), at 570 nm (τ = 470 fs), and at 600–610 nm ($\tau > 150$ ps). Multi-configuration time-dependent Hartree (MCTDH) [134] simulations taking account of spin–orbit couplings have been performed to understand the ultrafast decay dynamics of the Re-complex [135, 136].

The GP/SMF/SC-AFIR search was applied to all possible pairs between the five lowest lying singlet and triplet electronic excited states, i.e. S_2/S_1, S_2/T_3, S_2/T_2, S_2/T_1, S_1/T_3, S_1/T_2, and S_1/T_1. In total, seven MECIs and four MESXs (one S_2/S_1-MECI, four T_3/T_2-MECIs, two T_2/T_1-MECIs, two S_2/T_3-MESXs, and two S_1/T_2-MESXs) were found. Then, by the meta-IRC (mass-weighted steepest descent path) calculations starting from the MECI and MESX geometries, the most energetically preferable decay path was determined as $S_2 \rightarrow S_1 \rightarrow T_2 \rightarrow T_1$ via S_2/S_1-MECI, S_1/T_2-MESX, and T_2/T_1-MESX. A schematic of the decay path is shown in Figure 8.12a. In Figure 8.12b, the relative energies of the S_0, S_1, S_2, T_1, and T_2 states calculated at MECI, MESX, and local minimum geometries along the dominant decay path are plotted. It should be noted that $(S_2)_{min}$ and $(T_2)_{min}$ are located inside the S_2/S_1-CI and T_2/T_1-CI, respectively, and are identical to (S_2/S_1) and (T_2/T_1).

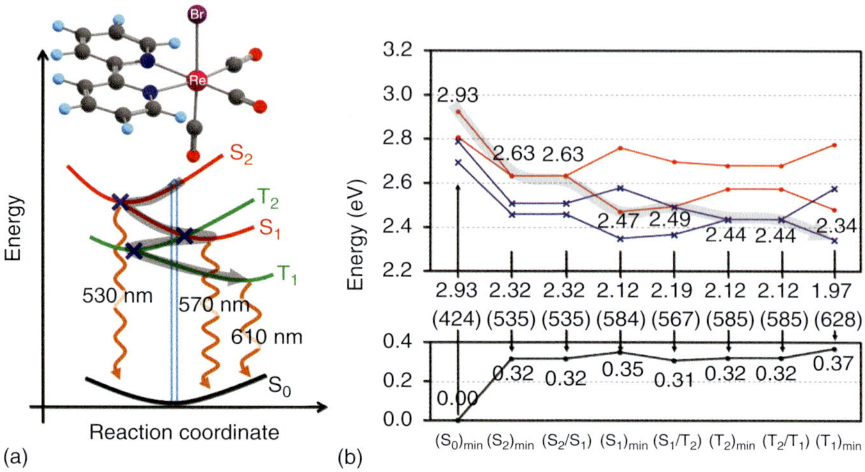

Figure 8.12 (a) Schematic of the ultrafast decay path of $ReBr(CO)_3$(bpy) and (b) variation of the potential energy (eV) for S_0 (black), S_1 (red), S_2 (red), T_1 (blue), and T_2 (blue), along the most preferable decay path. The state on which the system populates at each geometry is highlighted by the gray thick line. Relative energy values of S_0 and those of the state populated at each geometry are shown in electronvolts. The energy gap between S_0 and the state populated at each geometry, which corresponds to the emission energy, are also shown (emission wavelength in nanometers in parentheses) [53].

In Figure 8.12b, the state on which the system populates at each geometry is highlighted by the gray thick line. In addition, emission wavelength from the populated state at each geometry is presented in nanometers. Upon excitation, the system is expected to populate around $(S_2)_{min}$. The calculated emission wavelength from $(S_2)_{min}$ is 535 nm, which can be well compared to the first experimental peak at 530 nm, which decays in the lifetime 128 fs. Then, the system decays to $(S_1)_{min}$, and stays for a while in the region between $(S_1)_{min}$ and $(T_2)_{min}$. In this second region, the calculated emission wavelength from $(S_2)_{min}$ to $(T_2)_{min}$ are 584 and 585 nm, respectively. The emission from this region would correspond to the experimental luminescence peak at 570 nm with the lifetime 470 fs. Finally, the system reaches $(T_1)_{min}$, and stays on the T_1 PES for a long time. The calculated 628 nm emission from $(T_1)_{min}$ would thus correspond to the experimental peak at 600–610 nm with the long lifetime >150 ps. These qualitative discussions were also consistent with those extracted from MCTDH simulations using the reduced-dimensional model Hamiltonian [53].

8.3.12 Photochemical Ligand Substitution Reactions of *fac*-$[Re^I(bpy)(CO)_3PR_3]^+$

Similarly to the ultrafast excited-state dynamics of *fac*-$[Re^I(bpy)(CO)_3X]$ (X = halogen), the mechanism of the photochemical ligand substitution (PLS) reactions of *fac*-$[Re^I(bpy)(CO)_3PR_3]^+$ has been studied by the GP/SMF/SC-AFIR method [54]. The PLS reaction is useful to introduce a functional group into the axial position. The reaction occurs in the *fac*-[$Re^I(bpy)(CO)_3PR_3]^+$ system but *fac*-[$Re^I(bpy)(CO)_3X$] complex is photostable [137, 138]. In this study, the mechanisms of the CO ligand dissociation of *fac*-[$Re^I(bpy)(CO)_3P(OMe)_3]^+$ and *fac*-[$Re^I(bpy)(CO)_3Cl$] have theoretically been investigated as the dominant process of the PLS reaction.

In both complexes, the PLS reactivity can be determined by the topology of the T_1 PES, since the photoexcited complex is able to decay into the T_1 state by internal conversions and intersystem crossings with sufficiently low energy barriers. Thus, the T_1-MINs, T_1-TSs, and S_0/T_1-MESXs were searched by the SC-AFIR method using the B3LYP/SDD, cc-pVDZ level.

As illustrated in Figure 8.13, the T_1 state has a character of metal-to-ligand charge transfer (^{3}MLCT) around the FC region, and it changes to the metal-centered (^{3}MC) state as the Re—CO bond is elongated and bent. The equatorial CO ligand has a much higher energy barrier to leave than that of the axial CO, so that the axial CO ligand selectively dissociates in the PLS reaction. The SC-AFIR search reveals the CO dissociation pathway in photostable *fac*-[$Re^I(bpy)(CO)_3Cl$] as well: however, the dissociation barrier on the T_1 state is substantially higher than that in *fac*-[$Re^I(bpy)(CO)_3PR_3]^+$ and the S_0/T_1-MESXs are located before and below the barrier. On the other hand, no S_0/T_1-MESXs were found before and below the barrier in the *fac*-[$Re^I(bpy)(CO)_3PR_3]^+$ system. We emphasize that the lowest S_0/T_1-MESX was identified by the exhaustive GP/SMF/SC-AFIR search before making this discussion. The origins of reactivity and position-selectivity of the PLS reaction have successfully been elucidated by the AFIR method.

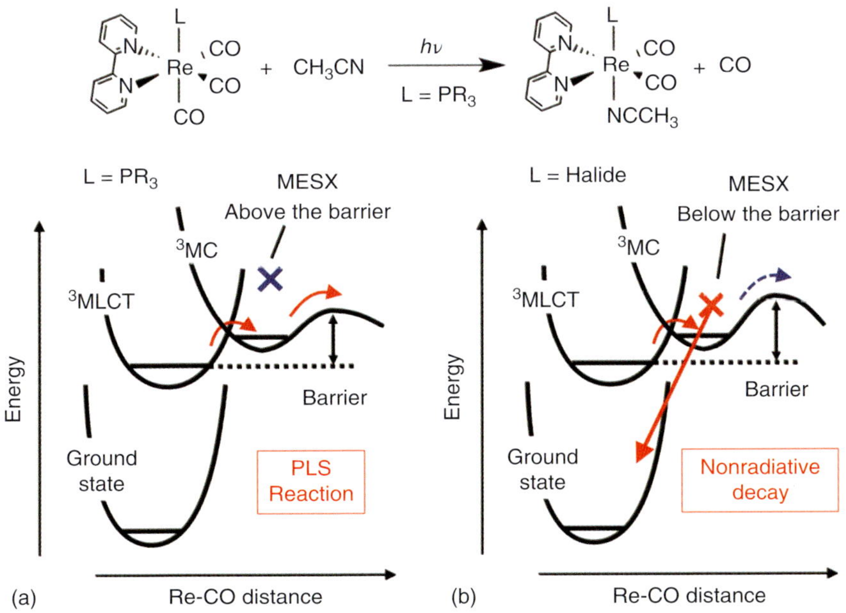

Figure 8.13 The reaction mechanisms for the PLS reaction proposed by the GP/SMF/SC-AFIR method [54].

8.4 Concluding Remarks

In this chapter, we described the AFIR method, which is an automated reaction path search method applied most extensively to quantum chemical PESs. The AFIR induces structural deformations in a molecular system systematically by applying positive or negative artificial force between two or more fragments. Two different algorithms of the AFIR method were introduced. The MC-AFIR algorithm has been used to sample many paths between two or more reactant molecules. In MC-AFIR, fragments to which the artificial force is applied are given reactant molecules or parts of them. The SC-AFIR algorithm has been employed to explore paths in a single system (molecule or complex) systematically. In the SC-AFIR, fragments are automatically generated in an entire system. The algorithm to explore MESX and MECI structures between two different PESs was further introduced by combining the AFIR method with the GP/SMF approach. By the GP/SMF/AFIR approach, the automated exploration of nonadiabatic pathways has been performed.

It was demonstrated with many application examples that the AFIR method has a wide application scope. In this chapter, its applications to organic reaction, multicomponent reaction, organometallic catalysis, cluster structure sampling, metal cluster catalysis, photoaddition reaction, nonradiative decay in molecules and organometallic complexes, and photoinduced ligand substitution in organometallic complex, were showcased, among many other applications. The function of finding out unexpected reaction paths possessed by the AFIR method has been very useful in these previous studies. Further applications to various reactions are also under progress. Further extension of the AFIR method for its application to enzyme reaction, crystal phase transition, surface reaction, etc. is also proceeding actively. The AFIR method would thus be expected as one of the tools to accelerate the development of future "molecular strategy."

Acknowledgments

This work was partly supported by a grant from Japan Science and Technology Agency with a Core Research for Evolutional Science and Technology (CREST) in the Area of "Establishment of Molecular Technology towards the Creation of New Functions" at Hokkaido University.

References

1 Koga, N. and Morokuma, K. (1991). *Chem. Rev.* 91: 823–842.
2 Niu, S. and Hall, M.B. (2000). *Chem. Rev.* 100: 353–406.
3 Siegbahn, P.E.M. and Blomberg, M.R.A. (2000). *Chem. Rev.* 100: 421–438.
4 Torrent, M., Solà, M., and Frenking, G. (2000). *Chem. Rev.* 100: 439–494.
5 Ziegler, T. and Autschbach, J. (2005). *Chem. Rev.* 105: 2695–2722.
6 Houk, K.N. and Cheong, P.H.-Y. (2008). *Nature* 455: 309–313.

7 Fukui, K. (1981). *Acc. Chem. Res.* 14: 363–368.
8 Maeda, S., Harabuchi, Y., Ono, Y. et al. (2015). *Int. J. Quant. Chem.* 115: 258–269.
9 Jensen, F. (2007). *Introduction to Computational Chemistry*, 2e. Chichester, UK: Wiley.
10 Schlegel, H.B. (2011). *WIREs Comput. Mol. Sci.* 1: 790–809.
11 Bondensgård, K. and Jensen, F. (1996). *J. Chem. Phys.* 104: 8025–8031.
12 Quapp, W., Hirsch, M., Imig, O., and Heidrich, D. (1998). *J. Comput. Chem.* 19: 1087–1100.
13 Černohorský, M., Kettou, S., and Koča, J. (1999). *J. Chem. Inf. Comput. Sci.* 39: 705–712.
14 Irikura, K.K. and Johnson, R.D. III (2000). *J. Phys. Chem. A* 104: 2191–2194.
15 Müller, E.M., de Meijere, A., and Grubmüller, H. (2002). *J. Chem. Phys.* 116: 897–905.
16 Ohno, K. and Maeda, S. (2004). *Chem. Phys. Lett.* 384: 277–282.
17 Maeda, S. and Ohno, K. (2005). *J. Phys. Chem. A* 109: 5742–5753.
18 Ohno, K. and Maeda, S. (2006). *J. Phys. Chem. A* 110: 8933–8941.
19 Ohno, K. and Maeda, S. (2008). *Phys. Scr.* 78: 058122/1–058122/8.
20 Maeda, S., Ohno, K., and Morokuma, K. (2009). *J. Phys. Chem. A* 113: 1704–1710.
21 Baker, J. and Wolinski, K. (2011). *J. Comput. Chem.* 32: 43–53.
22 Maeda, S., Ohno, K., and Morokuma, K. (2012). *Adv. Phys. Chem.* 2012: 268124/1–268124/13.
23 Zimmerman, P.M. (2013). *J. Comput. Chem.* 34: 1385–1392.
24 Rappoport, D., Galvin, C.J., Zubarev, D.Y., and Aspuru-Guzik, A. (2014). *J. Chem. Theory Comput.* 10: 897–907.
25 Kim, Y., Choi, S., and Kim, W.Y. (2014). *J. Chem. Theory Comput.* 10: 2419–2426.
26 Schaefer, B., Mohr, S., Amsler, M., and Goedecker, S. (2014). *J. Chem. Phys.* 140: 214102/1–214102/13.
27 Maeda, S., Taketsugu, T., Morokuma, K., and Ohno, K. (2014). *Bull. Chem. Soc. Jpn.* 87: 1315–1334.
28 Martínez-Núñez, E. (2015). *J. Comput. Chem.* 36: 222–234.
29 Zimmerman, P.M. (2015). *J. Comput. Chem.* 36: 601–611.
30 Bergeler, M., Simm, G.N., Proppe, J., and Reiher, M. (2015). *J. Chem. Theory Comput.* 11: 5712–5722.
31 Zhang, X.-J. and Liu, Z.-P. (2015). *Phys. Chem. Chem. Phys.* 17: 2757–2769.
32 Wang, L.-P., McGibbon, R.T., Pande, V.S., and Martinez, T.J. (2016). *J. Chem. Theory Comput.* 12: 638–649.
33 Habershon, S. (2016). *J. Chem. Theory Comput.* 12: 1786–1798.
34 Maeda, S. and Morokuma, K. (2010). *J. Chem. Phys.* 132: 241102/1–241102/4.
35 Maeda, S. and Morokuma, K. (2011). *J. Chem. Theory Comput.* 7: 2335–2345.
36 Maeda, S., Ohno, K., and Morokuma, K. (2013). *Phys. Chem. Chem. Phys.* 15: 3683–3701.

37 Maeda, S., Taketsugu, T., and Morokuma, K. (2014). *J. Comput. Chem.* 35: 166–173.

38 Maeda, S., Harabuchi, Y., Takagi, M. et al. (2016). *Chem. Rec.* 16: 2232–2248.

39 Maeda, S., Komagawa, S., Uchiyama, M., and Morokuma, K. (2011). *Angew. Chem. Int. Ed.* 50: 644–649.

40 Maeda, S. and Morokuma, K. (2012). *J. Chem. Theory Comput.* 8: 380–385.

41 Hatanaka, M., Maeda, S., and Morokuma, K. (2013). *J. Chem. Theory Comput.* 9: 2882–2886.

42 Uematsu, R., Maeda, S., and Taketsugu, T. (2014). *Chem. Asian J.* 9: 305–312.

43 Isegawa, M., Maeda, S., Tantillo, D.J., and Morokuma, K. (2014). *Chem. Sci.* 5: 1555–1560.

44 Zeng, G., Maeda, S., Taketsugu, T., and Sakaki, S. (2014). *Angew. Chem. Int. Ed.* 126: 4721–4725.

45 Uematsu, R., Yamamoto, E., Maeda, S. et al. (2015). *J. Am. Chem. Soc.* 137: 4090–4099.

46 Sameera, W.M.C., Maeda, S., and Morokuma, K. (2016). *Acc. Chem. Res.* 49: 763–773.

47 Sameera, W.M.C., Sharma, A.K., Maeda, S., and Morokuma, K. (2016). *Chem. Rec.* 16: 2349–2363.

48 Vong, K., Maeda, S., and Tanaka, K. (2016). *Chem. Eur. J.* 22: 18865–18872.

49 Maeda, S., Taketsugu, T., and Morokuma, K. (2013). *Z. Phys. Chem.* 227: 1421–1433.

50 Maeda, S., Harabuchi, Y., Taketsugu, T., and Morokuma, K. (2014). *J. Phys. Chem. A* 118: 12050–12058.

51 Harabuchi, Y., Taketsugu, T., and Maeda, S. (2015). *Phys. Chem. Chem. Phys.* 17: 22561–22565.

52 Murdock, D., Ingle, R.A., Sazanovich, I.V. et al. (2016). *Phys. Chem. Chem. Phys.* 18: 2629–2638.

53 Harabuchi, Y., Eng, J., Gindensperger, E. et al. (2016). *J. Chem. Theory Comput.* 12: 2335–2345.

54 Saita, K., Harabuchi, Y., Taketsugu, T. et al. (2016). *Phys. Chem. Chem. Phys.* 18: 17557–17564.

55 Harabuchi, Y., Taketsugu, T., and Maeda, S. (2016). *Chem. Lett.* 45: 940–942.

56 Yamazaki, K., Miyazaki, Y., Harabuchi, Y. et al. (2016). *J. Phys. Chem. Lett.* 7: 4001–4007.

57 Harabuchi, Y., Taketsugu, T., and Maeda, S. (2017). *Chem. Phys. Lett.* 674: 141–145.

58 Banerjee, A., Adams, N., Simons, J., and Shepard, R. (1985). *J. Phys. Chem.* 89: 52–57.

59 Elber, R. and Karplus, M. (1987). *Chem. Phys. Lett.* 139: 375–380.

60 Choi, C. and Elber, R. (1991). *J. Chem. Phys.* 94: 751–760.

61 Ayala, P.Y. and Schlegel, H.B. (1997). *J. Chem. Phys.* 107: 375–384.

62 Wein, E., Ren, W., and Vanden-Eijnden, E. (2002). *Phys. Rev. B* 66: 052301/1–052301/4.

63 Peters, B., Heyden, A., Bell, A.T., and Chakraborty, A. (2004). *J. Chem. Phys.* 120: 7877–7886.

64 Maeda, S., Saito, R., and Morokuma, K. (2011). *J. Phys. Chem. Lett.* 2: 852–857.
65 Maeda, S., Taketsugu, T., Ohno, K., and Morokuma, K. (2015). *J. Am. Chem. Soc.* 137: 3433–3445.
66 Koga, N. and Morokuma, K. (1985). *Chem. Phys. Lett.* 119: 371–374.
67 Manaa, M.R. and Yarkony, D.R. (1993). *J. Chem. Phys.* 99: 5251–5256.
68 Bearpark, M.J., Robb, M.A., and Schlegel, H.B. (1994). *Chem. Phys. Lett.* 223: 269–274.
69 Anglada, J.M. and Bofill, J.M. (1997). *J. Comput. Chem.* 18: 992–1003.
70 Ciminelli, C., Granucci, G., and Persico, M. (2004). *Chem. Eur. J.* 10: 2327–2341.
71 Keal, T.W., Koslowski, A., and Thiel, W. (2007). *Theor. Chem. Acc.* 118: 837–844.
72 Levine, B.G., Coe, J.D., and Martinez, T.J. (2008). *J. Phys. Chem. B* 112: 405–413.
73 Sicilia, F., Blancafort, L., Bearpark, M.J., and Robb, M.A. (2008). *J. Chem. Theory Comput.* 4: 257–266.
74 Maeda, S., Ohno, K., and Morokuma, K. (2010). *J. Chem. Theory Comput.* 6: 1538–1545.
75 Enders, D., Hüttl, M.R., Grondal, C., and Raabe, G. (2006). *Nature* 441: 861–863.
76 Passerini, M. and Simone, L. (1921). *Gazz. Chim. Ital.* 51: 181–189.
77 Schuler, F.W. and Murphy, G.W. (1950). *J. Am. Chem. Soc.* 72: 3155–3159.
78 Vance, R.L., Rondan, N.G., Houk, K.N. et al. (1988). *J. Am. Chem. Soc.* 110: 2314–2315.
79 Sumiya, Y., Nagahata, Y., Komatsuzaki, T. et al. (2015). *J. Phys. Chem. A* 119: 11641–11649.
80 Nagahata, Y., Maeda, S., Teramoto, H. et al. (2016). *J. Phys. Chem. B* 120: 1961–1971.
81 Hebrard, F. and Kalck, P. (2009). *Chem Rev* 109: 4272–4282.
82 Versluis, L., Ziegler, T., Baerends, E.J., and Ravenek, W. (1989). *J. Am. Chem. Soc.* 111: 2018–2025.
83 Versluis, L., Ziegler, T., and Fan, L. (1990). *Inorg. Chem.* 29: 4530–4536.
84 Huo, C.-F., Li, Y.-W., Beller, M., and Jiao, H. (2003). *Organometallics* 22: 4665–4677.
85 Huo, C.-F., Li, Y.-W., Beller, M., and Jiao, H. (2005). *Organometallics* 24: 3634–3643.
86 Rush, L.E., Pringle, P.G., and Harvey, J.N. (2014). *Angew. Chem. Int. Ed.* 126: 8816–8820.
87 Kobayashi, S., Sugiura, M., Kitagawa, H., and Lam, W.W.L. (2002). *Chem. Rev.* 102: 2227–2302.
88 Kobayashi, S. (1994). *Synlett* 689–701.
89 Yamamoto, E., Izumi, K., Horita, Y., and Ito, H. (2012). *J. Am. Chem. Soc.* 134: 19997–20000.
90 Gao, M., Lyalin, A., Maeda, S., and Taketsugu, T. (2014). *J. Chem. Theory Comput.* 10: 1623–1630.
91 Paternò, E. and Chiefi, G. (1909). *Gazz. Chim. Ital.* 39: 341–361.

92 Büchi, G., Inman, C.G., and Lipinsky, E.S. (1954). *J. Am. Chem. Soc.* 76: 4327–4331.

93 Zimmerman, H.E. (1966). *Science* 153: 837–844.

94 Freilich, S.C. and Peters, K.S. (1981). *J. Am. Chem. Soc.* 103: 6255–6257.

95 Griesbeck, A.G., Mauder, H., and Stadtmueller, S. (1994). *Acc. Chem. Res.* 27: 70–75.

96 Palmer, I.J., Ragazos, I.N., Bernardi, F. et al. (1994). *J. Am. Chem. Soc.* 116: 2121–2132.

97 Kutateladze, A.G. (2001). *J. Am. Chem. Soc.* 123: 9279–9282.

98 Griesbeck, A.G., Abe, M., and Bondock, S. (2004). *Acc. Chem. Res.* 37: 919–928.

99 Yabuno, Y., Hiraga, Y., Takagi, R., and Abe, M. (2011). *J. Am. Chem. Soc.* 133: 2592–2604.

100 D'Auria, M. (2012). *Tetrahedron* 68: 8699–8703.

101 Brogaard, R.Y., Schalk, O., Boguslavskiy, A.E. et al. (2012). *Phys. Chem. Chem. Phys.* 14: 8572–8580.

102 Olivucci, M., Ragazos, I.N., Bernardi, F., and Robb, M.A. (1993). *J. Am. Chem. Soc.* 115: 3710–3721.

103 Olivucci, M., Bernardi, F., Ottani, S., and Robb, M.A. (1994). *J. Am. Chem. Soc.* 116: 2034–2048.

104 Celani, P., Bernardi, F., Olivucci, M., and Robb, M.A. (1995). *J. Chem. Phys.* 102: 5733–5742.

105 Bernardi, F., Olivucci, M., and Robb, M.A. (1997). *J. Photochem. Photobiol., A* 105: 365–371.

106 Sakai, S. (1998). *Chem. Phys. Lett.* 287: 263–269.

107 Garavelli, M., Frabboni, B., Fato, M. et al. (1999). *J. Am. Chem. Soc.* 121: 1537–1545.

108 Izzo, R. and Klessinger, M. (2000). *J. Comput. Chem.* 21: 52–62.

109 Sakai, S. (2000). *Chem. Phys. Lett.* 319: 687–694.

110 Garavelli, M., Bernardi, F., Olivucci, M. et al. (2001). *J. Phys. Chem. A* 105: 11496–11504.

111 Sicilia, F., Blancafort, L., Bearpark, M.J., and Robb, M.A. (2007). *J. Phys. Chem. A* 111: 2182–2192.

112 Dick, B., Haas, Y., and Zilberg, S. (2008). *Chem. Phys.* 347: 65–77.

113 Levine, B.G., Ko, C., Quenneville, J., and Martinez, T.J. (2006). *Mol. Phys.* 104: 1039–1051.

114 Celani, P., Garavelli, M., Ottani, S. et al. (1995). *J. Am. Chem. Soc.* 117: 11584–11585.

115 Levine, B.G. and Martinez, T.J. (2009). *J. Phys. Chem. A* 113: 12815–12824.

116 West, A.C. and Windus, T.L. (2012). *Theor. Chem. Acc.* 131: 1251/1–1251/11.

117 Zilberg, S. and Haas, Y. (1999). *J. Phys. Chem. A* 103: 2364–2374.

118 Li, Q., Mendive-Tapia, D., Paterson, M.J. et al. (2010). *Chem. Phys.* 377: 60–65.

119 Blancafort, L. and Robb, M.A. (2012). *J. Chem. Theory Comput.* 8: 4922–4930.

120 Merchan, M., Gonzalez-Luque, R., Climent, T. et al. (2006). *J. Phys. Chem. B* 110: 26471–26476.

121 Perun, S., Sobolewski, A.L., and Domcke, W. (2006). *J. Phys. Chem. A* 110: 13238–13244.

122 Serrano-Perez, J.J., Gonzalez-Luque, R., Merchan, M., and Serrano-Andres, L. (2007). *J. Phys. Chem. B* 111: 11880–11883.

123 Zechmann, G. and Barbatti, M. (2008). *J. Phys. Chem. A* 112: 8273–8279.

124 Asturiol, D., Lasorne, B., Robb, M.A., and Blancafort, L. (2009). *J. Phys. Chem. A* 113: 10211–10218.

125 Lan, Z., Fabiano, E., and Thiel, W. (2009). *J. Phys. Chem. B* 113: 3548–3555.

126 Serrano-Andrés, L. and Merchán, M. (2009). *J. Photochem. Photobiol., C* 10: 21–32.

127 Barbatti, M., Aquino, A.J., Szymczak, J.J. et al. (2010). *Proc. Natl. Acad. Sci. USA* 107: 21453–21458.

128 Gonzalez-Luque, R., Climent, T., Gonzalez-Ramirez, I. et al. (2010). *J. Chem. Theory Comput.* 6: 2103–2114.

129 Yamazaki, S. and Taketsugu, T. (2012). *J. Phys. Chem. A* 116: 491–503.

130 Nakayama, A., Arai, G., Yamazaki, S., and Taketsugu, T. (2013). *J. Chem. Phys.* 139: 214304/1–214304/11.

131 Krauter, C.M., Mohring, J., Buckup, T. et al. (2013). *Phys. Chem. Chem. Phys.* 15: 17846–17861.

132 Heydova, R., Gindensperger, E., Romano, R. et al. (2012). *Chem. A* 116: 11319–11329.

133 Cannizzo, A., Blanco-Rodriguez, A.M., El Nahhas, A. et al. (2008). *J. Am. Chem. Soc.* 130: 8967–8974.

134 Meyer, H.D., Manthe, U., and Cederbaum, L.S. (1990). *Chem. Phys. Lett.* 165: 73–78.

135 Gourlaouen, C., Eng, J., Otsuka, M. et al. (2015). *J. Chem. Theory Comput.* 11: 99–110.

136 Eng, J., Gourlaouen, C., Gindensperger, E., and Daniel, C. (2015). *Acc. Chem. Res.* 48: 809–817.

137 Koike, K., Tanabe, J., Toyama, S. et al. (2000). *Inorg. Chem.* 39: 2777–2783.

138 Koike, K., Okoshi, N., Hori, H. et al. (2002). *J. Am. Chem. Soc.* 124: 11448–11455.

9

Computational Molecular Technology Toward Macroscopic Chemical Phenomena: Red Moon Methodology and Its Related Applications

Masataka Nagaoka[1,2,3,4], *Masayoshi Takayanagi*[1], *Norio Takenaka*[1,3], *Yuichi Suzuki*[1], *Kentaro Matsumoto*[1], *Nobuaki Koga*[1,2,4], *Sandhya Karakkadparambil Sankaran*[1], *Purushotham Uppula*[1], and *Yukichi Kitamura*[1]

[1] *Nagoya University, Graduate School of Informatics, Chikusa-ku, Nagoya, 464-8601, Japan*
[2] *Japan Science and Technology Agency, Core Research for Evolutional Science and Technology, Honmachi, Kawaguchi 332-0012, Japan*
[3] *Kyoto University, Elements Strategy Initiative for Catalysts and Batteries (ESICB), Katsura, Kyoto 615-8520, Japan*
[4] *Nagoya University, Future Value Creation Research Center, Chikusa-ku, Nagoya 464-8601, Japan*

9.1 Introduction

The recent development of new measurement techniques has made us realize how significant it is to consider chemistry by the number of molecules, not by the amount of substance in a mole, leading to the increasingly important role of computational chemistry as a molecular science [1]. On the other hand, when we try to treat with computational chemistry the diffusion and chemical reactions in "molecular aggregation states" where a huge number of atoms and molecules are gathered in condensation, the fact that these phenomena occur only very rarely has long made it restrictive to deal with them by still "time-consuming" first principles methods of computational chemistry [2]. Even with traditional quantum chemical (QC) calculations and classical molecular simulations, i.e. the first-generation "computational molecular technology (CMT)," it is difficult to determine the large-scale and long-term properties and stereochemical characteristics that depend on or originate in such macroscopic-scale equilibrium states that finally emerge after a number of super-rare events have occurred repeatedly.

Under the circumstances, for the purpose of atomistically dealing with the phenomena and to understand their characteristics in "molecular aggregation states," especially the (i) complex chemical reactions, (ii) stereospecificity, and (iii) aggregate structure, we have been developing new large space–time scale "computational molecular technologies (CMTs)," i.e. second-generation CMTs, beyond the category of first-generation CMTs, such as conventional computational chemistry [3]. In fact, facing directly onto the macroscopic chemical phenomena (MACHeP), by using those CMTs, it should become possible to design and create new environmentally and energetically useful functional

Molecular Technology: Materials Innovation, Volume 3, First Edition.
Edited by Hisashi Yamamoto and Takashi Kato.

materials and develop them to control their assemblies and composite materials for industrial usage [4].

Now, we are aiming at the following major goal:

I. To construct a number of CMTs for MACHeP simulations, in order to realize the molecular control of the following triple:
   (i) complex chemical reactions,
   (ii) stereospecificity, and
   (iii) aggregate structures.

Furthermore, in order to contribute to accomplishing the above goal, we are trying in parallel to achieve another goal:

II. To develop a suite of MACHeP simulators and to prepare the execution environments of the CMTs.

Undoubtedly, the achievement of these goals will have a great scientific and technological impact. More specifically, in order to solve such a difficult issue that the phenomena are connected to both big number (Avogadro's number order) and the huge space–time scale, a multiscale simulation technique has been prepared so far. In addition, because any phenomenon is inseparable from chemical reactions, i.e. a succession of super-rare events occurring in the "molecular aggregation state," we would like to establish "first-principles" CMT, being based on microscopic information of atoms and molecules. We are convinced that such a new CMT basis would make it possible for us to explain, at the multilevel resolution, the specificity of each complex chemical reaction system that each chemist is targeting for.

In this chapter, we review the essential aspects of Red Moon methodology, i.e. a new CMT that we have been developing for the above goals, and show its related application examples to clarify several practical chemical phenomena where complex chemical reactions occur at the same time. We believe the accomplishment of this CMT is considered an important mission for today and it is not possible until then for computational chemists to collaborate closely with experimental chemists.

9.2 Methodology

9.2.1 What Today's Chemists Want and Need to Consider

In Japan, such national projects are now running as novel algorithms of molecular dynamics (MD) simulations, and quantum chemical (QC) calculations are developed for biomolecular and macromolecular systems, and computer simulations are being carried out massively parallel to understand their properties and phenomena at atomic resolution [5, 6]. Some computer programs developed in these projects are directed toward the development of macromolecular all-electron calculation (such as NTChem, GELLAN, SMASH, FMO) and others toward that for large-scale- and long-term-oriented molecular simulation of molecular aggregation of biopolymers and functional materials (such as GENESIS, Modylas, OCTA) [5, 6]. Each one exerts tremendous power in analyzing the electronic, microscopic, or mesoscopic dynamic characteristics of the molecule. In particular, the importance and necessity of the molecular orbital (MO) method as a fundamental CMT, one of the first-generation CMTs that can predict very successfully the reactivity and the selectivity of elementary chemical reactions, is undeniable.

However, systematizing a number of CMTs has not been attempted to be able to predict both stereoselectivity of complex reaction systems in molecular aggregation state and specificity of the aggregation structure [3]. For this purpose, it is essential to deal with the ceaseless motions, in their indefinite structures as they are, of organic or inorganic molecules and macromolecules in solution and in biological systems. But, the present *ab initio* molecular dynamics (AIMD) method is restricted to be able to merely simulate them on such a space–time scale as at the most several million cubic Å ($\sim10^{-24}$ m^3) by several thousand microseconds ($\sim10^{-3}$ seconds). Furthermore, it is totally impossible to calculate such statistical quantities that microscopically super-rare phenomena such as chemical reactions that can just occur with very small probability, i.e., one ten billionth [10^{-10}] or less (the ratio of the number of active molecules to that of inactive ones), should play a decisive role, their becoming well-defined, due to the greatness of the Avogadro's number, in the several orders larger scale that is directly connected to the real world of materials that has further a few orders larger dimensions. In fact, this is completely a scientific challenge and technical requirement that today's chemists should deal with and then need to consider.

For the readers to get an idea of the big number of molecules, it is worth mentioning the typical model system for our MD simulation that will be used to study the ethylene polymerization later in Section 9.3.1 [7, 8]. As shown

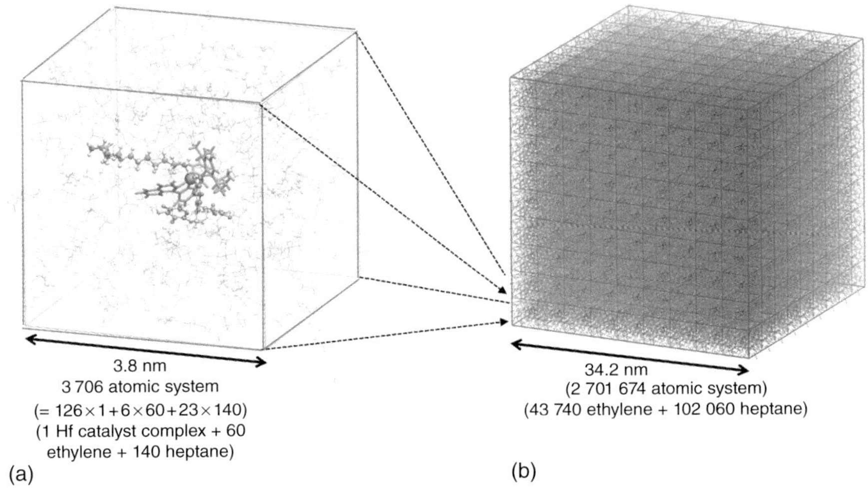

Figure 9.1 (a) A typical model system for the present MD simulation of ethylene polymerization by Hf catalyst, consisting of 1 catalyst ion pair, 60 ethylene monomers, and 140 heptane solvent molecules (3706 atomic system) (see Section 9.3.1). (b) A model system with minimally necessary size for the realistic MD simulation of the same reaction according to typical experimental condition at 400 K and 1.4 MPa, i.e. the concentration ratio [Hf]:[ethylene]:[heptane] = 1 : 40 000:160 000. "Minimally" means that at least one catalyst ion pair should be included in the whole model system. The volume (namely, total number of atoms [~3 million]) is 750 times larger than in (a) (~4000).

in the Figure 9.1a, according to the ability of our computing facilities, the system we prepared consisted of 1 catalytic ion pair, 60 ethylene monomers, and 140 heptane solvent molecules (3706 atomic system). However, in order to treat the ethylene polymerization reaction catalyzed by the Hf catalyst under experimental conditions at 400 K and 1.4 MPa, i.e. the concentration ratio [Hf]:[ethylene]:[heptane] = 1 : 40 000 : 160 000 [9], we need another model system consisting of 43 740 ethylene and 102 060 heptane monomers in addition to only one Hf catalyst ion pair (as shown in Figure 9.1b). The volume $40.0 \times 10^3\ nm^3$ (or, equivalently, the total number of atoms [~3 000 000]) is ~750 times larger than in the Figure 9.1a (~54.9 nm^3 [or ~4000]). As this is an interesting topic in catalytic synthesis of polymers, in order to study the chain transfer frequency between the catalysts, it would be natural to prepare at least two catalytic ion pairs, meaning that the model system should thus include twice as many numbers as that in the Figure 9.1b.

Under the circumstances, as a new entry to CMT, we have recently developed a new efficient hybrid Monte Carlo (MC)/molecular dynamics (MD) reaction method, a Rare-Event-Driving MethOdology Of Necessity (Red Moon), to realize a practical "atomistic" molecular simulation of large-scale and long-term chemical reaction systems (Figure 9.2) [10].

The research target of the "Red Moon" methodology aims at treating complex chemical reactions in the "molecular aggregation state" where a large number of molecules literally gather. For this purpose, the "Red Moon" methodology will be developed, with atomic and molecular information, to treat such chemical reactions that occur very rarely from the microscopic viewpoint (i.e. super-rare phenomena), then leading to a technology innovation. Realizing the stereochemical control of aggregation-state chemical reactions and the structural control of super-nano scale aggregates, new functional materials will be designed and

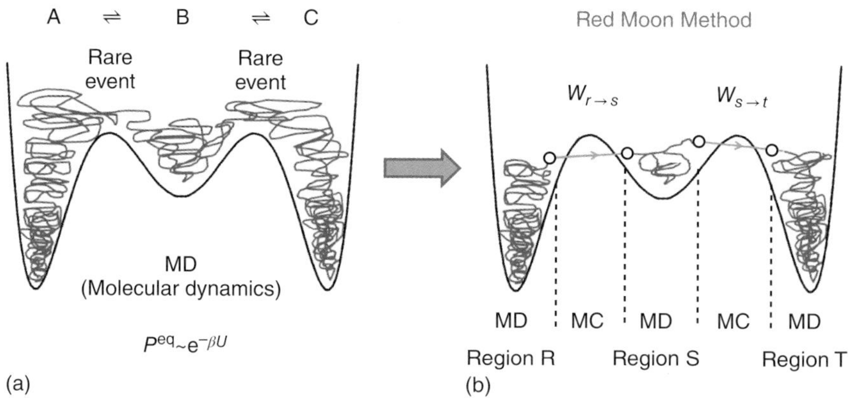

Figure 9.2 Schematic representation of Red Moon Method (a hybrid MC/MD reaction method). (a) The configurational distribution in equilibrium P^{eq} is proportional to the exponential factor $\exp[-\beta U]$ where U is the global potential function. (b) On the other hand, in right figure, the whole configurational distribution consists of regional ones. The connecting points of the two kinds of moves (open circles) are selected according to the criterion of chemical reaction. $W_{r\to s}$ and $W_{s\to t}$ are the transition probabilities from the configuration state r to s and from s to t, respectively. Source: Reproduced with permission from Nagaoka et al. [10]. Copyright 2013, Elsevier B.V.

created. We are dreaming to finally offer the general-purpose environment of the "CMT" of MACHeP simulation, certainly through collaboration with experimentalists [3, 10].

9.2.2 Red Moon Methodology – A Recent Computational Molecular Technology

9.2.2.1 Molecular Description of Complex Chemical Reaction Systems

Before explaining the Red Moon methodology itself, we introduce the total Hamiltonian in molecular number representation [10] and restrict the current discussion to a chemical system of complex chemical reactions that consists of four types of elementary chemical reactions relating to nine chemical species (i.e. a model type for the reaction scheme of the solid–electrolyte interface [SEI] film formation in sodium ion batteries in Section 9.4.2 [11]):

$$A_1 + A_2 \rightarrow A_3 \tag{9.1a}$$

$$A_4 + 3A_2 \rightarrow A_5 + 3A_6 \tag{9.1b}$$

$$A_3 \rightarrow A_7 + A_8 \tag{9.1c}$$

$$A_3 + A_3 \rightarrow A_9 \tag{9.1d}$$

This is because generalization is complicated if not impossible. The "microscopic" or "atomistic" Hamiltonian of the molecular assembly comprising these nine kinds of molecular species can be described by

$$H = H(N_{A_1}, N_{A_2}, N_{A_3}, N_{A_4}, N_{A_5}, N_{A_6}, N_{A_7}, N_{A_8}, N_{A_9}) \equiv H(\mathbf{L}_{i=1}^{9} N_{A_i}) \tag{9.2a}$$

$$= \sum_{i=1}^{9} H^{A_i} + \frac{1}{2} \sum_{i=1}^{9} \sum_{i=1}^{9} H^{A_i A_j} \tag{9.2b}$$

$$= \sum_{i=1}^{9} \sum_{\rho_i=1}^{N_{A_i}} H_{\rho_i}^{A_i} + \frac{1}{2} \sum_{i=1}^{9} \sum_{j=1}^{9} \sum_{\rho_i=1}^{N_{A_i}} \sum_{\sigma_j=1}^{N_{A_j}} H_{\rho_i \sigma_j}^{A_i A_j} \tag{9.2c}$$

where the notation means the line-up symbol

$$\mathrm{L}_{i=1}^{9} N_{A_i} \equiv N_{A_1}, N_{A_2}, \ldots, N_{A_9} \tag{9.3}$$

and i and j are used for the kinds of componential molecules; ρ_i and σ_j are the serial numbers of the ith and jth kinds of molecules, respectively, and N_{A_i} and N_{A_j} are the total numbers of molecules of kinds i and j, respectively. Further, H^{A_i} is the whole intramolecular Hamiltonian of the ith kind of molecules A_i and consists of a sum of N_{A_i} number of one-molecule Hamiltonians of the ith kind of molecule A_i, each denoted explicitly by $H_{\rho_i}^{A_i}$; $H^{A_i A_j}$ is the whole intermolecular Hamiltonian between the ith and jth kinds of molecules A_i and A_j, and consists of a sum of $\frac{1}{2} N_{A_i} \cdot N_{A_j}$ number of two-molecule Hamiltonians between the ith and jth kinds of molecules A_i and A_j, each denoted explicitly by $H_{\rho_i \sigma_j}^{A_i A_j}$. Assuming that componential molecules A_i consist of η_{A_i} numbers of atoms, respectively, the conservation of the total number of atoms (i.e. the conservation of mass) stands as

$$\sum_{i=A_1}^{A_9} N_i \cdot \eta_i = n (= \text{const.}) \tag{9.4}$$

In the above formulation, it should be noted that such chemical species as ions and radicals might also be called just "molecules" for simplicity.

The whole solution system, expressed by the "atomistic" canonical ensemble with constant NVT, is well described in a region R in equilibrium by the Hamiltonian

$$H_r \equiv H_r(\mathbf{N}^r) = H_r(\mathrm{L}_{i=1}^{9} N_{\mathrm{A}_i}^r) \tag{9.5}$$

in the molecule number representation $\mathbf{N}$ with the representative phase $\mathrm{P}_r(\mathbf{r}_r, \mathbf{p}_r)$. Then, if one of the chemical reactions in Eqs. (9.1a)–(9.1d), e.g. (9.1a), would occur with one molecule each of A_1 and A_2 creating just one A_3 molecule without any other reaction, the whole system then transfers to another state s (assumed in equilibrium) in the different phase space region, expressed well by the representative phase $\mathrm{P}_s(\mathbf{r}_s, \mathbf{p}_s)$ with the Hamiltonian

$$H_s \equiv H(\mathbf{N}^s) = H(\mathrm{L}_{i=1}^{9} N_{\mathrm{A}_i}^s) \tag{9.6a}$$

$$= H(N_{\mathrm{A}_1}^s, N_{\mathrm{A}_2}^s, N_{\mathrm{A}_3}^s, N_{\mathrm{A}_4}^s, N_{\mathrm{A}_5}^s, N_{\mathrm{A}_6}^s, N_{\mathrm{A}_7}^s, N_{\mathrm{A}_8}^s, N_{\mathrm{A}_9}^s) \tag{9.6b}$$

$$= H(N_{\mathrm{A}_1}^r - 1, N_{\mathrm{A}_2}^r - 1, N_{\mathrm{A}_3}^r + 1, N_{\mathrm{A}_4}^r, N_{\mathrm{A}_5}^r, N_{\mathrm{A}_6}^r, N_{\mathrm{A}_7}^r, N_{\mathrm{A}_8}^r, N_{\mathrm{A}_9}^r) \tag{9.6c}$$

$$= \sum_{i=1}^{9} \sum_{\rho_i=A_1}^{N_{A_i}'} H_{\rho_i} + \frac{1}{2} \sum_{i=1}^{9} \sum_{j=1}^{9} \sum_{\rho_i=1}^{N_i'} \sum_{\sigma_j=1}^{N_j'} H_{\rho_i \sigma_j} \tag{9.6d}$$

Here, one should notice that $N_{\mathrm{A}_1}^s = N_{\mathrm{A}_1}^r - 1$, $N_{\mathrm{A}_2}^s = N_{\mathrm{A}_2}^r - 1$, and $N_{\mathrm{A}_3}^s = N_{\mathrm{A}_3}^r + 1$, leaving the numbers of the other molecules totally unchanged.

The energy difference ΔH between these configuration states r and s before and after the reaction occurrence is

$$\Delta H_{rs} = H_s(\mathbf{r}_s, \mathbf{p}_s) - H_r(\mathbf{r}_r, \mathbf{p}_r) \tag{9.7}$$

As far as the NVT ensemble may be concerned,

$$\Delta H_{rs} = \Delta U_{rs} = U_s(\mathbf{r}_s) - U_r(\mathbf{r}_r) \tag{9.8}$$

where ΔU_{rs} is the energy difference of the reaction and U_r and U_s are the configurational potential energies of the whole system in the states r and s, respectively. However, the above treatment can be easily extended to other thermodynamic ensembles, with other chemical reactions.

9.2.2.2 Red Moon Method – A Rare Event-Driving Methodology of Necessity (Red Moon)

In this section, we first explain the Red Moon methodology rather conceptually. In the MD part of Red Moon approach [10], we adopt the usual NVT-MD simulation [12, 13] for spatial moves in a stable phase space region among R, S, and T (Figure 9.2), e.g. in region R, by numerically solving a large number of simultaneous Newtonian equations of motion:

$$\begin{cases} \dfrac{\mathrm{d}\mathbf{r}^{\mathrm{N}}}{\mathrm{d}t} = \dfrac{\partial H}{\partial \mathbf{p}^{\mathrm{N}}} \\[2ex] \dfrac{\mathrm{d}\mathbf{p}^{\mathrm{N}}}{\mathrm{d}t} = -\dfrac{\partial H}{\partial \mathbf{r}^{\mathrm{N}}} \end{cases} \tag{9.9}$$

so long as a certain pair of atoms meets the condition leading to a chemical reaction. In Eq. (9.9), H is the "microscopic" or "atomistic" Hamiltonian of a molecular assembly that is described by a set of atomistic phase variables ($\mathbf{r}^{\mathbf{N}}$, $\mathbf{p}^{\mathbf{N}}$). In Figure 9.2a,b, the curly curves represent the MD moves in phase space following Eq. (9.9), while the straight lines with arrows represent the MC moves (or transitions) of the system (panel (b)), whose dynamical moves would be extremely rare events. In panel (a), the configurational distribution in equilibrium p^{eq} is proportional to the exponential factor $\exp[-\beta U]$ where U is the global potential function. Panel (b) shows regional distributions, e.g. $p_r{}^{\text{eq}} \approx \exp[-\beta U_r]$ in region R. The connecting points of the two kinds of moves (open circles) are selected according to the criteria for chemical reaction occurrence. Note that they do not represent real connections since these points correspond to almost identical states in configuration space but not in momentum space and with different configurational gradients. $W_{r\to s}$ and $W_{s\to t}$ are the transition probabilities from state r to s and from s to t, respectively [10].

An MC part of the method can be now executed in the present "atomistic" canonical ensemble if a reasonably likely type of move between nonoverlapping phase space regions before and after a chemical reaction can be found. This can be accomplished by making (breaking) a chemical bond between a candidate pair of atoms belonging to two different molecules (a dissociative molecule) [10]. On rearranging these atoms, their original "atom types" in the reactant molecules (a dissociative molecule [or fragment]) that participate in a chemical reaction are legitimately switched to those suitable to the product species, meaning a change of the corresponding atomic potential parameters in the molecular mechanical (MM) force field [12, 14] together with a proper correction of the "zero" point of energy by the corresponding potential energy of reaction.

It is clear that a Markov chain giving a limiting distribution of $\exp[-\beta U]$ is then obtained [12, 13] by accepting a reactive move from state r to state s with the transition probability $W_{r\to s}$ (Figure 9.2) obtained by

$$W_{r\to s} = \min\{1, \exp[-\beta \Delta U_{rs}]\} \tag{9.10}$$

For this result, we impose the detailed balance (or microscopic reversibility) condition between the states r and s before and after the reaction

$$p_{r\to s} = p_{s\to r} \tag{9.11}$$

where the ps indicate the unweighted probability of a move, and then find

$$W_{r\to s} P_r^{\text{eq}} = W_{s\to r} P_s^{\text{eq}}. \tag{9.12}$$

For example, in a reaction such as in Eq. (9.1a), one can take for the r and s states with the molecular mixture compositions ($\mathbf{N}^r$) and ($\mathbf{N}^s$), respectively; P_r^{eq}, P_s^{eq} and $W_{r\to s}$ can then be regarded as

$$P_r^{\text{eq}} = P_r^{\text{eq}}(\mathbf{r}_r^{\mathbf{N}^r})(\equiv P_r^{\text{eq}}(\mathbf{n}^r, V, T)) \tag{9.13a}$$

$$P_s^{\text{eq}} = P_s^{\text{eq}}(\mathbf{r}_s^{\mathbf{N}^s})(\equiv P_s^{\text{eq}}(\mathbf{n}^s, V, T)) \tag{9.13b}$$

and

$$W_{r\to s} = W(\mathbf{r}_r^{\mathbf{N}^r}; \mathbf{r}_s^{\mathbf{N}^s}) \tag{9.14}$$

respectively.

9.2.2.3 Algorithmic Procedure of Red Moon Method

Here, we have explained more precisely the procedure in the Red Moon method assuming a reaction scheme (RS) that consists of a set of chemical reactions R_1, R_2, …, i.e. chemical reaction processes. The activation energy $\Delta E_a^{R_1}$, $\Delta E_a^{R_2}$, …, that each would occur necessarily with is obtained properly as the "free" energy of activation that is estimated in advance, within the continuum model [15], the free energy gradient method [16–18], or some experimental methods for each chemical reaction.

To summarize, the NVT-Red Moon simulation consists of a combination of the following types of processes:

1. Equilibrate the whole system through the NVT-MD simulation and select a configuration state r.
2. Generate reaction steps with the NVT-MC procedure and classical NVT-MD simulation, consisting of the following steps:
   (i) Search for a pair(s) of "reactive" atoms among reactant molecules with the whole system at a given configuration state r according to some criteria for whether possible chemical reactions R_1, R_2, … might occur. In the studies in some of the following sections, the criterion for a *candidate pair(s)* among their molecules has been simply taken to be whether the interatomic distance is less than the sum of their van der Waals (vdW) radii. However, it is occasionally more convenient and reasonable to determine the criterial distances for a candidate pair(s) of molecules on the basis of the chemical kinetic treatment. As a result, a set of the number of candidate pairs for each chemical reaction $\{N_{\text{cand}}^{R_1}, N_{\text{cand}}^{R_2}, \ldots\}$ that is able to generate the chemical reactions is obtained at a given configuration state r.
   (ii) Randomly select a chemical reaction R_i among R_1, R_2, … according to their corresponding relative weights of selection w^{R_1}, w^{R_2}, … , where a relative weight w^{R_i} for R_i is expressed by the product of $N_{\text{cand}}^{R_i}$ and $\exp(-\beta \Delta E_a^{R_i})$ as follows:
   $$w^{R_i} = N_{\text{cand}}^{R_i} \exp(-\beta \Delta E_a^{R_i}) \tag{9.15}$$
   Then, randomly select a candidate pair(s) among the candidate pairs for R_i. If there are no pairs for any chemical reactions, execute a short NVT-MD simulation instead of the following steps (iii) and (iv) to update the configuration state.
   (iii) Switch the MM parameters and, if necessary, the functional forms of MM force fields of the reactant molecules to those of the product ones, and virtually react them to generate a possible configuration state s, structurally optimizing the whole system to obtain the state s through a short NVT-MD simulation.
   (iv) Compute the energy change $\Delta U_{rs} (= U_s - U_r)$ (Eq. (9.8)) and accept (or reject) the reaction step according to the transition probability $W_{r \to s}$ (Eq. (9.10)) under the Metropolis scheme [12, 13, 19]. In the present study, ΔU_{rs} is approximately estimated as follows:
   $$\Delta U_{rs} = \Delta U_{rs}^{\text{MM}} + \Delta U_0^{\text{reac}} \tag{9.16}$$

where $\Delta U_{rs}^{\mathrm{MM}}$ is the naïve difference in total potential energies obtained in the original MM force field, meaning a change of the corresponding MM potential parameters and potential function forms together with a proper correction of the "zero" point of energy by the corresponding potential energy of reaction $\Delta U_0^{\mathrm{reac}}$.

3. If the molecular mixture composition might scarcely change, then stop. Otherwise return to process 2.

One cycle consisting of the above three different types of processes is called an "MC/MD cycle." By repeating the MC/MD cycle, we can stochastically realize a succession of various types of chemical reaction many times even if they occur competitively and simulate reasonably the atomistic dynamics of the whole complex chemical reaction system. The convergence of the Red Moon method is affected by the relative ratios of the three different types of processes, and the optimal choice depends on the system properties [10, 11, 20].

9.2.3 A Set-Up Using Conventional Computational Molecular Technology

9.2.3.1 Reaction Scheme – Quantum Chemistry (QC) and Experiment

To apply the Red Moon methodology to some chemical reaction system, we need to decide a RS, i.e. a reaction network consisting of the molecular structures at critical states, i.e. reactant, product, and transition states, and their energetics that should be plausible to describe necessarily and sufficiently the whole system concerned (see the RS of (9.1a)–(9.1d), for example). These reaction data may be sometimes easy to obtain from the beginning, but usually not so easy if the system includes many chemical species showing many kinds of mutual chemical reactions at the same time. Some data could be obtained from the experimental database [21]. However, even if we could assume some reasonable RS for the chemical reaction system concerned, we would often find that there exist no reliable MM force fields, especially for metal atoms important in polymerization reactions using metal complex catalysts. It is, therefore, inevitable to use extensive QC calculations [22] by a conventional CMT, e.g. density functional theory (DFT), to know the critical structures and their energetics, to set up the RS fully.

In this chapter, following the standard CMT protocol in recent QC treatment, structural optimizations are performed using DFT as implemented in, for example, the standard QC software suite Gaussian 09 [22]. The functional chosen is M06, a hybrid meta exchange-correlation functional parameterized for both transition metals and nonmetals [23]. In particular, for the Zr and the Hf atom, the LanL2DZ basis set is often used, which is a combination of the Los Alamos National Laboratory effective core potentials and the valence double zeta basis set along with f polarization functions [24, 25]. All other atoms are described by the appropriate basis set that varies from 6-31G(d,p) [26] to 6-31++G(d,p) [27] according to the molecular system concerned and the level of quality required. The transition states (TSs) are confirmed by checking whether they have only one imaginary mode of vibration, while the other optimized structures, reactant and product states, are confirmed with only real modes of vibration in the vibrational frequency analysis. Solvation effect is also treated using the SMD method [28] in single point calculations on all gas-phase optimized geometries, i.e. a continuum model with a dielectric constant ε, e.g. 2.3741 for toluene.

9.2.3.2 Molecular Mechanical (MM) Force Fields

For the dynamic studies in this chapter, we have basically employed AMBER 12 [29] in conjunction with the general AMBER force field (GAFF) [30] for the MM and MD simulation. The MM force field model considers the following contributions of energy components to the total potential energy [14, 30]:

$$V(\mathbf{R}) = \sum_{\text{bonds}} K_{ij}^{R}(R_{ij} - R_{ij}^{\text{eq}})^2 + \sum_{\text{angles}} K_{ijk}^{\theta}(\theta_{ijk} - \theta_{ijk}^{\text{eq}})^2 + \sum_{\text{dihedrals}} \frac{V_{ijkl}^{n}}{2}[1 + \cos(n\phi_{ijkl} - \gamma_{ijkl})] + \sum_{j=1}^{\text{atoms}} \sum_{i>j}^{\text{atoms}} \left[\frac{A_{ij}}{\rho_{ij}^{12}} - \frac{B_{ij}}{\rho_{ij}^{6}} + \frac{q_i q_j}{\varepsilon_r \rho_{ij}} \right] \tag{9.17}$$

In Eq. (9.17), **R** is the molecular structural variable as a whole, R_{ij}, θ_{ijk}, and ϕ_{ijkl} are the bond length between the interaction sites (usually atoms) i and j, the bond angle between the bonds ij and jk, and the dihedral angle between the surfaces ijk and jkl, respectively. Each variable with right superscript eq indicates its constant value in equilibrium. K_{ij}^{R}, K_{ijk}^{θ}, and V_{ijkl}^{n} are the force constants for R_{ij}, θ_{ijk}, and ϕ_{ijkl}, respectively. The factor n and γ_{ijkl} represent the periodicity and the phase shift of the dihedral potential, respectively. Further, ρ_{ij} is the nonbonded distance between the interaction sites (usually atoms) i and j, and A_{ij} and B_{ij} are the inner wall repulsion parameter and the van der Waals attraction parameter between the interaction sites i and j, respectively. q_i and q_j are the effective partial charges (in electron units) of electrostatic interaction sites i and j, respectively. ε_r is the dielectric constant.

While the force constants of the bond stretching, the bond angle bending and the dihedral angle bending, and the bond lengths and bond angles in equilibrium, etc. for many common molecules are available in GAFF [30] or the literature, there are often no force field parameters, namely, no MM force field parameters (hereafter simply MM parameters) available for several atoms in some specific molecules, e.g. transition metal atoms located at the active center of useful catalysts. Even in these cases, MM parameters for the common types of atoms in the molecules should be often modified by calibrating the parameters in GAFF. However, the MM parameters of such specific atoms as transition metals have to be newly developed, or modified by fitting the energy change to that obtained by DFT calculations. In addition, the effective partial charges are also determined using DFT calculations with a charge assignment method, e.g. a grid-based method (CHELPG, CHarges from Electrostatic Potentials using a Grid-based method [31]), the Mertz–Singh–Kollman method [32, 33], and the natural population analysis (NPA) [34].

9.3 Applications

In this section, we have shown four applications using not only a recently developed CMT, the Red Moon methodology, but also a number of conventional CMTs, i.e. an extensive combination usage of QC methods, and MM and MD simulation.

9.3.1 Ethylene Coordinative Chain Transfer Polymerization Mechanism on (Pyridylamide)Hf(IV) Catalyst

9.3.1.1 Active Site Opening Mechanism in Ion Pair of (Pyridylamide)Hf(IV) Catalyst: An Associative Mechanism

In the first application, we report a MD study on the ion pair of [{N^-,N,C^-}HfMe][MeB(C_6F_5)$_3$] [7], which is the active species of (pyridylamide)Hf(IV) complex for olefin polymerization. The model system is composed of an ion pair of catalyst and co-catalyst in an explicit organic solvent (heptane molecules) with and without ethylene monomers. MD simulations revealed an active site opening (ASO) mechanism, i.e. the associative active site opening (AASO), where the counter anion dissociates from the active site associatively with monomer coordination to the active site.

AASO also explains consistently the experimental fact that the activity of the catalyst increases when the counter anion has no Me groups. Further, we show that AASO is the major mechanism of ASO by using replica exchange molecular dynamics (REMD) method [35]. In addition, we discuss two important factors for the AASO: one is the structural condition of the ion pair and the other is the net stabilization energy. Finally, we conclude that both the ASO and the monomer coordination should be taken into account to accurately design the behavior of the ion pair and predict the activity.

9.3.1.2 Ion Pair Structure and Its Molecular Mechanical (MM) Force Fields

For elementary information on the MD simulation and the Red Moon methodology, i.e. a new CMT, we first need to confirm the ion pair structure by extensive calculations of DFT (also a conventional CMT). For the present system, we have obtained two important insights.

The first insight is on the attractive interactions in the ion pair that exist between the Hf atom of the cation and F atoms of the perfluorophenyl groups, and between the Hf atom and H atoms of the borate Me group. These interactions enable the anion to interact with the cation in different manners, leading typically to three different structures of the ion pair (Figure 9.3a–c). The second insight is that there are two stable relative positions of the cation and anion (Figure 9.4), i.e. "top structure" and "side structure," respectively. In the top structure, the anion is located above the active site as in the structures in Figure 9.4a. Thus the active site is completely closed (Figure 9.4a, $\theta(N_{pyr}-Hf-B)$ is around 180°). In the side structure, the anion is on the side of the cation and

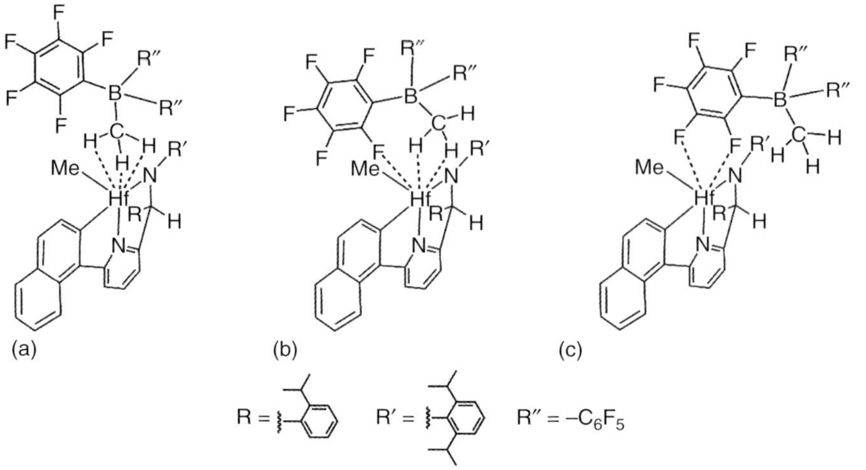

Figure 9.3 Three typical structures of the ion pair with different coordination manners. (a) The anion coordinates to the active site only with its Me group. (b) The anion coordinates with its Me group and an F atom. (c) The anion coordinates only with its F atoms. Source: Reproduced with permission from Matsumoto et al. [7]. Copyright 2016, American Chemical Society.

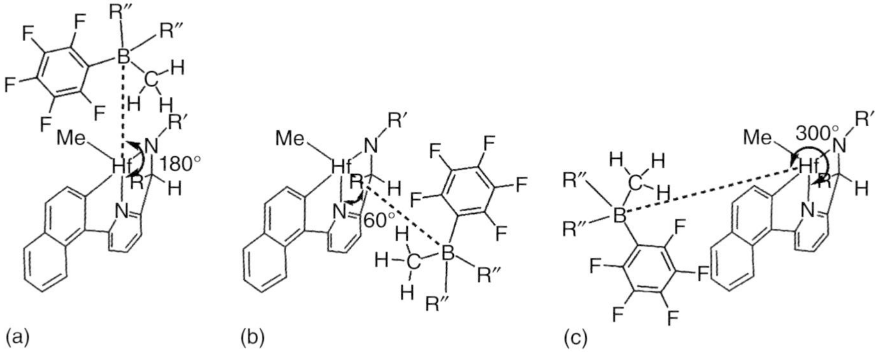

Figure 9.4 Typical ion pair structures with different relative positions of the cation and anion. (a) Top structure, (b) side structure, (c) an example of the relative positions ignored. Source: Reproduced with permission from Matsumoto et al. [7]. Copyright 2016, American Chemical Society.

the active site is open (Figure 9.4b, $\theta(N_{pyr}$–Hf–B) is around 60°). In terms of the ion pair concept of transition metal complexes, they are classified into the inner sphere ion pair (ISIP) and the outer sphere ion pair (OSIP), respectively [36]. Our DFT calculations revealed that the top structure is 15.3 kcal mol^{-1} more stable than the side structure on average. Thus, most of the ion pairs in solution have the top structure, and the active site is completely closed. We assumed that these two relative positions are essential for the ASO and ignored other relative positions. For an example of such ignored positions, the anion can be located at a different side of the cation (Figure 9.4c; $\theta(N_{pyr}$–Hf–B) is around 300°). The lack of direct interaction with the Hf atom will make such positions unstable.

From this information on the structure and energy, we are able to construct the reliable MM force field of the interactions between the cation and the anion, which are of importance to accurately simulate the ion pair's behavior. To this end, on the basic usage of the GAFF force field [29, 30], Eq. (9.17), we newly developed and calibrated an intramolecular MM parameter set of the Hf cation by fitting to DFT calculations, and then 6–12 Lennard–Jones (LJ) intermolecular interaction parameters between the Hf atom and borate anion atoms were calibrated to reproduce the interaction energy of the ion pair [7]. The obtained MM parameters were able to correctly simulate the interaction energy and structural dynamics of the ion pair [7]. In all the MD simulations, we adopt these developed MM force fields.

9.3.1.3 Propagation Reaction on the Active Site of (Pyridylamide)Hf(IV) Catalyst

In applying the Red Moon methodology, i.e. a new CMT, to simulate ethylene polymerization reaction, the model system consisting of 1 ion pair of catalyst and co-catalyst, 60 ethylene monomers, and 140 heptane molecules is located in the elementary cubic cell 38 Å × 38 Å × 38 Å (see also Figure 9.1a). These numbers follow from the same concentration as that in the previous work [37].

Hence, the Red Moon simulation was performed under the NVT condition with a temperature of 400 K [38]. Several insertion reactions occurred during the simulation, indicating the importance of the ion pair configuration. For example, at 700 MC/MD cycle for a typical simulation trajectory, the active site is occupied with the anion, and the configuration cannot be considered a candidate for a trial of elongation reaction (Figure 9.5a). At 742 MC/MD cycle, the anion is dissociated and the ethylene molecule gets closer to the active site and the condition for the reaction is satisfied (Figure 9.5b). At 747 MC/MD cycle, a reaction trial is examined and the ethylene molecule is inserted into the side of the Hf atom of the ethylene oligomer (Figure 9.5c). Finally, at 748 MC/MD cycle, a reaction trial of ethylene molecule insertion is accepted and the elongation reaction is completed (Figure 9.5d).

9.3.2 Propylene Polymerization Reaction Mechanism on C2 Symmetric $[H_2Si(Ind)_2ZrCH_3]^+$ and $[H_2Si(Ind)_2ZrCH_3]^+[CH_3B(C_6F_5)_3]^-$

The stereoregularity of propylene polymerization mechanism catalyzed by $[H_2Si\ (Ind)_2ZrCH_3]^+$ (Ind=indenyl) has been studied at the M06 level of theory [8]. There are four different approaches of propylene to the reactive catalyst, leading to four isomeric products due to the C_2-symmetry of silylene-bridged bis(indenyl) ligand of the catalyst. Consequently, numerous stereoisomers emerge due to various possibilities of propylene attack and orientation of polymer chain. A number of calculations of the first and second insertions with various conformers revealed the most favorable reaction pathway, showing that isotactic propagation is more favorable (3.5 kcal mol^{-1}) than syndiotactic propagation. The catalyst resting states have various agostic interactions between the CH bond and the Zr atom that stabilize the catalytic systems and play important roles in determining the favorable reaction pathway.

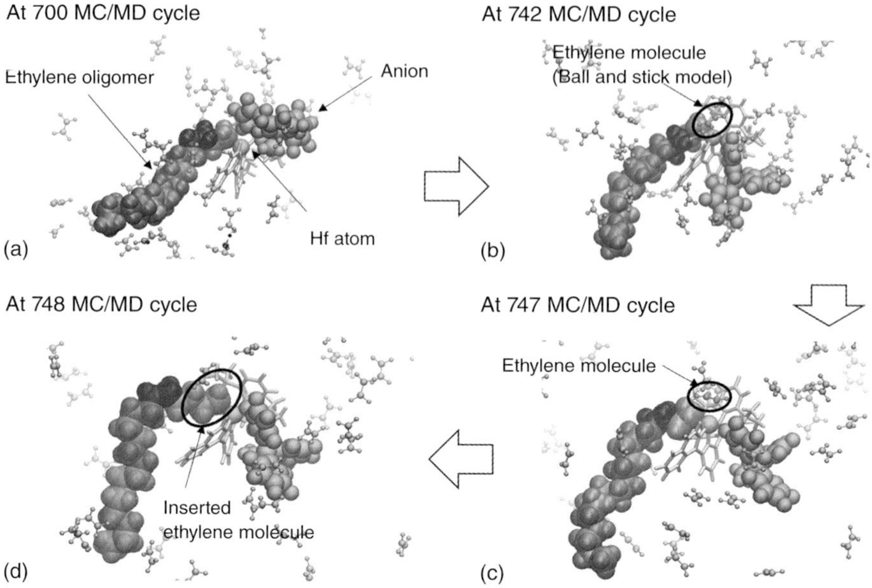

Figure 9.5 Typical insertion process in ethylene polymerization mechanism by Hf catalyst. (a) At 700 MC/MD cycle, because the active site is occupied with the anion, this is not a candidate for a trial of elongation reaction. (b) At 742 MC/MD cycle, the anion is dissociated and the ethylene molecule (in an oval) gets closer to the active site and the condition for reaction is satisfied. (c) At 747 MC/MD cycle, a reaction trial is examined and the ethylene molecule (in an oval) is inserted into the side of the Hf atom of ethylene oligomer. (d) At 748 MC/MD cycle, a reaction trial of ethylene molecule insertion is accepted and the elongation reaction is completed.

The influence of the counter anion (CA) $[CH_3B(C_6F_5)_3]^-$ on the reactivity of the catalyst was also studied. The results confirm that the trans orientation of the counter anion with respect to propylene is more favorable than its cis orientation and also clarify the most favorable reaction pathway in the first and second insertions. Because agostic interactions are involved in various aspects, the atoms-in-molecules (AIM) analysis has been used to find the bonding nature of agostic interactions as well as ion pair bonds. The overall results suggest that the rigidity of the ansa-zirconocene catalyst, the unique structure of C_2 symmetric ansa ligand, the influence of $[CH_3B(C_6F_5)_3]^-$, and the β agostic interaction may restrict the orientation of propylene attack to yield only isotactic polymerization and not syndiotactic polymerization.

9.3.2.1 Energetics of Propylene Insertion into Active Catalyst $H_2Si(Ind)_2ZrCH_3^+$ – Enantioselectivity and Regioselectivity

The present active catalyst $H_2Si(Ind)_2ZrCH_3^+$ is generated from the C_2-symmetric metallocene $H_2Si(Ind)_2ZrCl_2$ using methylaluminoxane (MAO) [39] or borates [40] as a co-catalyst. It is convenient to use $[CH_3B(C_6F_5)_3]^-$ as CA due to the flexible structure of MAO. Further, we restrict our discussion to consider the S form of $H_2Si(Ind)_2ZrCH_3^+$ of its possible mirror image R and S forms [41].

Four possible modes of propylene insertion into this active intermediate would produce four conformationally different intermediates based on the C_2 symmetry of silylene-bridged bis(indenyl) ligand and the orientation of propylene (Figure 9.6a). Thus, four different approaches of propylene generally lead to four different products (Figure 9.6b). On the basis of a number of extensive calculations for the first and second insertions by DFT, i.e. a conventional CMT, the energetics of the profiles clarified that the low-energy isotactic pathway is formed from one of the propylene primary insertions, considering all possible approaches of propylene for the formation of isotactic, syndiotactic, and atactic polymers [8]. It was understood that such syndiotactic products that might be also possible in the second and later insertion in the **D** approach are dormant due to a high-energy barrier in the first insertion (Table 9.1).

However, even in the above case, the orientations of the generating polymer and the CH_3 group of propylene should have less steric interaction, i.e. they are in the trans position. This process should continuously propagate for isotactic product formation. It is noticed that the overall mechanism could operate without site epimerization or back-skip mechanism, because it requires tail flipping, which we did not consider.

9.3.2.2 Reaction Mechanism of cis and trans Approach of Counter Anion $[CH_3B(C_6F_5)_3]^-$ on the Various Stereoisomers

Following the study in Section 9.3.2.1, we investigated the first and second insertions with CA in cis and trans positions with respect to the coordinating propylene (Figure 9.7). The profiles of solvent-effect-corrected relative Gibbs free energies for the first insertions with CA are summarized in Figure 9.8a, in which the reaction pathways are identified by the orientation of CA with respect to propylene.

In the inner sphere ion pair **IP1** ($[H_2Si(Ind)_2ZrCH_3]^+[CH_3B(C_6F_5)_3]^-$) (i.e. an initial catalyst, not shown in Figure 9.8), CA strongly coordinates to the Zr atom through the CH_3 group of CA with a distance 4.09 Å between the B and Zr atoms. This ion pair **IP1** should be rearranged to form two different outer sphere ion pairs **IP2** and **IP3** with CA in the trans and cis positions with respect to the propylene monomer, so that this propylene can coordinate to the Zr atom. Both **IP2** and **IP3** are less stable than **IP1** by 37.86 and 31.29 kcal mol^{-1}, respectively, **IP3** being more stable than **IP2** by 6.58 kcal mol^{-1} (Figure 9.8a). The large unstability of **IP2** and **IP3** compared to **IP1** clearly means that the change in the coordination manner of CA from the inner sphere to the outer sphere is rather difficult and should be the rate-determining step compared to the propagation step [42]. In general, it is shown that the first insertion in the trans pathway is more favorable than that in the cis pathway (Figure 9.8a). In particular, the **C** approach of propylene in the trans pathway with the lowest activation barrier, i.e. 3.41 kcal mol^{-1}, is hence the most favorable among those paths (Figure 9.8a and Table 9.2). These results indicate that some nonnegligible interaction may take place between CA and the Zr complex to affect the favorable reaction pathway and the nature of ansa ligand, and the presence or absence of CA can change significantly the reaction pathway of the insertion and stereochemistry of the system.

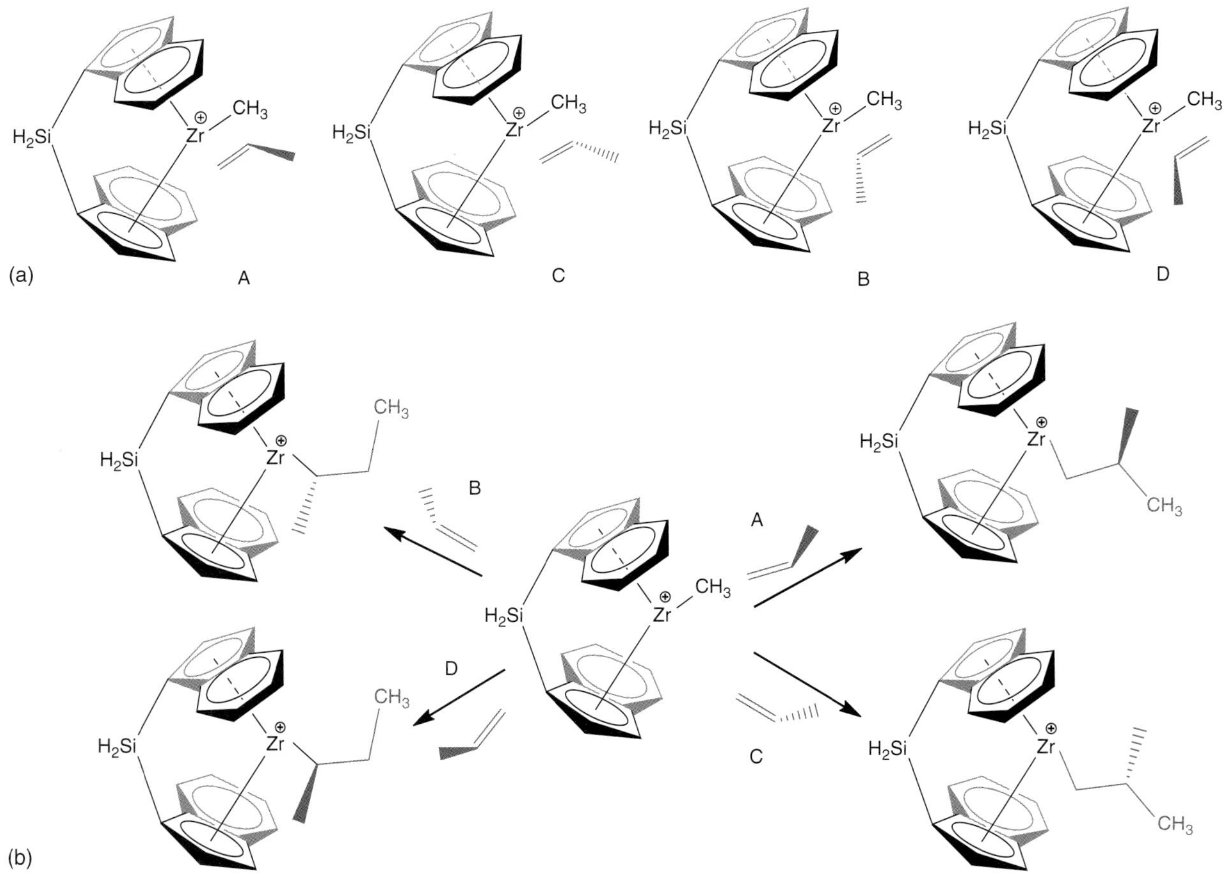

Figure 9.6 (a) Four different conformers of intermediate propylene π complex (two 1,2 insertions [or primary insertions] and two 2,1 insertions [or secondary insertions]). In the **A** and **D** approaches the propylene CH_3 group is pointing up, whereas that in the **B** and **C** approaches is pointing down. (b) Four possible modes in the first insertion of propylene that lead to four different products. The β carbon atom in the products in the **A** and **C** approaches has the same stereochemistry except the conformation. Source: Reproduced with permission from Sankaran et al. [8]. Copyright 2016, The Chemical Society of Japan.

Table 9.1 Solvent-effect-corrected relative Gibbs free energies for the first insertion in the **A**, **C**, **B**, and **D** pathways.

	Relative Gibbs free energy(kcal mol^{-1})			
	A	C	B	D
Int	0.00	0.00	0.00	0.00
R1	−13.90	−15.04	−13.27	−12.46
TS1	−6.10	−5.82	0.00	5.60
P1	−13.89	−14.84	−11.63	−7.20
Activation energy	7.80	9.22	13.27	18.06

Furthermore, by analyzing the second insertion paths with CA being in cis and trans positions and their resting catalysts **ReC1** prior to second insertion (Figure 9.8b), it can be said that in the trans pathways the **C** approach is more favorable than the other approaches. Finally, comparing all the paths in the second insertion with the careful calculation analysis on the basis of DFT, i.e. a conventional CMT, it can be finally suggested that the most favorable path follows the **C** approach with CA in the trans position [43].

9.3.2.3 Toward Propagation Reaction on the Active Site of the Catalyst Ion Pair $[(CH_3)_2Si(Ind)_2ZrCH_3]^+[CH_3B(C_6F_5)_3]^-$

Following the reaction mechanism discussed in Section 9.3.2.2, if the polymer propagation proceeds repeatedly with the most favorable insertion mode following the **C** approach with CA in the trans position, isotactic products would be yielded. However, in order to clarify the whole mechanism of stereoselective polymerization with CA, it is necessary to investigate how CA would move and change in coordination mode in the propagation steps. As explained in Section 9.2.1, to answer these questions, because static DFT calculations are too time-consuming some other simulation methods such as MD technique, a conventional established CMT, are necessary, together with the proper use of appropriate MM force fields.

For the purpose of treating the present polymerization, many MM parameters for $(CH_3)_2Si(Ind)_2$ were taken from the available reports [44–48]. Further, force constant parameters for bond stretching, angle bending, and dihedral rotation for propylene π complex have been determined using the QM results at the M06 level, together with determining other MM parameters for $(CH_3)_2Si(Ind)_2ZrCH_3(C_3H_6)^+$ with propylene in the **A** approach. The remaining transferable MM parameters were taken from the previous reports by Brintzinger and Bosnich [47, 49, 50]. We have comparatively used different charge assignment methods such as CHELPG [31], Merz–Kollman (MK) method [32, 33], and NPA [34] for the propylene π complex of metallocenes in order to study the robustness of different charges in MD simulation.

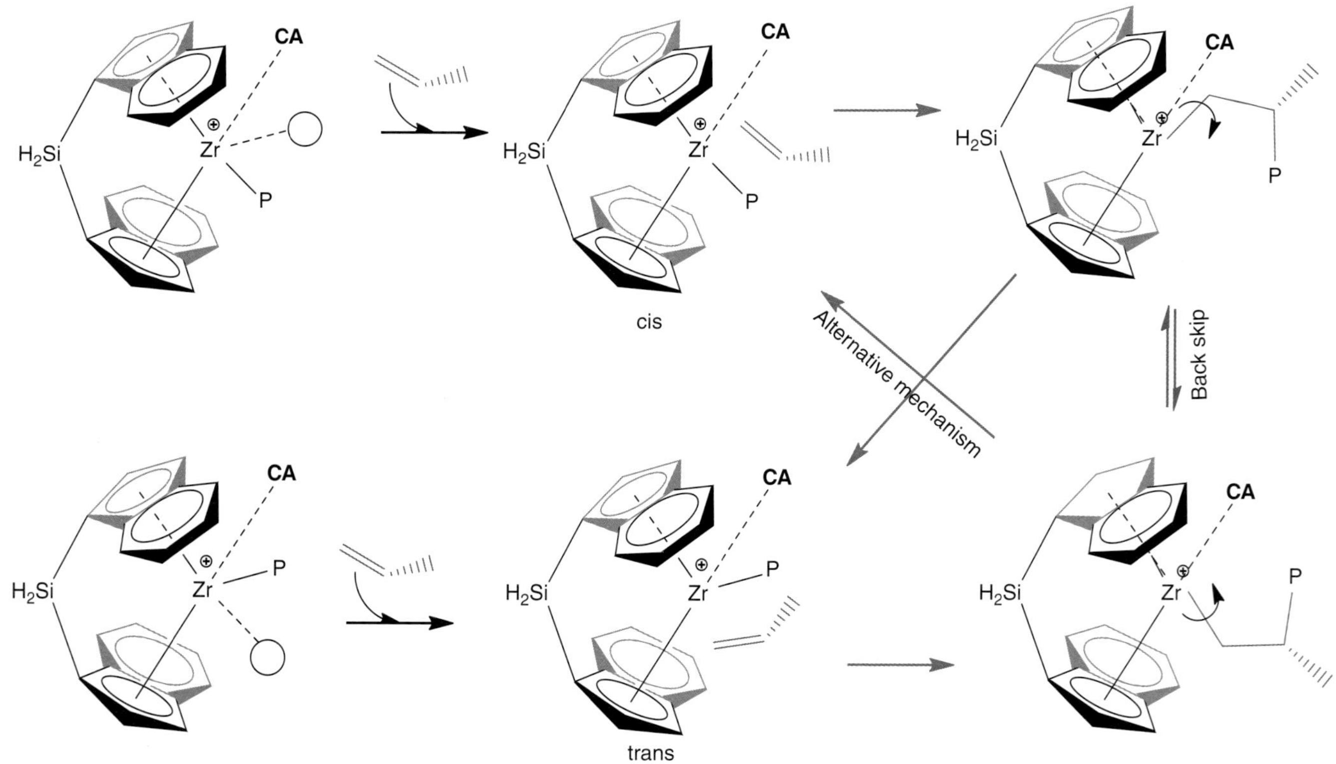

Figure 9.7 Reaction mechanism of first and second insertions of **A/C** approach in the presence of **CA** in cis and trans position. Back skip and alternative mechanisms are also denoted. Source: Reproduced with permission from Sankaran et al. [8]. Copyright 2016, The Chemical Society of Japan.

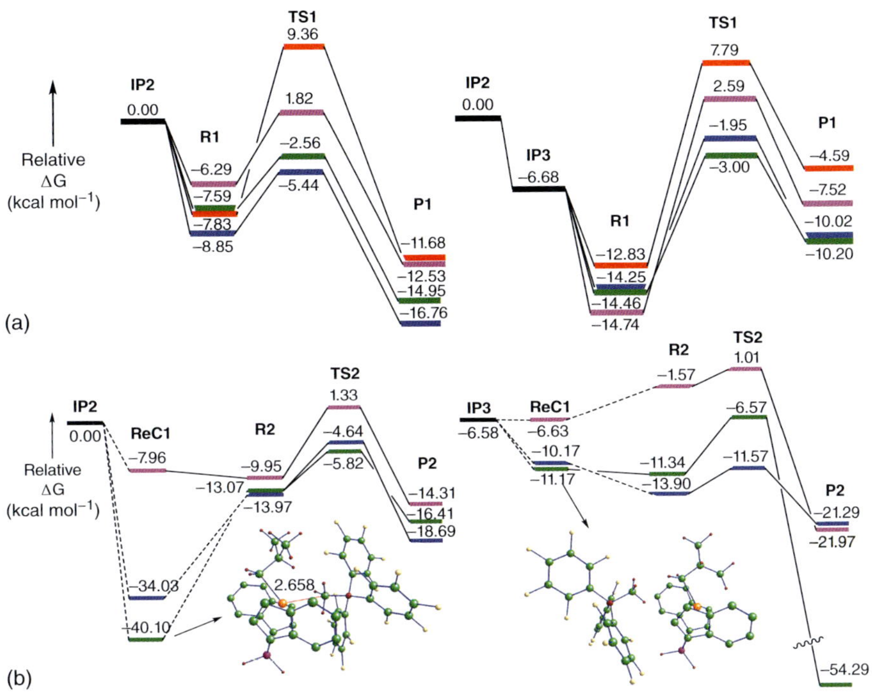

Figure 9.8 Solvent-effect-corrected free energy profile (a) for the first insertion of stereoisomeric reactions with trans (left) and cis (right) orientation of CA with respect to propylene. Green, pink, blue, and red lines represent **A**, **B**, **C**, and **D** approaches, respectively, and (b) for the second insertion via the **A**, **B**, and **C** pathways with CA being in the cis (left) and trans (right) position with respect to propylene. Green, pink, and blue lines represent **A**, **B**, and **C** approaches, respectively. Source: Reproduced with permission from Sankaran et al. [8]. Copyright 2016, The Chemical Society of Japan.

Table 9.2 Solvent-effect-corrected relative Gibbs free energies for the first insertion with the trans approach of CA (with different orientation) in the **A**, **C**, **B**, and **D** pathways.

	Relative Gibbs free energy (kcal mol^{-1})			
	A	**C**	**B**	**D**
IP2	0.00	0.00	0.00	0.00
R1	−7.59	−8.85	−6.29	−7.83
TS1	−2.56	−5.44	−1.82	9.36
P1	−14.95	−16.76	−12.53	−11.68
Activation energy	5.03	3.41	4.47	17.19

In this section, to show the importance of MM and MD treatments as traditional first-generation CMTs, we have focused on (i) the dynamic properties of π-coordinated propylene, especially its rotation, and (ii) the orientation of $[CH_3B(C_6F_5)_3]^-$ during the MD simulation of $[(CH_3)_2Si(Ind)_2ZrCH_3(C_3H_6)]^+$ $[CH_3B(C_6F_5)_3]^-$ (Ind = indenyl) ion pairs with propylene in all approaches. Actually, the MD simulations with the newly prepared MM parameters clarified the dynamic properties of the propylene complex ion pairs that were not found in static calculations of DFT, i.e. one of first-generation CMTs. Further, it was demonstrated that rotation and flipping of propylene occurs significantly during the MD simulations, which validates the experimental observation of misinsertion and stereoerror in polymerization and that the CA resides mostly in the outer sphere (with the average distance of 6.80 Å between the Zr atom and the B atom of the CA for the **A** approach of propylene, for example) with a "bent sandwich" and cis orientation with respect to the propylene for all conformers (Figure 9.9), so that $[(CH_3)_2Si(Ind)_2Zr(C_{13}H_{27})(C_3H_6)]^+[CH_3B(C_6F_5)_3]^-$ with an elongated polymer chain resides mainly in the **C** conformer. This will provide essential and crucial insights into the reaction mechanism of CA in isotactic polymerization. Such a study using Red Moon simulation to investigate the polymerization mechanism with CA using molecular simulation is now in progress.

9.3.3 Aromatic Polyamide Polymerization

The aromatic polyamide membrane, i.e. FT-30, is taken as the third example that has been used for reverse osmosis (RO) and is formed by polymerization reaction at the interface between the aqueous and organic phases, i.e. interfacial polycondensation (IP). The two constituent monomers of the membrane, *m*-phenylenediamine (MPD) and benzene 1,3,5-tricarboxylic acid chloride (TMC), are dissolved in those phases, respectively. Although the composition ratios of elements in the membrane have been observed to be dependent on its depth [51], the detailed microscopic structure of the membrane has not been fully characterized experimentally. Therefore, we have applied the Red Moon method to this membrane formation with the minimum bond convention (MBC), which is an effective treatment of intercellular chemical bonds under the periodic boundary condition (PBC) and allows two atoms belonging not only to the main cell but also to the main cell and an image cell to be able to form a chemical bond [3, 20]. Then, we have investigated the structural characteristics of the membrane at the atomistic level [20]. From the results it is interesting to note that the present simulations of the MPD/TMC mixing ratios 1 : 4 and 1 : 1 have shown good agreement with the corresponding experimental ones obtained by X-ray photoelectron spectroscopy (XPS) and Rutherford backscattering spectrometry (RBS), respectively [51].

9.3.3.1 Microscopic Clarification of the MPD/TMC Mixing Ratios for the Interfacial Polycondensation Reaction Process [20]

Table 9.3 shows the present results of the degree of polymer cross-linking (DPC) and the composition ratios of carbon, oxygen, and nitrogen of FT-30 membrane

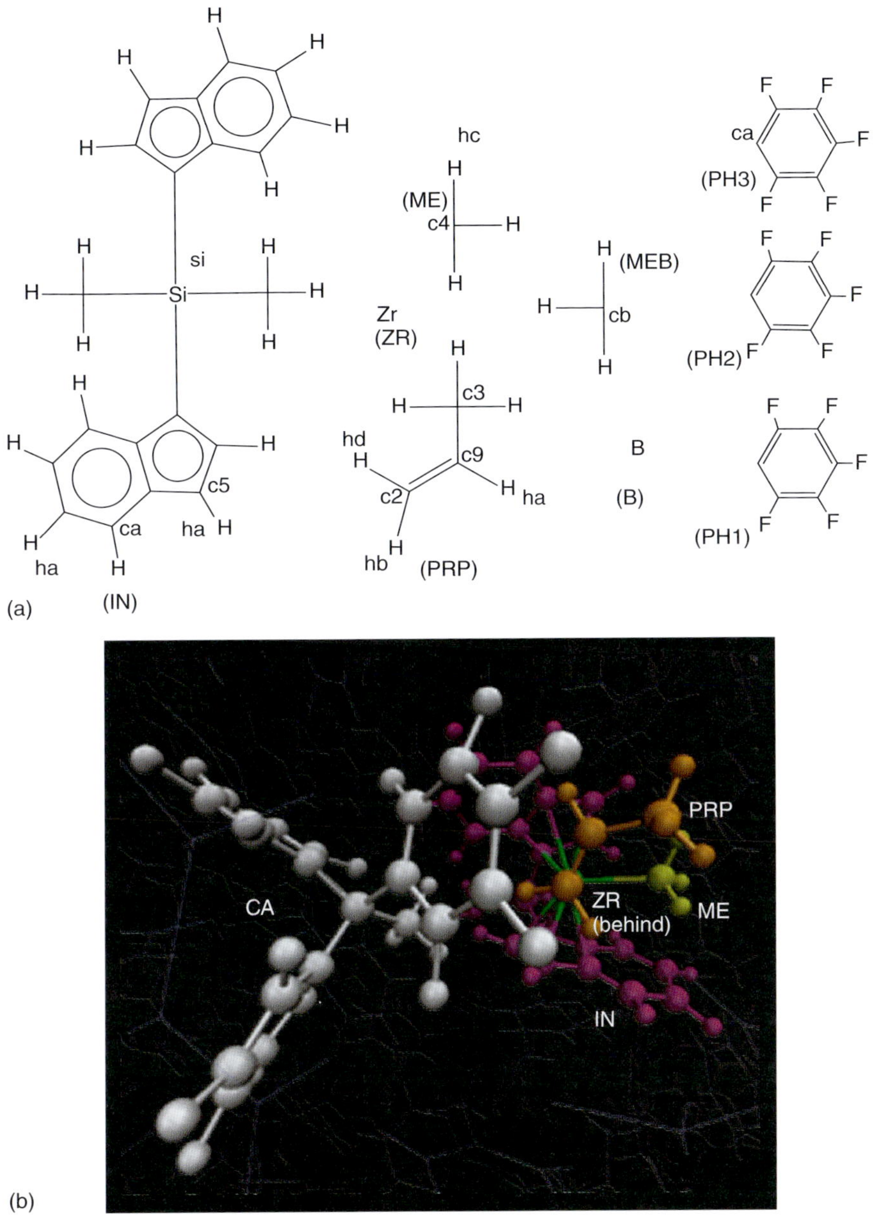

Figure 9.9 (a) Residues of $[(CH_3)_2Si(Ind)_2ZrCH_3(C_3H_6)]^+[CH_3B(C_6F_5)_3]^-$; PRP, ME, ZR, and IN represent the residue names. Atom types are also depicted in lowercase and those for the Zr and B atoms are zr and b, respectively. (b) Snapshot of the ion pair with propylene in the **A** approach at 2 ns in a typical MD trajectory. The residues PRP, ME, ZR, and IN are in orange, yellowish green, green, and violet. The CA is represented in silver color. The "bent sandwich" cis orientation of the CA with respect to propylene can be seen.

Table 9.3 Computational values of the degree of polymer cross-linking (DPC), the composition ratios of carbon, oxygen, and nitrogen, and their experimental values in the FT-30 membrane.

	Present[a)]				Experiment[b)]	
	Ratio 1 : 4	Ratio 1 : 1	Ratio 3 : 2	Ratio 4 : 1	XPS	RBS
DPC	94.3 ± 0.3	96.9 ± 0.2	99.0 ± 0.1	100.0 ± 0.0	91.7 ± 3.9	96.2 ± 0.2
%C	67.8 ± 0.1	71.6 ± 0.2	73.6 ± 0.1	74.9 ± 0.0	71.0 ± 1.6	71.0 ± 1.2
%O	25.7 ± 0.3	18.6 ± 0.3	14.2 ± 0.1	9.4 ± 0.0	26.2 ± 1.7	17.4 ± 0.4
%N	6.5 ± 0.1	9.9 ± 0.1	12.2 ± 0.0	15.6 ± 0.1	2.8 ± 1.7	11.5 ± 0.6

a) The calculated results were obtained as the averages of five sample models.
b) The experimental results were obtained by XPS and RBS [51].

and the corresponding experimental results. The present calculated values for each of the four MPD/TMC mixing ratios, 1 : 4, 1 : 1, 3 : 2, 4 : 1, were obtained as the averages of five bulk membrane models simulated from the different monomer model systems with each of the four ratios. It was found that the present results of the MPD/TMC mixing ratios 1 : 4 and 1 : 1 were in good agreement with the corresponding experimental results obtained by XPS and RBS, respectively [51].

Taking account of the actual mechanism of the IP reaction and experimental treatment of water permeation, unreacted acid chloride groups in the FT-30 membrane might be converted into carboxylic groups in the experimental situation [51, 52]. Therefore, we regarded the chlorine atoms of acid chloride groups as the oxygen atoms of carboxylic groups for the estimation of the present results. It may also be considered that the remaining monomers of MPD and TMC in the FT-30 membrane essentially flow out. Therefore, these monomers were deleted from the membrane models obtained and not included for the estimation.

To examine the correspondence between the present and experimental results, we reconsider the reaction mechanism of the IP in detail. As shown in Figure 9.10, an FT-30 membrane is experimentally formed by IP reaction between the aqueous and the organic phase in which MPD and TMC monomers are dissolved, respectively [53]. It was reported that the rate-determining step in the IP reaction is the diffusion of MPD monomers in the aqueous phase into the organic one [52, 54–58]. In fact, this polymerization reaction is promoted by the formation of amide bonds between MPD and TMC monomers, the former diffusing into the organic phase and encountering the latter [52, 54–58]. It can be expected, therefore, that the concentration of MPD monomer should become lower gradually from the aqueous to the organic phase (see Figure 9.10).

On the other hand, experimental observations [51, 59] have shown that the structure of an FT-30 membrane is divided into two regions: (i) the near-surface active (NSA) region (from the surface of the membrane to ~25 nm depth [59]) consisting of the coating layer and the partial active layer and (ii) the interior active (IA) region (from 25 to ~141 nm depth [49]) consisting of the active layer

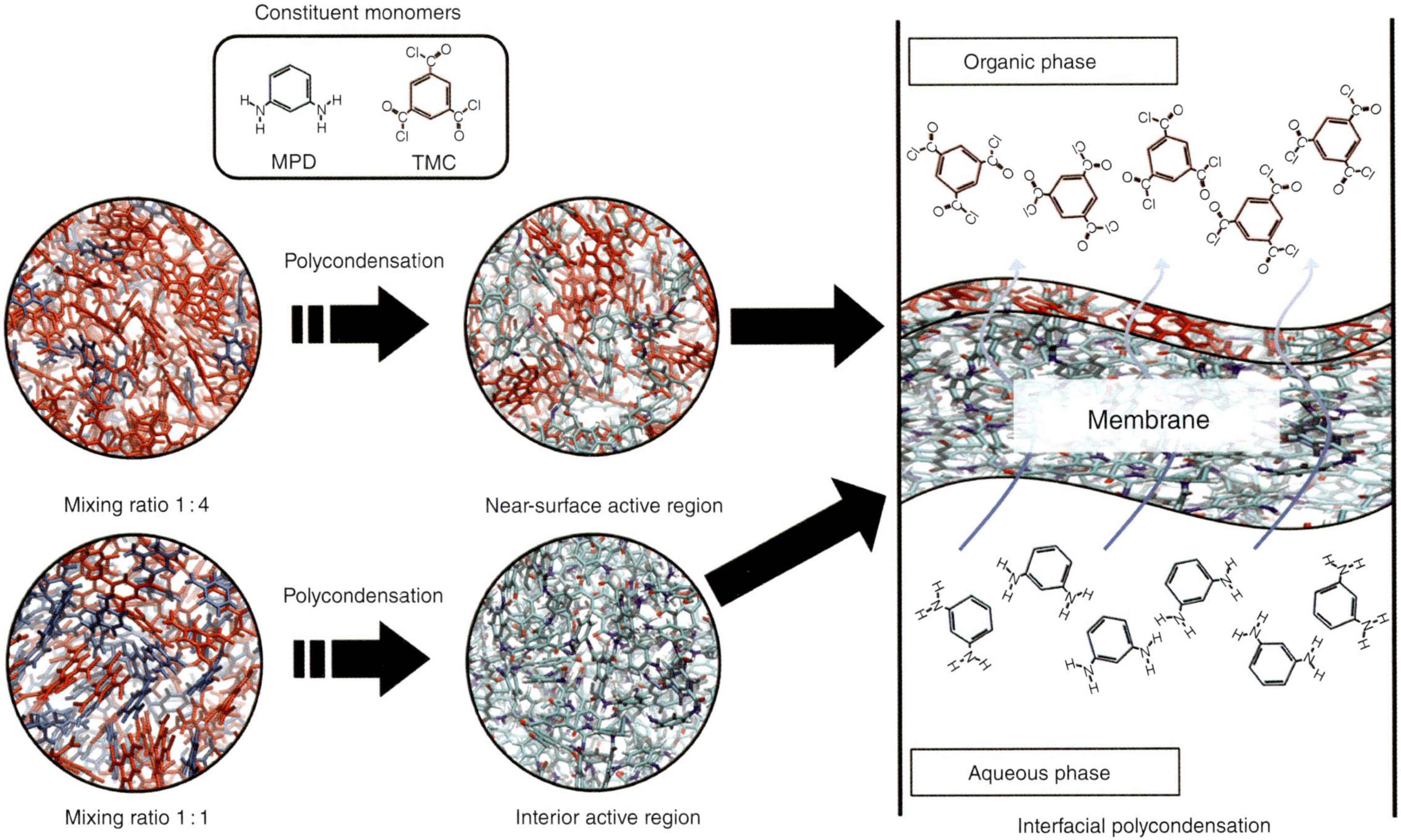

Figure 9.10 Schematic representation of the reaction mechanism of the interfacial polycondensation in the experimental treatment to produce FT-30 membrane. Source: Reproduced with permission from Suzuki et al. [20]. Copyright 2015, American Chemical Society.

including most of the membrane structure (see Figure 9.10). The difference in structural properties between these regions has been observed in some experiments [51, 59, 60]. According to the experimental study by Coronell et al. [51], in the FT-30 membrane the properties of the NSA region can be characterized by XPS, while the volume-averaged characteristics of the IA region can be quantified by RBS.

From the correspondence shown in Table 9.3, the reaction mechanism of the IP, and the properties of the experimental methods, we determine the MPD/TMC mixing ratio for the formation process of the FT-30 membrane. It seems that the NSA region would be formed at the boundary surface between the IA region and the organic phase (see Figure 9.10). In addition, the number density of MPD monomer in the NSA region should be lower than that of TMC monomer because the MPD concentration decreases from the aqueous to the organic phase. It is also expected that the number density of TMC monomer in the NSA region would be higher than that of MPD because of contact with the organic phase. On the other hand, since the IA region can be formed at the interface where MPD and TMC monomers are mixed in the same amounts, their number densities in the region should be equal. From the above results regarding the formation process of the FT-30 membrane, it can be concluded that the IA region is formed from the initial states with the MPD/TMC mixing ratio 1 : 1, and then the NSA region is formed with the MPD/TMC mixing ratio 1 : 4 at the boundary surface between the IA region and the organic phase (see Figure 9.10).

9.3.3.2 Water Permeability and Fidelity of the Membrane Model

Considering that the major part of the FT-30 membrane is the IA region [20], we made a model system consisting of a slab membrane model sliced out from the bulk model with MPD/TMC mixing ratio 1 : 1, and bulk water rearranged on both sides of the slab model (Figure 9.11a). We then performed a water diffusion (permeation) simulation through the slab model. To investigate the properties of water permeability, Figure 9.11b shows the number density profile of water molecules in the membrane interior region and bulk water model. The abscissa of Figure 9.11b is the z coordinate originating at the central position of all the atoms forming the whole slab model. The total mass density of constituent atoms in the membrane and the water molecules within a region of ±24.35 Å (corresponding to the total width 48.7 Å of the slab membrane model) (Figure 9.11a) was 1.39 g cm^{-3}, which is in good agreement with the experimental value of the hydrated FT-30 membrane (1.38 g cm^{-3}) [61].

To evaluate the properties of the water permeability of the membrane obtained, we further calculated the partition coefficient K defined by

$$K = \frac{\rho_{\text{mbrn}}}{\rho_{\text{bulk}}} \tag{9.18}$$

where ρ_{mbrn} and ρ_{bulk} are the water number densities in the membrane interior region and in bulk water region at the equilibrium state by MD simulation, respectively. To estimate the ρ_{mbrn} value, we determined a membrane interior region within ±10 Å to prevent the influence of the membrane structure where the intercellular amide bonds were cut artificially. Both ρ_{mbrn} and ρ_{bulk} values

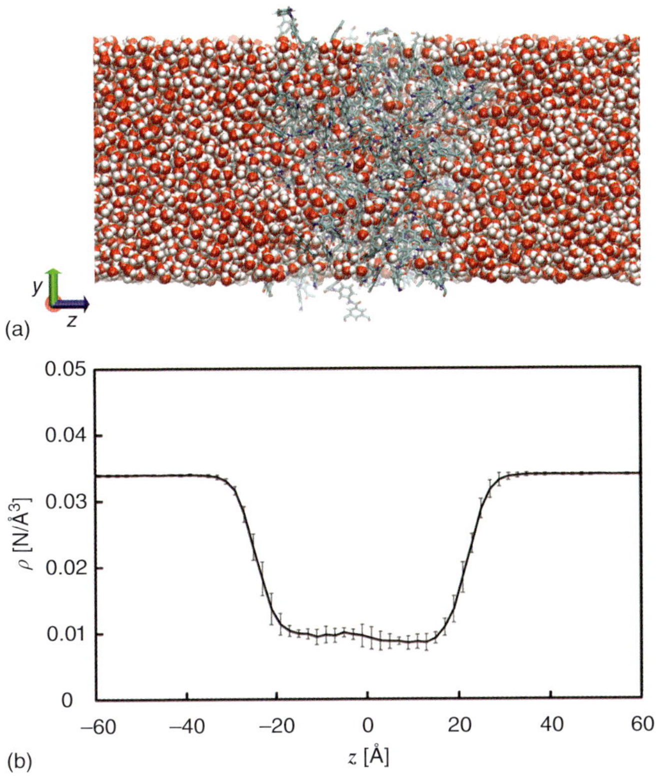

Figure 9.11 (a) A typical snapshot of the equilibrium state of the model system consisting of the slab membrane model and bulk water arranged on both sides of the slab model. The slab membrane model and all water molecules are expressed by ball and stick and sphere models, respectively. (b) Number density of the water molecule profile in the region of the slab membrane and bulk water as a function of *z* coordinate originating at the central position of all the atoms forming the whole slab model. Source: Reproduced with permission from Suzuki et al. [20]. Copyright 2015, American Chemical Society.

were then calculated by averaging the number densities in the membrane interior region and in the two regions of bulk water from +35 to 55 Å and from −55 to −35 Å, in Figure 9.11b, respectively.

As a result, we obtained 0.275 for K in good agreement with both the experimental value of 0.29 [61] and the recent theoretical ones (0.21 and 0.30) [62, 63]. The former value of 0.21 was theoretically calculated for the same system size as the present one, while the latter value 0.30 was calculated for a system that is eight times larger. From this fact, the present result may be considered to be sufficiently reasonable and, therefore, the present Red Moon method with the MBC to realize the formation of intercellular amide bonds should be valid enough to reproduce effectively the experimental properties in such simulation models with smaller system sizes.

In addition, it is meaningful to note that the standard error of partition coefficient K was ± 0.018 (sufficiently small) and, therefore, K can be recognized to be nearly independent of the initial configurations in the water diffusion simulations through slab membrane models.

The simulation results of water diffusion (permeation) through slab membrane models should be significantly affected by the spatial arrangement constructed. However, it was found that not only the DPC and the composition ratios but also the water number density of the hydrated membrane and the partition coefficient K corresponded to the experimental values quantitatively [20]. Thus, it should be concluded that the membrane model formed by Red Moon simulation fully reproduces the actual inhomogeneous spatial arrangement of the FT-30 membrane.

9.3.3.3 Characteristics of Red Moon Method and Its Possibility

Even if both PBC and MBC are applied legitimately to the model systems with the unit cell (48.7 Å × 48.7 Å × 48.7 Å), this size is insufficient in treating the target system corresponding to the actual material. Therefore, the model systems obtained by performing the Red Moon simulation can be considered identical to a partial sampling of the equilibrium structure at chemical equilibrium state. Nevertheless, since both the scale sizes of the actual NSA and IA regions are one order of magnitude larger than the size of the unit cell, the present model systems of aromatic polyamide membranes may be regarded as local equilibrium structures in their regions. Thus, it is meaningful to note that application of the present method to the FT-30 membrane formation was successful and the membranes obtained could be utilized to simulate a stable spatial structure in the local equilibrium state under a nonequilibrium stationary state of permeation.

In the future, by performing saltwater diffusion (permeation) simulation through an RO membrane such as FT-30 formed by the present methods, we will evaluate the free energy profiles of both water and salt ions for the membrane systems. In this way, we would try to reveal the microscopic reasons for physicochemical phenomena that are not well understood, i.e. how water can permeate membranes with a high flux while salt ions are largely rejected. The atomistic reaction simulation using the Red Moon method with MBC should become a unique approach complementary to experiments that will support the design of new materials with high performance such as RO membranes.

9.4 Lithium Ion and Sodium Ion Batteries

For the fourth example, the Red Moon method was applied to investigate the formation process of solid–electrolyte interphase (SEI) film in lithium ion batteries (LIBs) [11] and sodium (Na) ion batteries (NIBs) [64]. The current LIB performance with liquid electrolytes is strongly dependent on the formation of stable SEI film on the anode surface, formed as a result of reduction of the electrolyte [65]. The Red Moon simulations were executed in 1.1 mol l^{-1} $LiPF_6$ ethylene carbonate (EC)- or propylene carbonate (PC)-based electrolytes with the carbon anode. In the SEI films thus obtained, it was found that the inorganic salts (e.g. Li_2CO_3) are accumulated closer to the anode, while the organic ones (e.g. Li_2BDC, Li_2DMBDC) mainly exist in the outer layer as seen in the experimental observations [66–68]. We also found that the dense EC-based SEI film develops so as to protect the electrolyte from reduction, although the PC-based one becomes sparse and cannot protect the electrolyte (Figures 9.12 and 9.13). It was understood that the SEI film formation shows strong sensitivity to the small structural difference of electrolyte molecules at the microscopic

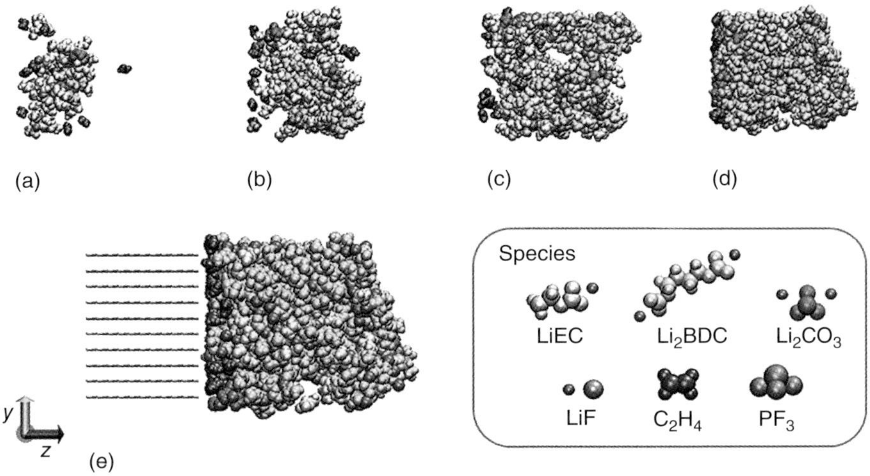

Figure 9.12 The SEI film formation process in the EC-based electrolyte. Five typical snapshots of the changes of the aggregation states of reaction products during the Red Moon simulation (bulk EC not shown) in the (a) 200 MC/MD cycle, (b) 400 MC/MD cycle, (c) 800 MC/MD cycle, (d) 1600 MC/MD cycle, and (e) 2000 MC/MD cycle. Source: Reproduced with permission from Takenaka et al. [11]. Copyright 2014, American Chemical Society.

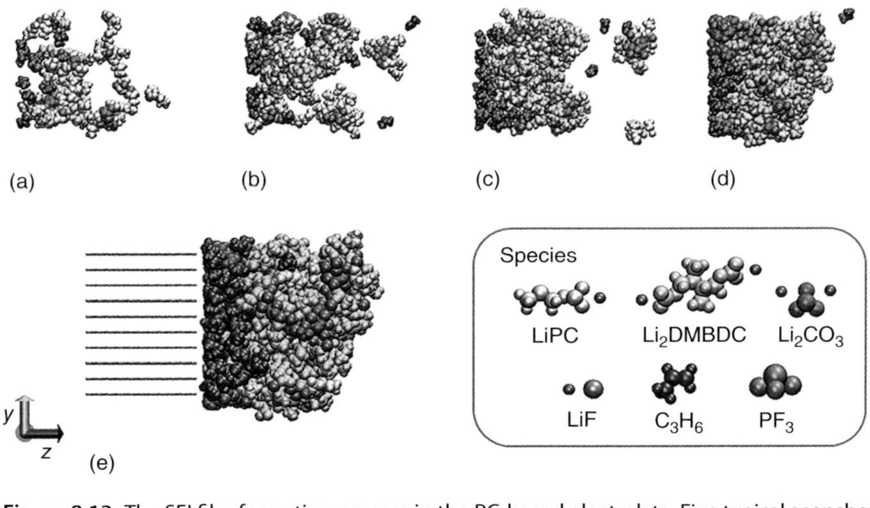

Figure 9.13 The SEI film formation process in the PC-based electrolyte. Five typical snapshots of the changes of the aggregation states of reaction products during the Red Moon simulation (bulk PC not shown) in the (a) 200 MC/MD cycle, (b) 400 MC/MD cycle, (c) 800 MC/MD cycle, (d) 1600 MC/MD cycle, and (e) 2000 MC/MD cycle. Source: Reproduced with permission from Takenaka et al. [11]. Copyright 2014, American Chemical Society.

level. On the other hand, since the cost of Li increases by the year, NIBs have recently attracted special attention as a substitute for Li ion batteries and the Red Moon simulations in PC-based electrolyte were performed with and without fluoroethylene carbonate (FEC) additives [64] because it is known as an efficient electrolyte additive to improve SEI film formation in NIBs.

9.4.1 Strong Sensitivity to Small Structural Difference of Electrolyte Molecules on the Solid–Electrolyte Interphase (SEI) Film Formation

In Figures 9.12 and 9.13, the five snapshots show changes of the aggregation states of reaction products in the EC- and PC-based electrolyte, respectively, displaying how the constituents of electrolyte solutions grow into the SEI films during the Red Moon simulations. As shown in Figure 9.12, the solvent EC molecules were continually reduced to form the LiEC complexes on the graphite surface until the reaction products developed into the stable SEI film. During the SEI film formation processes, the LiEC complexes were further reduced on the graphite surface and dissociated to form C_2H_4 gases and Li_2CO_3 complexes. At the same time, the Li_2BDC dimers were formed via their radical polymerization. Additionally, a few PF_6^- anions were reduced on the graphite surface, and dissociated to form PF_3 gases and LiF complexes. Finally, these reaction products were linked to each other by the ionic bond with the sharing Li^+ cations, and formed the stable SEI film. However, in the PC-based electrolyte, it was found that some reaction products drift away into the bulk region of the electrolyte solution during the SEI film formation process (Figure 9.13). Since the structural difference between the solvent EC and PC molecules is due to the existence of the methyl group, its existence in solvent PC molecules and the PC-originating reaction products should prevent them from clumping together to make the SEI film formation processes unstabilized, enhancing the products drifting into the bulk region of the electrolyte solution.

9.4.2 Microscopic Additive Effect on Solid–Electrolyte Interphase (SEI) Film Formation in Sodium Ion Batteries

NIBs are appealing alternatives to LIBs due to the increase in the cost of Li [69–74]. The NIB performance with liquid electrolytes strongly depends on a stable SEI film on the anode surface. FEC is known as an efficient electrolyte additive to improve the SEI film formation in NIBs [70]. However, the molecular mechanism of such additive effect is still unknown. To investigate this mechanism, assuming the RS consisting of a list of elementary reaction processes necessary and sufficient to simulate the SEI film formation as shown in Figure 9.14, a number of atomistic reaction simulations in PC-based electrolyte with and without FEC additives were performed using Red Moon methodology, and the smaller irreversible capacity and the smoother SEI film surface in FEC-added electrolyte were obtained as in the experimental observations [64]. Further, this study showed for the first time that intact FEC molecules can improve SEI film formation so as to enhance the network formation of organic reaction products such as NaPC and Na_2DMBDC owing to the large electronegativity of their

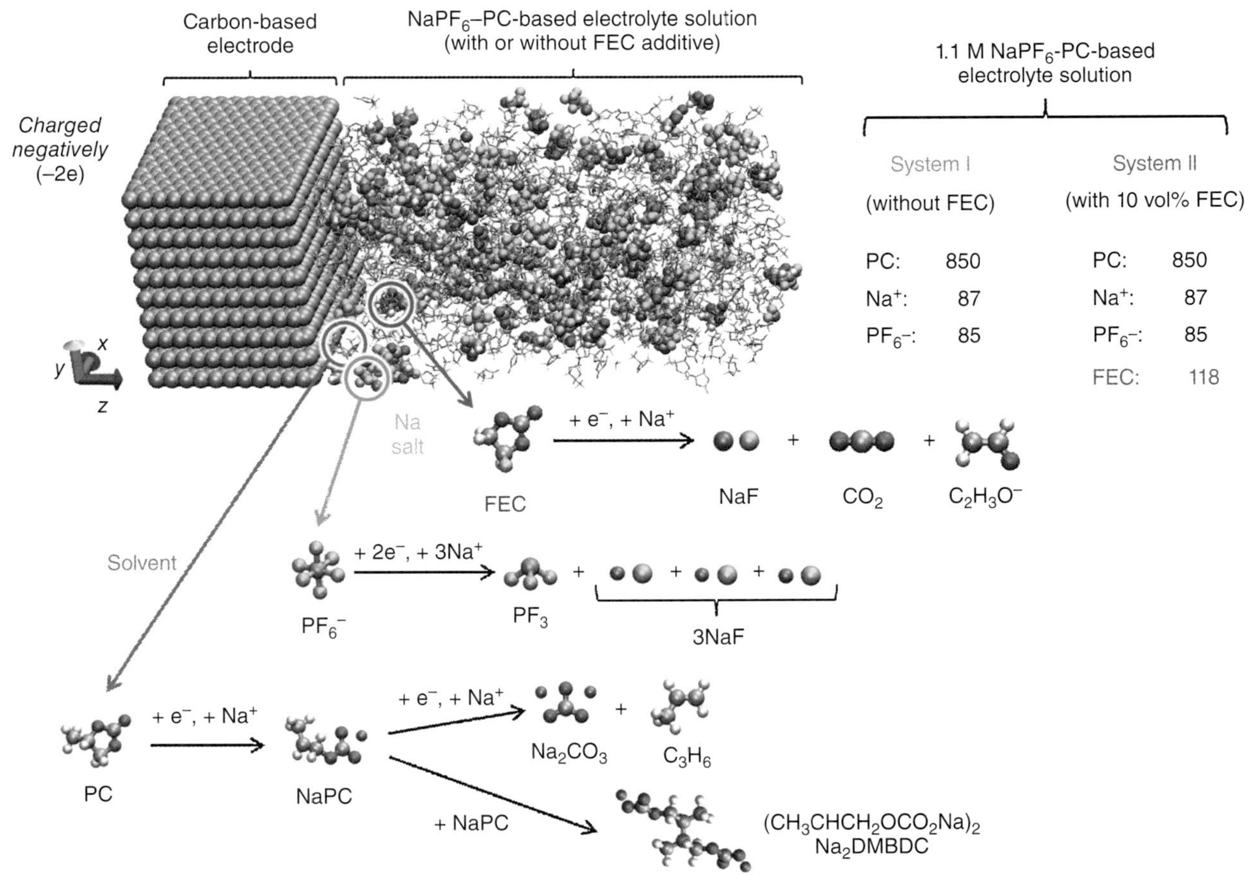

Figure 9.14 Model simulation system and reaction scheme for Red Moon simulations in the FEC-added NaPF$_6$–PC electrolyte solution. Here, the solvent PC molecules are depicted by a stick model, while the remaining ones are depicted by a ball model (full vdW radii). Source: Reproduced partially with permission from Takenaka et al. [64]. Copyright 2015, American Chemical Society.

fluorine atoms (Figure 9.15) [64]. Specifically, by adding FECs, the film thickness was found to decrease by 32% within the present approximation. Such drastic change is originating in the suppression of the unstable growth of the SEI film (Figure 9.16), which must be one of the most important effects of FEC additive at the microscopic level. These new microscopic insights should provide an important guiding principle for use in designing the best electrolytes to develop high-performance NIBs.

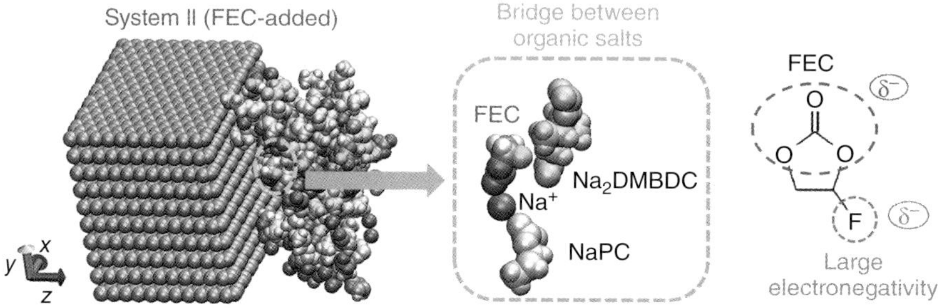

Figure 9.15 Typical snapshot of SEI film formation process at 500 MC/MD cycle in FEC-added electrolyte solution. Source: Reproduced partially with permission from Takenaka et al. [64]. Copyright 2015, American Chemical Society.

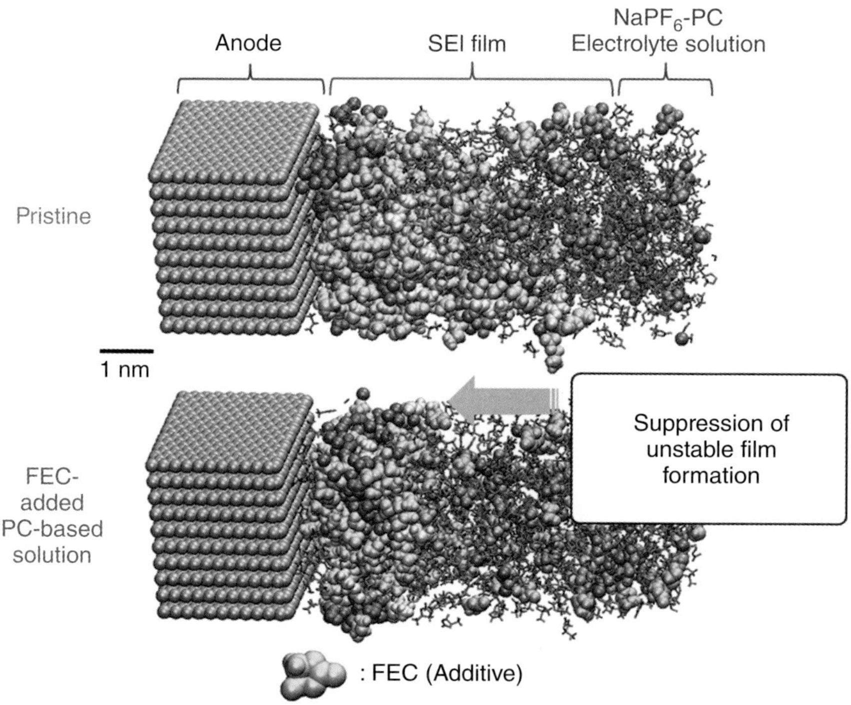

Figure 9.16 Structural characteristic of SEI film formed by FEC addition. FEC additive makes the SEI film compact and suppresses the unstable film formation. Source: Reproduced partially with permission from Takenaka et al. [64]. Copyright 2015, American Chemical Society.

9.5 Summary and Conclusions

In this chapter, after introducing our motivation from the viewpoint of what today's chemists want and need to consider, we have first reviewed in general the Red Moon methodology that is an effective hybrid MC/MD reaction method with a rare event-driving mechanism. It was elucidated that the mechanism simply drives such rare events as chemical reactions and successfully provides meaningful information regarding not only the molecular compositions in chemical equilibrium but also "atomistic" spatial distributions of heterogeneous chemical systems in nonequilibrium thermodynamic states.

The Red Moon methodology can be said to be one of the second-generation CMTs for a practical "atomistic" molecular simulation of large-scale chemical reaction systems. In this chapter, however, we have also touched concisely a number of conventional first-generation CMTs, namely, the QC methodology of density functional theory and MM and MD methods. Actually, the extensive use of these CMTs is essential to construct the RSs that should be prepared before applying the Red Moon methodology.

In the latter half of the chapter, we have demonstrated four typical applications using these first- and second-generation CMTs. As the first and second applications, we have dealt with two radical polymerization reaction mechanisms; one is the ethylene coordinative polymerization reaction mechanism on (Pyridylamide)Hf(IV) catalyst and the other is the propylene polymerization reaction mechanism on $[H_2Si(Ind)_2ZrCH_3]^+$ catalyst cation and $[H_2Si(Ind)_2ZrCH_3]^+$ $[CH_3B(C_6F_5)_3]^-$ catalysts ion pair. In the former, the active site-opening mechanism is discussed using the MM and MD simulations, while in the latter, the energetics of propylene insertion into active catalyst is thoroughly analyzed by DFT calculations. In the third application, the Red Moon methodology was applied to the interfacial polycondensation reactions of aromatic polyamide polymerization. In the fourth application, the SEI film formations in lithium ion and sodium ion batteries were treated to show their characteristic changes depending on the variety of components of electrolytes.

In conclusion, the Red Moon method is promising as a new CMT to study complex chemical reaction systems for their reaction mechanisms, various properties of products, and their stereochemistry. By only slightly changing the way that the product state s is generated, the framework of the present method can be easily transferred to chemical reaction systems with the forces obtained by *ab initio* electronic structure calculations, including those by quantum mechanical/molecular mechanical (QM/MM) methods [10, 18].

We are now trying to deal with and to study more extensive chemical systems from the viewpoint of the Red Moon methodology by extending the protocol of Red Moon methodology to include more general rare events other than chemical reactions themselves. We believe this methodology would become another CMT in the near future. We are also dreaming of or even looking forward to these CMTs spreading over as a standard tool not only to pure scientists but also to industrial chemists.

Acknowledgments

This work was supported by the Core Research for Evolutional Science and Technology (CREST) of the Japan Science Technology Agency (JST) "High Performance Computing for Multi-scale and Multi-physics Phenomena" and "Establishment of Molecular Technology towards the Creation of New Functions" [4]; by a Grant-in-Aid for Science Research from the Ministry of Education, Culture, Sport, Science, and Technology (MEXT) in Japan; and also by the MEXT programs "Elements Strategy Initiative to Form Core Research Center" (since 2012) [6] and Priority Issue on Post-K computer (Development of new fundamental technologies for high-efficiency energy creation, conversion/storage, and use). The calculations were partially performed using several computing systems at the Information Technology Center in Nagoya University.

References

1 Schreiner, P.R., Allen, W.D., Orozco, M. et al. (2014). *Computational Molecular Science.* Wiley.

2 (a) Schleyer, P.v.R., Allinger, N.L., Clark, T. et al. (eds.) (1998). *Encyclopedia of Computational Chemistry.* Wiley. (b) Leszczynski, J., Kaczmarek-Kedziera, A., Puzyn, T. et al. (eds.) (2017). *Handbook of Computational Chemistry,* 2e. Springer.

3 Research Project (JST-CREST) (2013). Establishment of computational molecular technology towards macroscopic chemical phenomena -molecular control of complex chemical reactions, stereospecificity and aggregate structures. http://www.mt.jst.go.jp/en/researchers/masataka_nagaoka.html (accessed 25 March 2017).

4 Research Area of JST-CREST (2013). Establishment of Molecular Technology towards the Creation of New Functions. http://www.mt.jst.go.jp/en/ (accessed 25 March 2017).

5 RIKEN (2010). Advanced Institute for Computational Science (AICS). http://www.riken.jp/en/research/labs/aics/ (accessed 25 March 2017).

6 New Energy and Industrial Technology Development Organization (NEDO) (2016). News release. http://www.nedo.go.jp/english/news/AA5en_100081.html (accessed 25 March 2017).

7 Matsumoto, K., Sankaran, S.K., Takayanagi, M. et al. (2016). *Organometallics* 35: 4099.

8 Sankaran, S.K., Koga, N., and Nagaoka, M. (2016). *Bull. Chem. Soc. Jpn.* 89: 1093.

9 Zhibin, G. (ed.) (2009). *Metal Catalysts in Olefin Polymerization.* Berlin: Springer.

10 Nagaoka, M., Suzuki, Y., Okamoto, T., and Takenaka, N. (2013). *Chem. Phys. Lett.* 583: 80.

11 Takenaka, N., Suzuki, Y., Sakai, H., and Nagaoka, M. (2014). *J. Phys. Chem. C* 118: 10874.

12 Allen, M.P. and Tildesley, D.J. (1987). *Computer Simulation of Liquids.* Oxford: Clarendon.

13 Heerman, D.W. (1990). *Computer Simulation Methods in Theoretical Physics*, 2e. Berlin: Springer.
14 Rappe, A.K. and Casewit, C.J. (1997). *Molecular Mechanics across Chemistry.* University Science Books.
15 Tomasi, J. and Persico, M. (1994). *Chem. Rev.* 94: 2027.
16 Okuyama-Yoshida, N., Nagaoka, M., and Yamabe, T. (1998). *Int. J. Quantum Chem.* 70: 95.
17 Nagaoka, M., Okuyama-Yoshida, N., and Yamabe, T. (1998). *J. Phys. Chem. A* 102: 8202.
18 Kitamura, Y., Takenaka, N., Koyano, Y., and Nagaoka, M. (2016). Free energy gradient method and its recent related developments: free energy optimization and vibrational frequency analysis in solution. *Quantum Modeling of Complex Molecular Systems*, Challenges and Advances in Computational Chemistry and Physics (ed. J.-L. Rivail, M. Ruiz-Lopez and X. Assfeld). Springer.
19 Metropolis, N., Rosenbluth, A.W., Rosenbluth, M.N. et al. (1953). *J. Chem. Phys.* 21: 1087.
20 Suzuki, Y., Koyano, Y., and Nagaoka, M. (2015). *J. Phys. Chem. B* 119: 6776.
21 Binran, K. (2004) Pure Chemistry 5th Ed.; Applied Chemistry 6th Ed). *Handbook of Chemistry.* Tokyo: Maruzen.
22 Frisch, M.J., Trucks, G.W., Schlegel, H.B. et al. (2009). *Gaussian 09 (Revision D.01).* Wallingford, CT: Gaussian, Inc.
23 Zhao, Y. and Truhlar, D.G. (2008). *Theor. Chem. Acc.* 120: 215.
24 Hay, P.J. and Wadt, W.R. (1985). *J. Chem. Phys.* 82: 270.
25 Ehlers, A.W., Böhme, M., Dapprich, S. et al. (1993). *Chem. Phys. Lett.* 208: 111.
26 Ditchfield, R., Hehre, W.J., and Pople, J.A. (1971). *J. Chem. Phys.* 54: 724.
27 Clark, T., Chandrasekhar, J., Spitznagel, G.W., and Schleyer, P.V.R. (1983). *J. Comput. Chem.* 4: 294.
28 Marenich, A.V., Cramer, C.J., and Truhlar, D.G. (2009). *J. Phys. Chem. B* 113: 6378.
29 Case, D.A., Darden, T.A., Cheatham, T.E. et al. (2012). *AMBER 12.* San Francisco, CA: University of California.
30 Wang, J., Wolf, R.M., Caldwell, J.W. et al. (2004). *J. Comput. Chem.* 25: 1157.
31 Breneman, C.M. and Wiberg, K.B. (1990). *J. Comput. Chem.* 11: 361.
32 Besler, B.H., Merz, K.M., and Kollman, P.A. (1990). *J. Comput. Chem.* 11: 431.
33 Singh, U.C. and Kollman, P.A. (1984). *J. Comput. Chem.* 5: 129.
34 Reed, A.E., Weinstock, R.B., and Weinhold, F. (1985). *J. Chem. Phys.* 83: 735.
35 Sugita, Y. and Okamoto, Y. (1999). *Chem. Phys. Lett.* 314: 141.
36 Macchioni, A. (2005). *Chem. Rev.* 105: 2039.
37 Ahmadi, M. and Nasresfahani, A. (2015). *Macromol. Theory Simul.* 24: 311.
38 Arriola, D.J., Carnahan, E.M., Hustad, P.D. et al. (2006). *Science* 312: 714.
39 Kaminsky, W. and Laban, A. (2001). *Appl. Catal. A* 222: 47.
40 Chen, E.Y.X. and Marks, T.J. (2000). *Chem. Rev.* 100: 1391.
41 Kaminsky, W. and Polym, J. (2004). *J. Polym. Sci. A: Polym. Chem.* 42: 3911.
42 Laine, A., Coussens, B.B., Hirvi, J.T. et al. (2015). *Organometallics* 34: 2415.
43 Chan, M.S.W. and Ziegler, T. (2000). *Organometallics* 19: 5182.
44 Bosnich, B. (1994). *Chem. Soc. Rev.* 23: 387.

45 Doman, T.N., Landis, C.R., and Bosnich, B. (1992). *J. Am. Chem. Soc.* 114: 7264.
46 Doman, T.N., Hollis, T.K., and Bosnich, B. (1995). *J. Am. Chem. Soc.* 117: 1352.
47 Rappé, A.K., Casewit, C.J., Colwell, K. et al. (1992). *J. Am. Chem. Soc.* 114: 10024.
48 Udo, H., Roland, M., Markus, K., and Gerhard, E. (1994). *Organometallics* 13: 2380.
49 Brintzinger, H.H., Prosenc, M.H., Schaper, F. et al. (1999). *J. Mol. Struct.* 485–486: 409–419.
50 Nicole, S., Frank, S., Katrin, S. et al. (2000). *Organometallics* 19: 3597.
51 Coronell, O., Marinas, B.J., and Cahill, D.G. (2011). *Environ. Sci. Technol.* 45: 4513.
52 Kim, S.H., Kwak, S.Y., and Suzuki, T. (2005). *Environ. Sci. Technol.* 39: 1764.
53 FILMTEC Reverse Osmosis Membranes Technical Manual, Form No. 609-000710109; Dow Water Solutions: Midland, MI.
54 Freger, V. (2003). *Langmuir* 19: 4791.
55 Enkelmann, V. and Wegner, G. (1976). *Mechanism of Interfacial Makromol. Chem.* 177: 3177.
56 Janssen, L.J.J.M. and te Nijenhuis, K. (1992). *J. Membr. Sci.* 65: 59.
57 Karode, S.K., Kulkarni, S.S., Suresh, A.K., and Mashelkar, R.A. (1997). *Chem. Eng. Sci.* 52: 3243.
58 Ji, J., Dickson, J.M., Childs, R.F., and McCarry, B.E. (2000). *Macromolecules* 33: 624.
59 Coronell, O., Marinas, B.J., Zhang, X., and Cahill, D.G. (2008). *Environ. Sci. Technol.* 42: 5260.
60 Tang, C.Y., Kwon, Y.N., and Leckie, J.O. (2007). *J. Membr. Sci.* 287: 146.
61 Mi, B., Cahill, D.G., and Marinas, B.J. (2007). *J. Membr. Sci.* 291: 77.
62 Harder, E., Walters, D.E., Bodnar, Y.D. et al. (2009). *J. Phys. Chem. B* 113: 10177.
63 Luo, Y., Harder, E., Faibish, R.S., and Roux, B. (2011). *J. Membr. Sci.* 384: 1.
64 Takenaka, N., Sakai, H., Suzuki, Y. et al. (2015). *J. Phys. Chem. C* 119: 18046.
65 Xu, K. (2014). *Chem. Rev.* 114: 11503.
66 Aurbach, D., Ein-Ely, Y., and Zaban, A. (1994). *J. Electrochem. Soc.* 141: L1.
67 Aurbach, D., Zaban, A., Schecheter, A. et al. (1995). *J. Electrochem. Soc.* 142: 2873.
68 Peled, E., Golodnitsky, D., and Ardel, G. (1997). *J. Electrochem. Soc.* 144: L208.
69 Komaba, S., Murata, W., Ishikawa, T. et al. (2011). *Adv. Funct. Mater.* 21: 3859.
70 Komaba, S., Ishikawa, T., Yabuuchi, N. et al. (2011). *ACS Appl. Mater. Interfaces* 3: 4165.
71 Slater, M.D., Kim, D., Lee, E., and Johnson, C.S. (2013). *Adv. Funct. Mater.* 23: 947.
72 Yabuuchi, N., Kubota, K., Dahbi, M., and Komaba, S. (2014). *Chem. Rev.* 114: 11636.
73 Kitajyou, A., Yamaguchi, J., Hara, S., and Okada, S. (2014). *J. Power Sources* 247: 391.
74 Choi, N.S., Yew, K.H., Lee, K.Y. et al. (2006). *J. Power Sources* 161: 1254.

10

Multi-timescale Measurements with Energetic Beams for Molecular Technology

Masaki Hada[1] *and Taiki Hoshino*[2]

[1] *Okayama University, 3-1-1 Tsushimanaka, Kita-ku, Okayama 700-8530, Japan*
[2] *RIKEN SPring-8 Center, 1-1-1 Kouto, Sayo-cho, Sayo-gun, Hyogo 679-5148, Japan*

10.1 Introduction

The multi-timescale measurement methodology with energetic beams, e.g. neutrons, electrons, X-ray, and laser, will be introduced in this chapter. One of the challenges in science is to record atoms or molecules in motion to enable the direct observation of chemical reactions or responses as they occur. In general, the dynamics of molecules or molecular-assembled systems during the reactions or responses can be detected by time-domain, frequency-domain, or time-correlation measurements. The concept of the former measurements is rather simple; phenomena induced by triggers are detected by probes. In this sense, one might not need special techniques for time-domain measurements in the second, minute, hour, or longer timescale. Fast electronic detection or electronic synchronization of the triggers and probes can be useful to follow the faster dynamics of materials under stimuli on the milliseconds to nanoseconds (10^{-3} to 10^{-9} seconds) timescale. However, to understand even faster dynamics, e.g. on the picosecond to femtosecond (10^{-12} to 10^{-15} seconds) timescale, which would be considered the fastest motion of atoms or molecules that occurs, one needs to perform the so-called "pump–probe" measurements. Optically synchronized pulsed triggers and probes whose durations are shorter than the target time range are used for pump–probe experiments. In the first part of this chapter, we focus on the recent progress of time-domain measurements on the picosecond-to-femtosecond timescale. Raman spectroscopy is one of the representative frequency-domain measurements, where the light wave resonates with the atomic or molecular vibration modes. Thus, the dynamics of the molecular vibration can be also indirectly obtained with frequency-domain measurements. Time-correlation measurements have an advantage in studying the dynamics in systems where the average structure remains constant. In "photon correlation spectroscopy (PCS) technique," the dynamics of the systems are studied by analyzing the temporal correlations of the scattered intensities of coherent light. Using visible laser as the illumination probe, PCS, which is also called dynamic light scattering (DLS), is widely used in laboratories

Molecular Technology: Materials Innovation, Volume 3, First Edition.
Edited by Hisashi Yamamoto and Takashi Kato.

for studying the dynamics of systems on length scales from micrometers down to sub-micrometers. On the other hand, "X-ray photon correlation spectroscopy" (XPCS) is a technique using coherent X-rays, which has been developed at third-generation synchrotron radiation facilities, and with which the dynamics on timescales ranging from $\sim 10^{-6}$ to 10^3 seconds and length scales from 10^{-9} to 10^{-3} m are available. XPCS is a powerful technique to study molecular dynamics, which has been advancing with the recent development of X-ray sources. In the latter part of this chapter, we will present an overview of the studies using the XPCS technique and its recent progress.

10.2 Time-Domain Measurements

10.2.1 Femtosecond Optical Pump–Probe Studies

Almost all pump–probe experiments are based on femtosecond lasers. Femtosecond lasers are commercially available with pulse lengths as short as 10–100 fs, 1–10 W of average power, and repetition rates in the 1 kHz to 100 MHz range. The most common laser systems are based on titanium-doped sapphire (Ti:sapphire) gain media [1], and recently more thermally stable ytterbium-doped potassium-gadolinium tungstate (Yb:KGW) or ytterbium-doped potassium-yttrium tungstate (Yb:KYW) gain media are available. The femtosecond laser pulses are used for both the trigger and the probe in the simplest pump–probe experiments, i.e. optical pump–probe experiments. Figure 10.1 illustrates the schematics of a typical optical pump–probe setup. The femtosecond optical pulse from the laser is separated by a beam splitter into two arms, the pump and the probe arms. Different nonlinear media and optical components are usually used to modify the fundamental wavelength of the

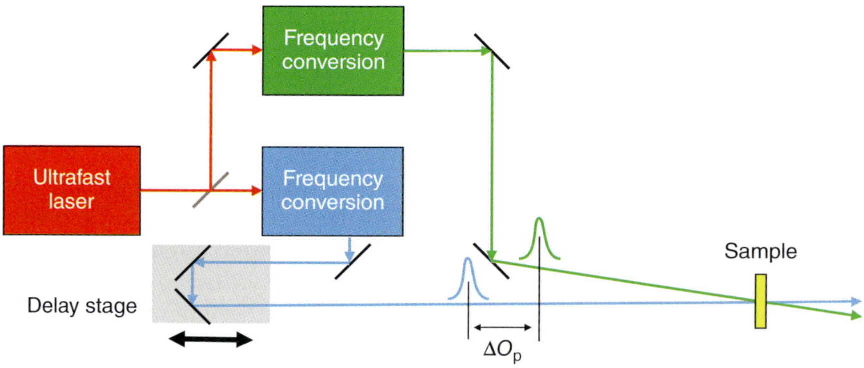

Figure 10.1 General schematics of an optical pump–probe setup. The optical pulse from ultrafast laser is split into two arms by a beam splitter. Different optical components are used to modify the fundamental frequency of the laser according to the experimental requirements. The optical path difference between the probe and pump arms can be modified through the motion of a delay stage. ΔO_p is the optical path difference between the pump and probe pulses. The time delay τ is obtained as the optical path difference divided by the speed of the light.

laser according to the electronic properties of the sample under study and the experimental requirements. The pump arm usually carries most of the energy and the optical pump pulse is responsible for triggering the dynamics of interest. A probe pulse monitors the changes in the material of interest. The time delay (τ) between the probe and pump pulses is given by the optical path difference (ΔO_p) between the pulses as they reach the sample position:

$$\tau = \frac{\Delta O_p}{c} \tag{10.1}$$

where c is the speed of light. A temporal increment of 100 fs translates into an optical path increment of only 30 or 15 μm in the motion of a translation stage as shown in Figure 10.1.

In the following section, we introduce the molecular dynamics revealed by femtosecond optical pump–probe experiments [2]. The examples observed by the femtosecond pump–probe experiments range from fundamental science, such as the dynamics of photoinduced phase transition [3, 4], photochemical [5–7] or biological reactions in photosynthesis materials, photoactive chemicals, and proteins [8, 9] to application-oriented science, such as the creation and annihilation of excitons in solar cells and charge-transfer processes in catalysts. Typical time-evolution spectra obtained in femtosecond pump–probe experiments on photofunctional proteins in the liquid phase are shown in Figure 10.2 [10], where the horizontal and vertical axes present time delay and delta-transmission ($\Delta T/T$), sometimes delta-absorption ($\Delta A/A$), or delta-reflection ($\Delta R/R$). Delta-transmission is calculated to be $(T'(\tau) - T)/T$, where T is the transmission without pump light and $T'(\tau)$ is the transmission at the time delay of τ. The time-evolution spectrum has a steep increase (when the sample becomes more transparent) or decrease (darker) just after the photoexcitation, and then it recovers to zero with certain decay time constants. The delta-absorption, delta-transmission, and delta-reflection often differ as

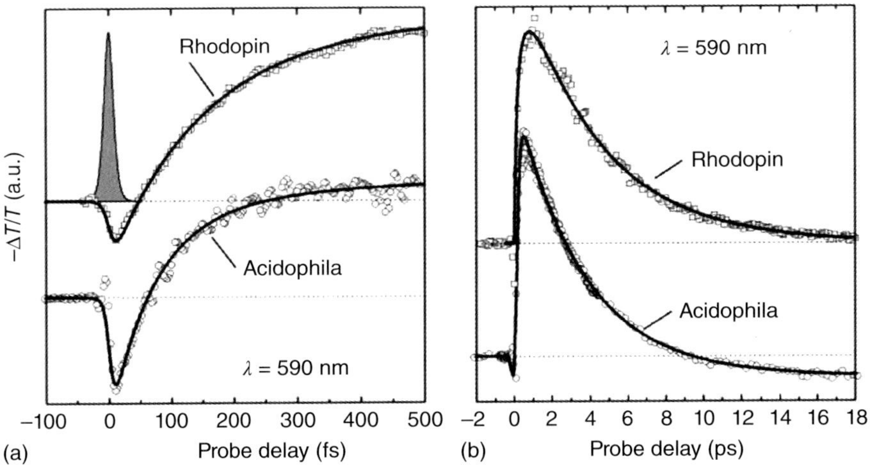

Figure 10.2 The typical time-evolution spectrum (Delta-transmission) obtained in femtosecond pump–probe experiments on photofunctional proteins in solution phase [10].

functions of the probe wavelength, which presents the photoexcited transient nature of the samples, as shown in Figure 10.3. To fully understand the dynamics of the photoinduced phase transition, photochemical, or biological reactions from the femtosecond pump–probe spectra, the assistance of theoretically based model calculations is required; thus, femtosecond optical (especially near-infrared (near-IR), visible-to-ultraviolet (UV) region) pump–probe experiments provide occasionally indirect evidence of the dynamics in materials. To overcome this problem, there are efforts to develop time-resolved mid-IR vibrational spectroscopy setups or Raman vibrational spectroscopy setups with time-resolution of sub-picoseconds to a few picoseconds. These vibrational spectroscopies provide more direct transient structural information [11, 12].

10.2.2 Femtosecond X-ray and Electron Pulse Sources

Other energetic beams such as X-ray and electron can be used as pulsed probes for the picosecond-to-femtosecond timescale. The ultrashort pulses of the wave particles of interest with the proper (<1 Å) wavelength or de Broglie wavelength (λ) enable us to directly access information about the motion of atoms or molecules in the systems. The structural information (d) is given by the well-known Bragg's rule ($2d \sin\theta = n\lambda$) [13–15], where θ is Bragg's angle, or by the intensity of the diffraction peaks ($I(\boldsymbol{K})$) from the calculation of the crystal structural factors ($F(\boldsymbol{K})$):

$$I(\boldsymbol{K})^2 = |F(\boldsymbol{K})|^2 |G(\boldsymbol{K})|^2 \tag{10.2}$$

$$F(\boldsymbol{K}) = \sum_{j=1} f_j(\boldsymbol{K}) T_j(\boldsymbol{K}) \exp(2\pi \boldsymbol{K} \cdot \boldsymbol{r}_j) \tag{10.3}$$

where $\boldsymbol{K}$, $G(\boldsymbol{K})$, $f_j(\boldsymbol{K})$, and $T_j(\boldsymbol{K})$ are the wave vector, Laue function, atomic scattering factor, and temperature factor of the jth atom, respectively. The atomic position in the unit cell is represented as $\boldsymbol{r}_j$. The high-flux X-ray pulsed probes are usually produced from synchrotron or linear accelerator sources. With <100-ps time-resolution synchrotron sources, these have provided insight into reaction intermediates, conformational changes, and crystalline rearrangement [16–19]. However, their nominal temporal resolution is determined by the duration of each stored electron bunch, typically approximately 30 ps [20–22]. The X-ray free electron lasers (XFELs) based on linear accelerators such as linac coherent light source (LCLS), SPring-8 Angstrom Compact free electron LAser (SACLA), and the upcoming European and Swiss XFELs provide very high X-ray flux with femtosecond pulse duration. The huge progress on the generation of ultrashort X-ray pulses reflects the relevance of achieving atomic- or molecular-level views of complex phenomena. There is also a variety of ultrafast diffraction techniques based on femtosecond electron sources. The effective brightness of table-top femtosecond electron sources can be very close to that of XFELs once the ratio of their respective scattering cross sections is taken into account, i.e. electrons scatter about 10^4 to 10^5 times more efficiently than X-rays and are approximately 10^3 times less damaging. In the following section, we will introduce the basic experimental methodology of femtosecond hard X-ray diffraction/spectroscopy and electron diffraction experiments.

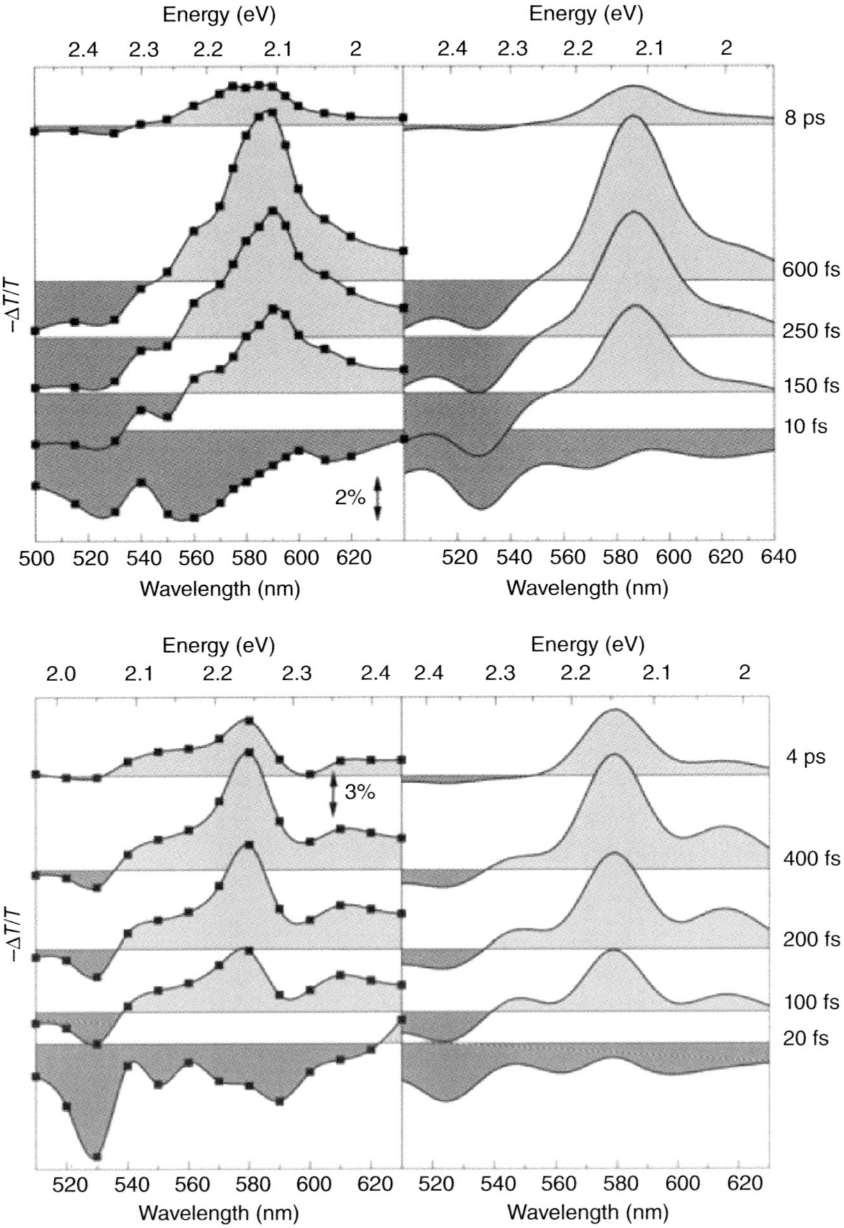

Figure 10.3 Delta-transmission of photofunctional proteins in solution phase as a function of the probe wavelength. Source: Cerullo et al. 2007 [10]. Reprinted with permission of the Royal Society of Chemistry.

Figures 10.4–10.6 show several kinds of X-ray sources in a synchrotron facility. In a synchrotron, single or multiple bunches of electrons are accelerated to the energy of several Giga-electron-volt and sent to a low-emittance storage ring. The electron bunch in the storage ring is compensated for the energy loss induced by the synchrotron emission using a radio frequency (RF) cavity. The RF cavity increases the electrical field as the electron bunch arrives; therefore, the head group of the electron bunch receives less energy than the tail group of the bunch. Thus, the RF cavity not only supplies energy to the electron bunch but also simultaneously compresses the pulse duration of the electron bunch to approximately 30 ps. The compressed electron bunch undergoes undulation by magnetic arrays with alternating polarity, and emits synchrotron hard X-ray radiation (10^4 to 10^5 X-rays per pulse with a bandwidth of 0.1%) along the traveling direction. The pulse duration of the generated X-ray is approximately the same as that of the electron bunch. The X-ray pulses are selected by a mechanical chopper for the pump–probe experiments (Figure 10.4) [23]. The interaction between the electron bunch and a femtosecond optical pulse makes it possible

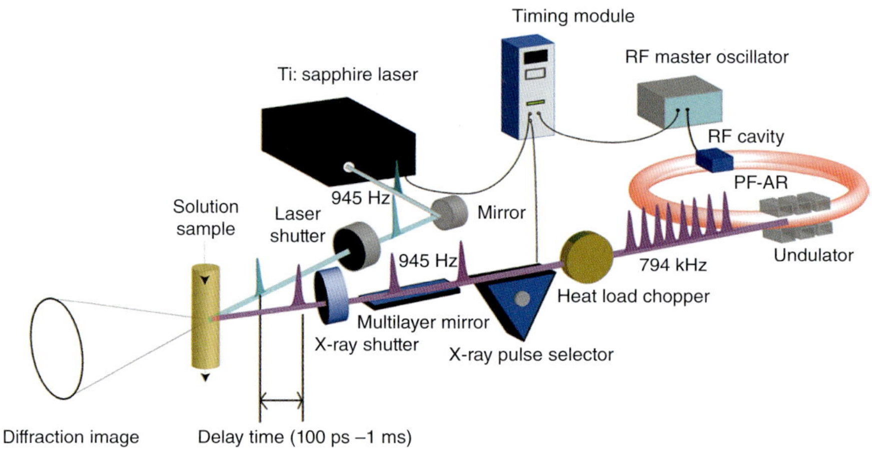

Figure 10.4 Drawings of the synchrotron X-ray diffraction-based time-resolved X-ray diffractometer. Source: Ichiyanagi et al. [23]. Copyright 2009, International Union of Crystallography, with permission.

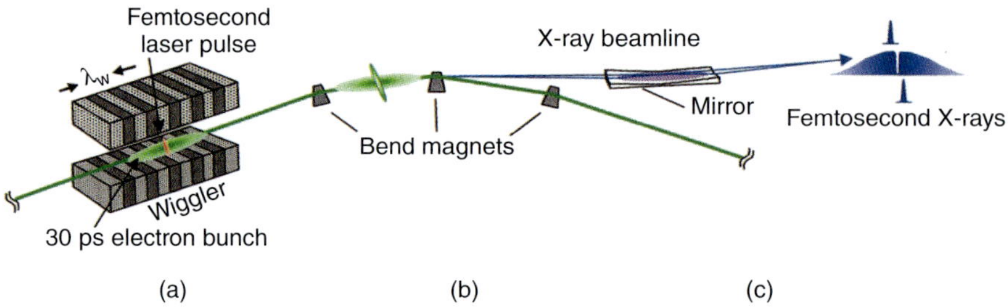

Figure 10.5 Laser-slicing X-ray method to produce the ultrashort pulsed X-ray. Source: Schoenlein et al. 2000 [24]. Reprinted with permission of the American Association for the Advancement of Science (AAAS).

to produce an ultrashort X-ray pulse with a duration of ~100 fs (Figure 10.5), which is called a slicing technique [24, 26–29]. This technique has overcome this barrier by producing approximately 100-fs X-ray pulses. Slicing techniques have proved very useful to detect ultrafast molecular motion, despite their low X-ray flux (10^2 to 10^3 X-rays per pulse). The XFELs are distinguished by their high brightness (~10^{12} X-rays per pulse), high coherence, and short pulse (10–100 fs) with a repetition rate of 100 Hz [30–33]. The coherence of the X-ray is referred to more in Section 10.3. Visible or IR free-electron lasers are assembled with high reflectors on both sides of the undulator; however, there is no production of high reflectors for hard X-ray. Instead of using high reflectors, in XFEL, long-aligned undulators are applied, and laser saturation is achieved with a single pass through the long-aligned undulators, which is well known as "self-amplified stimulated emission" of high-energy, high-peak-current, and ultrashort pulsed electron bunches (Figure 10.6) [25]. These pulsed X-rays generated from a synchrotron or linear accelerator are electrically or optically synchronized with femtosecond pump laser pulses and produced for spectroscopic or diffraction experiments [34–36].

Ultrashort pulsed electrons are usually generated from femtosecond laser irradiation on metal or semiconductor surfaces via the photoelectric effect [37–40]. A schematic of the simplest electron diffraction setup with DC acceleration is shown in Figure 10.7. Femtosecond electron diffraction setups are based on a photocathode under strong DC or AC electric fields, magnetic lenses

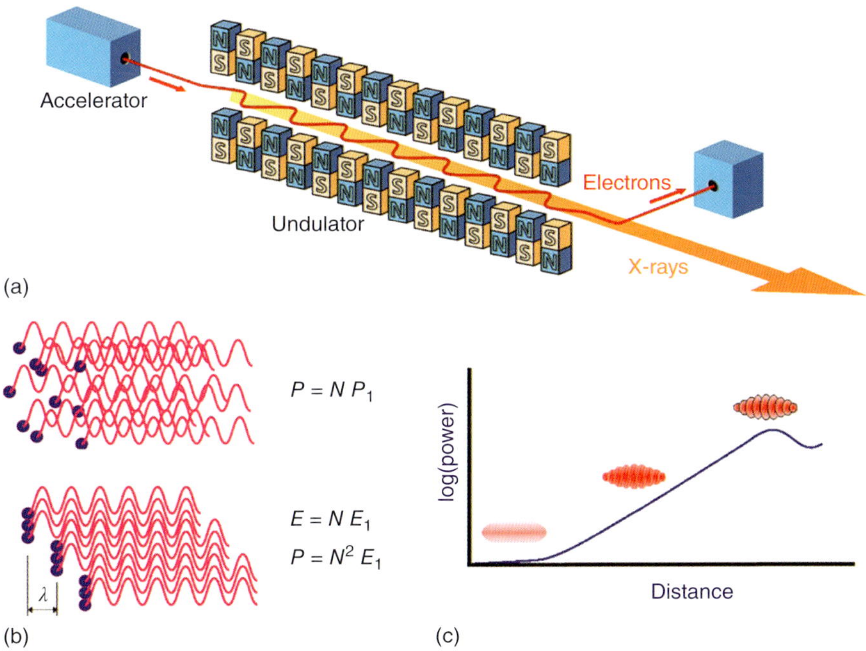

Figure 10.6 Schematic of the X-ray free electron laser emission (a) by the self-amplified stimulated emission (b,c). Source: Patterson and Abela 2010 [25]. Reprinted with permission of the Royal Society of Chemistry.

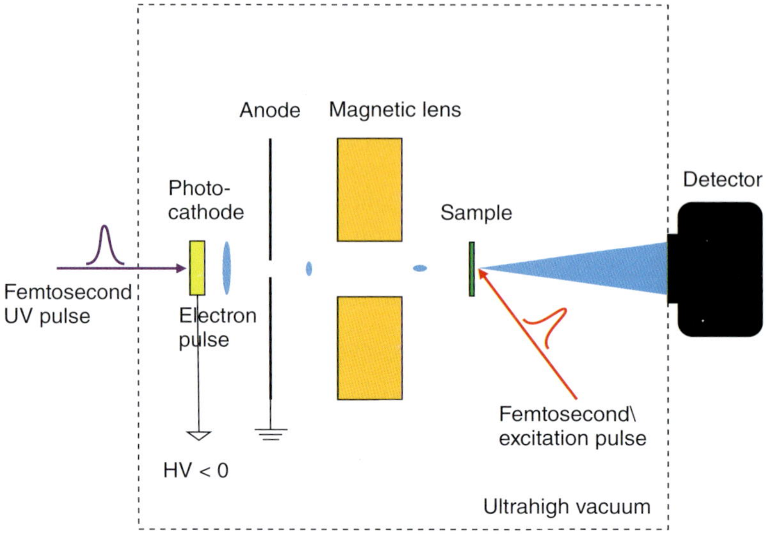

Figure 10.7 Femtosecond electron diffraction layout. An fs-optical pulse generates, via the photoelectric effect, an fs-electron pulse. The photocathode is held at negative potential, typically 30–100 kV with respect to ground. Electrons are therefore accelerated through the cathode–anode potential drop. A magnetic lens is used to collimate the electron beam before it scatters off the irradiated sample's area to generate a diffraction pattern at the screen. Again, optical pump and electron probe pulses can be arbitrarily delayed with respect to each other by placing a translation stage in one of the arms as shown in Figure 10.1. Diffraction patterns are recorded by a large area electron detector.

for electron beam collimation and focusing, and large-area detectors to record diffraction patterns. Electrons are photoemitted by a 20-nm thick Au film. The Au film is deposited on a sapphire or fused-silica substrate, which is transparent to $\lambda = 266\,\text{nm}$ (the third harmonic of a Ti:sapphire laser). The photocathode is held at negative high voltage, typically between 30 and 100 kV with respect to a grounded anode aperture. The photocathode-to-anode distance is less than 1 cm. After being accelerated and collimated by a magnetic lens, the electron pulse scatters off the sample to produce a diffraction pattern on a phosphor screen, which is recorded by a large-area camera. The main challenges in the design of DC-electron guns are to obtain ultrashort and bright electron pulses at the sample position. Electrons are charged particles. They repel each other, and in the case of dense electron bunches, strong space-charge effects quickly lead to electron pulse broadening. This is probably the most limiting factor in obtaining sub-picosecond electron probe pulses. The first attempts to solve this problem were based on either the use of low-density electron pulses [41–44] or compact electron sources in which the propagation distance was minimized [45–52]. Other techniques to obtain ultrashort electron pulses include electron pulse compression techniques. Static magnetic field or RF synchronized with the electron emission is able to modulate the time profile of the electron bunches [53–55]. In spite of the complicated setups, high brightness and ultrashort pulsed electrons can be obtained with electron compression techniques.

10.2.3 Structural Dynamics Revealed by X-ray or Electron Probes

The big advantage of X-ray probes compared with electron probes is their high coherence length. The synchrotron-based pulsed X-ray has a micro-to-millimeter coherence length, which is even sufficient to resolve the atomic or molecular structural dynamics in protein crystals. In 2003, F. Schotte et al. performed time-resolved X-ray diffraction measurements on a protein crystal (myoglobin mutant) with pulsed synchrotron X-rays at the duration of approximately 150 ps [16]. Figure 10.8 shows the myoglobin molecule before (magenta) and 100 ps after (green) photoexcitation. As one can see from the figure, three large-scale displacements near the CO-binding site (large yellow arrows) are accompanied by more subtle correlated rearrangements throughout the entire protein. Even faster dynamics of the molecules in protein crystals can be observed with ultrashort pulsed X-ray (pulse duration of 10–100 fs) from

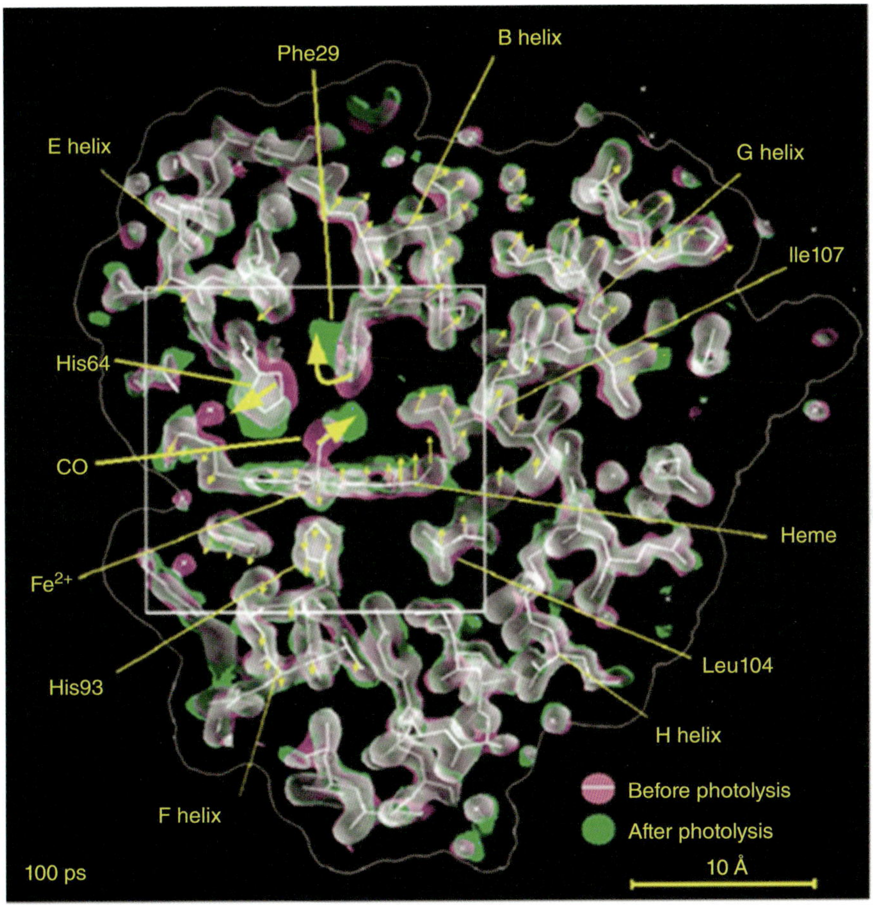

(a)

Figure 10.8 The motion of the molecules in myoglobin crystal revealed by picosecond time-resolved X-ray diffraction measurements. Source: Schotte et al. 2003 [16]. Reprinted with permission of the American Association for the Advancement of Science (AAAS).

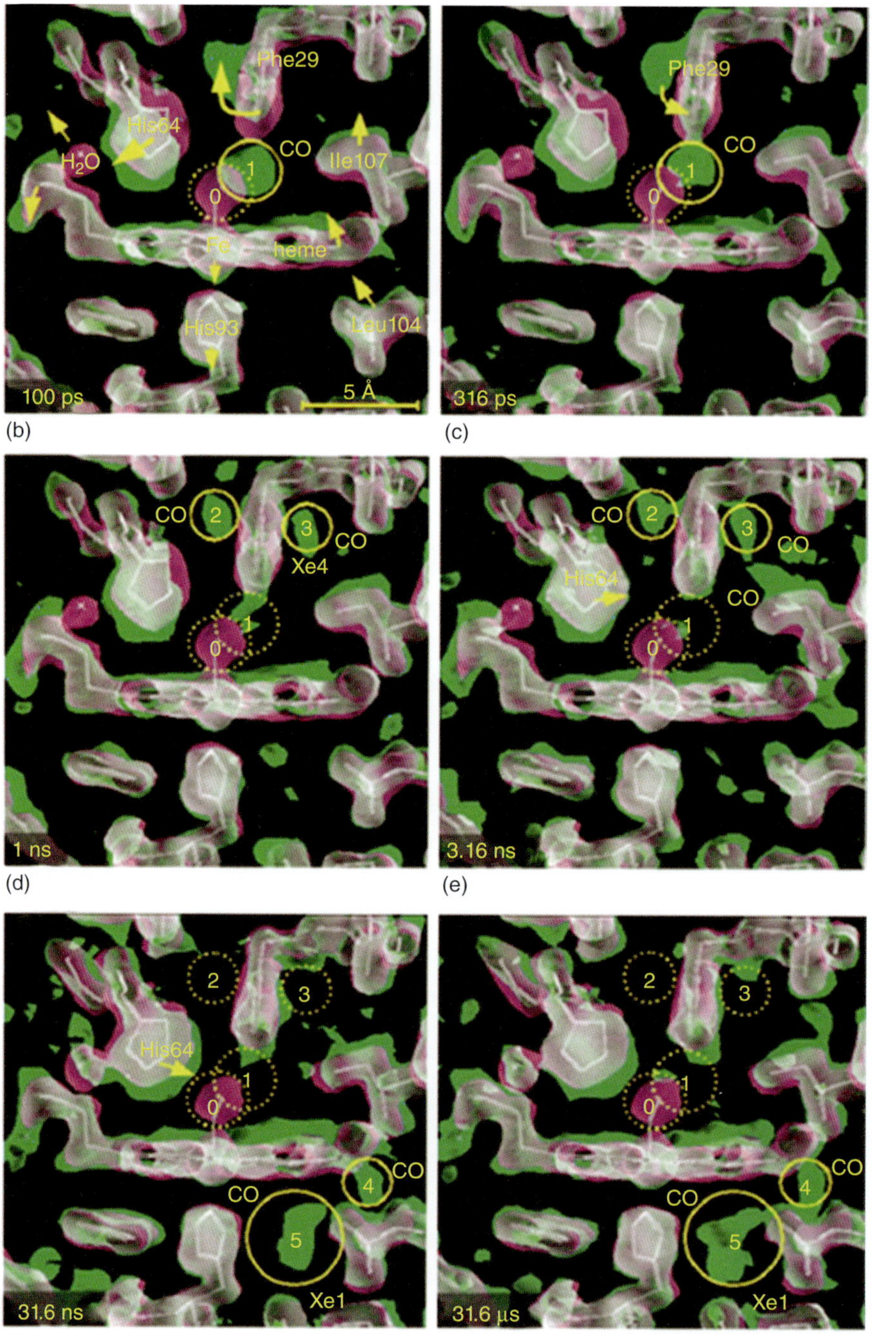

(b) (c) (d) (e) (f) (g)

Figure 10.8 (*Continued*)

XFEL. One of the big issues of the XFEL is damage on the organic or biological samples. Owing to the extremely high brightness of the X-ray pulse from XFEL, it induces fatal damage to the protein crystal; therefore, one should change the sample by a shot of the X-ray from XFEL. Very recently, continuous flow jet [56] or chip loading [57] methodologies have been developed for quick sample exchange. Ultrafast time-resolved X-ray diffraction measurements can even be applied to gas- or liquid-phase materials, which will provide quite important insights for fundamental physics and chemistry research.

On the other hand, electron probes also have an advantage in strong scattering. The X-ray and electron probes mainly interact with the inner shell electrons and the potential created by the electrons and atoms in the materials, respectively. The latter has a 10^4 to 10^5 times stronger scattering cross section than the former, resulting in the requirement of freestanding ultrathin (~100 nm) samples for electron probes. It is challenging to fabricate these freestanding ultrathin samples. However, it is also beneficial to use such ultrathin samples for time-resolved measurements because typical optical penetration depths of photofunctional materials are in the range 10–100 nm, which matches with the requirements of sample thickness for the electron probes. Taking advantage of this, several molecular dynamics of organic materials have been revealed with time-resolved electron diffraction. In 2013, M. Gao et al. reported the first molecular dynamics in an organic material ($(EDOTTF)_2PF_6$ (EDOTTF: ethylenedioxy-tetrathiafulvalene)) using time-resolved electron diffraction measurements with ultrabright and ultrashort electron pulses [58]. Figure 10.9 shows structures of $(EDOTTF)_2PF_6$ in low-temperature and high-temperature phases [59]. This material undergoes an insulator-to-metal phase transition

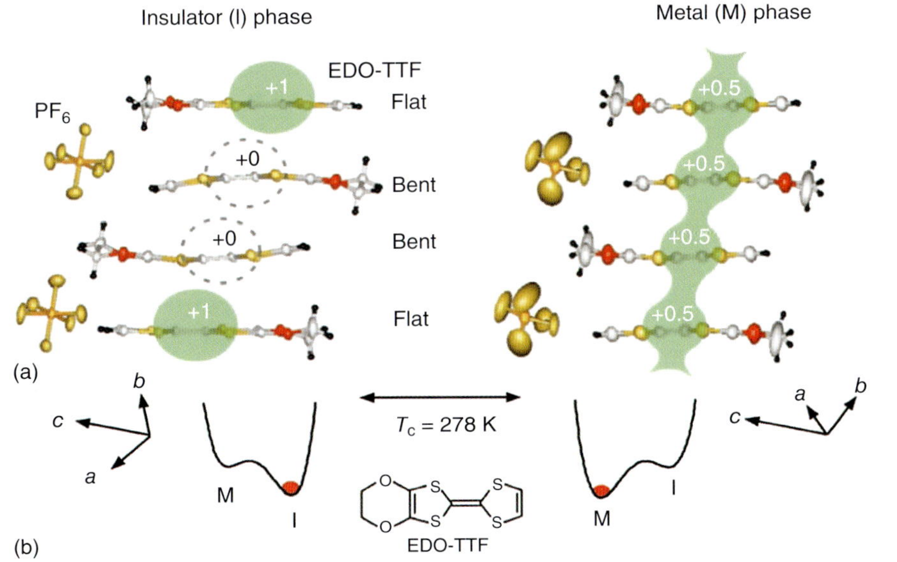

Figure 10.9 Structures of EDO-TTF PF_6 in the insulator and metal phase. Source: Mueller et al. 2015 [57]. Reprinted with permission of the American Association for the Advancement of Science (AAAS).

at a temperature of 278 K. It includes four cations EDOTTF, each of which has 0.5 positive charges, and two anions PF_6 with a negative charge in the unit cell. As shown in Figure 10.9, the two centered EDOTTFs take a bent shape in the ground state and they get flattened in the high-temperature phase. The charges are localized because of these bend shaped EDOTTFs; therefore, $(EDOTTF)_2\ PF_6$ is insulating and metallic in the low- and high-temperature phase, respectively. According to the static structural measurements and the femtosecond optical pump–probe experiments, one may have expected ultrafast bend-to-flat motions of the centered EDOTTFs; however, time-resolved electron diffraction measurements have revealed faster sliding motions (~1 ps) of the other two EDOTTFs and relatively slower bend-to-flat motions (10–100 ps) of the two centered EDOTTFs. Thus, one cannot predict the atomic or molecular motions only with static information and femtosecond time-resolved optical pump–probe experiments, and time-resolved X-ray and electron diffraction will need to be performed to understand the real structural dynamics during the photoinduced phase transition, photochemical, and photobiological reactions. There is also a report about a famous photochromic material, diarylethene, which undergoes ring-opening and ring-closing isomerization with UV and visible photoexcitation [60]. More recently, photoinduced dynamics of an even larger scaled and more complicated organometallic compound ($Me_4P\ (Pt(dmit)_2)_2$) with more than 100 atoms included in a unit cell (Figure 10.10) have been revealed by electron diffraction measurements (Figure 10.11) [61]. The unit cell of this compound is more than a few nanometers, which is comparable

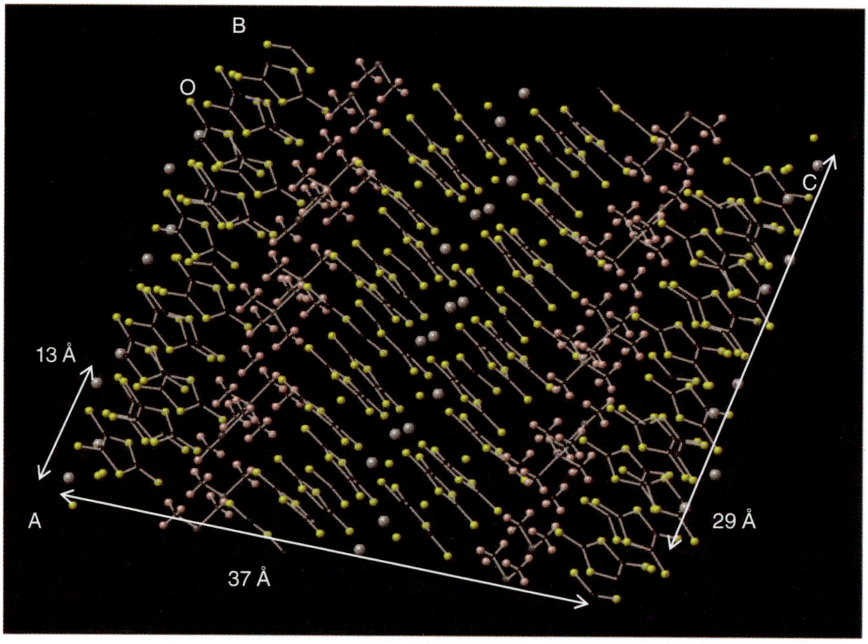

Figure 10.10 A unit cell of $Me_4P\ (Pt(dmit)_2)_2$ at the ground state. The size of the unit cell is 29 Å × 13 Å × 37 Å.

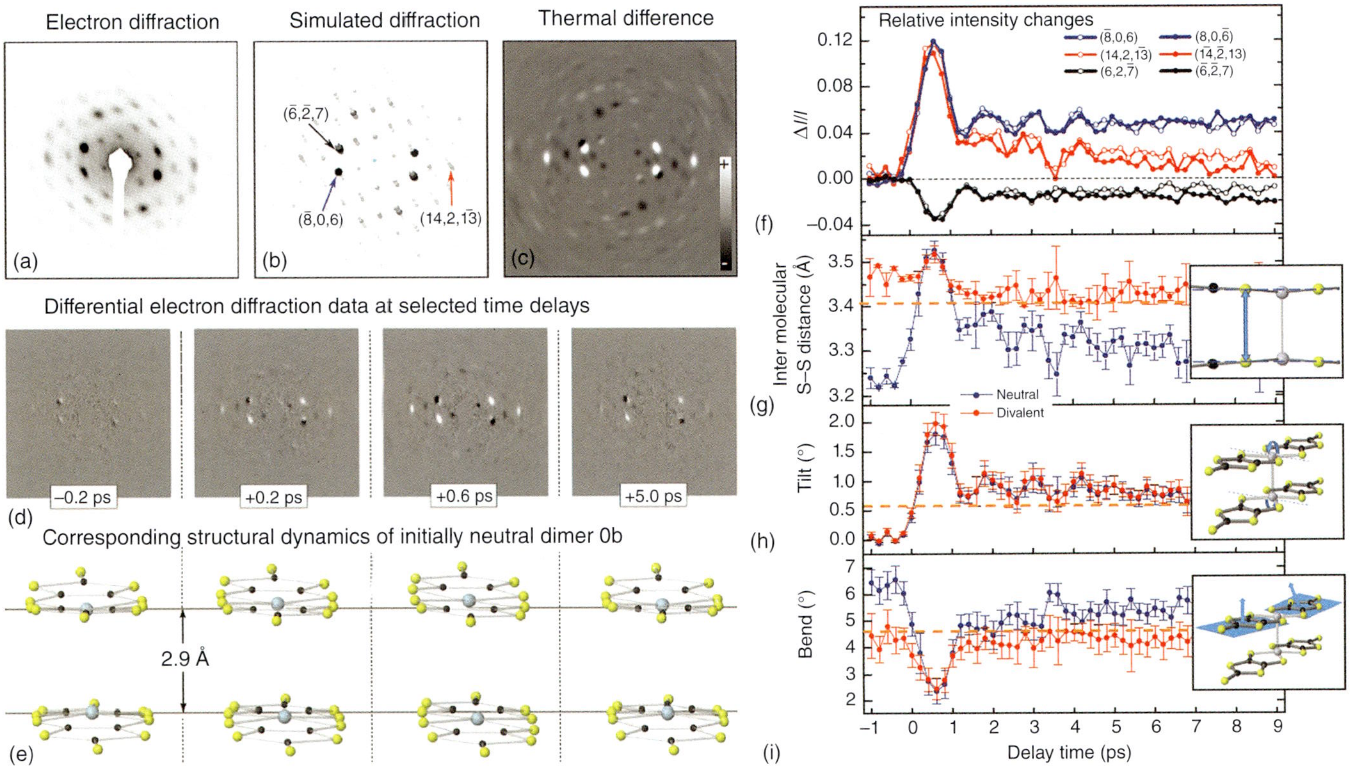

Figure 10.11 Molecular dynamics of the photoexcited Me_4P $(Pt(dmit)_2)_2$. Source: Ishikawa et al. 2015 [61]. Reprinted with permission of the American Association for the Advancement of Science (AAAS).

to those of small protein crystals. It is suggested that the molecular dynamics of protein crystals could be accessed by table-top time-resolved electron diffraction measurements with a slight increase of the coherence length in near future.

10.3 Time-Correlation Measurements

10.3.1 Introduction of X-ray Photon Correlation Spectroscopy

Here, we introduce the XPCS technique, which can investigate dynamic features in the molecular scale by examining the time correlation of coherent X-ray scattered photons. As shown in Figure 10.12, there are various dynamic aspects of physical phenomena over a wide range of spatial and temporal scales [62]. In Figure 10.12, some of the polymer dynamics are depicted as representative molecular motion. Scattering methods are powerful tools to reveal such a hierarchical structure, because they can adjust the spatial or time scale targets by changing the scattering angle or incidental energy. Furthermore, we can simultaneously obtain dynamics at many spatial and time scales. DLS mainly measures relatively large (>μm) and slow (>μs) dynamics corresponding to the repetition or cooperative segment motion of polymers. On the other hand, neutron spin echo, quasielastic or inelastic neutron scattering, and inelastic X-ray scattering cover the nanoscale (~10^{-1} to 10^{2} nm) and fast time range (~10^{-14} to 10^{-8} seconds). The main region accessed by XPCS is an intermediate

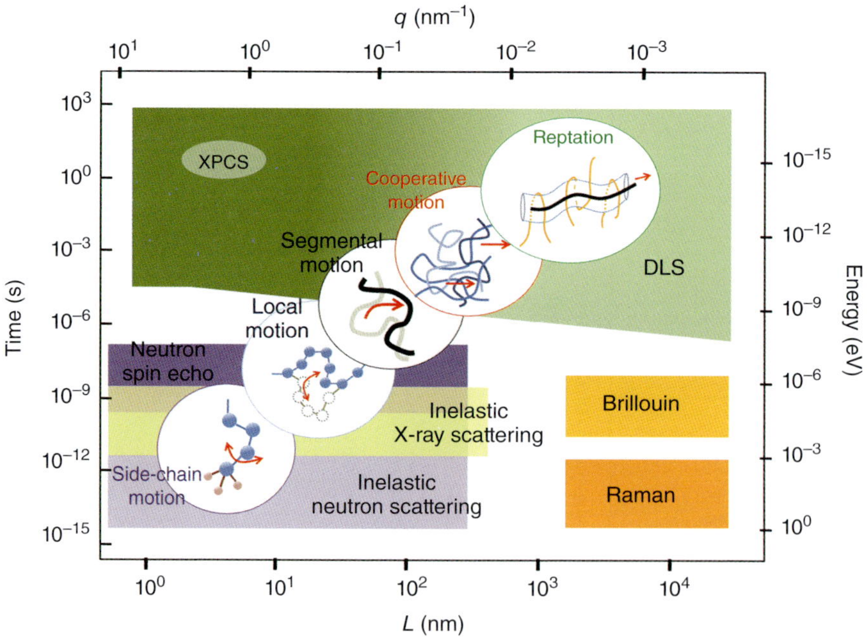

Figure 10.12 Time and length scale of various scattering techniques and hierarchal dynamical structure in polymers. Source: Reprinted from Hosino et al. 2013 [62].

region between the two aforementioned cases, which corresponds to cooperative motion or segmental motion of polymers.

10.3.2 Principle of XPCS

XPCS experimentally accesses various microscopic dynamic phenomena using partially coherent X-rays [63–68]. When an arbitrary arrangement of scatterers is illuminated by coherent radiation, the scattered intensity shows a so-called speckle pattern, which reflects the instantaneous configuration of the scatterer. Movement of the scatterer causes a corresponding change in the speckle pattern, which therefore contains information on the dynamics of the system. The principle of XPCS is similar to the principle of DLS performed using a visible laser light source. XPCS has two main advantages compared to DLS. The first is that shorter wavelength X-rays can study length scales down to the nanometer range. The second is that multiple scattering is often negligible even if the sample is opaque owing to high X-ray transmittance. Taking advantage of this, research on concentrated colloidal suspensions has been vigorously undertaken, and the understanding of hydrodynamic interactions of diffusion movement has been advanced [69–73].

Although DLS experiments are performed with a fully coherent light source, a visible laser, the XPCS light source is usually not fully coherent. Here, we consider the coherence of X-rays obtained from a third-generation synchrotron radiation source, which is a chaotic light source with a Gaussian intensity distribution. The transverse coherence length is given by $\xi_t = \lambda R/(2\pi\sigma)$, where λ, R, and σ are the wavelength, distance from the light source, and beam size of the light source, respectively [74, 75]. For example, at BL19LXU in SPring-8, which is a high-brilliance beamline with a 27-m-long undulator, $\xi_t \sim 16 \times 132$ $(H \times V)$ μm^2 is obtained from $\lambda = 1.55$ Å, $R \sim 75$ m, and $\sigma \sim 113 \times 14$ $(H \times V)$ μm^2.

Since the irradiated area of the sample should be approximately equal to the transverse coherence length, the beam size is cut to less than 20 μm × 20 μm with slits or a pinhole, and as a result, the XPCS beam flux is 0.1% of the raw beam. Therefore, the XPCS measurement has been performed on the high-brilliance beam lines of the third-generation synchrotron radiation source. The longitudinal coherence length, given by $\xi_l = \lambda(\lambda/\Delta\lambda)$ where $\lambda/\Delta\lambda$ is the bandwidth, relates to the maximum measurable scattering angle 2θ. The path length difference (PLD) derived from the positional difference of the scatterers within the scattering volume should satisfy the relationship $\xi_l >$ PLD. Therefore, it is necessary to satisfy the relation between the beam size and the sample thickness l: $\xi_l > 4l(\sin^2\theta)$. When using a monochromatic light beam, data of about $2\theta < 25$ mrad, i.e. approximately $q < 1$ nm^{-1}, which is a typical value obtained by a monochromator, are available. In small angle X-ray scattering (SAXS) geometry, the PLD is inherently small, and thus XPCS measurements with quasi-monochromatic X-ray sources have often been performed.

In general, XPCS measurements and data evaluation are performed in a process as shown in the schematic diagram in Figure 10.13. First, the temporal variation $I(\boldsymbol{q}, t)$ of the speckle intensity obtained by coherently irradiating the sample is

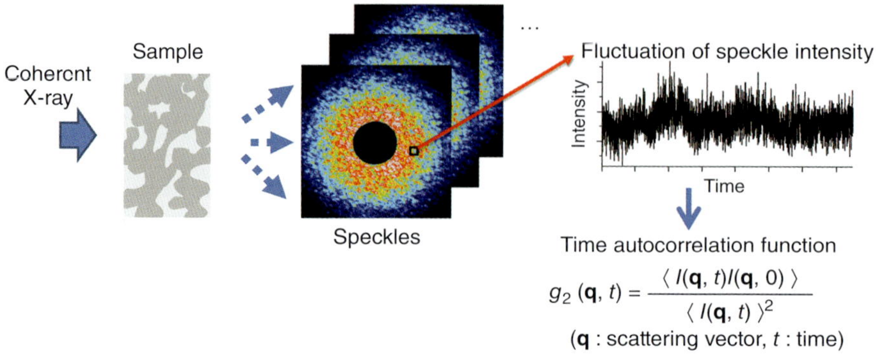

Figure 10.13 Schematic diagram of an XPCS measurement.

detected with various scattering vectors $\boldsymbol{q}$. Second, the autocorrelation function is calculated by

$$g_2(\boldsymbol{q}, t) = \langle I(\boldsymbol{q}, t) I(\boldsymbol{q}, 0) \rangle / \langle I(\boldsymbol{q}) \rangle^2 \tag{10.4}$$

where the brackets mean time-averaging. $g_2(\boldsymbol{q}, t)$ is frequently expressed in terms of the intermediate scattering function $S(\boldsymbol{q}, t)$:

$$g_2(\boldsymbol{q}, t) = \beta(\boldsymbol{q}, t) \left| \frac{S(\boldsymbol{q}, t)}{S(\boldsymbol{q})} \right|^2 + 1 \tag{10.5}$$

where $S(\boldsymbol{q})$ is the static structure factor, and $\beta(\boldsymbol{q}, t)$ is the contrast determined by the optical setup, which varies between 0 and 1.

In order to detect the intensity fluctuation of the speckle pattern, the XPCS detector must have a spatial resolution comparable to the speckle size and a time resolution faster than the scatterers' dynamics. Various detectors have been developed and used. Up to now, a considerable number of experiments have been performed with various types of 2D X-ray detectors, such as the directly illuminated X-ray charge coupled device (CCD) [76], the indirectly illuminated X-ray CCD [77], and photon counting 2D detectors including the Medipix-2 detector [78], MAXIPIX detector [79], and PILATUS detector [80]. Since the PILATUS detector has a relatively large pixel size (172 μm), it is necessary to increase the distance between the sample and the detector in order to detect the speckle fluctuation. The PILATUS detector was also used with a grid mask resolution enhancer placed in front of the detection pixel to detect fluctuations in speckle intensity [81]. A smaller pixel size (75 μm) detector, EIGER, is already in use as the next-generation PILATUS detector. This detector operates at frame rates up to 3 kHz [82].

10.3.3 Example

Dynamical studies of various condensed matter systems have been performed using XPCS techniques. As a representative case, two basal systems will be presented here: particle dispersed systems and fluid surfaces. Since these systems have interesting dynamical phenomena at the spatiotemporal scales corresponding to the area covered by XPCS using third-generation radiations, they have been continuously studied from an early stage.

10.3.4 Particle Diffusion

Particles in a fluid are thermally agitated and their dynamical behavior is strongly influenced by the surrounding fluid. The particles in a Newtonian fluid exhibit simple Brownian motion when there are no interactions between particles. The mean squared displacement (MSD) of the NPs linearly increases with time, $\langle x^2(t)\rangle \propto t$. However, if the surrounding fluid is non-Newtonian, the MSD follows another power law, $\langle x^2(t)\rangle \propto t^\alpha$, with $\alpha \neq 1$. If $0 < \alpha < 1$, the particle behavior is subdiffusive. Subdiffusive behavior is commonly observed in constrained systems, such as living cells [83, 84], actin networks [85], and crowded polymer systems [86–88]. In contrast, if $\alpha > 1$, the NP behavior is hyperdiffusive or superdiffusive. This behavior is observed in soft solids, such as colloidal gels [89, 90] and glass-forming systems [78, 91, 92].

If the particles move in a Brownian manner, $S(\boldsymbol{q}, t)$ is expressed by

$$S(\boldsymbol{q}, t) = \exp(-Dq^2t) \tag{10.6}$$

where D is the diffusion coefficient, and $q = |\boldsymbol{q}|$. If the particles are dilute and the particle–particle interactions are negligible, $g_2(\boldsymbol{q}, t)$ can be expressed by

$$g_2(\boldsymbol{q}, t) = 1 + \exp(-2\Gamma t) \tag{10.7}$$

where

$$\Gamma = Dq^2 \tag{10.8}$$

If the surrounding matrix is a non-Newtonian fluid, the diffusive motion of the particles does not follow simple Brownian motion and $g_2(\boldsymbol{q}, t)$ deviates from Eq. (10.7). In such cases, $g_2(\boldsymbol{q}, t)$ can be frequently expressed in stretched or compressed exponential form:

$$g_2(\boldsymbol{q}, t) = A\ \exp[-2(\Gamma t)^\gamma] + 1 \tag{10.9}$$

The compressed exponential form, $\gamma > 1$, is observed if the diffusive motion is hyperdiffusive (or superdiffusive, ballistic). In contrast, if the motion is subdiffusive, the stretched exponential form, $\gamma < 1$, is observed.

For some of the cases, a continuous-time random walk (CTRW) model can be applied to analyze the details of particle behavior. Caronna et al. [78] investigated the dynamic properties of silica nanoparticles suspended in a supercooled glass-forming liquid (1,2-propanediol) by XPCS. They observed the crossover from Brownian to hyperdiffusive behavior at $1.26T_g$ as the temperature decreased, where T_g is the calorimetric glass transition temperature of the liquid. In the system of polystyrene grafted silica nanoparticles dispersed in a polystyrene matrix, the nanoparticles exhibited non-Brownian behavior, and as the temperature increased, the crossover from hyperdiffusion to subdiffusion occurred at $1.25T_g$, as representative autocorrelation functions shown in Figure 10.14 [93]. Hyperdiffusive behavior is caused by the dynamical heterogeneity of the polymer matrix associated with the glass transition. When

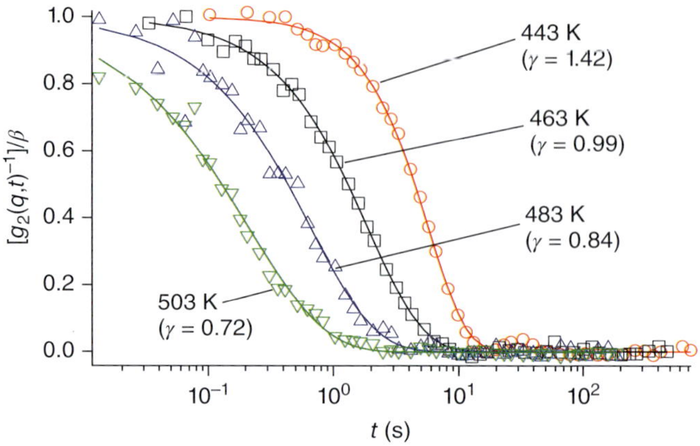

Figure 10.14 Normalized autocorrelation functions (symbols) measured at $q = 2.15 \times 10^{-2}$ nm^{-1} for the PS-grafted nanoparticles dispersed in a PS matrix at different temperatures. Solid lines are fitting curves for Eq. (10.9) [93].

the temperature was higher than $1.25T_g$, the interaction between the graft polymer and the polymer matrix became relatively significant and the dynamical behavior of the nanoparticles changed. In both cases, they directly fit $g_2(\boldsymbol{q}, t)$ by the expressions deduced from the CTRW model, and allow successful discussion of the microscopic motion.

Some studies have investigated the diffusion movement of particles in a restricted space. For example, there are studies under shear conditions [94, 95] and studies of the movement of particles in thin films [96, 97]. Koga et al. studied the dynamics of markers in PS thin films, and successfully evaluated the surface viscosity and the internal viscosity independently using a resonance-enhanced X-ray scattering technique in grazing incidence geometry [98].

10.3.5 Surface Fluctuation of Fluids

A fluid surface in an equilibrium state is macroscopically smooth and flat, but microscopically fluctuates owing to the thermal motion of the molecules and repeated generation and damping of waves. These waves are called capillary waves, and waves of various wavelengths travel with their respective frequencies.

If the amplitude is sufficiently small compared to the wavelength, such as a wave caused by thermal fluctuation, its behavior can be described by a linearized Navier–Stokes equation. In the case of a liquid with a low viscosity, such as water, the wave exhibits a damped oscillating behavior, but when the viscosity is high, such as a polymer fluid, it shows overdamped behavior without oscillation. A transition from oscillating to overdamped behavior can also occur by changing the wave number, and observations have been reported by XPCS [98] and DLS [99, 100]. Overdamped behavior with high viscosity will be focused on in the following text.

The frequency-domain spectrum of the surface fluctuation due to thermal fluctuation is expressed from the fluctuation dissipation theorem using the dynamic susceptibility that characterizes the linear response of the surface height to the periodic external force in the vertical direction as follows [101]:

$$S_{zz}(k,\omega) = 2k_{\rm B}T\frac{\mathrm{im}[\chi_{zz}(k,\omega)]}{\omega} \tag{10.10}$$

Here, k is the wave number of the surface wave, ω is the frequency, $k_{\rm B}$ is the Boltzmann constant, and T is the temperature. Frequencies and relaxation rates are obtained as poles in the lower half of the complex ω plane. In the present case, only the overdamped mode appears, and thus the spectrum in the time domain is expressed as the sum of the overdamped mode at the pole as follows:

$$S_{zz}(k,t) = S_{zz}(k)\sum_{m} a_{\rm m}(k)\mathrm{e}^{-\Gamma_{\rm m}(k)t} \tag{10.11}$$

Here, $S_{zz}(k) = k_{\rm B}T\chi_{zz}(k,0)$ is the total power of the spectrum, and $a_{\rm m}$ is the mode intensity expressed by

$$a_{\rm m}(k) = \frac{\mathrm{Res}[\chi_{zz}^{(ij)}(k,-i\Gamma_{\rm m})]}{\chi_{zz}^{(ij)}(k,0)\Gamma_{\rm m}} \tag{10.12}$$

at which the sum is $\sum_m a_{\rm m}(k) = 1$.

XPCS at grazing incidence (GI-XPCS) is effective for measuring the surface fluctuation of a fluid. Surface dynamics have been actively studied by GI-XPCS measurements since the observation example of the high-viscosity liquid (glycerin) capillary wave reported in 2001 [102, 103]. For example, Kim et al. observed the overdamping spectrum of a capillary wave on a polystyrene thin film surface [104].

In a homogeneous high-viscosity fluid thin film, Eq. (10.11) is known to be dominant in one relaxation mode and is expressed by the following exponential function:

$$g_2(q_{//},t) = \beta(q_{//})\exp[-2\Gamma(q_{//})t] + 1 \tag{10.13}$$

where $q_{//}$ is the in-plane component of the scattering vector. From the linearized Navier–Stokes equation, the dispersion relation under the boundary condition where the velocity at the interface of the substrate in the thin film of thickness d is zero is expressed by the following equation:

$$\Gamma(q_{//}) = \frac{\gamma}{2\eta}q_{//}\frac{\sinh(q_{//}d)\cosh(q_{//}d) - q_{//}d}{\cosh(q_{//}d)^2 + (q_{//}d)^2} \tag{10.14a}$$

Equation (10.14a) can be rewritten as the following expression by multiplying both sides by d, and Γd is expressed as $q_{//}d$ function:

$$\Gamma(q_{//})d = \frac{\gamma}{2\eta}q_{//}d\frac{\sinh(q_{//}d)\cosh(q_{//}d) - q_{//}d}{\cosh(q_{//}d)^2 + (q_{//}d)^2} \tag{10.14b}$$

Equation (10.14b) is effective in evaluating capillary waves of thin films of different thickness. In Figure 10.15a, $\Gamma d - q_{//}d$ is plotted for PS ($M_n = 25$k) thin film of thickness 44, 89, and 155 nm and polyhedral oligomeric silsesquioxane (POSS)-functionalized polystyrene (PS-POSS) ($M_n = 25$k) of thicknesses 44, 79, and 144 nm are shown. A theoretical line obtained by substituting the surface

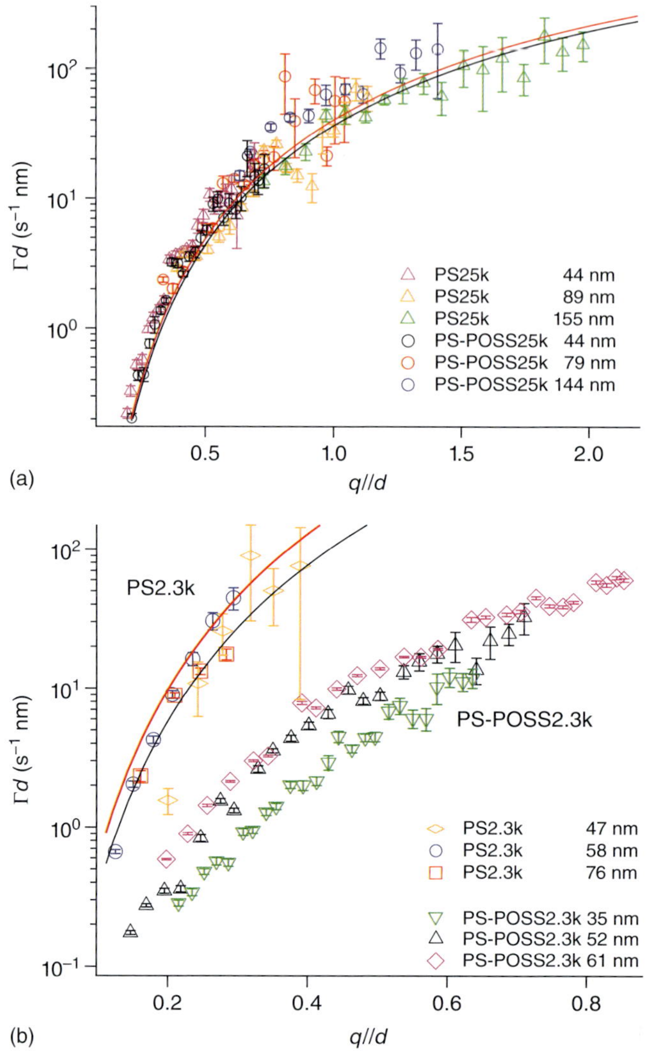

Figure 10.15 GI-XPCS results for PS25k and PS-POSS25k thin films at 413 K. (a) *qd* dependence of Γd for thin films of various thicknesses. The red solid line and the black line are calculated from Eq. (10.14b) with the viscosity and surface tension obtained in the laboratory for PS25k and PS-POSS25k, respectively. (b) *qd* dependence of Γd for thin films of various thicknesses. The red solid line and the black line are calculated from Eq. (10.14b) with the viscosity and surface tension obtained in the laboratory for PS2.3k and PS-POSS2.3k, respectively. Source: Hoshino et al. 2016 [105]. Reprinted with permission from Elsevier.

tension and viscosity independently measured in the laboratory into Eq. (10.14b) is also drawn in Figure 10.15a. At this molecular weight, plots of different thicknesses of PS thin film and PS-POSS thin film are on almost the same line and it can be interpreted that these PS and PS-POSS thin films have the same physical properties.

It is also interesting when the Eqs. (10.14a) and (10.14b) do not hold. The unique properties of the thin film can be determined by the dynamic behavior of the capillary wave on the thin film. Wang et al. measured surface fluctuations of branched polymer thin films, and the viscosity calculated from the capillary wave spectrum of the thin film was 100 times higher than that of the bulk [106]. Jiang et al. considered the dynamics of bilayer membranes as representative examples of heterogeneous thin films and calculated the surface and interfacial dynamics of polymeric bilayer films based on the fluctuation–dissipation theorem [107].

For the PS-POSS thin film, it was also found that the behavior of the capillary wave cannot be interpreted in Eqs. (10.14a) and (10.14b) when the molecular weight is small [105]. Figure 10.15b shows $\Gamma d - q_{//}d$ plots obtained from capillary waves of PS ($M_n = 2.3$k) thin films and PS-POSS ($M_n = 2.3$k) thin films with various thicknesses at 393 K. The PS thin film plots coincide well with the theoretical line obtained from Eq. (10.14b), whereas the plots of the PS-POSS thin films are far from the theoretical line obtained from Eq. (10.14b). In the literature, by analyzing the dispersion relation of capillary waves by introducing a three-layer model in which the viscosities at the surface and at the substrate interface are different from that of the bulk, the existence of high-viscosity layers at the surfaces and at the substrate interfaces of PS-POSS ($M_n = 2.3$k) thin films is suggested.

The correlation functions described above are obtained by homodyne experiments, and therefore the phase information of the scattered electric fields is lost. To examine the movement of several types of scatterers for example, phase information is indispensable. In such cases, in DLS, heterodyne experiments have been conducted, in which the reference light is mixed into the scattered light at the detector [108, 109]. On the other hand, heterodyne XPCS experiments are not very common. However, quite a few experiments have been carried out by mixing reference X-rays with scattered X-rays at the detector. In GI geometry, specularly reflected beams play the role of reference light by interfering with fluctuating diffuse reflected beams [110–112]. In transmission geometry experiments, a more controlled heterodyne measurement is demonstrated by placing a static reference sample immediately upstream of the main sample [113, 114]. Recently, also in GI geometry, Ulbrandt et al. obtained heterodyne signals by mixing scattering signals from the surface and bulk during real-time studies of sputter deposition of a-Si and a-WSi2 films [115]. They successfully controlled the homodyne mixing ratio by controlling the X-ray penetration and escape depths.

10.3.6 Summary and Perspective

Samples measured by XPCS so far are varied. However, in XPCS, it is necessary to measure speckle fluctuations with temporal resolution; thus, it is required that the scattering intensity from the sample be sufficiently strong. Typical samples that have been measured by XPCS are shown in the distribution diagram in Figure 10.16. Various particle dispersion systems, such as silica particles, metal particles, polymer particles, and magnetic particles, have been measured and various phenomena have been observed since it is relatively easy to obtain strong scattering from particles. Fluid surfaces can also give enough scattering intensity by making X-rays incident on the surface at an angle smaller than the total

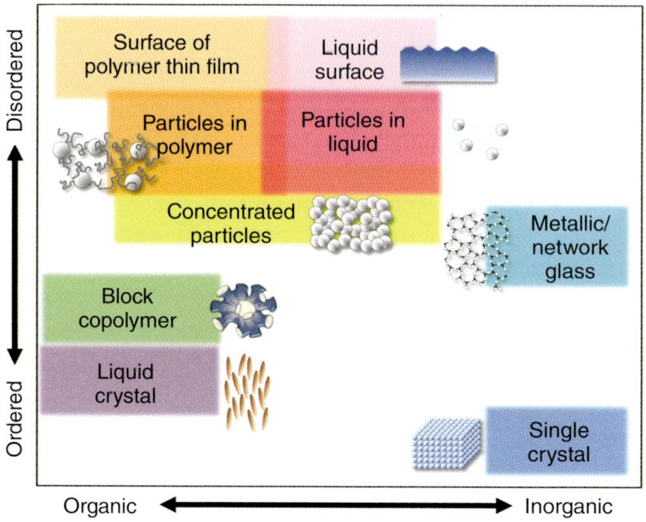

Figure 10.16 Diagram of typical samples measured by XPCS.

reflection critical angle, and there are many examples of observing the capillary waves of the liquid surface and polymer thin film surface. Among soft materials, there are some examples of studying the fluctuation of domains and lamellae from fluctuations of speckle around the scattering vector, which can yield strong scattering in the orderly systems, such as block copolymer [116, 117] and liquid crystal [118].

There is much research on glass transition, since the timescale targeted by XPCS is relatively slow. In discussing such nonequilibrium phenomena, analysis using two-time correlation functions has been actively conducted, and not only relatively large dynamics, such as in the domain of alloys [119] and particle motions in gel systems [120, 121], but also dynamics on the atomic scale of metallic glass relaxation phenomena, have been observed [122, 123].

Various dynamics measurement examples have also been reported for the single crystal, which is the most ordered system: a quasihexagonal reconstruction of the Au(001) surface and the diffusion of single atoms interpreted by the nearest neighbor jumps [124], intermetallic alloy $Cu_{90}Au_{10}$, and avalanche dynamics caused by the incubation time effect, along with slow dynamics at the transformation temperature in an athermal martensitic shape memory alloy ($Au_{50.5}Cd_{49.5}$) [125]. Regarding research on magnetism, observing a spin density wave or charge density wave [126], fluctuation of the domains of a spiral antiferromagnet [127], and spin reorientation [128] by obtaining fluctuation of magnetic scattering by irradiating X-rays near the absorption edge were reported.

As we have seen, XPCS has been used to measure various soft and hard condensed materials, but in spite of that, this is overwhelmingly more research on soft materials. One reason is that with soft materials, it is easier to design the samples and conditions according to the applicable temporal and spatial scales of XPCS. In general, however, the organic material that occupies the majority of soft materials is not optimal for XPCS, as it has low X-ray irradiation resistance and low X-ray scattering ability owing to its low electron density. With the future development of X-ray technologies, such as light sources and detectors, the application to inorganic materials will be greatly expanded if wide-angle, high-speed measurements become possible.

In order to observe speckle fluctuation at high-q, a light source with high coherent flux is necessary. For example, for XPCS using XFEL, sequential measurements using pulse shots at regular time intervals are reported in LCLS and SACLA [129, 130]. In these cases, the time/spatial scales have not exceeded the area covered by third-generation synchrotron radiation. As a technique that positively exploits short pulse characteristics, the split-delay method is being developed aiming at fast timescale dynamics reaching tens of femtoseconds, and high-speed dynamics measurements in the time domain are expected [131, 132]. However, there could be a gap between the timescale (femtosecond-to-nanosecond) that is considered to be targeted by split-delay and the timescale that the current XPCS covers, and for unified interpretation of dynamics, continuous development of XPCS using a stable light source is indispensable. The low-emittance ring light source currently under development around the world is expected to greatly advance dynamics research by XPCS.

References

1 Silfvast, W.T. (2004). *Laser Fundamentals*. Cambridge University Press.
2 Zewail, A.H. (1999). *Femtochemistry*. Nobel Lecture.
3 Nasu, K. (ed.) (2004). *Photoinduced Phase Transition*. World Scientific Publishing.
4 Koshihara, S., Tokura, Y., Takeda, K., and Koda, T. (1992). *Phys. Rev. Lett.* 68: 1148.
5 Shapiro, S.L. (ed.) (1977). *Ultrashort Light Pulses*, Topics in Applied Physics. Springer.
6 Fleming, G.R. (1982). *Adv. Chem. Phys.* 49: 1.
7 Powell, R.C. and Soos, Z.G. (1975). *J. Lumin.* 11: 1.
8 Barber, J. (1978). *Pep. Prog. Phys.* 41: 1158.
9 Campillo, A.J. and Shapiro, S.L. (1978). *Photochem. Photobiol.* 28: 975.
10 Cerullo, G., Manzoni, C., Lüera, L., and Pollia, D. (2007). *Photochem. Photobiol. Sci.* 6: 135–144.
11 Sahoo, S.K., Umapathy, S., and Parker, A.W. (2011). *Appl. Spectrosc.* 65: 1087–1115.
12 Mukuta, T., Fukazawa, N., Murata, K. et al. (2014). *Inorg. Chem.* 53: 2481–2490.
13 Bragg, W.L. (1922). *The Diffraction of X-rays by Crystals*. Nobel Lecture.
14 von Laue, M. (1915). *Concerning the Detection of X-ray Interferences*. Nobel Lecture.
15 Shmueli, U. (2001). *International Tables for Crystallography: Reciprocal Space (2008)*, 3e. Springer.
16 Schotte, F., Lim, M., Jackson, T.A. et al. (2003). *Science* 300: 1944.

17 Collet, E., Lemee-Cailleau, M.-H., Cointe, M.B.-L. et al. (2003). *Science* 300: 612.
18 Schotte, F., Cho, H.S., Kaila, V.R.I. et al. (2012). *Proc. Natl. Acad. Sci. U. S. A.* 109: 19256.
19 Ihee, H., Lorenc, M., Kim, T.K. et al. (2005). *Science* 309: 1223.
20 Seeck, O.H., Deiter, C., Pflaum, K. et al. (2011). *J. Synchrotron Radiat.* 19: 30.
21 Abela, R., Joho, W., Milton, S., and Rivkin, L. (1992). *Rev. Sci. Instrum.* 63: 1606.
22 Stock, S.R., Mini, S.M., and Perry, D.L. (2000). *Applications of Synchrotron Radiation Techniques to Materials Science V*. Cambridge University Press.
23 Ichiyanagi, K., Sato, T., Nozawa, S. et al. (2009). *J. Synchrotron Radiat.* 16: 391–394.
24 Schoenlein, R.W., Chattopadhyay, S., Chong, H.H.W. et al. (2000). *Science* 287: 2237.
25 Patterson, B.D. and Abela, R. (2010). *Phys. Chem. Chem. Phys.* 12: 5647–5652.
26 Cavalleri, A., Wall, S., Simpson, C. et al. (2006). *Nature* 442: 664.
27 Beaud, P., Johnson, S.L., Streun, A. et al. (2007). *Phys. Rev. Lett.* 99: 174801.
28 Johnson, S.L., Beaud, P., Milne, C.J. et al. (2008). *Phys. Rev. Lett.* 100: 155501.
29 Beaud, P., Johnson, S.L., Vorobeva, E. et al. (2009). *Phys. Rev. Lett.* 103: 155702.
30 Shintake, T. (2003). *Nucl. Instrum. Methods Phys. Res., Sect. A* 507: 382.
31 Patterson, B.D., Abela, R., Braun, H.-H. et al. (2010). *New. J. Phys.* 12: 035012.
32 Ding, Y., Brachmann, A., Decker, F.J. et al. (2009). *Phys. Rev. Lett.* 102: 5–8.
33 Ishikawa, T., Aoyagi, H., Asaka, T. et al. (2012). *Nat. Photonics* 6: 540–544.
34 Beaud, P., Caviezel, A., Mariager, S.O. et al. (2014). *Nat. Mater.* 13: 923–927.
35 Minitti, M.P., Budarz, J.M., Kirrander, A. et al. (2015). *Phys. Rev. Lett.* 114: 255501.
36 Kim, K.H., Kim, J.G., Nozawa, S. et al. (2015). *Nature* 518: 385–389.
37 Zewail, A.H. (2010). *Science* 328: 187–193.
38 Miller, R.J.D. (2014). *Science* 343: 1108–1116.
39 Miller, R.J.D., Ernstorfer, R., Harb, M. et al. (2010). *Acta Crystallogr., Sect. A: Found. Crystallogr.* 66: 137.
40 Hada, M., Pichugin, K., and Sciaini, G. (2013). *Euro. Phys. J. Special Topics* 222: 1093–1123.
41 Aidelsburger, M., Kirchner, F.O., Krausz, F., and Baum, P. (2010). *Proc. Natl. Acad. Sci. U. S. A.* 107: 19714.
42 Gahlmann, A., Park, S.T., and Zewail, A.H. (2008). *Phys. Chem. Chem. Phys.* 10: 2894.
43 Baum, P. and Zewail, A.H. (2006). *Proc. Natl. Acad. Sci. U. S. A.* 103: 16105.
44 Baum, P., Yang, D.-S., and Zewail, A.H. (2007). *Science* 318: 788.
45 Siwick, B.J., Green, A.A., Hebeisen, C.T., and Miller, R.J.D. (2005). *Opt. Lett.* 30: 1057.
46 Hebeisen, C.T., Ernstorfer, R., Harb, M. et al. (2006). *Opt. Lett.* 31: 3517.
47 Hebeisen, C.T., Sciaini, G., Harb, M. et al. (2008). *Opt. Express* 16: 3334.

48 Sciaini, G., Harb, M., Kruglik, S.G. et al. (2009). *Nature* 458: 56.
49 Ernstorfer, R., Harb, M., Hebeisen, C.T. et al. (2009). *Science* 323: 1033.
50 Harb, M., Ernstorfer, R., Hebeisen, C.T. et al. (2008). *Phys. Rev. Lett.* 100: 155504.
51 Eichberger, M., Schaefer, H., Krumova, M. et al. (2010). *Nature* 468: 799.
52 Hada, M., Zhang, D., Pichugin, K. et al. (2014). *Nat. Commun.* 5: 3863.
53 van Oudheusden, T., Pasmans, P.L.E.M., van der Geer, S.B. et al. (2010). *Phys. Rev. Lett.* 105: 264801.
54 Gao, M., Jean-Ruel, H., Cooney, R.R. et al. (2012). *Opt. Express* 20: 12048.
55 Chatelain, R.P., Morrison, V.R., Godbout, C., and Siwick, B.J. (2012). *Appl. Phys. Lett.* 101: 081901.
56 Chapman, H.N., Fromme, P., Barty, A. et al. (2011). *Nature* 470: 73–77.
57 Mueller, C., Marx, A., Epp, S.W. et al. (2015). *Struct. Dyn.* 2: 054302.
58 Gao, M., Lu, C., Jean-Ruel, H. et al. (2013). *Nature* 496: 343–346.
59 Chollet, M., Guerin, L., Uchida, N. et al. (2005). *Science* 307: 86–89.
60 Jean-Ruel, H., Gao, M., Kochman, M.A. et al. (2013). *J. Phys. Chem. B* 117: 15894–15902.
61 Ishikawa, T., Hayes, S.A., Keskin, S. et al. (2015). *Science* 350: 1501–1505.
62 Hoshino, T., Tanaka, Y., Jinnai, H., and Takahara, A. (2013). *J. Phys. Soc. Jpn.* 82: 021014.
63 Sutton, M. (2008). *C.R. Phys.* 9: 657.
64 Grübel, G., Madsen, A., and Robert, A. (2008). *Soft Matter Characterization* (ed. R. Borsali and R. Pecora), 953. Netherlands, Dordrecht: Springer.
65 Sinha, S.K., Jiang, Z., and Lurio, L.B. (2014). *Adv. Mater.* 26: 7764.
66 Leheny, R.L., Rogers, M.C., Chen, K. et al. (2015). *Curr. Opin. Colloid Interface Sci.* 20: 261.
67 Nogales, A. and Fluerasu, A. (2016). *Euro. Poly. J.* 81: 494.
68 Bikondoa, O. *X-Ray Photon Correlation Spectroscopy for the Characterization of Soft and Hard Condensed Matter: Datasheet from: "X-ray and Neutron Techniques for Nanomaterials Characterization"*, Springer Materials https://doi.org/10.1007/978-3-662-48606-1_3. Berlin Heidelberg: Springer-Verlag.
69 Tsui, O.K.C. and Mochrie, S.G.J. (1998). *Phys. Rev. E* 57: 2030.
70 Lurio, L.B. et al. (2000). *Phys. Rev. Lett.* 84: 785.
71 Banchio, A.J. et al. (2006). *Phys. Rev. Lett.* 96: 138303.
72 Riese, D.O., Wegdam, G.H., Vos, W.L. et al. (2000). *Phys. Rev. Lett.* 85: 5460.
73 Gapinski, J., Patkowski, A., Banchio, A.J. et al. (2009). *J. Chem. Phys.* 130: 084503.
74 Goodman, J.W. (1985). *Statistical Optics*. New York, NY: Wiley.
75 Mandel, E.W.L. (1995). *Optical Coherence and Quantum Optics*. Cambridge, UK: Cambridge University Press.
76 Lumma, D., Lurio, L.B., Mochrie, S.G.J., and Sutton, M. (2000). *Rev. Sci. Instrum.* 71: 3274.
77 Shinohara, Y., Imai, R., Kishimoto, H. et al. (2010). *J. Synchrotron Radiat.* 17: 737.
78 Caronna, C., Chushkin, Y., Madsen, A., and Cupane, A. (2008). *Phys. Rev. Lett.* 100: 055702.

79 Orsi, D., Cristofolini, L., Baldi, G., and Madsen, A. (2012). *Phys. Rev. Lett.* 108: 105701.
80 Westermeier, F., Autenrieth, T., Gutt, C. et al. (2009). *J. Synchrotron Radiat.* 16: 687.
81 Hoshino, T., Kikuchi, M., Murakami, D. et al. (2012). *J. Synchrotron Radiat.* 19: 988.
82 Johnson, I., Bergamaschi, A., Buitenhuis, J. et al. (2012). *J. Synchrotron Radiat.* 19: 1001.
83 Golding, I. and Cox, E.C. (2006). *Phys. Rev. Lett.* 96: 098102.
84 Guigas, G., Kalla, C., and Weiss, M. (2007). *Biophys. J.* 93: 316.
85 Wong, I.Y., Gardel, M.L., Reichman, D.R. et al. (2004). *Phys. Rev. Lett.* 92: 178101.
86 Omari, R.A., Aneese, A.M., Grabowski, C.A., and Mukhopadhyay, A. (2009). *J. Phys. Chem. B* 113: 8449.
87 Sprakel, J., van der Gucht, J., Stuart, M.A.C., and Besseling, N.A. (2007). *Phys. Rev. Lett.* 99: 208301.
88 Guo, H., Bourret, G., Lennox, R.B. et al. (2012). *Phys. Rev. Lett.* 109: 055901.
89 Cipelletti, L., Manley, S., Ball, R.C., and Weitz, D.A. (2000). *Phys. Rev. Lett.* 84: 2275.
90 Chung, B., Ramakrishnan, S., Bandyopadhyay, R. et al. (2006). *Phys. Rev. Lett.* 96: 228301.
91 Bellour, M., Knaebel, A., Harden, J.L. et al. (2003). *Phys. Rev. E Stat. Nonlin. Soft Matter Phys.* 67: 031405.
92 Guo, H., Bourret, G., Corbierre, M.K. et al. (2009). *Phys. Rev. Lett.* 102: 075702.
93 Hoshino, T., Murakami, D., Tanaka, Y. et al. (2013). *Phys. Rev. E: Stat. Nonlinear Soft Matter Phys.* 88: 032602.
94 Burghardt, W.R., Sikorski, M., Sandy, A.R., and Narayanan, S. (2012). *Phys. Rev. E: Stat. Nonlinear Soft Matter Phys.* 85: 021402.
95 Rogers, M.C., Chen, K., Andrzejewski, L. et al. (2014). *Phys. Rev. E: Stat. Nonlinear Soft Matter Phys.* 90: 062310.
96 Narayanan, S., Lee, D.R., Hagman, A. et al. (2007). *Phys. Rev. Lett.* 98: 185506.
97 Koga, T., Li, C., Endoh, M.K. et al. (2010). *Phys. Rev. Lett.* 104: 066101.
98 Madsen, A., Seydel, T., Sprung, M. et al. (2004). *Phys. Rev. Lett.* 92: 096104.
99 Ohmasa, Y., Hoshino, T., Osada, R., and Yao, M. (2008). *Chem. Phys. Lett.* 455: 184.
100 Hoshino, T., Ohmasa, Y., Osada, R., and Yao, M. (2008). *Phys. Rev. E* 78: 061604.
101 Jackle, J. (1998). *J. Phys. Condens. Matter* 10: 7121.
102 Tolan, M., Seydel, T., Madsen, A. et al. (2001). *Appl. Surf. Sci.* 182: 236.
103 Seydel, T., Madsen, A., Tolan, M. et al. (2001). *Phys. Rev. B* 63: 073409.
104 Kim, H., Rühm, A., Lurio, L.B. et al. (2003). *Phys. Rev. Lett.* 90: 068302.
105 Hoshino, T., Nojima, S., Sato, M. et al. (2016). *Polymer* 105: 487.
106 Wang, S.-f., Yang, S., Lee, J. et al. (2013). *Phys. Rev. Lett.* 111: 068303.
107 Jiang, Z., Kim, H., Mochrie, S.G.J. et al. (2006). *Phys. Rev. E* 74: 011603.

108 Berne, B.J. and Pecora, R. (1976). *Dynamic Light Scattering: with Applications to Chemistry, Biology, and Physics.* New York, NY: Wiley.
109 Langevin, D. (1992). *Light Scattering by Liquid Surfaces and Complementary Techniques* (ed. M. Dekker), 41. New York, NY: Surfactant Science Series.
110 Gutt, C., Ghaderi, T., Chamard, V. et al. (2003). *Phys. Rev. Lett.* 91: 076104.
111 de Jeu, W.H., Madsen, A., Sikharulidze, I., and Sprunt, S. (2005). *Physica B* 357: 39.
112 Sikharulidze, I. and de Jeu, W.H. (2005). *Phys. Rev. E: Stat. Nonlinear Soft Matter Phys.* 72: 011704.
113 Livet, F., Bley, F., Ehrburger-Dolle, F. et al. (2006). *J. Synchrotron Radiat.* 13: 453.
114 Ehrburger-Dolle, F., Morfin, I., Bley, F. et al. (2012). *Macromolecules* 45: 8691.
115 Ulbrandt, J.G., Rainville, M.G., Wagenbach, C. et al. (2016). *Nat. Phys.* 12: 794.
116 Falus, P., Borthwick, M.A., Narayanan, S. et al. (2006). *Phys. Rev. Lett.* 97: 066102.
117 Jang, W.-S., Koo, P., Sykorsky, M. et al. (2013). *Macromolecules* 46: 8628.
118 Sikharulidze, I., Dolbnya, I.P., Fera, A. et al. (2002). *Phys. Rev. Lett.* 88: 115503.
119 Malik, A., Sandy, A.R., Lurio, L.B. et al. (1998). *Phys. Rev. Lett.* 81: 5832.
120 Madsen, A., Leheny, R.L., Guo, H. et al. (2010). *New J. Phys.* 12: 055001.
121 Hernández, R., Criado, M., Nogales, A. et al. (2015). *Macromolecules* 48: 393.
122 Ruta, B., Chushkin, Y., Monaco, G. et al. (2012). *Phys. Rev. Lett.* 109: 165701.
123 Ruta, B., Baldi, G., Chushkin, Y. et al. (2014). *Nat. Commun.* 5: 3939.
124 Pierce, M.S., Chang, K.C., Hennessy, D. et al. (2009). *Phys. Rev. Lett.* 103: 165501.
125 Muller, L., Waldorf, M., Gutt, C. et al. (2011). *Phys. Rev. Lett.* 107: 105701.
126 Shpyrko, O.G., Isaacs, E.D., Logan, J.M. et al. (2007). *Nature* 447: 68.
127 Chen, S.W., Guo, H., Seu, K.A. et al. (2013). *Phys. Rev. Lett.* 110: 217201.
128 Seu, K.A., Roy, S., Turner, J.J. et al. (2010). *Phys. Rev. B* 82: 012404.
129 Carnis, J., Cha, W., Wingert, J. et al. (2014). *Sci. Rep.* 4: 6017.
130 Lehmkühler, F., Kwaśniewski, P., Roseker, W. et al. (2015). *Sci. Rep.* 5: 17193.
131 Roseker, W., Franz, H., Schulte-Schrepping, H. et al. (2011). *J. Synchrotron Radiat.* 18: 481.
132 Osaka, T., Hirano, T., Sano, Y. et al. (2016). *Opt. Express* 24: 9187.

11

Single Molecule Magnet for Quantum Information Process'

Tadahiro Komeda[1], *Keiichi Katoh*[2], *and Masahiro Yamashita*[2,3,4]

[1] *Tohoku University, Institute of Multidisciplinary Research for Advanced Materials (IMRAM, Tagen), 2-1-1, Katahira, Aoba, Sendai 980-0877, Japan*
[2] *Tohoku University, Department of Chemistry, Graduate School of Science, Aramaki-Aza-Aoba, Aoba-Ku, Sendai 980-8578, Japan*
[3] *Tohoku University, WPI Research Center, Advanced Institute for Materials Research, 2-1-1 Katahira, Aoba-ku, Sendai 980-8577, Japan*
[4] *School of Materials Science and Engineering, Nankai University, Tianjin 300350, China*

11.1 Introduction

The use of the spin degree of freedom of an electron or a nucleus in electronics is developing rapidly. A spintronic device combines the conductance of electrons and spin degree of freedom of electrons for possible applications in information technology and signal processing, whose uses include information storage and logic devices [1, 2]. Quantum computing using the entanglement of spins has been envisioned as a promising technology for advanced information processing. The use of molecules as a material that realizes such advanced electronic devices using spin degree of freedom is being discussed actively [1, 3]. Some of the superior characteristics of the molecules are reasons for this application. First, a molecule can be a building block for the bottom-up approaches, where the molecules form a certain structure in a self-assembled manner. This might solve the problems of the current top-down lithographic technology, which is facing limitations due to the design rules of electronic devices that are already much smaller than the wavelength of light. It has been argued that the lithography size might be limited to a few tens of nanometers [4]. Second, the molecules can easily change their conformation with outer stimulations such as light, electric and magnetic fields, temperatures, and so on. This nature might be used to control the electronic and spin properties. Third, the molecules are composed of light elements, such as carbon and nitrogen, whose spin–orbit coupling is small and the spin coherent length is long, which is favorable for spin transport.

Moreover, it is needless to say that the size of the molecules is genuinely on the atomic size level. Wires for the electron conductance can be made of molecule chains. However, atom-sized magnets are more difficult to form than wires. For classical magnets of ferromagnetic materials, spins interact with each other to stabilize the energy. In this many-body effect, they obtain larger

Molecular Technology: Materials Innovation, Volume 3, First Edition.
Edited by Hisashi Yamamoto and Takashi Kato.

energy stabilization if the number of participating spins increases [5]. In the miniaturization of magnets, magnetic anisotropy energy, which corresponds to the energy barrier to flip the magnetization, decreases and the spontaneous flip of the magnetization becomes a problem. From a chemist's point of view, this can be solved by increasing the anisotropic energy of the spin by tuning the surrounding ligands. This idea has been realized in the synthesis of single molecule magnets (SMMs) that act as magnets of nanometer scale with magnetic hysteresis of an atomic scale below a certain temperature [6, 7]. The successful synthesis has been followed by proposals of possible applications [3]. SMMs are complexes consisting of metal center(s) with organic ligands. The ligand design enables the spins of the metal ions to couple with other spins within the molecule.

Initial SMM research has been concentrated mainly on examining polymetallic 3d-metal complexes, mainly Mn^{2+} ions [8]. The strategy to achieve a high energy barrier to the reversal of the magnetization (U_{eff}) has been to place many anisotropic centers together to create high spin multiplicity (S) as well as high magnetic anisotropy (D) [9].

The research on SMMs accelerated upon finding that double-decker lanthanide phthalocyanine ($LnPc_2$) complexes show good SMM properties with remnant magnetization at single-ion level. After the report of pioneering works by Ishikawa et al. [10], many types of SMMs were reported [11] that bear wider hysteretic loops and energy barriers up to 2 orders of magnitude higher than that observed in the 3d-counterparts [12, 13]. More active discussions of SMMs as building blocks for molecular spintronic or quantum information technologies have been made.

In this chapter, we review recent studies of SMM molecules used in electronic devices using the spin degree of freedom [14]. In such devices, SMMs are combined with metal electrodes, and therefore the characterization of the molecule on surfaces is mandatory. In Section 11.2, we examine recent developments in the synthesis and magnetic characterization of lanthanide double-decker SMMs. AC susceptibility measurement and micro superconducting quantum interference device (μ-SQUID) measurement reveal SMM properties of Ln double-decker molecules and magnetic relaxation through direct tunneling (DT) and quantum tunneling of magnetization (QTM). By tuning the ligand and crystal packing, it has been found that Ln–Ln interactions can be changed, slowing the magnetic relaxation process, especially QTM. In Section 11.3, we present examples of SMMs applied in electronic devices using the spin properties of SMMs. Spintronic devices are fabricated combining $TbPc_2$ derivative molecules and single-walled carbon nanotubes (SWCNTs) in which the SWCNT works as a quantum dot and the SMMs adsorbed at CNT as second quantum dots. The device demonstrates that the conductance can be changed by flipping the spins of SMMs. The experiments executed in mK region can read out the information of the nucleus spin state of SMMs. In Section 11.4 we review recent scanning tunneling microscope (STM) characterization of the SMM film deposited on metal surfaces. STM is a powerful technique not only for the topographic observation of the SMM molecules but also for the characterization of electronic and spin states of SMMs on surfaces. The spin state of SMMs is characterized by observing Kondo effect, which is a many-body effect caused by shielding of an isolated spin by conduction electrons.

11.2 Synthesis and Magnetic Properties of Double-Decker SMM

In this section, we present the developments in the synthesis of SMMs of double-decker phthalocyanate Ln complex. In 2003, Ishikawa et al. reported that the Tb^{III}–Pc double-decker complex $(TBA)^+[TbPc_2]^-$ (Tb^{III} has a ground state with $S = 3$, $L = 3$, and $g = 3/2$; TBA^+ = tetrabutylammonium cation) behaves as an SMM with a long magnetization relaxation time (τ), which originates from the ligand field (LF) [10, 15]. Magnetic properties of rare earth metal ions are strongly related to the charge density distribution. The complexes exhibit a uniaxial magnetic anisotropy along the easy axis, and the separation of ground-state multiplets is caused by the LF at the axial position. The energy gap between the ground and first excited states is associated with the energy barrier for the reversal of the magnetization (ΔE). The larger barrier corresponds to a slow relaxation of the magnetization. The mechanism is different from those of 3d metal cluster SMMs [6]. Consequently, the direction and magnitude of the LF have a large effect on the magnetic properties. The LF potential around a Tb^{III} ion ($4f^8$) with a total angular momentum (J) of 6 splits the ground multiplets so that the lowest sublevel has the largest J_z value.

The structure of $TbPc_2$ is shown in Figure 11.2a,b. The electronic ground state of the Tb^{III} ion splits into seven multiplets with angular momenta (expressed as $|J_z\rangle$) of $|0\rangle$, $|\pm1\rangle$, $|\pm2\rangle$, $|\pm3\rangle$, $|\pm4\rangle$, $|\pm5\rangle$, $|\pm6\rangle$ by the LF of Pc ligands. As a consequence, there is an energy gap between the $|\pm5\rangle$ and $|\pm6\rangle$ levels of $\sim400\ cm^{-1}$, which is correlated to ΔE (see Figure 11.2c). Furthermore, the J of $|\pm6\rangle$ couples with the nuclear spins of the Tb^{III} ions ($I = \pm3/2, \pm1/2$), and the ground state splits into eight states (I–J coupling), which causes QTM [16] (see Figure 11.1a). In addition, a contraction of the square-antiprismatic (SAP) coordination environment has an influence on the LF of the Ln^{III} ions, which affects the SMM properties (Figure 11.1b). It has been reported that when considering the LF potential (V), the V_{SAP} for an ideal SAP coordination geometry with D_{4d} symmetry is different from the V_{SP} for SP (square-prismatic) coordination geometry with D_{4h} symmetry [20]. The differences in the characteristics of SMMs are thought to be due to the presence or absence of particular off-diagonal terms of LF.

Lanthanoid double-decker complexes ($LnPc_2$) consist of a Ln^{III} ion and two Pc ligands, each having a formal charge of −2 with a closed shell π electron system. It is known that one-electron oxidation of the anion radical can be caused at the ligand, resulting in a neutral complex with an open shell π electron system. Therefore, $LnPc_2$ has two spin systems: a delocalized unpaired π electron on the Pc ligands and the Ln^{III} ion with 4f electrons. Deep green crystals of the products "without" CH_2Cl_2 are obtained by using $CHCl_3$ as the solvent [21]. $TbPc_2$ has a Tb^{III} ion sandwiched between two Pc ligands with eight isoindole-nitrogen (N_{iso}) donor atoms. The eight Tb–N_{iso} bond distances, based on the ionic radius (r_{ion}) of an octacoordinated Tb^{III} ion, change in the range of 0.240–0.243 nm [21]. The twist angle (φ) between the two rings has been determined to be 41° with a SAP coordination environment around the metal center (Figure 11.1b,c). Both Pc rings adopt a saucer shape and are slightly distorted. However, one ring deviates

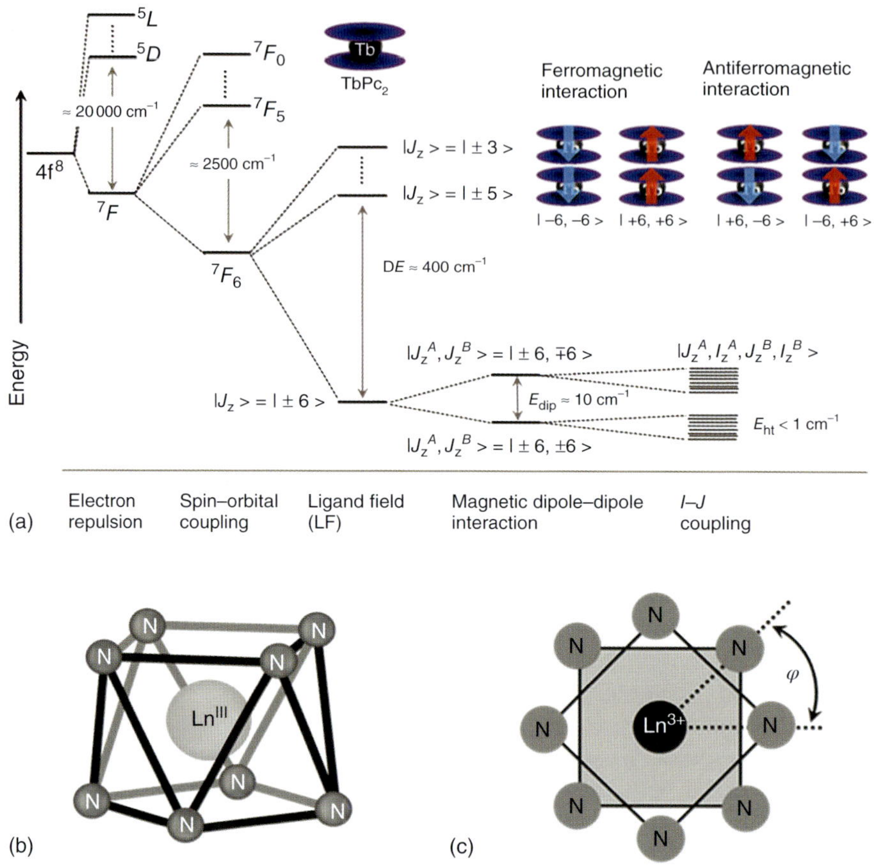

Figure 11.1 (a) Schematic representation of the electronic structure of a Tb^{III} ion ($4f^8$) and $TbPc_2$ complexes. Magnetic dipole–dipole interactions are important for the Tb^{III} dinuclear systems. (b) Schematic illustration of the square-antiprismatic (SAP) coordination environment of $LnPc_2$. (c) Twist angle φ in an SAP in multiple-decker Ln^{III}–Pc complexes. Source: Adapted with permission from Katoh et al. 2016 [19].

more from planarity than the other. The crystal structures are similar to those of previously described Ln^{III}–Pc complexes with solvent molecules [22]. The size and height of the molecule are estimated to be ~1.6 and ~0.4 nm, respectively, on the basis of the distance between the two hydrogen atoms at both ends and between the hydrogen atoms of the two Pc rings, respectively.

Here we examine the details of the QTM process. Although the SMM behavior of the molecule of $TbPc_2$ can be examined by measuring AC susceptibility, as shown in Figure 11.2d, insight into the QTM of the TBA[$TbPc_2$] complex has been achieved by employing a μ-SQUID array, which is a powerful tool for the understanding of the magnetic properties of TBA[$TbPc_2$]. Hysteresis curves associated with SMMs and quantum tunneling steps, including the ones triggered by hyperfine interactions of the electronic spin of the f-electrons with the nuclear spin of the Tb^{III} ion, are obtained. First, the SMM nature of the complex is corroborated through magnetic hysteresis measurements of a single crystal of

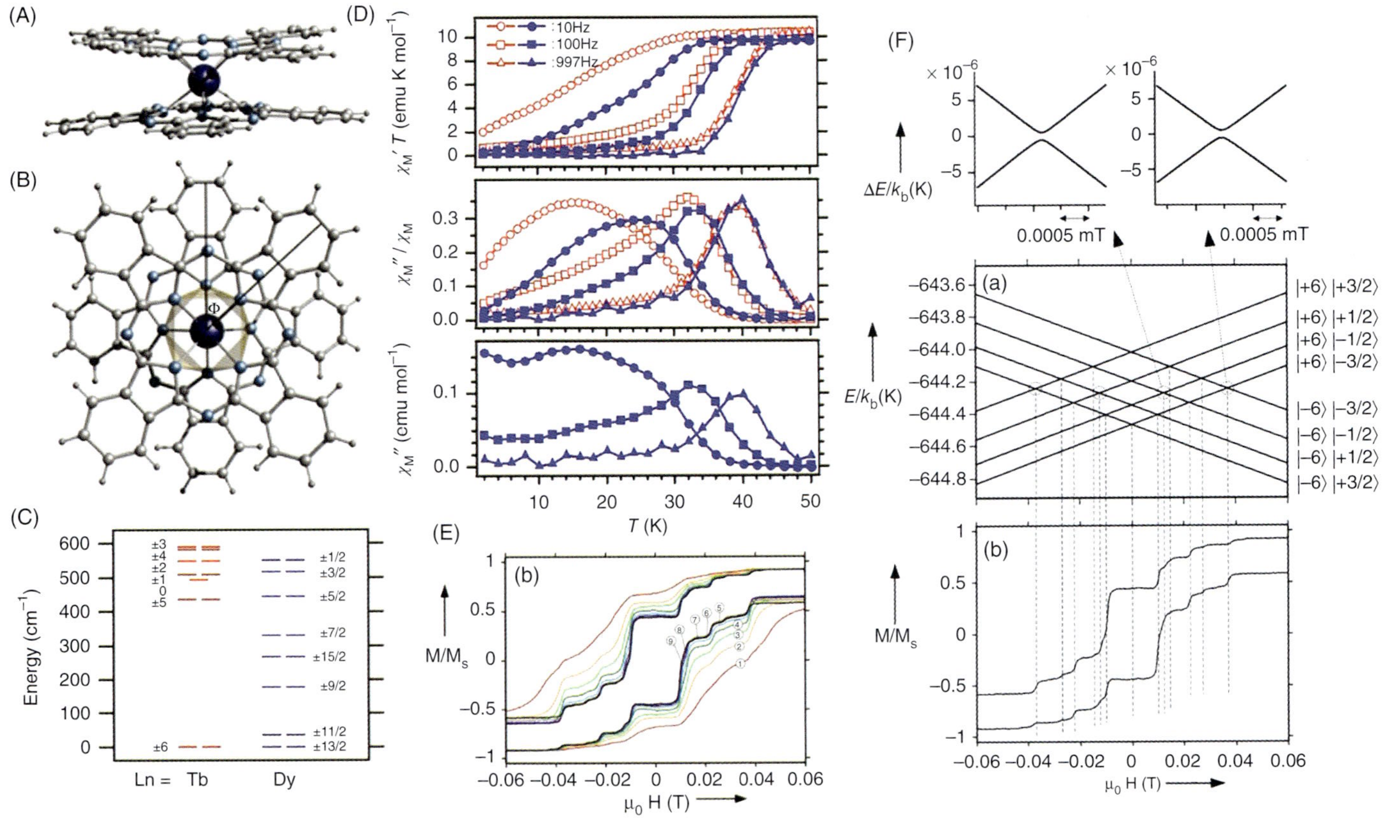

Figure 11.2 Crystal structure of [$TbPc_2$] (a) side and (b) top view. Color for elements: H, dark gray; C, gray; N, cyan; Tb, dark blue. The TBA counterion was omitted for clarity. (c) Energy diagram for TBA[$LnPc_2$] (Ln = Tb^{III} and Dy^{III}) determined by Ishikawa, and (d) dynamic magnetic behavior of TBA[$TbPc_2$] sample (open symbols) and diluted into TBA[YPc_2] (filled symbols), in a zero DC field and an oscillating field of 3.5 G at different frequencies. Source: (a–d) Adapted with permission from Ishikawa et al. 2003 [10] and Ishikawa et al. 2005 [16]. (e) Hysteresis loops for a single crystal of diluted TBA[$Tb_{0.02}Y_{0.98}Pc_2$] at different scan rates at $T = 40$ mK with the field applied along the easy axes of magnetization Source: Adapted with permission from Ganivet et al. 2013 [17]. (f) Schematic representation of the split magnetic ground state $J = 6$ into four states by hyperfine interactions with the nuclear spin $I = \pm 3/2$. Source: Adapted with permission from Ganzhorn et al. 2013 [18].

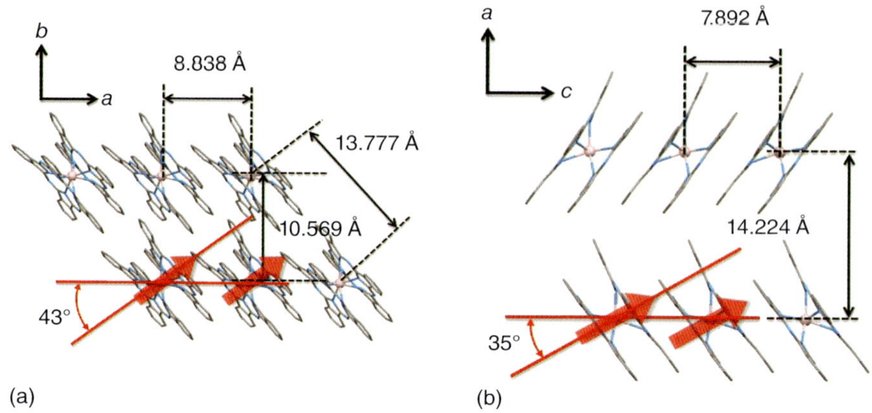

Figure 11.3 Molecular packing of $TbPc_2$ (a) for bare crystal viewed from the *c* axis, (b) in CH_2Cl_2-stabilized crystal viewed from the *b* axis. The values in the figure are the intermolecular Tb^{III}–Tb^{III} distances. Hydrogen atoms were omitted for clarity. Source: Adapted with permission from Yamabayashi et al. 2017 [24].

diluted TBA[$TbPc_2$] in TBA[YPc_2] (2 : 48) at mK temperatures with the magnetic field aligned along the easy axes of magnetization (Figure 11.2e) [16]. Several quantum tunneling events are observed that are associated to quantum tunneling between the states of the J_z multiplets ($J_z = \pm 6$) and the nuclear spin of the Tb ion ($I = \pm{}^3/_2$). The QT steps arise from a combination of the J_z states (± 6) with each J_z state of the nuclear spin, i.e. $m_J = \pm 6$ with $m_I = \pm{}^3/_2$ and $\pm{}^1/_2$ and a quadrupolar term, which leads to 13 level crossings, corresponding to a step in the hysteresis loop determined by the μ-SQUID experiment (Figure 11.2f).

The first example of crystal field modification was realized in an oxidized version of the original subject of Ishikawa's study, i.e. TBA[$TbPc_2$], which upon oxidation becomes neutral $TbPc_2$. AC susceptibility studies show an increase in the out-of-phase $\chi''(\nu)$ up to 10 K, which is higher than that observed in the TBA[$TbPc_2$] [23]. This upward shift is associated to a π-radical ($S = ½$) delocalized over the Pc ligands and the $J = 6$ ground state of the Tb ion. Similarly, Ganivet et al. have shown that upon peripheral substitution of Pc groups on homo- and heteroleptic complexes, enhancement of the U_{eff} due to ligand field effects can be achieved [17].

Recently, Yamashita and coworkers have reported that the relaxation by QTM can be greatly suppressed by changing the crystal packing of the $TbPc_2$ molecules [24]. This has been confirmed by comparing the bare crystal (without any crystal solvent molecules) and that with CH_2Cl_2. In the latter, CH_2Cl_2 remains in the crystal, and the structure shows that the Tb–Tb distance is shorter than that of the bare crystal. In addition, it has been shown that the magnetic dipole alignment of the Tb ions is closer to a collinear configuration in the CH_2Cl_2-stabilized crystal (see Figure 11.3a,b). Magnetic measurements show that the activation barrier for spin reversal is similar for both cases, although the relaxation time is 1 order of magnitude slower for the CH_2Cl_2-stabilized crystal than it is for the bare crystal. The slower relaxation rate should be related to the suppression of QTM process (see Figure 11.4a,b). This suggests that the collinear alignment of the Tb

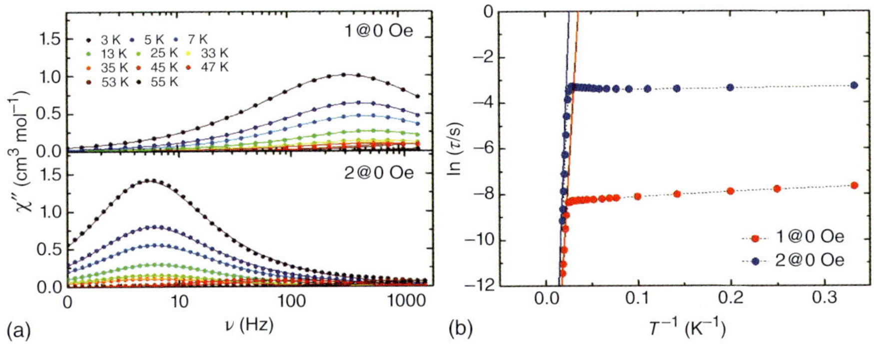

Figure 11.4 (a) χ_M'' vs. ν plot for **1** (bare crystal) and **2** (CH_2Cl_2 stabilized crystal) in a zero field. The solid lines were fitted by using the generalized Debye model. (d) Arrhenius plots for **1** and **2**. The solid lines were fitted by using the Arrhenius equation. The dashed lines are guides for eyes. Source: Adapted with permission from Yamabayashi et al. 2017 [24].

spins contributes to making the off-diagonal term in the LF smaller, and the rate of QTM is slowed. Meanwhile, the ferromagnetic coupling between Tb spins in the bare crystal does not contribute to the suppression of the QTM process. These findings imply that molecular packing, especially the relative angle between the spins of Ln, can be used to tune the QTM process.

11.3 Device Applications of SMM for Spintronic Operations

In this section, we examine several electronic devices that demonstrate a tuning of the electronic conductance by using the spin of SMM molecules. To combine SMMs with electronic devices, the choice of SWCNT and graphene devices is rational, partially because of the sensitivity of these devices due to a large surface/volume ratio compared to conventional field effect transistors (FETs). Moreover, there is a stock of knowledge concerning the attachment of organic molecules to carbon sp^2 surfaces. Kyatskaya et al. have reported the magnetic properties of $TbPc_2$ molecules when attached to SWCNT. In the experiment, terbium phthalocyanine ($TbPc_2^*$), a molecule comprising a heteroleptic double-decker, is used where one of the Pc ligands is substituted with three hexyl groups and one 4-(4-Pyren-1-ylbutoxy) group [25]. This is because the hexyl and pyren have good π–π stacking with sp^2 carbon layered materials. In the magnetic property measurements, the authors mentioned that it is crucial to form pure diamagnetic SWCNTs, since metal atoms are often used as a catalyst for the growth of CNT and can be a problem for magnetic measurements. They measure AC susceptibility for $TbPc_2^*$ molecules on SWCNTs, which shows a clear maximum at a certain frequency and indicates that the SMM property of the $TbPc_2$ can be preserved on SWCNT. In addition, μ-SQIID results measured at 40 mK show a hysteresis loop [25].

This experiment is followed by a conductance measurement with a $TbPc_2^*$–SWCNT device equipped with source, drain, and gate electrodes,

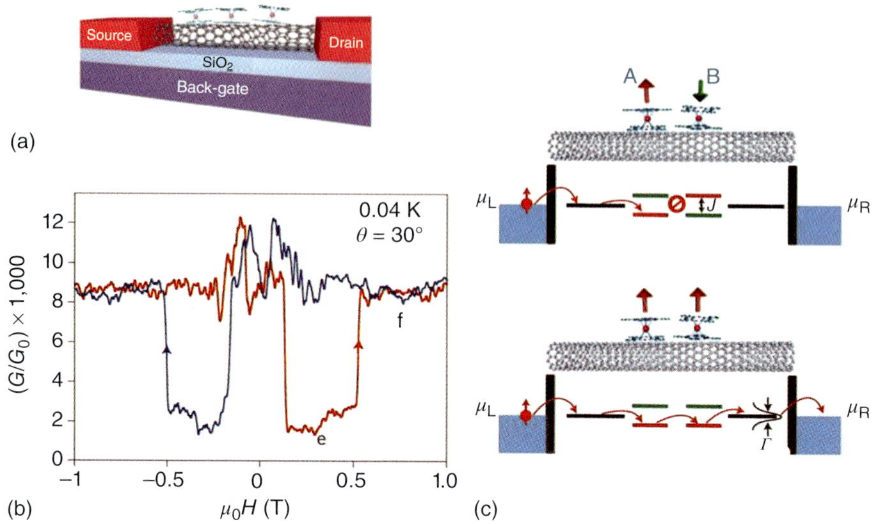

Figure 11.5 Spintronic device combining $TbPc_2^*$ and SWCNT. (a) Schematic representation of the molecular spin valve. (b) Zero-bias conductance vs. magnetic field. The red curve corresponds to the conductance under increasing field (−1 to +1 T) and the blue curve under decreasing field. The last jumps around ±500 mT are attributed to a direct relaxation process of a single molecule. H, high conductance; L, low conductance. (c) A model of electron conductance with a relation to the spin states of two $TbPc_2^*$ molecule. When the spins are antiparallel, the electron flow through the SWCNT is hindered. Source: Adapted with permission from Urdampilleta et al. 2011 [26].

which demonstrates supramolecular spin valve behavior (see Figure 11.5a) [26]. Owing to the sharp and discrete electronic states of the SWCNT, the FET device of $TbPc_2^*$–SWCNT shows a Coulomb blockade in which the SWCNT behaves as a quantum dot. The $TbPc_2^*$ present on the SWCNT acts as a secondary quantum dot.

Experimentally the zero-bias conductance is measured as a function of the magnetic field, as shown in Figure 11.5b. The red curve corresponds to the conductance in an increasing field (−1 to +1 T) and the blue curve in a decreasing field [26]. The conductance jumps around the zero field are attributed to a quantum tunneling of molecules due to tunnel splitting. The last jumps around ±500 mT are attributed to a direct relaxation process of a single molecule showing high and low conductance.

Owing to the coupling between $TbPc_2^*$ and the SWCNT, new spin polarized electronic states, which play a critical role in the spin-valve behavior, appear. This can be explained using a model in which two $TbPc_2^*$ molecules work as a pair. With the presence of the spin of the $TbPc_2^*$ molecule, the hybridized state of $TbPc_2^*$ and SWCNT is spin polarized. If the spin directions of the two molecules are parallel with each other, the created hybridized states have an identical spin direction for the stable states. In this case, the electron with a certain spin direction injected from the left electrode can be transferred smoothly to the right electrode hopping the two dots formed by the paired $TbPc_2^*$. However, if the two molecules have an antiparallel spin direction, the hopping is hampered by the crossing of the levels due to the flipping of the spins. These are illustrated in the upper and lower panels of Figure 11.5c, which correspond to the high and low conductance states, respectively. It can be summarized that the observed magnetoresistance properties of the SMM-SWCNT can serve as an effective tunneling barrier, triggered by the magnetic arrangement of $TbPc_2^*$ interacting with the nonmagnetic quantum dot, i.e. SWCNT.

It should be also noted that the presence of the unpaired π orbital of the Pc ligand plays an important role in mixing the Tb $J = 6$ state with that of SWCNT. The authors also show a strong angular dependence of the conductance change, which should be attributed to the direction of the easy axis of the magnetization to be perpendicular to $TbPc_2$.

The combination between $TbPc_2$ and SWCNT can also give an intriguing spin–phonon coupling [27]. It is demonstrated that magnetization has a dominant single relaxation pathway, that is, one-dimensional phonons associated to the mechanical motion of the SWCNT. As shown in Figure 11.6a, a direct

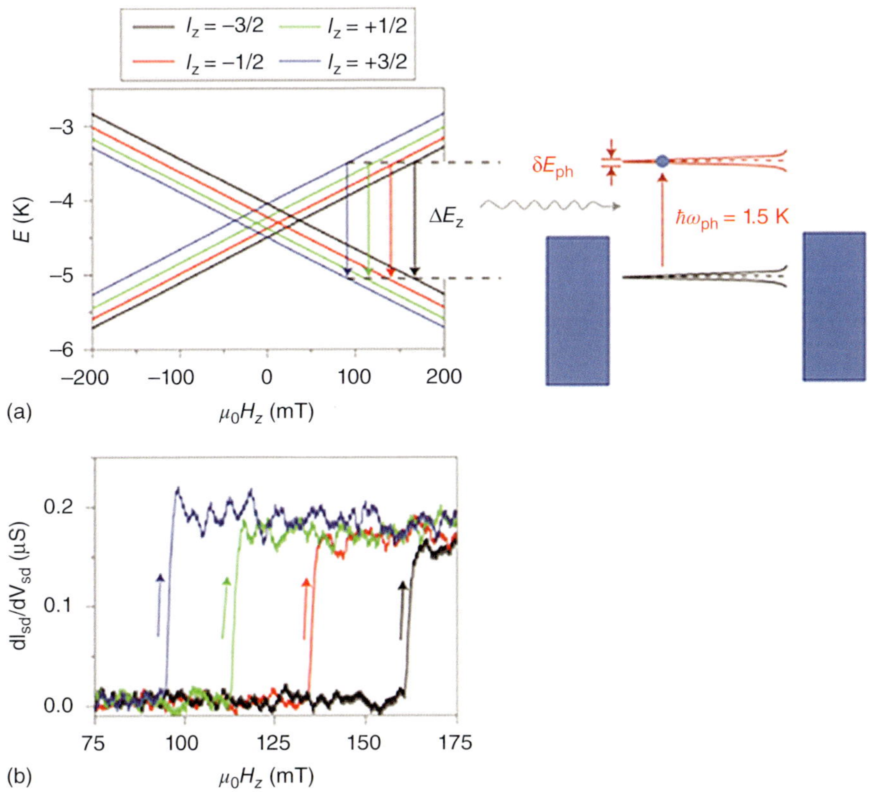

Figure 11.6 Spin–phonon coupling of a $TbPc_2$-SMM and an SWCNT. (a) The magnetization reversal of the $TbPc_2^*$ via direct transition results in excitation of the electron into a vibrational state in the CNT resonator. Four different transitions corresponding to the four nuclear spin states as depicted by the colored arrows and a transition energy $\Delta E_z = \hbar\omega_{ph} = 1.5$ K. (b) Differential conductance vs. magnetic field $\mu_0 H_z$ (parallel to the easy axis of the SMM). The switching field depends on the nuclear spin state occupied before magnetization reversal. The arrows depict the scan direction. Source: Adapted with permission from Ganzhorn et al. 2013 [27].

tunneling process is expected to occur between the corresponding sublevels of the nuclear spin states while dissipating the energy to the phonon.

For the four nucleus spin states, the energy differences corresponding to the states of $J_z = 6$ and -6 are all different. Thus, if a single phonon is considered for energy dissipation, the direct tunneling transitions are expected to occur at different magnetic fields for the four nucleus states. This is illustrated in Figure 11.6a with a corresponding phonon with an energy of 1.5 K. In the actual experiment, the differences can be detected by the four different onsets in the plot of dI/dV vs. magnetic field shown in Figure 11.6b. This experiment demonstrates that readout of the nucleus spin states by conduction measurement is possible.

An intriguing difference in the spin and conductance behavior of the $TbPc_2^*$ can be seen when the molecule is attached to the SWCNT, the two ends of which are suspended by two metal electrodes and the middle portion is not touching the substrate [28]. For such a device, QTM is greatly suppressed if compared to the case where the entire SWCNT is touching the substrate. This can be seen in the magnetization reversal histogram against the magnetic field shown in Figure 11.7a. No quantum tunneling can be seen around $B = 0$, meaning the process is much suppressed (see Figure 11.7b). This is due to the restricted conservation of angular momentum and energy on the suspended SWCNT. This demonstrates that the suppression of the QTM might be possible by tuning the coupling between the molecule and the substrate.

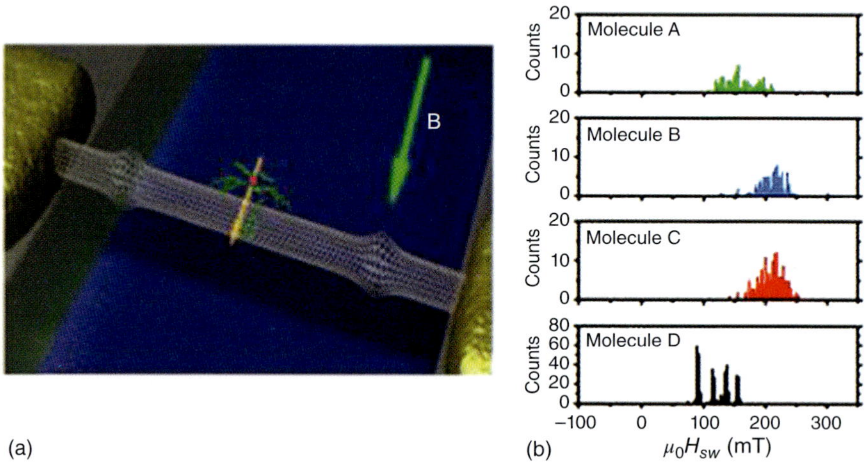

Figure 11.7 Suppression of the QTM in suspended SWCNT-$TbPc_2^*$-SMM resonator. (a) Representation of SWCNT-$TbPc_2^*$-SMM resonator. (b) Magnetization reversal histograms on four different SWCNT-$TbPc_2^*$-SMM resonators. The histograms are recorded at $T = 20$ mK and a magnetic field sweep rate of 50 mT s^{-1}. No QTM is observed around zero magnetic field, but a 100% probability for a magnetization reversal via a direct transition of the electronic spin state of the Tb^{III} ion. Source: Adapted with permission from Ganzhorn et al. 2016 [28].

11.4 Scanning Tunneling Microscopy (STM) and Spectroscopy (STS) Phthalocyaninato–Terbium(III) Multiple-Decker Complexes

STM has played important roles in the studies of molecules adsorbed on surfaces [29, 30], which can give topographic information of an individual molecule with a sub-molecule resolution. Much of the recent developments of STM, however, have been devoted to reveal the physical properties of the molecule for chemical identification.

Scanning tunneling spectroscopy (STS) measurement is one of the most widely used techniques for such a purpose, by which the density of states (DOS) of the molecule at a specific site of the molecule can be deduced. This is achieved by measuring the dI/dV curve of the tunneling current (I) and the gap voltage (V), normally by using lock-in amplifier technique. Electronic states of a molecule such as the highest occupied molecular orbital (HOMO) and the lowest unoccupied molecular orbital (LUMO) appear as peaks in a dI/dV plot.

Characterization of the spin properties of magnetic molecules after film formation is also crucial. Although electron paramagnetic resonance (EPR)/NMR technique has a good resolution to distinguish small changes in the magnetic properties, it cannot be used for film characterization due to a problem of sensitivity. Instead, there is a rapid development of spin detection techniques that have an atomic scale resolution. Among these, we see spin-polarized STM [31–34], inelastic tunneling spectroscopy (IETS) of spin flipping [35–37], and Larmor precession detection in the high-frequency component of the tunneling current [38–41]. Detection of the Kondo resonance is another good technique. The Kondo effect, caused by the interactions between conduction electrons and localized spin [42], appears as a sharp feature near the Fermi level in the dI/dV curve of an atomic scale device or in the tunneling spectroscope. The detection of the Kondo resonance using STM started with measuring a single magnetic atom absorbed on nonmagnetic surfaces [43–48], followed by the reports for magnetic molecules [49–72].

The Kondo resonance of a magnetic molecule is illustrated in Figure 11.8a, where the spin center of the molecule is placed in the gap between the STM tip and the substrate. For a magnetic molecule, the electron in a singly occupied molecular orbital (SOMO) often becomes the spin center, which interacts with the conduction electrons to form a Kondo resonance. The SOMO state, which is specified by the energy ε, should show the increase in the energy due to the on-site Coulomb energy U when it is doubly occupied. A sharp increase in the DOS at the Fermi level occurs if the Kondo resonance is formed (Figure 11.8b).

Nagaoka et al. showed the following formula for the T dependence of the Kondo peak width ($\Gamma(T)$):

$$\Gamma(T) = 2\sqrt{(\pi k_B T)^2 + 2(k_B T_K)^2} \tag{11.1}$$

where k_B is the Boltzmann constant and T_K is the Kondo temperature [73].

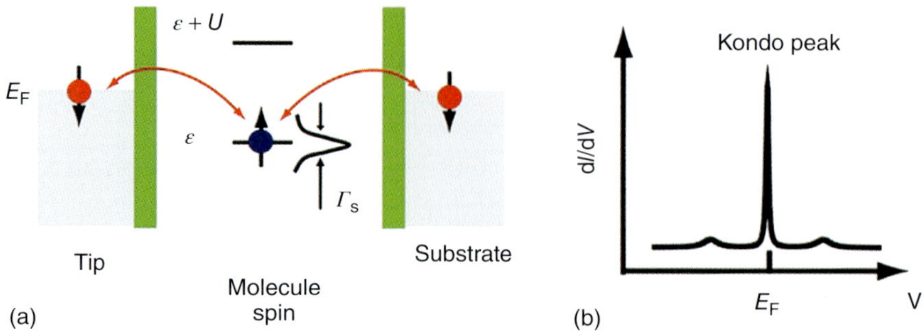

Figure 11.8 (a) Schematic model of the Kondo resonance. ε, U, and Γ_s represent the energy level, Coulomb repulsion energy, and the peak width of the singly occupied molecular orbital (SOMO) state, respectively. (b) Schematic of a typical STS plot of the Kondo resonance.

T_K can be expressed as follows:

$$T_K \propto \exp(-\pi|\varepsilon|/m\Gamma_s) \tag{11.2}$$

where Γ_s is the peak width of the electronic state of the spin center after hybridization with the surface, ε is its energy level relative to E_F, and m is the degeneracy of the orbital [74, 75]. Γ_s decreases when the distance between the spin and the substrate are separated, owing to a decrease in the hybridization between the molecule and the substrate phthalocyaninato–terbium(III) multiple-decker complexes.

11.4.1 Film Formation on Metal Substrates

For an application of Tb–Pc complexes to electrical devices, it is necessary to form a film on the metal or semiconductor substrates. Thus, characterization of the molecules after being transferred to the substrates is necessary. In order to analyze the film properties, molecules are transferred to a substrate using a sublimation method from a heated Ta boat in an ultrahigh vacuum (UHV). Au(111) and Ag(111) surfaces, which are cleaned in UHV chambers prior to deposition, are used as substrates [76, 77]. For the STM/STS experiments, the sample is kept at $T_s \approx 4.7$ K. The dI/dV spectra are obtained using a lock-in amplifier with a modulation voltage of 4 mV (1 mV) superimposed on the tunneling bias voltage for wide-range (near the Fermi level) STS measurements. Studying the films at cryogenic temperatures improves the STM capabilities for molecular characterization for a variety of reasons. First, the motion of the molecules is frozen at low T_s. Second, the resolution of STS spectra is improved at lower T_s because the tunneling electrons are in thermal equilibrium with the sample T. The peak width is expressed by the following equation:

$$W = \sqrt{\left(5.4\frac{k_B T}{e}\right)^2 + {W_1}^2} \tag{11.3}$$

where k_B the Boltzmann constant and W_1 is an intrinsic width of the peak [78]. The increase in the resolution is nearly linear with a decrease in the sample T.

To understand the STM and STS results, a comparison with calculations is required. First-principles calculations are performed using VASP code with a plane wave basis set and projector-augmented-wave (PAW) potentials to describe the valence electron behavior [79, 80]. In addition, a generalized gradient Perdew–Burke–Ernzerhof (PBE) exchange-correlation potential is used [81]. Structure optimization is repeated until the forces are less than 0.05 eV $Å^{-1}$. Owing to the absence of dispersion forces in the local and semi-local exchange-correlation approximations, the molecule–surface distance for weak bonding, such as van der Waals interactions, is still controversial. Thus, whether charge transfer from the substrate to the molecule occurs or not is still unclear. Nevertheless, the results of the calculations on the adsorbed molecules with van der Waals interactions are sufficient to understand the spin behavior when compared with the results calculated for molecules in vacuum. The gold surface is modeled as a five-atom thick slab.

Before an understanding of the electronic and spin states of the molecules on surfaces can be obtained, it is mandatory to determine the adsorption structures. The information should include both the bonding configuration with respect to the metal surface and the lattice structures of the molecule films.

11.4.2 Bonding Configuration of Pc Molecule

Since our target molecules have a common configuration where Pc ligands sandwich Ln atoms, it is useful to examine previous reports of the adsorption of the Pc molecules on surfaces, which have been performed for variety of systems [82–89]. We first see the case of a H_2Pc molecule adsorbed on Au(111) surface [90].

Komeda et al. have reported an STM image of H_2Pc molecules on Au(111) surface, as shown in Figure 11.9a, which is obtained at $T_s \sim 4.7$ K. The cross-like protrusions correspond to H_2Pc molecules. Several other single-decker Pc metal complexes have been reported to show similar STM images [82–89]. Although we see the cross-like skeleton in the structure model of the Pc ligand, the appearance of the cross-shaped protrusion is not only due to the electronic structure of the Pc but also due to hybridization with the electronic state of the Au(111) surface. This can be examined by comparing the results with those from density functional theory (DFT) calculations, where the Pc's molecular orbitals near the Fermi level show eight protrusions rather than four [91, 92].

The symmetry lines connecting the centers of the two diagonal phenyl rings are often used to determine the azimuthal rotation angle of each molecule with respect to the substrate's crystal direction. We often see cases where one of the two symmetry lines of a molecule is parallel to the close-packed directions of the substrate (A1, A2 or C in Figure 11.9b). It is assumed that the system can gain larger van der Waals interaction energy by this configuration due to a larger overlap between the molecule and the substrate [90].

An example of the STM image of a monolayer film of H_2Pc molecules is shown in Figure 11.9c, in which an ordered array of molecules together with long-range undulation, reflecting the herring-bone structure of Au(111), can be seen.

There, cross-like protrusions are observed as an internal structure of the H_2Pc molecule (see Figure 11.9d). The lattice of the film can be defined with two unit

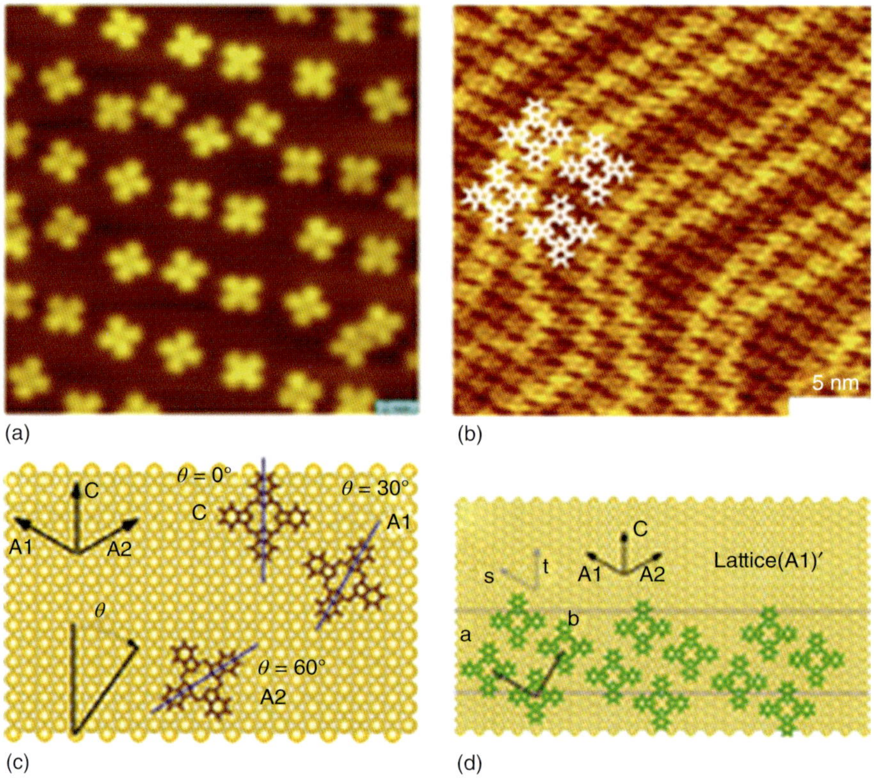

Figure 11.9 (a) STM image of adsorbed H_2Pc molecules. (b) Schematics of Au(111) lattice and the azimuthal rotation of adsorbed Pc molecule in the film. A1, A2, and C represent crystallographically equivalent close-packed directions of the Au(111) substrate. Pc center lines, depicted by blue lines in three Pc molecules, are parallel to A1, A2, and C. (c) STM image of H_2Pc at 1ML coverage for a large area. (d) Schematic of the lattice configuration of H_2Pc film, showing unit vectors of the molecular lattice, **a** and **b**, and of the substrate, **s** and **t**. The lengths of the vectors of the substrate are expanded four times larger than the actual lengths for better visibility. Source: Adapted with permission from Komeda et al. 2011 [90].

vectors, **a** and **b**, which are perpendicular to each other. **a** is parallel to $[\bar{1}10]$ with a length of $5a$, and **b** is parallel to $[11\bar{2}]$ with a length of $3\sqrt{3}a$ (a is the nearest neighbor distance of Au atoms of $\sim$0.288 nm).

Similar lattices have been reported for various combinations of MPc molecules and substrates, which include H_2Pc on a highly oriented pyrolytic graphite (HOPG) surface [93] and FePc [84, 86], CoPc [82, 83, 87, 88], NiPc [86], CuPc [83, 85, 87], and SnPc [89] on metal surfaces. In these, the azimuthal rotation angles of the Pc molecules and the lattices of the films with respect to the metal surfaces are similar to those mentioned for H_2Pc on Au(111).

11.4.3 Molecule Films of Double-Decker Phthalocyaninato–Tb(III) Complexes: Bonding Configuration and Film Structure

Here the bonding configuration of the double-decker Pc molecules is determined using STM observation. $TbPc_2$ molecules were deposited on Au(111) surfaces by a thermal evaporation method with the sample temperature being kept at room temperature during the deposition. We observed two types of molecules, as shown in Figure 11.10a. One type has a cross-shaped structure of a height of 0.14 nm, which can be seen in the upper part of Figure 11.10b. The symmetry lines of the cross are along the [$\bar{1}$10] direction or its equivalents. These features are almost identical with those of the H_2Pc molecule, and therefore, we consider

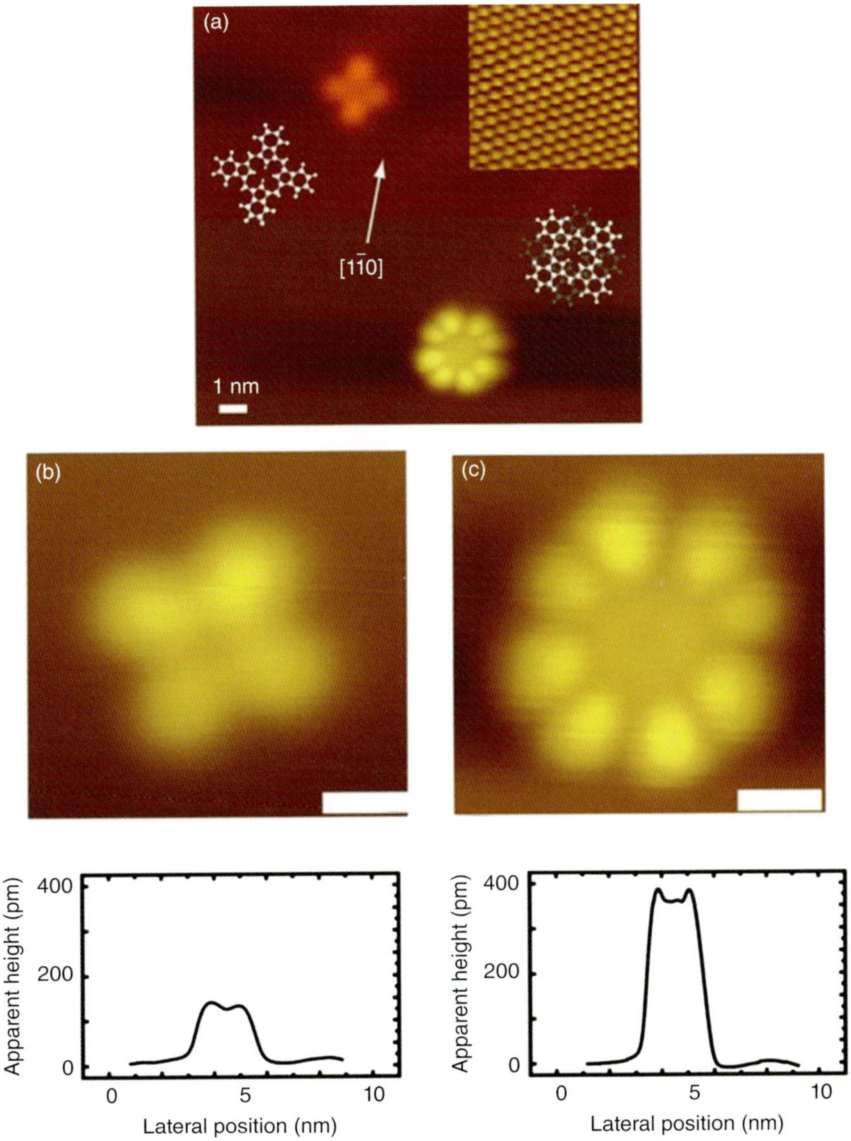

Figure 11.10 STM images of $TbPc_2$. (a) Molecules of single-decker (upper) and double-decker (lower) Pc. An atomic image of Au(111), whose [$\bar{1}$10] direction is marked by the arrow, is shown in the inset. Schematic model of the Pc plane is superimposed. Topographic image and cross-section view of the (b) TbPc and (c) $TbPc_2$ molecules. Bars indicate the length of 1 nm. Source: Adapted with permission from Katoh et al. 2011 [94].

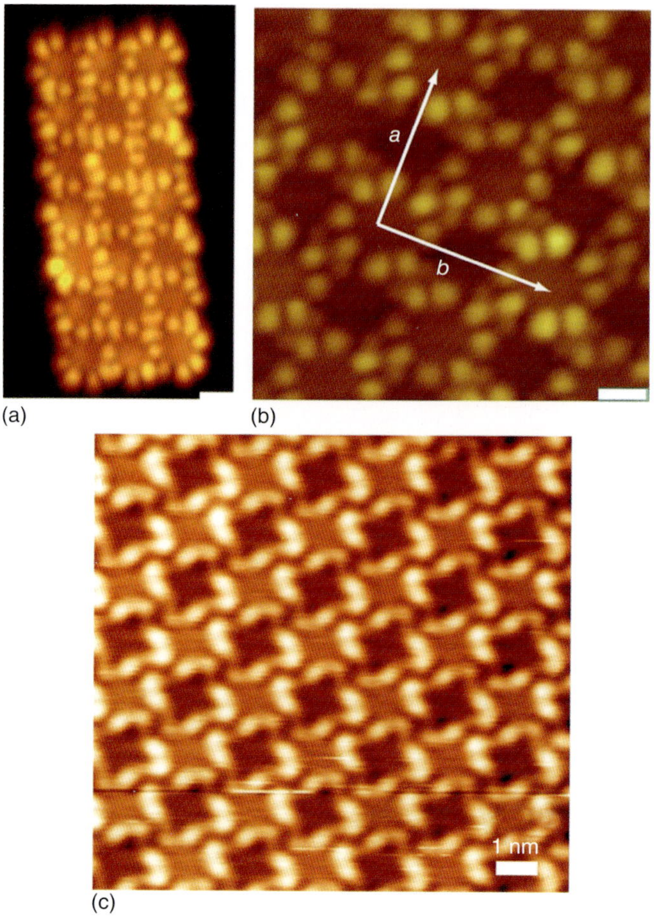

Figure 11.11 (a) $TbPc_2$ film composed of 21 molecules ($V_s = 0.8$ V and $I_t = 0.3$ nA). (b) A magnified image of (a). Checkerboard arrangement of bright and dark molecules, the latter of which is almost invisible with high contrast between them ($V_s = -0.8$ V and $I_t = 0.3$ nA). (c) Same checkerboard contrast variation observed for $TbPc_2$ on Ag(111). Source: Adapted with permission from Katoh et al. 2016 [19].

that this species corresponds to either TbPc or Pc, which are formed by a cracking of the $TbPc_2$ molecule during the thermal deposition process. However, the cross-like molecules are a minority species on the surface.

The other type shows eight protruded lobes, which can be seen in the lower part of Figure 11.10a. The center is slightly depleted from the eight protruded lobes (see Figure 11.10c). The height of the film from the substrate is ~0.4 nm. This should correspond to the $TbPc_2$ molecule transferred to the surface without cracking.

When the coverage of the $TbPc_2$ increases, the eight-lobe molecules form a film that has a pseudo-square lattice. A similar film is formed on both Au(111) and Ag(111) surfaces, and the one on the latter is shown in Figure 11.11a [95]. The film height is similar to that of the isolated molecule of ~0.4 nm. When the lattice image is magnified, we see a checkerboard-type contrast of the image for the film in which bright and dark alternate. The images are shown for the films on Au(111) and Ag(111) surfaces in Figure 11.11b and 11.11c, respectively. If we consider a lattice of the molecule film using a unit cell containing one bright and one dark molecule, the unit cell can be defined using two vectors, **a** and **b**, which are perpendicular to each other. **a** is parallel to $[\bar{1}10]$ with a length of $10a$ (a is the nearest neighbor distance of Au atoms) and **b** is parallel to $[11\bar{2}]$ with a length of $6\sqrt{3}a$. Thus, if the bright–dark contrast is ignored, the lattice is identical. Although it is not possible to observe the lower Pc molecules, it is thought that they adsorb on Au(111) surface in the same manner as H_2Pc molecules do.

In this section, we consider how the STM image of the $TbPc_2$ molecule is formed from the bonding and electronic structures. A magnified STM image of an isolated $TbPc_2$ molecule (Figure 11.12a) has been compared with a model of the molecule (Figure 11.12b), and the charge distribution of HOMO, which is calculated using the DFT method (Figure 11.12c). Since the lower Pc is too far from the tip to contribute to the tunneling current, only the HOMO level of the upper Pc ligand should be examined. We see that the distribution of the STM protrusions of the molecule is similar to that of the HOMO level. The eight lobes are formed since each phenyl ring is flanked by two protrusions.

However, the HOMO distribution is similar both for the single-decker Pc and $TbPc_2$, even though different STM images of the cross and the eight lobes are observed for the former and the latter, respectively. This is mainly due to the difference in the magnitude of the hybridization with the substrate. For the double-decker molecule, the effect of hybridization with the substrate is negligible for the upper ligand, and thus the distribution of the HOMO level of the Pc ligand can be clearly observed.

Next, we discuss the checkerboard image contrast observed in the $TbPc_2$ film. A magnified image of two neighboring molecules is shown in Figure 11.12d, which has a bright (B) or a dark (D) contrast. The eight bright sites of B are marked as 1–8. However, only four tiny lobes (2, 4, 6, and 8) are visible for D. In addition, the angles between the symmetry line of the molecule and the crystal orientation of the substrate are different for the B and D molecules. With respect to the vertical line along the $[01\bar{1}]$ direction, the symmetry line is rotated by 75° for B and by 60° for D. Thus, the azimuthal angle of the upper Pc of B and D is rotated by 15°. An atomic model is shown in Figure 11.12e.

In Figure 11.12e, we show a model where the lower Pc (blue) is assumed to be aligned along the $[01\bar{1}]$ direction and the upper Pc ligand (red) is rotated by 45° with respect to the lower Pc. The symmetry line connecting the two phenyl groups of the upper Pc, **c**, makes an angle of 75° to the vertical line, **v**. This agrees with the observed angle for the B molecule shown in Figure 11.12d. Consequently, the symmetry line for the D molecule is well explained by assuming the azimuthal rotation angle of 30° to the lower Pc. The alternating azimuthal rotation angle is caused by a large steric repulsion among the upper Pc ligands in case all upper Pcs have the same azimuthal rotation angle of 45° to the lower Pcs.

The observed heights of the B and D molecules are different in the STM image. As stated above, the height of a molecule in an STM image is a combination of the

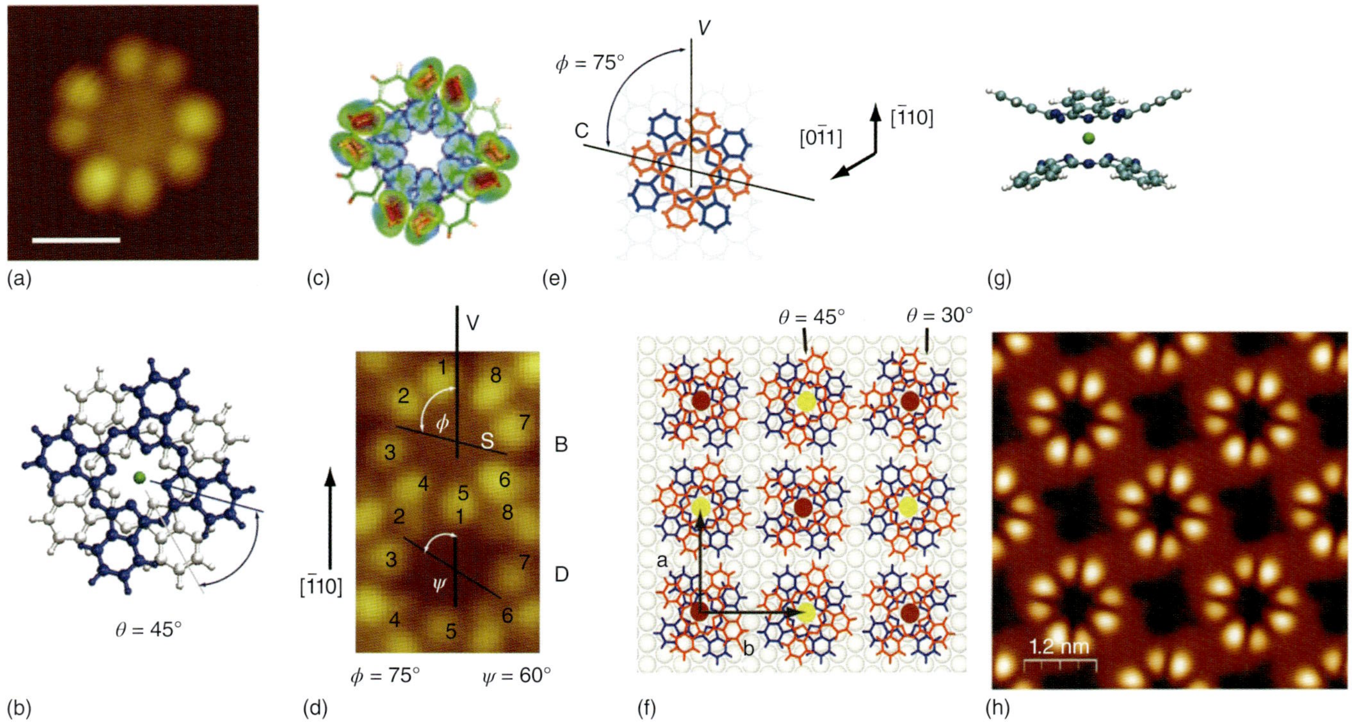

Figure 11.12 An isolated molecule of $TbPc_2$: (a) STM image of Au(111), (b) top view of the model, (c) distribution of HOMO level, and (d) STM images of a $TbPc_2$ film. Bright (marked B) and dark (marked D) molecules are aligned along the [$\bar{1}$10] direction (the vertical line, **v**). Symmetry line of the molecule is marked as **s**. For a bright (dark) molecule, the angle φ (ψ) between **s** and **v** reads ~75° (~60°). (e) Atomic scale model of the bright molecule in (d). (f) Model of film of $TbPc_2$ molecules. (g) Side view model after structural optimization using VASP. (h) Simulated STM image of $TbPc_2$ film. A molecule with $\theta \approx 45°$ appears as bright eight lobes, while $\theta \approx 30°$ is almost invisible. Source: Adapted with permission from Komeda et al. 2011 [96].

physical height and DOS of the molecule. Thus, a comparison with a theoretical model is required.

The optimized results of the simulation for the model are shown in Figure 11.12f. A side view is shown in Figure 11.12g. In the simulated molecules, all Pc ligands adopt a bent configuration. In addition, for the B molecule, the angle of the bend of the upper Pc ligand is steeper than that of the D molecule. This makes the former molecule taller than the latter and should be the origin of the height difference between B and D. The image is dominated by the contribution from the HOMO state, which gives more contrast than do higher lying orbitals because it is more compact. The simulated STM image shown in Figure 11.12h agrees with the experimental STM image.

So far, we have assumed a flat-lying adsorption configuration of the Pc ligand on the surface. However, it is known that they are tilted from the flat-lying positions when multiple Pc planes are deposited [97]. This might be a manifestation of the polymorphic nature of the phthalocyaninato molecules in which polymorph phases have been reported [98–100]. This is also true for a film of CoPc molecules on Au(111) surface, where a tilt of ~3° has been found in the second layer [88]. Thus, it is not surprising that the upper Pc of a $TbPc_2$ molecule tilts. Actually, this tilting affects the spin property of the molecule, which will be discussed later.

11.4.4 Hetero-ligand Double-Decker Molecule

As discussed in the synthesis section, SMM properties, such as the blocking temperature, can be controlled by changing the ligands. We have synthesized a hetero-ligand double-decker molecule, 2,3-naphthalocyaninato phthalocyaninato Tb^{III} double-decker complex (TbNPcPc). The structure of TbNPcPc molecule is shown in Figure 11.13a, in which Pc and NPc ligands sandwich a Tb^{III} ion [61]. In the model, the smaller Pc and larger NPc ligands stack parallel to each other. As can be seen in the side view, Pc and NPc ligands adopt a bent configuration with respect to each other [61]. Top views of both ligands are shown in Figure 11.13b. The difference between the Pc and NPc ligands is the four phenyl rings marked with an "O" in the figure extending from "I". The adsorbed TbNPcPc molecule has two distinct electronic states since the symmetry is lost after the molecule is adsorbed on the surface.

In addition, the adsorption of the hetero-ligand molecule is significantly different from that of homo-ligand molecules. This is because there are two types of molecules upon adsorption for hetero-ligand molecules, depending on which ligand attaches to the substrate. This difference in adsorption induces chirality in the molecule, although the molecule itself is not chiral.

The Tb(NPcPc) molecule can be deposited onto Au(111) surface with a sublimation method. The images are shown in Figure 11.13c,d, in which Pc-up and NPc-up molecules can be distinguished. On the basis of the length of the diagonal (~1.5 and ~2.0 nm for Pc and NPc, respectively), the images in Figure 11.13c,d are assigned to be Pc-up and NPc-up molecules, respectively. This is further confirmed via comparisons with simulated STM images created using DFT calculations. The distributions of local DOS (LDOS) of Pc-up and NPc-up molecules are shown in Figure 11.13e,f with the marks of "I" and "O," respectively.

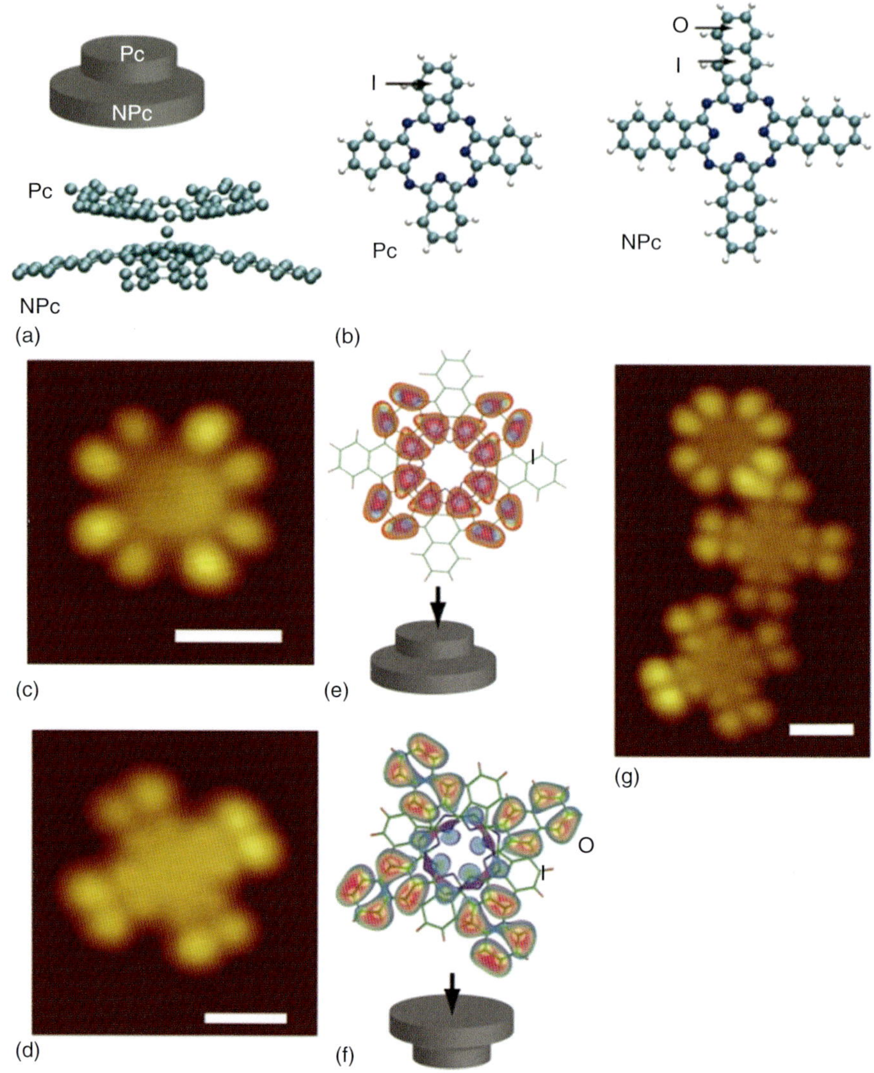

Figure 11.13 (a) and (b) Models of the TbNPcPc double-decker molecules. "I" and "O" in (b) specify the position of the outer phenyl rings. (c–f) STM image and DFT-simulated LDOS mapping of isolated Pc-up ((c), (e)) and NPc-up ((d), (f)) molecules. (g) STM image of aggregated TbNPcPc molecules. White bars in (c), (d), and (g) correspond to a length of 1 nm. Source: Adapted with permission from Komeda et al. 2013 [61].

In Figure 11.13g, there are one Pc-up and two NPc-up molecules. The separation between Pc-up and NPc-up molecules has been determined to be ~1.7 nm. Although the two molecules cannot physically touch each other, they look overlapped in the STM image. The lobes of the Pc-up molecules are more protruded than those of the NPc-up molecules. This is consistent with the LDOS calculation done using a DFT method. The results are shown in Figure 11.13e,f where the

DOS for the occupied states within ~0.8 eV from the Fermi level of the Pc-up molecules is more enhanced than those for the NPc-up molecules.

When TbNPcPc molecules are deposited on Au(111) whose temperature is kept at ~80 °C, an ordered chain structure such as that shown in Figure 11.14a appears. Interestingly, the chain is composed only of NPc-up molecules, as illustrated in Figure 11.14c. A model of the adsorption configuration is shown in Figure 11.14b. The chain is parallel to the direction expressed as $-4\mathbf{s}+9\mathbf{t}$ using the unit vectors of substrate **s** and **t**, which is slightly tilted from the $[11\bar{2}]$ direction. The chain is preferentially formed at the fcc sites between the soliton walls of Au(111) reconstructed surface [90], and the centers of the molecules are separated by ~2.2 nm. The symmetry line is tilted by ~75° from the $[11\bar{2}]$ direction, which can be explained by a model where the lower NPc symmetry line is parallel to the $[1\bar{1}0]$ direction and upper and lower ligands are rotated 45° from each other, which is the same as that used for the $TbPc_2$ modeling.

Further deposition of Tb(NPcPc) molecules at the substrate temperature of ~80 °C leads to the formation of a well-ordered two-dimensional (2D) film, which is shown in Figure 11.14d. We identified two different molecules in the image, which are marked with red and blue circles. The lattice vectors **u** and **v** of the film are determined in relation to **s** and **t**:

$$\begin{pmatrix} u \\ v \end{pmatrix} = \begin{pmatrix} 9 & 0 \\ -4 & 9 \end{pmatrix} \begin{pmatrix} \mathbf{s} \\ \mathbf{t} \end{pmatrix}$$

assuming that molecules are at the origin and at the center of the unit rhombohedra $2.5\mathbf{s}+4.5\mathbf{t}$ in the unit cell.

The STM images of the blue and red molecules correspond to those of an isolated Pc-up and NPc-up molecules, respectively. The lobes corresponding to the "O" phenyl rings are missing for NPc-up molecule, probably because the outer parts cannot be observed when they are close to a Pc-up molecule. A schematic model of the ordered film is shown in Figure 11.14e, in which NPc-up and Pc-up molecules alternate. The nearest distance between the molecules is ~1.7 nm, which is smaller than that observed in the 1D chain.

11.4.5 Triple-Decker Molecule

Here, we see the STM topography images of the film of Y_2Pc_3 molecules [101]. A triple-decker molecule is necessary to investigate the metal–metal interaction within the molecule. A molecular model of Y_2Pc_3 optimized by using a VASP calculation is shown in Figure 11.15a, where both the top and bottom Pc ligands are bent with respect to the center Pc. Again the molecule can be transferred to the Au(111) surface by a sublimation method. A well-ordered film on Au(111) surface is formed when the molecules are deposited while keeping the substrate at room temperature. The lattice of the molecules is visible along $[\bar{1}10]$, which is the close-packed direction of Au(111).

The cross-section plot of the film along the white line in Figure 11.15b is indicated in Figure 11.15c. The height is ~0.55 nm above the bare Au substrate, which is higher than the film of the double-decker Pc (~0.4 nm). However, it does not reflect the physical height difference since STM traces the isosurface of the LDOS.

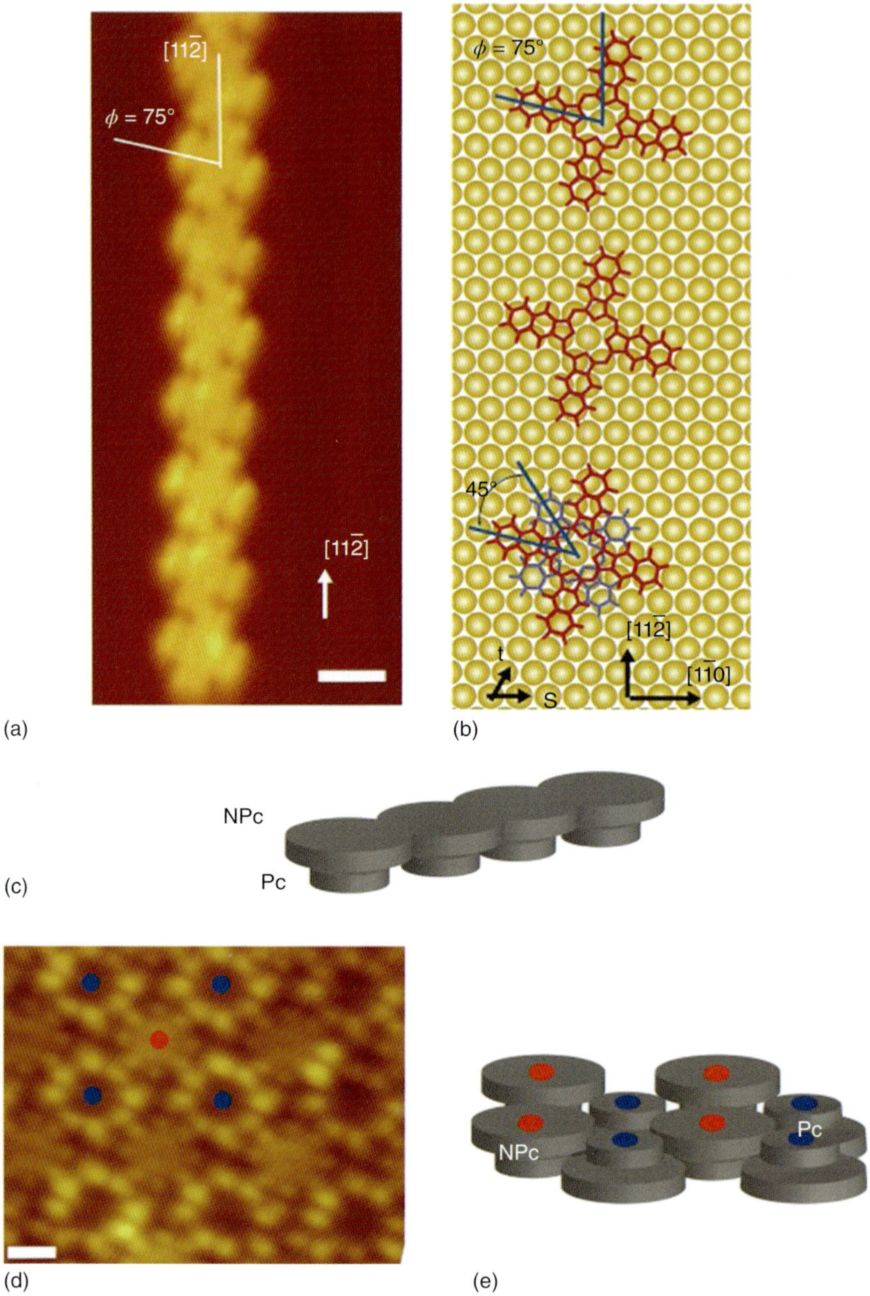

Figure 11.14 (a) STM image of NPc-up chain. White bar = 1 nm. (b) and (c) Schematic models of the structure of the chain in (a). (d) STM image of a TbNPcPc film. Blue and red dots correspond to Pc-up and NPc-up molecules, respectively. White bar = 1 nm. (e) Schematic model of the film of (d). Source: Adapted with permission from Komeda et al. 2013 [61].

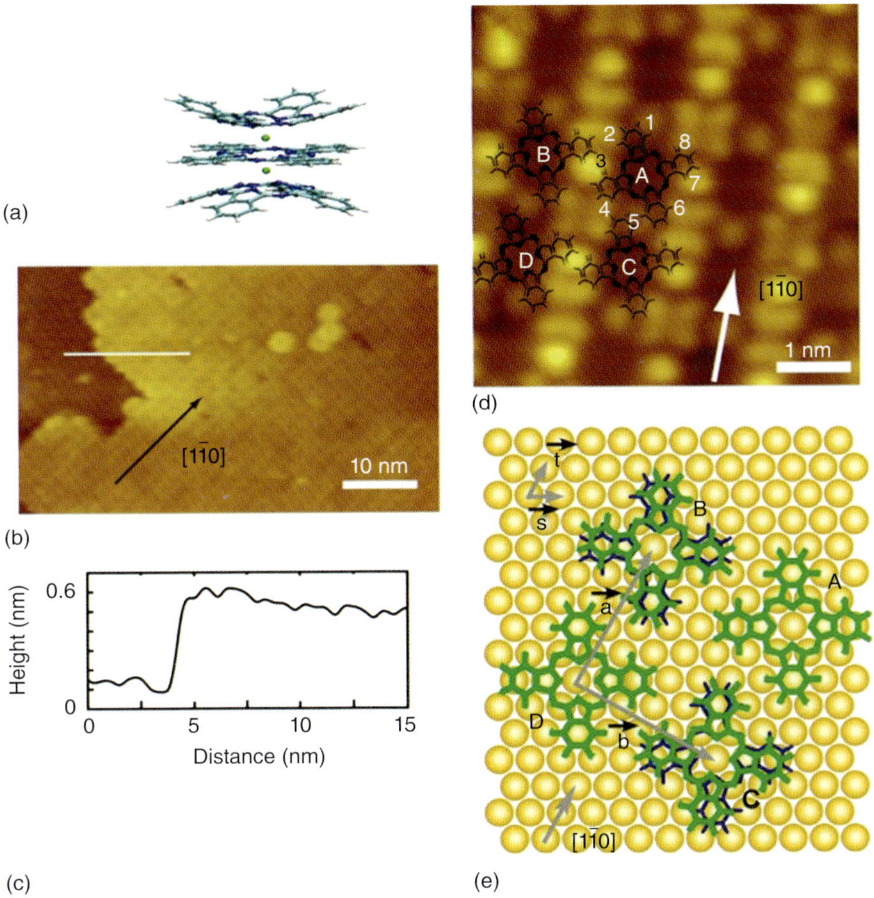

Figure 11.15 (a) Side view of a triple-decker Y_2Pc_3 molecule. (b) STM image of monolayer film of Y_2Pc_3 molecules adsorbed on Au(111). Both film and Au bare surface are seen ($45 \times 27\ nm^2$, $I_t = 0.4$ nA, $V = -0.8$ V, $T = 4.7$ K). (c) Cross section plot along the white line in (b). (d) Magnified images of Y_2Pc_3 film on Au(111) for occupied state image ($5.2 \times 5.2\ nm^2$, $I_t = 0.4$ nA, $V = -0.8$ V). Models of top Pc of Y_2Pc_3 molecule are superimposed for molecules A–D. Numbers 1–8 correspond to bright spots originating from molecule A. (e) A schematic model of Y_2Pc_3 on Au(111). Yellow circles are Au atoms. Top and bottom Pc ligands are shown in green and blue, respectively, the latter of which are eclipsed by the former in A and C. **s** and **t** are unit vectors of the gold substrates. Source: Adapted with permission from Isshiki et al. 2010 [101].

Thus, the triple-decker Y_2Pc_3 molecules can be transferred to the surface by the thermal evaporation process without cracking.

A magnified STM image of the film in an occupied state (I_t of 0.4 nA, V_s of −0.8 V) is shown in Figure 11.15d. The four molecules marked A–D in Figure 11.15d form a pseudo-square lattice unit of the Y_2Pc_3 film.

The lattice is the same one proposed for H_2Pc and double-decker $TbPc_2$ molecules. As discussed above, this suggests that the lower Pc ligand, which attaches to the substrate, should have the same bonding configuration as that of the H_2Pc molecule.

In order to consider the rotations of the middle and top Pc ligands, we examined the reported XRD measurements for triple-decker molecules, such as Bi_2Pc_3 [102], $Cd_2(OAc)_3$ (OAc = 1,4,8,11,15,18,22,25-octakis(hexyl)phthalocyanine) [103], and $Tb_2(obPc)_3$ (obPc = dianion of 2,3,9,10,16,17,23,24-octabutoxyphthalocyanine) [94]. From these studies, we can deduce a common Pc stacking configuration for the triple-decker Pc molecules. For the Bi_2Pc_3 molecule, the middle Pc ligand is rotated by 38.3° from the top and bottom Pc ligands. The top Pc is slightly (~0.5°) rotated from the lower Pc. Similar rotation angles among the three Pc ligands can be seen for the other two molecules, and the twist angles are reported to be ~34.2° and ~32° for the $Cd_2(OAc)_3$ and $Tb_2(obPc)_3$, respectively [103].

These examples suggest that the relative azimuthal angle between the top and bottom Pc ligands is small. As shown above, the $TbPc_2$ molecule shows bright spots at the right and left borders of each phenyl ring. Thus, models of the top Pc ligand are superimposed on the image of Figure 11.15d. An atomic scale model with respect to the Au substrate is shown in Figure 11.15e. For molecules A and D, the top Pc is aligned with the lower Pc. This configuration reproduces the observed STM image. However, for molecules B and C, the top Pc has to be rotated by ~8° in the clockwise direction. This discrepancy in the azimuthal angle appears as a checkerboard contrast modulation, similar to the bright–dark sequence of the $TbPc_2$ film.

Although it is impossible to directly access the rotation angle of the middle Pc ligand, it is thought that their angles alternate between 45° and 30° as in the case of $TbPc_2$ molecules. This is due to a large steric repulsion among the middle Pc ligands of the molecules in case the azimuthal rotation angle does not alternate. Thus, the rotation of the middle Pc ligand contributes to the appearance of checkerboard-type variation in the rotational angle of the top ligands.

11.4.6 Double- and Triple-Decker Pc and Kondo Behavior

Here, we compare STS near the Fermi level observed for $TbPc_2$, YPc_2, Y_2Pc_3, and TbNPcPc molecules [61, 96]. First, the $TbPc_2$ molecule has a two-spin system: one involves the central Tb^{III} ion with $J = \pm 6$ and the other involves one unpaired electron in the π orbital of the Pc ligand, both of which can be the spin center for the Kondo resonance.

In Figure 11.16, we show the dI/dV plot of the Kondo resonance for the $TbPc_2$ molecule on Au(111) [96]. A clear zero bias peak (ZBP) is observed for the bright molecules ($\theta \approx 45°$) with the tip positioned on one of the lobes (labeled I). When the tip is placed over the center of the molecule, however, no ZBP is observed (labeled II). This is clearly seen in the mapping image of Figure 11.16b, in which the changes in the dI/dV spectra are shown as the color difference (upper panel) for the tip positions along the symmetry line of $TbPc_2$ (lower panel).

This mapping shows that the Kondo peak is observed strongly at the ligand position and nearly disappears above the Tb position. A similar spatial distribution of the Kondo resonance has been observed for CuPc/Ag(100) in which the spin center is attributed to the unaired π orbital [65]. Although the unpaired electron in the ligand is formed due to a charge transfer from the Ag(100) substrate to

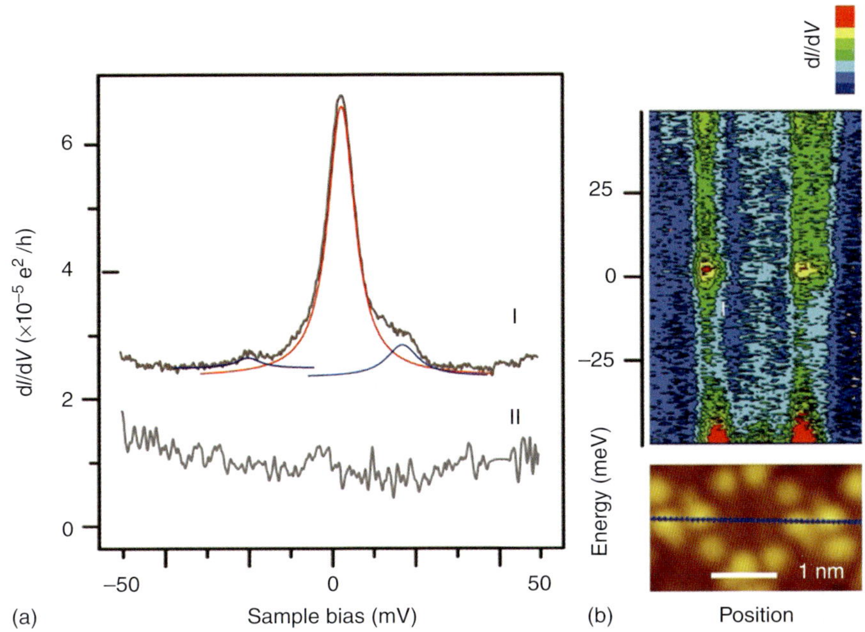

Figure 11.16 (a) dI/dV spectra measured at the ligand position, I, and at the center of the molecule, II. (b) Change in the Kondo peak height when the tip position is moved along the blue line in the STM topography panel. The different colors correspond to the dI/dV intensity, described below the figure. Source: Adapted with permission from Komeda et al. 2011 [96].

CuPc, it is not for the $TbPc_2$ molecule on Au(111). Rather, the molecule of $TbPc_2$ has an intrinsic radical ligand due to the existence of the trivalent Tb^{III} ion and the divalent Pc^{2-} ligand. Thus, a radical is formed due to molecular design instead of charge transfer from the substrate.

It is important to identify the 3D spatial distribution of the unpaired electron in the molecule. However, since the STM can detect the phenomena only for the upper ligand, the distribution of the spin must be calculated. Such a plot is shown in Figure 11.17, where higher spin-polarization sites are concentrated only in the upper Pc. The upper and lower Pcs are not equivalent upon adsorption, which is due to the dipole moment created by the Au surface. Thus, it is reasonable that even the spin distribution is not symmetric for the two ligands. It can be concluded that the Kondo resonance of $TbPc_2$ is formed by the unpaired π states in the upper Pc.

T_K of $TbPc_2$/Au(111) is determined to be ~31 K, while the ones observed for CuPc and NiPc are both ~27 K. This similarity is along the scenario that the Kondo resonance is formed by the spin center of π radical in both cases. However, the T_K should be sensitive to the coupling between the spin and the substrate. This can be seen in Eq. (11.2), where the coupling strength is expressed with the parameter of Γ_s. The spin center is distributed in the upper Pc of $TbPc_2$, which is much farther separated from Au(111) substrate than those of CuPc and NiPc. If the coupling is determined only by the distance between the spin center and the

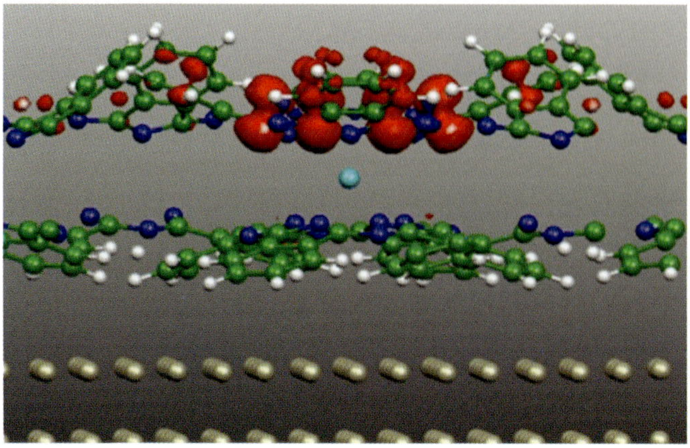

Figure 11.17 Calculated spin polarization for $TbPc_2$ molecules adsorbed on an Au(111) surface. Source: Adapted with permission from Komeda et al. 2011 [96].

substrate, T_K for the $TbPc_2$ molecule should be much lower than that for CuPc and NiPc considering the exponential dependence on Γ_s. However, the actual interaction between the two is stronger than that expected from the distance.

This can be explained as follows. The HOMO and SOMO states of $TbPc_2$ are formed as the bonding and antibonding orbitals of a Pc ligand. This can be seen in Figure 11.18a, where the HOMO and SOMO levels for $\theta = 0°$ are shown. The use of $\theta = 0°$ for the calculations is for simplicity of discussion. Identical orbitals located at each Pc are hybridized in-phase and out-of-phase for the HOMO and SOMO levels, respectively. The charge distributed between the two Pcs can be clearly seen in the side view of the HOMO level. The SOMO state is a case for an antibonding coupling between the two. It is clear that the unpaired π orbital of the upper Pc is a part of the molecular orbital extended over the entire molecule.

This argument is valid also for $\theta = 45°$. When the upper Pc is rotated by 45°, the lobes in the inner circles become in-phase for the SOMO state and out-of-phase for the HOMO state (see Figure 11.18a). Since the HOMO is defined as a more stable state than the SOMO, the HOMO state of $\theta = 0°$ configuration becomes the SOMO for $\theta = 45°$ configuration (level-crossing). Thus, the interaction of the two Pcs is similar for both $\theta = 0°$ and $\theta = 45°$.

The coupling between the two Pcs increases the interaction between the upper Pc and the Au(111) substrate. This can be examined by checking the partial DOS (PDOS) of the upper Pc. The model used for the PDOS calculations is shown in Figure 11.18b, and the calculated PDOS is shown in Figure 11.18c. We compare the cases where the molecule is placed in the vacuum and is adsorbed on Au(111), the results of which are shown in plots I and II, respectively. It is obvious that the HOMO and SOMO orbitals are broadened in plot II than those in plot I, which is caused by the interaction with the substrate. Moreover, in a new peak (marked M) for plot II, a large contribution from the Au substrate is confirmed by analyzing the atom orbital components. Thus, peak M is formed by the hybridization with the metal substrate, indicating a fairly large contribution from the substrate even in the upper Pc. This supports a strong interaction between the upper Pc and the

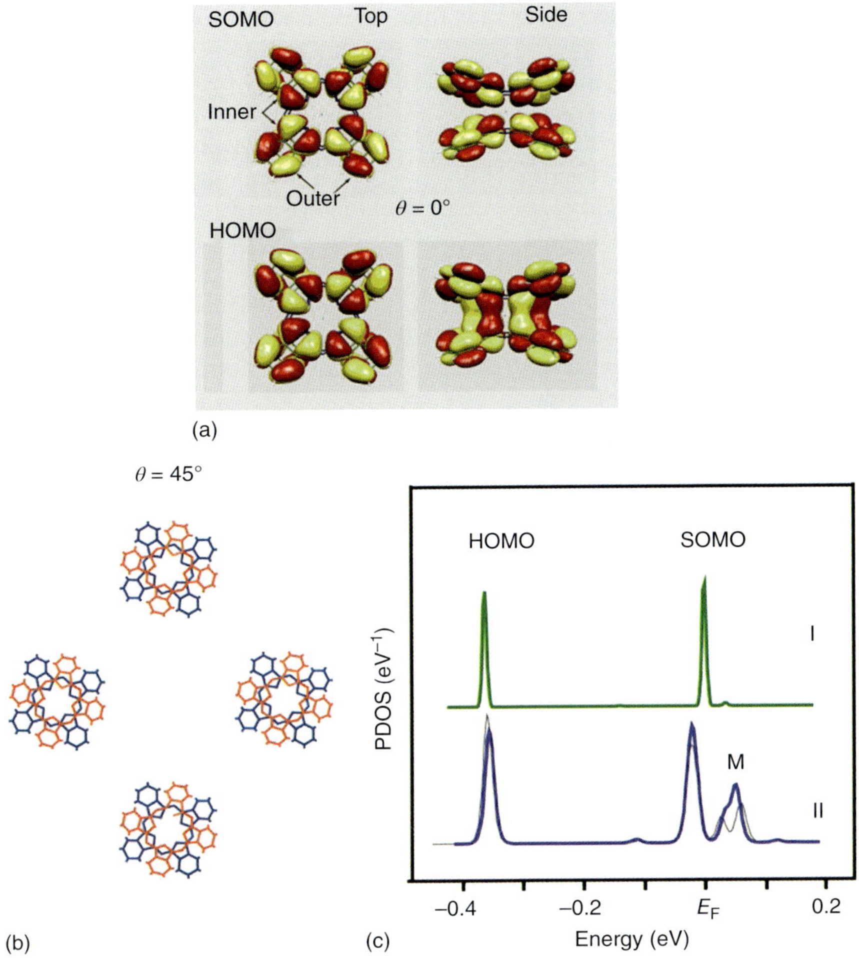

Figure 11.18 (a) SOMO and HOMO for $\theta = 0°$, and the colors represent their phases. Top view shows eight lobes both for the outer circles and inner circles of the molecule. In the side view, bonding between the two Pc ligands can be seen in HOMO. Note that the difference in the SOMO and HOMO are due to the phase differences of the upper and lower Pc ligands. (b) Model used for an isolated $\theta = 45°$ molecule. (c) PDOS of the carbon atom in the phenyl of the upper Pc ligand. I and II are calculated without and with an Au(111) substrate. Source: Adapted with permission from Komeda et al. 2014 [104].

substrate despite the large distance between them, and the interaction is formed mainly by the hybridization between the upper and lower Pcs' orbitals.

Although the $TbPc_2$ molecule has spins both in the Pc ligand and the Tb^{III} ion, only the unpaired electron in the ligand contributes to the appearance of the Kondo resonance. The absence of the Kondo resonance originated from the Tb^{III} spin, which might be explained by the low T_K. In the case of CuPc on Ag(100), a spin is present on the Cu^{II} ion. However, T_K is so low that the resonance cannot be detected at the temperature of ~4 K [65]. The T_K value involving the Tb^{III} ion will be lower because the spin of Tb^{III} ion is decoupled from the substrate due to the presence of the lower Pc ligand. Thus, it could not be observed using the current instrument.

This is consistent with the observation that the Kondo peaks appear only at the ligand position for $TbPc_2$ on Au(111). If charge transfer from the substrate fills an orbital with an unpaired spin, the Kondo resonance will disappear. In fact, Vitali et al. have reported that for $TbPc_2$/Cu(111) no Kondo peak is observed in the STS plot [105]. The $TbPc_2$ molecule should interact with a Cu(111) surface more strongly than with an Au(111) surface, causing charge transfer, which fills the SOMO.

In order to further confirm the origin of the Kondo resonance, we check the Kondo feature of other molecules, double-decker YPc_2 and triple-decker Y_2Pc_3 molecules [101]. The measured dI/dV spectra near the Fermi level are shown in Figure 11.19. A sharp Kondo feature appears for YPc_2, whereas there is no Kondo feature for Y_2Pc_3. This is rational following the model of the Kondo resonance caused by the π-radical of the ligand. The Y ion(s) in both YPc_2 and Y_2Pc_3 molecules are in +3 oxidation states. Thus, in the double-decker YPc_2 molecule, there is a single unpaired electron in the π orbital of the Pc ligand and no spin on the Y ion. In the triple-decker Y_2Pc_3 molecule, two Y ions have a total oxidation state of +6 but have neither d nor f spins. The sum of the oxidation states of the three Pc ligands is −6, meaning that Y_2Pc_3 is neutral and has no spin system. Thus, a Kondo peak is observed for YPc_2, whereas no Kondo peak is observed for Y_2Pc_3. DFT calculations on YPc_2 on Au(111) revealed that the spin is dominantly distributed in the upper Pc, which is similar to the case of $TbPc_2$ on Au(111).

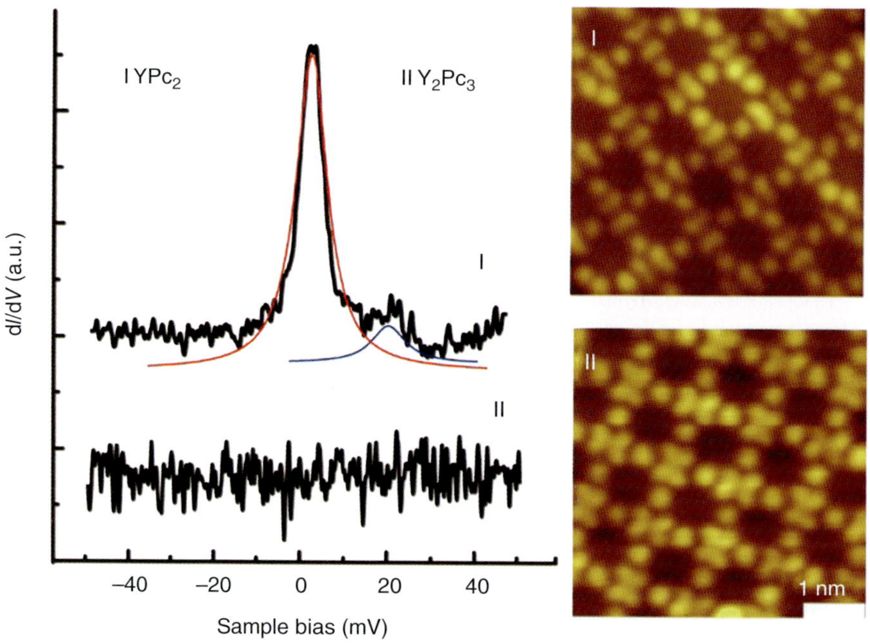

Figure 11.19 Comparison of the dI/dV curves for (I) YPc_2 and (II) Y_2Pc_3. Corresponding STM images are shown on the right-hand side ($V_s = 0.8$ V and $I_t = 0.3$ nA). Source: Adapted with permission from Komeda et al. 2011 [96].

In the unoccupied side of the Kondo peak, a side peak at the energy of ~19 meV is observed for both $TbPc_2$ and YPc_2. Such side peaks of the Kondo feature have been observed for other molecules and carbon nanotubes and have been attributed to magnetic [106, 107] and vibrational [51, 53] excitations. The analysis of the side peaks should be beneficial to probe the molecule's characteristics.

An example where the side peaks are caused by singlet–triplet spin excitation can be seen in the report of CuPc molecule on Ag(100) [65]. This is a two-spin system: one in an unpaired π orbital of the ligand, and the other is from the Cu^{II} ion. These two spins can be coupled either in an antiparallel manner, $S = 0$, or parallel manner, $S = 1$. When the tip is placed on the ligand position, the spin excitation appears from $S = 0$ to $S = 1$ and appears as a side peak around the Kondo peak. The spectra recorded over Cu and Ni ions show inelastic excitation of the vibrational modes of CuPc and NiPc, which cause distortion of the metal–N bond without an accompanying Kondo feature.

In the case of YPc_2, magnetic excitation is not possible since the Y^{III} ion in YPc_2 molecule does not have a spin. Thus, the most plausible mechanism is the excitation of the vibrational mode of the molecule. In order to determine the possibility of the vibrational excitation, the peak energy is compared with those in previous reports on the vibrational modes of the double-decker Pc molecules measured using high-resolution electron energy loss spectroscopy (HREELS). However, so far no HREELS data is available for double-decker Pc molecules to the best of our knowledge. An HREELS study of CuPc on Au(100), in which ν(Cu–N) is observed at 18.5 meV, has been reported [108]. It should be noted that the Y^{III} ion is located out of the Pc plane, whereas the Cu^{II} ion is located in the Pc plane. The effect of the out-of-plane location of the metal ion on the metal–N stretching mode can be estimated by examining the data for a nonplanar MPc complex. Phthalocyaninato tin (SnPc) is a nonplanar molecule, and from HREELS data, ν(Sn–N) is at ~18 mV [109]. Thus, the effects of the out-of-plane location of the Y^{II} ion together with the difference in the mass of the metal ion is limited to the peak energy, and ν(Y–N) should appear at ~18 mV for YPc_2. Although there is a possibility of spin excitation for $TbPc_2$, the side peak for $TbPc_2$ is assigned to a vibrational excitation of the same mode based on the energy position. At this moment, the side peaks of YPc_2 and $TbPc_2$ accompanying the Kondo peak are correlated to the excitation of the vibrational mode of ν(metal–N).

11.4.7 Ligand Effect on the Kondo Behavior

The properties of metal complexes can be tuned by changing the ligand of the molecules. The tunability of the ligand is a beneficial characteristic of magnetic molecules over inorganic materials. Here we see how the Kondo resonance changes with changes in the ligand of the double-decker MPc molecules.

In the dI/dV spectra for TbNPcPc molecules, a Kondo resonance is observed near the Fermi level. Similar ZBPs are observed in the spectra of both Pc-up (Figure 11.20a) and NPc-up molecules (Figure 11.20b) when the STM tip is placed over the lobes of the ligands. The origin of the Kondo peak is attributed to an unpaired spin in a π orbital for both molecules, which is supported by the electronic structure determined using spin-resolved DFT calculations, as shown

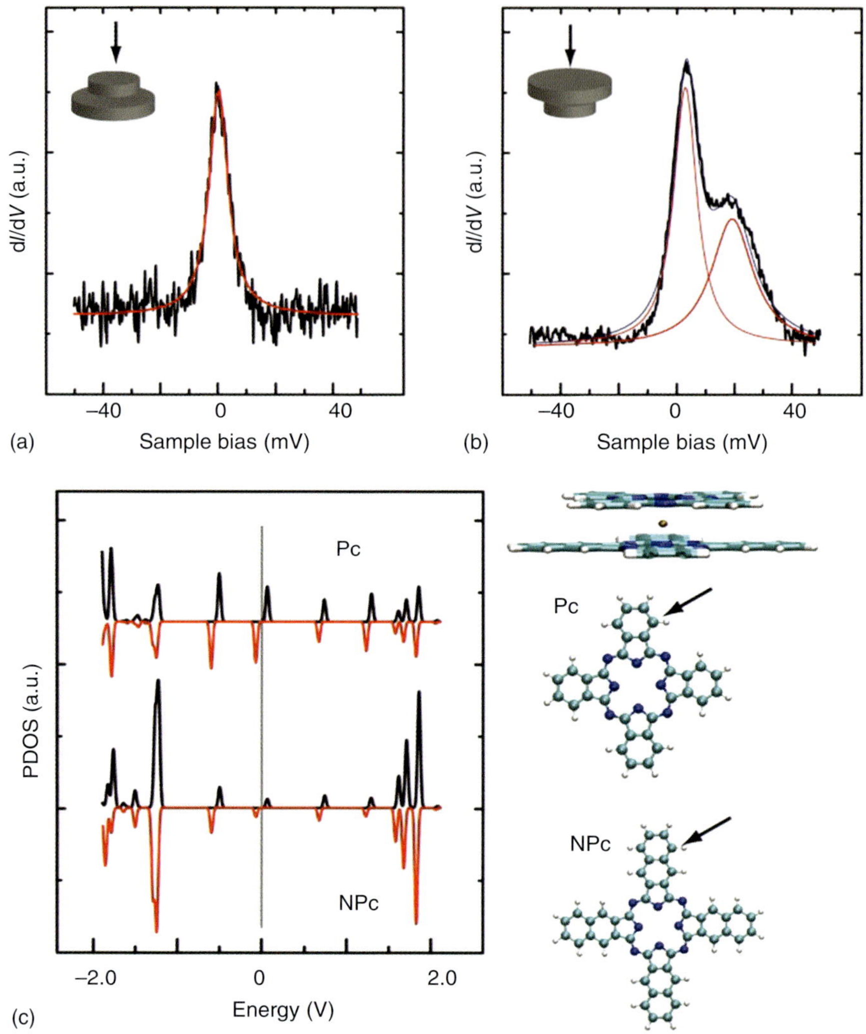

Figure 11.20 dI/dV spectra over isolated (a) Pc-up and (b) NPc-up molecules. Solid lines are fitted using a Lorentzian function. (c) Spin-resolved PDOS vs. energy plots of the carbon atoms of Pc and NPc ligands of TbNPcPc, indicated by arrows. Red and black lines for major and minor spin components. Source: Adapted with permission from Komeda et al. 2013 [61].

in Figure 11.20c. At the Fermi level, a singly occupied state for both molecules is observed, which supports the existence of spins on both ligands.

However, the widths of the Kondo peaks for Pc-up and NPc-up molecules observed at the lobes are different from each other. Figure 11.20a,b shows Kondo peaks fitted with Lorentzian functions. The widths of the main peaks have been determined to be ~8 and ~12 mV for Pc-up and NPc-up molecules, respectively. The widths can be converted to T_K by using Eq. (11.1). The values of T_K have been calculated to be ~32 and ~48 K for Pc-up and NPc-up molecules, respectively.

The mechanism for the difference in T_K values for NPc-up and Pc-up molecules is discussed on the basis of the distance between the upper ligand and the Au substrate. For the $TbPc_2$ molecule, the unpaired π orbital is distributed in the vacuum-side ligand [96]. Similar DFT calculations on the NPc-up and Pc-up molecules show that the unpaired π orbital is distributed in the vacuum-side ligand for both cases. As already mentioned, this molecule has a bent configuration. The absolute height of the vacuum-side ligand from the substrate cannot be determined, since DFT calculation cannot calculate the dispersion force of the van der Waals interactions. However, it is possible to estimate the height difference between the NPc-up and Pc-up from the results of the DFT calculations on an isolated molecule in vacuum. The NPc ligand is larger than the Pc ligand, and both have a similar bent structure. Thus, the vacuum-side ligand of the Pc-up molecule is higher than that of the NPc-up molecule by ~1 Å.

Analysis using Eq. (11.2) indicates that the increase in the spin–substrate distance makes T_K lower due to the expected decrease in Γ_s. This explains why the T_K value for the Pc-up molecule is lower than that for the NPc-up molecule since the ligand in the former case is ~1 Å higher than it is in the latter case. The side peak is clearer, and its intensity is higher than that for $TbPc_2$ and YPc_2 molecules, which originates from the inelastic process accompanying the Kondo resonance. It has been shown that the yield of the IETS peak increases when the molecule is decoupled from the metal substrate. In experiments using copper phthalocyaninato molecules adsorbed on an ultrathin Al_2O_3 film grown on a NiAl(110) surface, the vibronic states of individual molecules are observed as a series of equally spaced features, which are absent when the bare surface is used [110, 111]. For magnetic excitation, the spin-flip IETS experiment using oxidized surfaces of NiAl [35] and Cu surfaces [37] shows a high yield of the IETS features.

For double-decker MPc_2 molecules, the lower Pc plays a role in decoupling the interaction between the upper ligand and the substrate. Therefore, the yield of the excitation of the vibration mode should be higher. This effect seems to be enhanced for the TbNPcPc molecule than for $TbPc_2$ and YPc_2 molecules because the upper Pc is more decoupled from the substrate for the TbNPcPc case due to the larger ligand (and its bent configuration) than it is for $TbPc_2$ and YPc_2 molecules.

11.4.8 Molecular Ordering and Kondo Resonance

It is of great interest to control the nature of a spin by changing its environment via the distance from and coordination numbers of neighboring spins. This can be achieved by manipulating the shape of a self-assembled molecule film on a surface. Iancu et al. have reported that the spin–electron coupling of a hexagonal assembly of TBrPP-Co molecules can be controlled by manipulating the nearest neighbor molecules [58]. The value of T_K increases from 105 to 170 K with a decrease in the number of nearest neighbor molecules from six to zero. They explain that the scattering of surface electrons by the molecules located at the edges of the molecular layer reduces the spin–electron coupling strength for the molecules inside the layer, resulting in the reduction of T_K. It has been claimed that the Cu(111) surface state is removed by the adsorbed molecule. Moreover, it has been theoretically shown that the local DOS of the substrate, $\rho_{m,}$ is reduced with the growth of the molecular assembly [112].

The variation of the Kondo resonance with chain formation has been investigated for Co-(5,5-Br_2-Salophen) molecules [52]. The 1D molecular chain is formed by removing Br atoms of each molecule and forming covalent bonds between the molecules. A Fano-dip of the Kondo resonance is detected over the Co ions. The Co ions are in a low spin state ($S = 1/2$) with three occupied and one partially occupied singlets. The $d_z{}^2$ is the HOMO, and d_{xz} is the LUMO, which carries the spin of the Co ion. It has been found that T_K increases with the number of the molecules included in the chain and varies with odd–even changes in the molecule number.

On the carbon atom at the end of the basic unit after the removal of Br atom, an unpaired electron remains and is coupled to the localized d_{xz} electron of the Co in a ferromagnetic manner. Accordingly, antiferromagnetic ordering between neighboring Co ions in the chains is formed. This is because the unpaired electrons of the carbon atoms couple in an antiferromagnetic manner in the new covalent C—C bond, causing antiferromagnetic ordering between the spins of the Co ions.

In an ordered chain of the hetero double-decker molecules, complex behaviors of the Kondo resonance can be seen (STM in Figure 11.21a) [61]. dI/dV spectra of the chain molecules are shown in Figure 11.21b, which are obtained at the lobe position of the edge molecule of the chain marked I in Figure 11.21a and at that of the inner molecule marked II in Figure 11.21a. In the spectrum obtained over an edge molecule (plot I in Figure 11.21b), a peak similar to that obtained over an isolated molecule is observed. However, the peak width is broader, and T_K (~60 K) is higher [61]. In addition, the Kondo resonance obtained over the molecule inside of the chain changes into a Fano-dip. In the rest of this section, the broadening of the Kondo peak at the edge and the shape change into the Fano-dip are discussed.

These changes have been examined for several systems. Wahl et al. have reported the manipulation of Kondo peaks for a Co atom on Cu(100) by controlling the coupling of the spin with the surrounding cobalt adatoms [48]. Although the compact dimer Co suppresses the Kondo resonance due to strong ferromagnetic coupling between Co atoms, the spectrum on the next nearest neighbor dimer shows a distinct Kondo resonance at the Fermi level, whose peak width is broadened compared to that of the isolated Co adatom. The broadening is attributed to antiferromagnetic coupling between the two Co adatoms. Tsukahara et al. have shown that the Kondo resonance for a FePc molecule on Au(111) changes with the formation of a two-dimensional lattice. A sharp Kondo peak appears in the single-impurity regime, which broadens and splits as the lattice builds up. This is related to Rudermann–Kittel–Kasuya–Yosida (RKKY) coupling between the molecular spins [37, 64]. In addition, distortion of the orbitals due to the intermolecular interaction changes the width of the Kondo resonance [65]. The broadening of the Kondo resonance observed at the edge molecule of the 1D chain of TbNPcPc molecules is due to spin–spin interactions and changes in the molecule configuration.

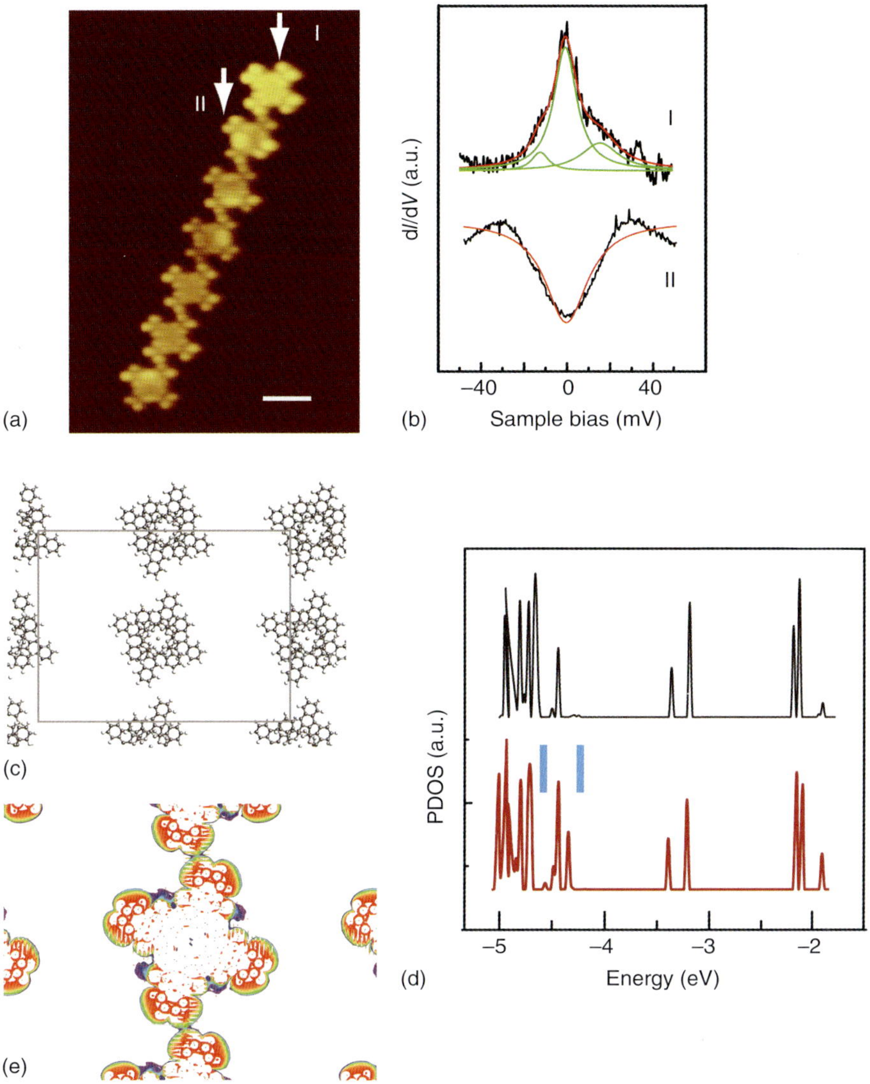

Figure 11.21 (a) High-resolution STM image of the NPc-up chain. White bar = 1 nm. (b) dI/dV spectra measured at I and II in (a). (c) Model of 1D chain of NPc-up molecules used for DFT calculations (free standing). (d) PDOS vs. energy plots for isolated NPc-up molecule (upper, black) and for NPc-up 1D chain molecule (lower, red). (e) Plot of the isosurface of integrated PDOS in the energy range between two blue lines of (d). Source: Adapted with permission from Komeda et al. 2013 [61].

Next, we discuss the changes in the spectral shape of the Kondo resonance with the formation of the chain structure. The spectra obtained over the molecules inside of the chain (plot II in Figure 11.21b) show a dip around the Fermi level. By fitting plot II, Γ has been determined to be 12 mV. The T_K of ~48 K, estimated from $\Gamma = k_B T_K$, is similar to that of an isolated NPc-up molecule.

The Kondo resonance changes from a peak to a Fano-dip due to changes in the chemical environment. For FePc/Au(111), there is interference in the two paths of the tunneling current, i.e. tip–Fe–substrate and tip–substrate. In the on-top configuration, which has a higher bonding energy, a Fano-dip forms due to an increase in the contribution of the tip–substrate tunneling. For CoPc, an increase in contribution of the tip–ligand–substrate pathway causes a Fano-dip.

For TbNPcPc, a Fano-dip occurs due to the formation of the linear chain. Several mechanisms for the conversion from the peak to Fano-dip can be considered. First, the change in the bonding configuration of the molecule occurs upon formation of the molecule chain. When the bonding between the molecule and substrate becomes stronger, a Fano-dip can appear. However, DFT calculations indicate that the formation of the chain causes the molecules to decouple from the substrate, meaning that it is not the mechanism for the Fano-dip formation [61].

Second, minute changes in the orbital due to the intermolecular interactions can cause a Fano-dip. It has been shown that the van der Waals interactions can dramatically change and even quench the Kondo interaction of ligand spins in MPc molecules [65]. Electronic state calculations for a freestanding chain of NPc molecules have been performed. The model used for the calculations is shown in Figure 11.21c, in which the NPc molecules are separated by 2.2 nm parallel to the chain direction and 3.5 nm perpendicular to it. The isolated molecule is modeled by expanding the distance from 2.2 nm to a longer distance.

The results of the calculations are shown in Figure 11.21d. There is a noticeable change in the energy range between the blue bars in Figure 11.21d. More states are identified for a chain in comparison to those of a single molecule because the single molecule state is split into multiple peaks when molecule–molecule interactions occur. The calculated integrated local DOS for the energy region between the blue lines is shown in Figure 11.21e. The isosurface clearly shows the overlapping of two NPc ligands.

Third, there is possible quantum interference of the tunneling current. It has been theoretically shown that the linear chain of quantum dots can exhibit a Fano-dip feature in the conductance due to interference in the tunneling current [113]. In the case of the linear chain of TbNPcPc molecules, the π orbital of the ligand is responsible for the Kondo resonance and, thus, can interfere with the current, causing a Fano-dip.

Further investigation is necessary to completely understand the mechanism for the conversion of the Kondo resonance from a peak to a Fano-dip upon formation of a molecule chain. In particular, the variations in the Kondo feature by controlling the number of the molecules in the chain and modifying the shape of the molecular assembly of a 2D lattice must be examined. Such studies will help reveal the mechanism of how the spin–spin interaction splits the Fano-dips, which has been reported by Tsukahara et al. [37, 64].

11.5 Summary and Future Scope

We have reviewed recent reports about SMM for its application in molecule spintronic devices. The developments in molecular designs and synthesis including the control of axial ligand fields, introducing exchange bias and dipole–dipole interaction have changed the activation barrier of the spin flipping and the rate of QTM. It is apparent that formation of an interface between electrode and SMM also has a large effect on these parameters. For example, spin–phonon interactions accelerate the QTM rate. In addition, the metal–orbital hybridization quenches some unpaired π orbital of Pc ligands, and changes the dipole–dipole coupling of 4f atoms.

Some of these are analyzed and controlled through fabricating actual spintronic devices, which combines the SMMs with SWCNT FET. The device demonstrates that a pair of $TbPc_2$-SWCNT hybridized spin-dependent electronic states works as a filter for the conduction electron; the conductance is high or low depending on whether the spins are parallel or antiparallel. The device at 40 mK clearly shows a conductance change corresponding to QTM, which can distinguish the nucleus spin state. Moreover, this device shows a specific phonon mode of the SWCNT preferentially coupled to the spin-flip of $TbPc_2$, giving an energy passivation path. Furthermore, if the SWCNT is suspended by two electrodes instead of being placed on the substrate, the rate of QTM is greatly suppressed, indicating the importance of the QTM process.

STM characterization of the adsorption of SMM on surfaces can reveal not only the topographic information of the adsorption and film configuration but also the electronic spin states of the adsorbed molecules. A $TbPc_2$ molecule adsorbed on Au(111) shows a flat-lying configuration, whose lattice is similar to the pseudo square lattice observed for a wide variety of metal phthalocyanate molecules, which indicates that the film structure is basically determined by the lower Pc. However, in the case of the upper Pc, two types of contrast are found to form a checkerboard-type arrangement. The contrast variation originates from a structural difference in the two molecules: azimuthal rotation angles (θ) of 45° and 30° for the bright and dark molecules, respectively. Such a rotation is formed to reduce the steric repulsion of the upper Pcs, which appears when all molecules have 45° configuration. The contrast is formed by the structure difference in the two molecules. Upper and lower Pc ligands show a saucer-like shape and the 45° (bright) molecule has a steeper bend angle of the upper Pc ligand than that of the 30° (dark) molecule, which makes the bright molecule higher in physical height. Triple-decker Y_2Pc_3 molecules also make a film with the identical lattice, but no alternative contrast variation among the molecules has been observed.

Next, we examined the STM observation of spin states of the $TbPc_2$ molecule and focus on the observation of the Kondo resonance, which is formed by a shielding of an isolated spin by conduction electrons. For the $TbPc_2$ molecule, the Kondo resonance formed by the unpaired π spin of the ligand is observed, whereas that formed by Tb 4f spin cannot be detected directly. This might be caused either by the low T_K for the Tb 4f Kondo state or by the shielding of the 4f state by the upper Pc. However, this indicates the strong coupling between the unpaired π orbital and the conduction electron. The Kondo resonance of the unpaired π orbital is sensitive to the relative rotational angle between the two Pc ligands; unpaired π spin almost disappears for the dark molecule ($\theta = 30°$), which is due to the variation of the SOMO level with the azimuthal rotation angle. The rotation of the relative azimuthal angle of the upper and lower Pc ligands can be induced by manipulation of the atom with the use of the tunneling electrons. The structure of the double-decker Pc can provide freedom of rotation of the azimuthal angle of the two Pc ligands, which can be a model for a molecular machine that mimics the rotational motion of the biomolecule. In addition to chemical methods to invoke the rotation, an injection of tunneling electrons from the STM tip into the $TbPc_2$ molecule can rotate the azimuthal angle of the vacuum-side Pc. The rotation accompanies a change in the spin properties of the unpaired π electron of the ligand, demonstrating the manipulation of a single spin with current.

Acknowledgments

The authors acknowledge all researchers and students who have been part of the work described in this review. This work was financially supported by a Grant-in-Aid for Scientific Research (grant No. 20225003, MY and grant No. 24750119, KK and grant No. 15K05467, KK and grant No. (B) (16H04191), TK), and Innovative Areas "Molecular Architectonics" (grant No. 25110006) from the Ministry of Education, Culture, Sports, Science, and Technology (MEXT) and M. Yamashita thanks for supporting the 111 Project (B18030) from China.

References

1 Sanvito, S. (2011). Molecular spintronics. *Chem. Soc. Rev.* 40: 3336–3355.
2 Žutić, I., Fabian, J., and Sarma, S.D. (2004). Spintronics: fundamentals and applications. *Rev. Mod. Phys.* 76: 323–410.
3 Bogani, L. and Wernsdorfer, W. (2008). Molecular spintronics using single-molecule magnets. *Nat. Mater.* 7: 179–186.
4 Troiani, F. and Affronte, M. (2011). Molecular spins for quantum information technologies. *Chem. Soc. Rev.* 40: 3119–3129.
5 Stoner, E.C. and Wohlfarth, E.P. (1991). A mechanism of magnetic hysteresis in heterogeneous alloys. *IEEE Trans. Magn.* 27: 3475–3518.
6 Sessoli, R., Gatteschi, D., Caneschi, A., and Novak, M.A. (1993). Magnetic bistability in a metal-ion cluster. *Nature* 365: 141–143.
7 Sessoli, R., Tsai, H.L., Schake, A.R. et al. (1993). High-spin molecules: [Mn12O12(O2CR)16(H 2O)4]. *J. Am. Chem. Soc.* 115: 1804–1816.
8 Bagai, R. and Christou, G. (2009). The *Drosophila* of single-molecule magnetism: $[Mn_{12}O_{12}(O_2CR)_{16}(H_2O)_4]$. *Chem. Soc. Rev.* 38: 1011–1026.
9 Christou, G., Gatteschi, D., Hendrickson, D.N., and Sessoli, R. (2000). Single-molecule magnets. *MRS Bull.* 25: 66–71.
10 Ishikawa, N., Sugita, M., Ishikawa, T. et al. (2003). Lanthanide double-decker complexes functioning as magnets at the single-molecular level. *J. Am. Chem. Soc.* 125: 8694–8695.
11 Sorace, L., Benelli, C., and Gatteschi, D. (2011). Lanthanides in molecular magnetism: old tools in a new field. *Chem. Soc. Rev.* 40: 3092–3104.
12 Blagg, R.J., Ungur, L., Tuna, F. et al. (2013). Magnetic relaxation pathways in lanthanide single-molecule magnets. *Nat. Chem.* 5: 673–678.

13 Gregson, M., Chilton, N.F., Ariciu, A.M. et al. (2015). A monometallic lanthanide bis(methanediide) single molecule magnet with a large energy barrier and complex spin relaxation behaviour. *Chem. Sci.* 7: 155–165.

14 Moreno Pineda, E., Komeda, T., Katoh, K. et al. (2016). Surface confinement of $TbPc_2$-SMMs: structural, electronic and magnetic properties. *Dalton Trans.* 45: 18417–18433.

15 Ishikawa, N., Sugita, M., Okubo, T. et al. (2003). Determination of ligand-field parameters and f-electronic structures of double-decker bis(phthalocyaninato)lanthanide complexes. *Inorg. Chem.* 42: 2440–2446.

16 Ishikawa, N., Sugita, M., and Wernsdorfer, W. (2005). Quantum tunneling of magnetization in lanthanide single-molecule magnets: bis(phthalocyaninato)terbium and bis(phthalocyaninato)dysprosium anions. *Angew. Chem. Int. Ed.* 44: 2931–2935.

17 Ganivet, C.R., Ballesteros, B., De La Torre, G. et al. (2013). Influence of peripheral substitution on the magnetic behavior of single-ion magnets based on homo- and heteroleptic TbIII bis(phthalocyaninate). *Chem. Eur. J.* 19: 1457–1465.

18 Ganzhorn, M., Klyatskaya, S., Ruben, M., and Wernsdorfer, W. (2013). Carbon nanotube nanoelectromechanical systems as magnetometers for single-molecule magnets. *ACS Nano* 7: 6225–6236.

19 Katoh, K., Komeda, T., and Yamashita, M. (2016). The frontier of molecular spintronics based on multiple-decker phthalocyaninato Tb^{III} single-molecule magnets. *Chem. Rec.* 16: 987–1016.

20 Sakaue, S., Fuyuhiro, A., Fukuda, T., and Ishikawa, N. (2012). Dinuclear single-molecule magnets with porphyrin-phthalocyanine mixed triple-decker ligand systems giving SAP and SP coordination polyhedra. *Chem. Commun.* 48: 5337–5339.

21 Katoh, K., Yoshida, Y., Yamashita, M. et al. (2009). Direct observation of lanthanide(III)-phthalocyanine molecules on Au(111) by using scanning tunneling microscopy and scanning tunneling spectroscopy and thin-film field-effect transistor properties of Tb(III)- and Dy(III)-phthalocyanine molecules. *J. Am. Chem. Soc.* 131: 9967–9976.

22 De Cian, A., Moussavi, M., Fischer, J., and Weiss, R. (1985). Synthesis, structure, and spectroscopic and magnetic properties of lutetium(III) phthalocyanine derivatives: $LuPc_2.CH_2Cl_2$ and $[LuPc(OAc)(H_2O)_2].H_2O.2CH_3OH$. *Inorg. Chem.* 24: 3162–3167.

23 Ishikawa, N., Sugita, M., Tanaka, N. et al. (2004). Upward temperature shift of the intrinsic phase lag of the magnetization of bis(phthalocyaninato) terbium by ligand oxidation creating an $S = 1/2$ spin. *Inorg. Chem.* 43: 5498–5500.

24 Yamabayashi, T., Katoh, K., Breedlove, B., and Yamashita, M. (2017). Molecular orientation of a terbium(III)-phthalocyaninato double-decker complex for effective suppression of quantum tunneling of the magnetization. *Molecules* 22: 999.

25 Kyatskaya, S., Mascarós, J.R.G., Bogani, L. et al. (2009). Anchoring of rare-earth-based single-molecule magnets on single-walled carbon nanotubes. *J. Am. Chem. Soc.* 131: 15143–15151.

26 Urdampilleta, M., Klyatskaya, S., Cleuziou, J.P. et al. (2011). Supramolecular spin valves. *Nat. Mater.* 10: 502–506.
27 Ganzhorn, M., Klyatskaya, S., Ruben, M., and Wernsdorfer, W. (2013). Strong spin-phonon coupling between a single-molecule magnet and a carbon nanotube nanoelectromechanical system. *Nat. Nanotechnol.* 8: 165–169.
28 Ganzhorn, M., Klyatskaya, S., Ruben, M., and Wernsdorfer, W. (2016). Quantum Einstein-de Haas effect. *Nat. Commun.* 7: 11443.
29 Frommer, J. (1992). Scanning tunneling microscopy and atomic force microscopy in organic chemistry. *Angew. Chem. Int. Ed.* 31: 1298–1328.
30 Rosei, F., Schunack, M., Naitoh, Y. et al. (2003). Properties of large organic molecules on metal surfaces. *Prog. Surf. Sci.* 71: 95–146.
31 Heinze, S., Bode, M., Kubetzka, A. et al. (2000). Real-space imaging of two-dimensional antiferromagnetism on the atomic scale. *Science* 288: 1805–1808.
32 Wachowiak, A., Wiebe, J., Bode, M. et al. (2002). Direct observation of internal spin structure of magnetic vortex cores. *Science* 298: 577–580.
33 Wiesendanger, R. (2009). Spin mapping at the nanoscale and atomic scale. *Rev. Mod. Phys.* 81: 1495–1550.
34 Wiesendanger, R., Güntherodt, H.J., Güntherodt, G. et al. (1990). Observation of vacuum tunneling of spin-polarized electrons with the scanning tunneling microscope. *Phys. Rev. Lett.* 65: 247–250.
35 Heinrich, A.J., Gupta, J.A., Lutz, C.P., and Eigler, D.M. (2004). Single-atom spin-flip spectroscopy. *Science* 306: 466–469.
36 Hirjibehedin, C.F., Lutz, C.P., and Heinrich, A.J. (2006). Spin coupling in engineered atomic structures. *Science* 312: 1021–1024.
37 Tsukahara, N., Noto, K.I., Ohara, M. et al. (2009). Adsorption-induced switching of magnetic anisotropy in a single iron(II) phthalocyanine molecule on an oxidized Cu(110) surface. *Phys. Rev. Lett.* 102: 167203.
38 Komeda, T. and Manassen, Y. (2008). Distribution of frequencies of a single precessing spin detected by scanning tunneling microscope. *Appl. Phys. Lett.* 92: 212506.
39 Manassen, Y., Hamers, R.J., Demuth, J.E., and Castellano, A.J. (1989). Direct observation of the precession of individual paramagnetic spins on oxidized silicon surfaces. *Phys. Rev. Lett.* 62: 2531–2534.
40 Manassen, Y., Terovanesyan, E., Shachal, D., and Richter, S. (1993). Electron-spin resonance-scanning tunneling microscopy experiments on thermally oxidized Si(111). *Phys. Rev. B* 48: 4887–4890.
41 Sainoo, Y., Isshiki, H., Shahed, S.M.F. et al. (2009). Atomically resolved Larmor frequency detection on Si(111)-7 × 7 oxide surface. *Appl. Phys. Lett.* 95: 082504.
42 Kondo, J. (1968). Effect of ordinary scattering on exchange scattering from magnetic impurity in metals. *Phys. Rev.* 169: 437.
43 Knorr, N., Schneider, M.A., Diekhoner, L. et al. (2002). Kondo effect of single Co adatoms on Cu surfaces. *Phys. Rev. Lett.* 88: 096804.
44 Li, J., Schneider, W.D., Berndt, R., and Delley, B. (1998). Kondo scattering observed at a single magnetic impurity. *Phys. Rev. Lett.* 80: 2893–2896.

45 Madhavan, V., Chen, W., Jamneala, T. et al. (1998). Tunneling into a single magnetic atom: spectroscopic evidence of the Kondo resonance. *Science* 280: 567–569.

46 Manoharan, H.C., Lutz, C.P., and Eigler, D.M. (2000). Quantum mirages formed by coherent projection of electronic structure. *Nature* 403: 512–515.

47 Neel, N., Kroger, J., Limot, L. et al. (2007). Conductance and Kondo effect in a controlled single-atom contact. *Phys. Rev. Lett.* 98: 016801.

48 Wahl, P., Diekhoner, L., Wittich, G. et al. (2005). Kondo effect of molecular complexes at surfaces: ligand control of the local spin coupling. *Phys. Rev. Lett.* 95: 166601.

49 Cho, S., Lim, J.M., Hiroto, S. et al. (2009). Unusual interchromophoric interactions in β,β' directly and doubly linked corrole dimers: prohibited electronic communication and abnormal singlet ground states. *J. Am. Chem. Soc.* 131: 6412–6420.

50 Choi, T., Badal, M., Loth, S. et al. (2014). Magnetism in single metalloorganic complexes formed by atom manipulation. *Nano Lett.* 14: 1196–1201.

51 Choi, T., Bedwani, S., Rochefort, A. et al. (2010). A single molecule Kondo switch: multistability of tetracyanoethylene on Cu(111). *Nano Lett.* 10: 4175–4180.

52 DiLullo, A., Chang, S.-H., Baadji, N. et al. (2012). Molecular Kondo chain. *Nano Lett.* 12: 3174–3179.

53 Fernandez-Torrente, I., Franke, K.J., and Pascual, J.I. (2008). Vibrational Kondo effect in pure organic charge-transfer assemblies. *Phys. Rev. Lett.* 101: 217203.

54 Fu, Y.S., Ji, S.H., Chen, X. et al. (2007). Manipulating the Kondo resonance through quantum size effects. *Phys. Rev. Lett.* 99: 256601.

55 Fu, Y.S., Xue, Q.K., and Wiesendanger, R. (2012). Spin-resolved splitting of Kondo resonances in the presence of RKKY-type coupling. *Phys. Rev. Lett.* 108: 087203.

56 Gao, L., Ji, W., Hu, Y.B. et al. (2007). Site-specific Kondo effect at ambient temperatures in iron-based molecules. *Phys. Rev. Lett.* 99: 106402.

57 Iancu, V., Deshpande, A., and Hla, S.W. (2006). Manipulating Kondo temperature via single molecule switching. *Nano Lett.* 6: 820–823.

58 Iancu, V., Deshpande, A., and Hla, S.W. (2006). Manipulation of the Kondo effect via two-dimensional molecular assembly. *Phys. Rev. Lett.* 97: 266603.

59 Jiang, Y., Zhang, Y.N., Cao, J.X. et al. (2011). Real-space imaging of Kondo screening in a two-dimensional O_2 lattice. *Science* 333: 324–328.

60 Kim, H., Chang, Y.H., Lee, S.-H. et al. (2013). Switching and sensing spin states of co-porphyrin in bimolecular reactions on Au(111) using scanning tunneling microscopy. *ACS Nano* 7: 9312–9317.

61 plus.1emKomeda, T., Isshiki, H., Liu, J. et al. (2013). Variation of Kondo peak observed in the assembly of heteroleptic 2,3-naphthalocyaninato phthalocyaninato Tb(III) double-decker complex on Au(111). *ACS Nano* 7: 1092–1099.

62 Li, Z., Li, B., Yang, J., and Hou, J.G. (2010). Single-molecule chemistry of metal phthalocyanine on noble metal surfaces. *Acc. Chem. Res.* 43: 954–962.

63 Liu, J., Isshiki, H., Katoh, K. et al. (2013). First observation of a Kondo resonance for a stable neutral pure organic radical, 1,3,5-triphenyl-6-oxoverdazyl, adsorbed on the Au(111) surface. *J. Am. Chem. Soc.* 135: 651–658.

64 Minamitani, E., Tsukahara, N., Matsunaka, D. et al. (2012). Symmetry-driven novel Kondo effect in a molecule. *Phys. Rev. Lett.* 109: 086602.
65 Mugarza, A., Krull, C., Robles, R. et al. (2011). Spin coupling and relaxation inside molecule-metal contacts. *Nat. Commun.* 2: 490.
66 Müllegger, S., Rashidi, M., Fattinger, M., and Koch, R. (2013). Surface-supported hydrocarbon Pi radicals show Kondo behavior. *J. Phys. Chem. C* 117: 5718–5721.
67 Perera, U.G.E., Kulik, H.J., Iancu, V. et al. (2010). Spatially extended Kondo state in magnetic molecules induced by interfacial charge transfer. *Phys. Rev. Lett.* 105: 106601.
68 Stróecka, A., Soriano, M., Pascual, J.I., and Palacios, J.J. (2012). Reversible change of the spin state in a manganese phthalocyanine by coordination of CO molecule. *Phys. Rev. Lett.* 109: 147202.
69 Tsukahara, N., Shiraki, S., Itou, S. et al. (2011). Evolution of Kondo resonance from a single impurity molecule to the two-dimensional lattice. *Phys. Rev. Lett.* 106: 187201.
70 Wegner, D., Yamachika, R., Zhang, X. et al. (2009). Tuning molecule-mediated spin coupling in bottom-up-fabricated vanadium-tetracyanoethylene nanostructures. *Phys. Rev. Lett.* 103: 087205.
71 Zhang, Y.H., Kahle, S., Herden, T. et al. (2013). Temperature and magnetic field dependence of a Kondo system in the weak coupling regime. *Nat. Commun.* 4: 2110.
72 Zhao, A.D., Hu, Z.P., Wang, B. et al. (2008). Kondo effect in single cobalt phthalocyanine molecules adsorbed on Au(111) monoatomic steps. *J. Chem. Phys.* 128: 234705.
73 Nagaoka, K., Jamneala, T., Grobis, M., and Crommie, M.F. (2002). Temperature dependence of a single Kondo impurity. *Phys. Rev. Lett.* 88: 077205.
74 Hewson, A.C. (1993). *The Kondo Problem to Heavy Fermions*. Cambridge University Press.
75 Schlottmann, P. (1989). Some exact results for dilute mixed-valent and heavy-fermion systems. *Phys. Rep.-Rev. Sec. Phys. Lett.* 181: 1–119.
76 Zhang, Y.F., Isshiki, H., Katoh, K. et al. (2009). A low-temperature scanning tunneling microscope investigation of a nonplanar dysprosium-phthalocyanine adsorption on Au(111). *J. Phys. Chem. C* 113: 14407–14410.
77 Zhang, Y.F., Isshiki, H., Katoh, K. et al. (2009). Low-temperature scanning tunneling microscopy investigation of bis(phthalocyaninato)yttrium growth on Au(111): from individual molecules to two-dimensional domains. *J. Phys. Chem. C* 113: 9826–9830.
78 Hallback, A.-S., Oncel, N., Huskens, J. et al. (2004). Inelastic electron tunneling spectroscopy on decanethiol at elevated temperatures. *Nano Lett.* 4: 2393–2395.
79 Kresse, G. and Furthmuller, J. (1996). Efficiency of ab-initio total energy calculations for metals and semiconductors using a plane-wave basis set. *Comput. Mater. Sci.* 6: 15–50.
80 Kresse, G. and Joubert, D. (1999). From ultrasoft pseudopotentials to the projector augmented-wave method. *Phys. Rev. B* 59: 1758–1775.
81 Perdew, J.P., Burke, K., and Ernzerhof, M. (1996). Generalized gradient approximation made simple. *Phys. Rev. Lett.* 77: 3865–3868.

82 Barlow, D., Scudiero, L., and Hipps, K.W. (2003). Scanning tunneling microscopy of 1, 2, and 3 layers of electroactive compounds. *Ultramicroscopy* 97: 47–53.
83 Barlow, D.E. and Hipps, K.W. (2000). Scanning tunneling microscopy and spectroscopy study of vanadyl phthalocyanine on Au(111): the effect of oxygen binding and orbital mediated tunneling on the apparent corrugation. *J. Phys. Chem. B* 104: 5993–6000.
84 Cheng, Z.H., Gao, L., Deng, Z.T. et al. (2007). Epitaxial growth of iron phthalocyanine at the initial stage on Au(111) surface. *J. Phys. Chem. C* 111: 2656–2660.
85 Chizhov, I., Scoles, G., and Kahn, A. (2000). The influence of steps on the orientation of copper phthalocyanine monolayers on Au(111). *Langmuir* 16: 4358–4361.
86 Lu, X. and Hipps, K.W. (1997). Scanning tunneling microscopy of metal phthalocyanines: d6 and d8 cases. *J. Phys. Chem. B* 101: 5391–5396.
87 Lu, X., Hipps, K.W., Wang, X.D., and Mazur, U. (1996). Scanning tunneling microscopy of metal phthalocyanines: d7 and d9 cases. *J. Am. Chem. Soc.* 118: 7197–7202.
88 Takada, M. and Tada, H. (2004). Low temperature scanning tunneling microscopy of phthalocyanine multilayers on Au(111) surfaces. *Chem. Phys. Lett.* 392: 265–269.
89 Walzer, K. and Hietschold, M. (2001). STM and STS investigation of ultrathin tin phthalocyanine layers adsorbed on HOPG(0001) and Au(111). *Surf. Sci.* 471: 1–10.
90 Komeda, T., Isshiki, H., and Liu, J. (2010). Metal-free phthalocyanine (H_2Pc) molecule adsorbed on the Au(111) surface: formation of a wide domain along a single lattice direction. *Sci. Technol. Adv. Mater.* 11: 054602.
91 Takada, M. and Tada, H. (2005). Direct observation of adsorption-induced electronic states by low-temperature scanning tunneling microscopy. *Ultramicroscopy* 105: 22–25.
92 Takada, M. and Tada, H. (2005). Scanning tunneling microscopy and spectroscopy of phthalocyanine molecules on metal surfaces. *Jpn. J. Appl. Phys., Part 1* 44: 5332–5335.
93 Nilson, K., Ahlund, J., Brena, B. et al. (2007). Scanning tunneling microscopy study of metal-free phthalocyanine monolayer structures on graphite. *J. Chem. Phys.* 127: 114702.
94 Katoh, K., Kajiwara, T., Nakano, M. et al. (2011). Magnetic relaxation of single-molecule magnets in an external magnetic field: an ising dimer of a terbium(III)-phthalocyaninate triple-decker complex. *Chem. Eur. J.* 17: 117–122.
95 Ara, F., Qi, Z.K., Hou, J. et al. (2016). A scanning tunneling microscopy study of the electronic and spin states of bis(phthalocyaninato)terbium(III) ($TbPc_2$) molecules on Ag(111). *Dalton Trans.* 45: 16644–16652.
96 Komeda, T., Isshiki, H., Liu, J. et al. (2011). Observation and electric current control of a local spin in a single-molecule magnet. *Nat. Commun.* 2: 217.
97 Smykalla, L., Shukrynau, P., and Hietschold, M. (2012). Investigation of ultrathin layers of bis(phthalocyaninato)lutetium(III) on graphite. *J. Phys. Chem. C* 116: 8008–8013.

98 Heutz, S., Bayliss, S.M., Middleton, R.L. et al. (2000). Polymorphism in phthalocyanine thin films: mechanism of the $\alpha \rightarrow \beta$ transition. *J. Phys. Chem. B* 104: 7124–7129.
99 Hoshino, A., Takenaka, Y., and Miyaji, H. (2003). Redetermination of the crystal structure of alfa-copper phthalocyanine grown on KCl. *Acta Crystallogr., Sect. B: Struct. Sci.* 59: 393–403.
100 Mason, R., Williams, G.A., and Fielding, P.E. (1979). Structural chemistry of phthalocyaninato-cobalt(II) and manganese(II). *J. Chem. Soc., Dalton Trans.* 676–683.
101 Isshiki, H., Liu, J., Katoh, K. et al. (2010). Scanning tunneling microscopy investigation of tris(phthalocyaninato)yttrium triple-decker molecules deposited on Au(111). *J. Phys. Chem. C* 114: 12202–12206.
102 Benihya, K., Mossoyan-Déneux, M., Hahn, F. et al. (2000). Synthesis, crystal structure and spectral characterization, of a new phase of tris(phthalocyaninato)dibismuth(III), $Bi_2(Pc)_3$. *Eur. J. Inorg. Chem.* 2000: 1771–1779.
103 Chambrier, I., Hughes, D.L., Swarts, J.C. et al. (2006). First example of a di-cadmium tris-phthalocyanine triple-decker sandwich complex. *Chem. Commun.* 3504–3506.
104 Komeda, T., Isshiki, H., Liu, J. et al. (2014). Variation of Kondo temperature induced by molecule-substrate decoupling in film formation of bis(phthalocyaninato)terbium(III) molecules on Au(111). *ACS Nano* 8: 4866–4875.
105 Vitali, L., Fabris, S., Conte, A.M. et al. (2008). Electronic structure of surface-supported bis(phthalocyaninato) terbium(III) single molecular magnets. *Nano Lett.* 8: 3364–3368.
106 Paaske, J., Rosch, A., Wolfle, P. et al. (2006). Non-equilibrium singlet-triplet Kondo effect in carbon nanotubes. *Nat. Phys.* 2: 460–464.
107 Sasaki, S., De Franceschi, S., Elzerman, J.M. et al. (2000). Kondo effect in an integer-spin quantum dot. *Nature* 405: 764–767.
108 Auerhammer, J.M., Knupfer, M., Peisert, H., and Fink, J. (2002). The copper phthalocyanine/Au(100) interface studied using high resolution electron energy-loss spectroscopy. *Surf. Sci.* 506: 333–338.
109 Salomon, E., Angot, T., Papageorgiou, N., and Layet, J.M. (2005). Self-assembled monolayer of tin-phthalocyanine on InSb(0 0 1)-(4 × 2)/c(8 × 2). *Surf. Sci.* 596: 74–81.
110 Nazin, G.V., Wu, S.W., and Ho, W. (2005). Tunneling rates in electron transport through double-barrier molecular junctions in a scanning tunneling microscope. *Proc. Natl. Acad. Sci. U.S.A.* 102: 8832–8837.
111 Qiu, X.H., Nazin, G.V., and Ho, W. (2004). Vibronic states in single molecule electron transport. *Phys. Rev. Lett.* 92: 206102.
112 Aguiar-Hualde, J.M., Chiappe, G., Louis, E., and Anda, E.V. (2007). Kondo effect in transport through molecules adsorbed on metal surfaces: from Fano dips to Kondo peaks. *Phys. Rev. B* 76: 155427.
113 Büsser, C.A., Moreo, A., and Dagotto, E. (2004). Conductance dip in the Kondo regime of linear arrays of quantum dots. *Phys. Rev. B* 70: 035402.

12

Molecular Technology for One- and Two-Dimensional Materials on Surfaces

Shigeki Kawai[1] *and Kazukuni Tahara*[2]

[1] *National Institute for Materials Science, International Center for Materials Nanoarchitectonics, 1-1, Namiki, Tsukuba, Ibaraki 305-0044, Japan*
[2] *Meiji University, Department of Applied Chemistry, School of Science and Technology, 1-1-1 Higashimita, Tama-ku, Kawasaki 214-8571, Japan*

12.1 General Introduction

In this chapter, on-surface chemical reaction studied with local scanning probe microscopy (SPM) will be described. This field rapidly progressed since the seminal invention of scanning tunneling microscopy (STM) and atomic force microscopy (AFM). Here, two extreme cases will be discussed: in ultrahigh vacuum (UHV) at low temperature (LT) and at the liquid/solid interfaces at ambient temperature. First, a brief description of the STM and AFM background will be introduced. Recently, the resolution of AFM was drastically improved by employing a functionalized tip, and one can readily see inner structures of molecules. Naturally, this technique has been applied in studies of the on-surface chemistry. The study of an on-surface reaction with high-resolution AFM just opened up so that many innovative works are foreseen; yet, it is certainly beneficial to summarize. Another important aspect relates to the study at the liquid/solid interface. Recently, on-surface chemistry at the organic liquid/solid interfaces became a very popular research field owing to its easy and inexpensive experimental approach. Furthermore, the dynamics during chemical reactions facilitated by the presence of a supernatant liquid phase can be directly probed. These interfacial techniques are also definitely worth summarizing in this chapter. Finally, summary and perspectives will be described.

12.1.1 Scanning Tunneling Microscopy

The invention of STM opened up nanotechnology, where controlling and using single atoms and molecules is the main focus [1]. In STM, a bias voltage is applied between the tip and the sample. When the junction is less than a few angstroms thick, the tunneling current flows. Since the distance dependence is extremely strong, atomic-scale imaging is readily performed. This microscope was first applied for studies of solid surfaces and then for molecules adsorbed on the

Molecular Technology: Materials Innovation, Volume 3, First Edition.
Edited by Hisashi Yamamoto and Takashi Kato.

solid surfaces. Similar to solution phase chemistry, studies on the reaction and manipulation on crystalline surfaces have tremendously attracted the interest of many researchers [2]. For instance, van der Waals forces [3, 4], dipole–dipole interaction [5], hydrogen bond interaction [6], and halogen bond interaction [7] are used to form one- or two-dimensional (1D or 2D) nanostructures (Figure 12.1a–d). Once these intermolecular interactions are well controlled by judiciously designed molecules, a large structure extended over a terrace is achievable. Since complex electronic couplings between metal substrate and molecule are responsible for the tunneling current, observed STM images are often peculiar. Therefore, density functional theory calculations are usually conducted to get a deep insight into the geometry and electronic property. Nevertheless, if a molecule is electrically decoupled from a metal substrate by inserting a thin insulator film such as NaCl, the spatial distribution and the energy of the lowest unoccupied and highest occupied molecular orbitals can directly be investigated (Figure 12.1e) [8].

12.1.2 Atomic Force Microscopy in Ultrahigh Vacuum Environment

AFM was invented by Binnig et al. in 1986 [9]. In the first AFM setup, an atomically sharp diamond tip attached to the end of a gold foil cantilever was used. The interaction force caused between the tip and the sample is detected via the quasi-static deflection of the cantilever, the so-called contact mode. The tip is laterally scanned over the surface, keeping a constant deflection. Since the imaging is based on the detection of forces, any kind of surface such as metal, semiconductor, or insulator, can readily be measured with high resolution. Particularly, AFM has become the most powerful tool in the field of biology nowadays. Since the interaction includes any kind of force such as van der Waals, electrostatic, magnetic, and chemical forces, the property of the tip significantly affects the observations. Among these forces, the chemical force is responsible for the atomic-scale contrast in UHV measurements. However, large background forces, arising from the macroscopic tip shape, prevent stable imaging since a soft cantilever typically used in the contact mode cannot sustain the atomic-scale junction. Therefore, the tip apex accidentally contacts the sample surface, the

Figure 12.1 Engineering intermolecular interactions to assemble molecules. (a) Dipole–dipole interaction between 5,10,15,20-tetrakis-(3,5-di-*tertiary*butylphenyl) porphyrin molecules with the cyanophenyl substitution. Source: Adapted with permission from Yokoyama et al. [5]. Copyright 2001, Macmillan Publishers Ltd: [Nature]. (b) O· · ·H and N· · ·H hydrogen bond interactions between perylene tetra-carboxylic diimide and melamine. Source: Adapted with permission from Theobald et al. [6]. Copyright 2003, Macmillan Publishers Ltd: [Nature]. (c) Van der Waals interactions among alkyl chains. Source: Adapted with permission from Tahara et al. [4]. Copyright 2006, American Chemical Society. (d) C—Br· · ·Br halogen bond interactions. Source: Adapted with permission from Shang et al. [7]. Copyright 2015, Macmillan Publishers Ltd: [Nature Chemistry]. (e) Direct observation of molecular orbitals by STM. Pentacene molecule is electronically decoupled from a metal substrate by a thin insulator film. Source: Adapted with permission from Repp and Meyer [8]. Copyright 2005, the American Physical Society.

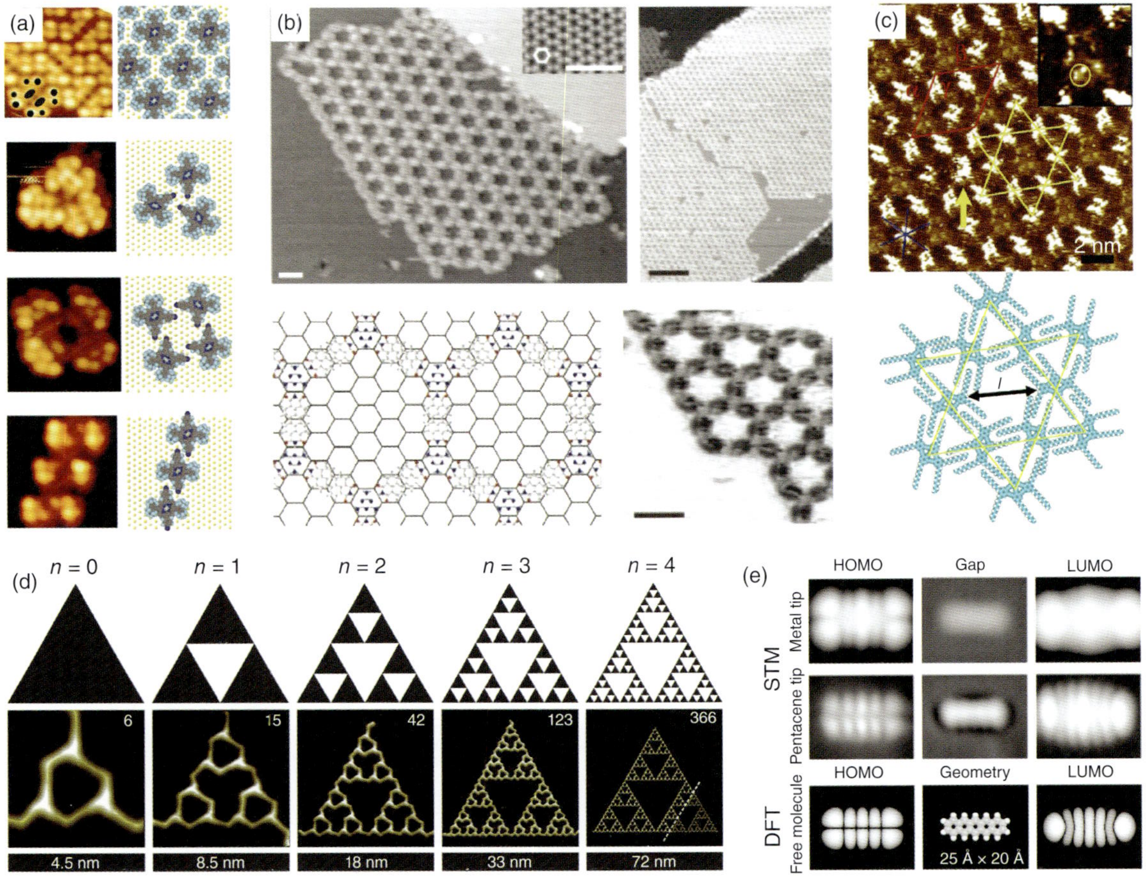
(a)
(b)
(c)
2 nm
l
(d)
n = 0
n = 1
n = 2
n = 3
n = 4
6
15
42
123
366
4.5 nm
8.5 nm
18 nm
33 nm
72 nm
(e)
HOMO
Gap
LUMO
STM
Metal tip
Pentacene tip
DFT
Free molecule
HOMO
Geometry
LUMO
25 Å × 20 Å

so-called jump-into-contact instability. Once this happens, the atomic-scale structure of the tip apex no longer stays constant during scanning. Using a stiff cantilever can prevent this issue; yet, the detection sensitivity drops due to the relation formula described as $dZ_{def} = F/k$, where dZ_{def} is the deflection of the cantilever, F is the interaction force, and k is the stiffness of the cantilever. To improve the force sensitivity, usage of the dynamic response of the cantilever was introduced [10]. Albrecht et al. developed a frequency modulation technique also called noncontact atomic force microscopy (NC-AFM) [11], in which the cantilever oscillates on its mechanical resonance. The resonance frequency shifted by the interaction force is detected by a phase-lock looped circuit and is used for controlling the tip–sample separation. Since the first systematic observation on Si(111)-7 × 7 with atomic resolution in 1995 by Giessibl [12], NC-AFM also became a powerful tool in nanotechnology [13]. The attractive part of the chemical interaction is usually used for atomic-scale imaging at a constant frequency shift mode. However, in contrast to STM, NC-AFM imaging is rather challenging as it took almost one decade since the invention of AFM. The difficulty relates to the fact that the tip apex usually has to come very close to the surface compared to STM measurements. Although the tip is still in the noncontact regime, a number of local energy minima lead to frequent tip change by the interaction force [14]. Nevertheless, once the interaction force and the tip apex structure are well controlled, atom manipulation can be conducted on semiconductor surfaces (Figure 12.2a) [15, 20] and metal surfaces [21] and even on insulator surfaces (Figure 12.2b) [16].

Another important aspect in NC-AFM is the capability to measure the atomic-scale force quantitatively. If the distance dependence of the frequency shift is measured at a certain atomic site, the interaction force and potential can be extracted with a numerical algorithm [22]. This technique, the so-called force spectroscopy, was first demonstrated on the Si(111)-7 × 7 by Lantz et al. [23]. Since then, force spectroscopy has been used for force measurement on alkali halides [24] and metal surfaces [25] as well as noble gases [26] with a force resolution of better than 10 pN. Furthermore, by comparing maximum attractive

Figure 12.2 AFM atom manipulation and molecular imaging at RT. (a) NC-AFM image of the rearranged single Sn atoms in Ge(111)-c(2 × 8) for the construction of "an atom inlay". Source: Adapted with permission from Sugimoto et al. [15]. Copyright 2005, Macmillan Publishers Ltd: [Nature Materials]. (b) NC-AFM image of NaCl(001), in which 20 Br anions are arranged as a shape of "Swiss cross." Source: Adapted with permission from Kawai et al. [16]. Copyright 2014, Macmillan Publishers Ltd: [Nature Communications]. (c) Chemical structure of di-*tert*-butyl-phenyl porphyrin and (d) the corresponding frequency shift map. Source: Adapted with permission from Loppacher et al. [17]. Copyright 2003, American Physical Society. (e) NC-AFM image of self-assembled cyanoporphyrin molecules on KBr(001). (f) Assemblies consisting of several parallel π–π stacked wires grow on flat terraces along the [18] direction. (g) High-resolution NC-AFM images of a monolayer consisting of several parallel porphyrin wires and (h) the porphyrin wires as well as the substrate. (inset) Fourier analysis showing periodicities for the molecules and the substrate. (j) Proposed molecular arrangement (π–π stacked). Source: Adapted with permission from Maier et al. [19]. Copyright 2008, John Wiley and Sons.

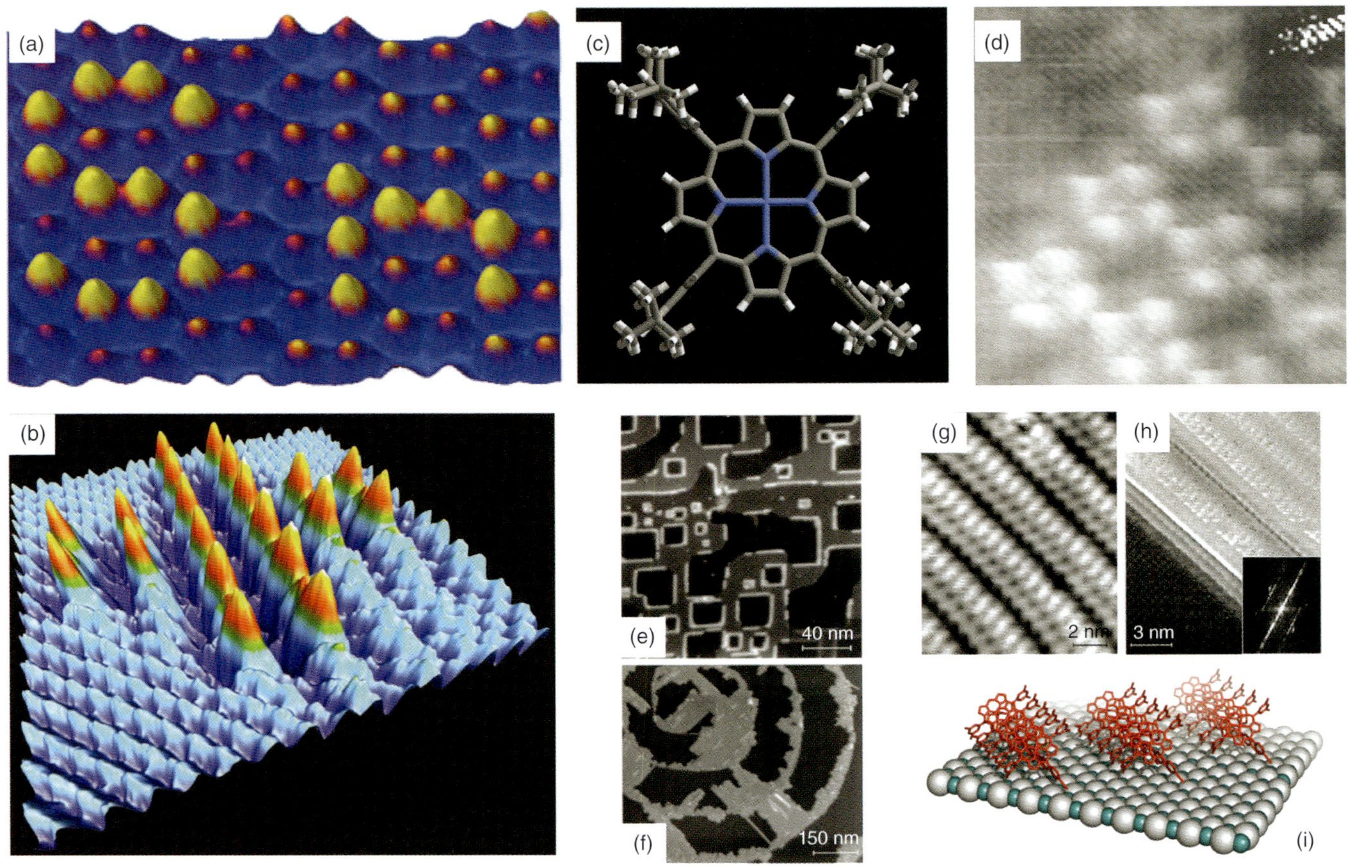
(a)
(b)
(c)
(d)
(e)
40 nm
(f)
150 nm
(g)
2 nm
(h)
3 nm
(i)

forces, Sugimoto et al. demonstrated discrimination of group 14 atoms (Si, Sn, and Pb), embedded in a Si(111) surface [27].

Besides these important improvements in imaging and force spectroscopy, molecules on surfaces also became one of the main topics in the field of NC-AFM. In 1998, Gotsmann et al. reported that AFM can indeed image individual perylene-3,4,9,10-tetracarboxylic-3,4,9,10-dianhydride molecules adsorbed on Ag(110) [28]. Molecules usually bind to surfaces weakly so that they can be moved by the interaction force [17], leading to a peculiar contrast. Therefore, a certain assumption is required to interpret sub-nanometer scale features (Figure 12.2c,d). Nevertheless, NC-AFM is only a tool to provide detailed information on individual molecules adsorbed on insulator surfaces in real space. Nony et al. found that chloro-[subphthalocyaninato] boron-(III) molecules can be stabilized inside nanoscale pits created on a KBr(001) surface even at room temperature due to a large dipole moment arising from the chlorine atom at the center [29]. A cyano group can also be used for anchoring molecules on the alkali halide surface (Figure 12.2e–i) [19]. In this way, the topics studied with NC-AFM at RT are often more oriented to a structural analysis of molecular self-assemblies and their applications.

12.1.3 High-Resolution Imaging with a Functionalized Tip Measurement at Low Temperature in Ultrahigh Vacuum Environment

Recently, the resolution of NC-AFM was drastically improved by the functionalized tip (Figure 12.3) [30]. A metal tip was terminated with a carbon monoxide molecule picked up from the surface. Since the reactivity of the tip significantly reduces, the tip apex can come close enough to a molecule on a surface to detect the repulsive interaction. Otherwise, the strong attractive interaction force with a metal tip often induces instability in imaging. Since the repulsive component of the interaction force has a stronger Z distance dependence, the spatial

Figure 12.3 High-resolution imaging with a functionalized tip at LT, following the seminal work [30]. (a) STM image of $C_{16}H_{10}N_2O_2$, which was *priori* identified with high-resolution mass spectrometry, yet the structure was unknown. (b) NC-AFM image (raw data) and (c) the same image with the molecular model of cephalandole overlaid as an eye guide. (d) Low-pass filtered 3D representation. Source: Adapted with permission from Gross et al. [31]. Copyright 2010, Macmillan Publishers Ltd: [Nature Chemistry]. (e) Chemical structure of free-base naphthalocyanine and (f) the corresponding NC-AFM image. Source: Adapted with permission from Mohn et al. [32]. Copyright 2012, Macmillan Publishers Ltd: [Nature Nanotechnology]. (g) AFM image of the self-assembled molecules condensed by weak hydrogen bond interactions. Source: Adapted with permission from Kawai et al. [33]. Copyright 2013, American Chemical Society. (h) The atom positions of both Cu layers at a step edge determine the hcp sites and fcc sites. The adsorption sites are given with respect to the centers of the C6 rings. (i) Olympicene on the hcp 30° site. (j) Radical on the fcc 30°. (k) Ketone on fcc 30° sites. Scale bars, 0.5 nm. Source: Adapted with permission from Schuler et al. [34]. Copyright 2013, American Physical Society. (l)-(o) NC-AFM images show intermediates along the reaction pathway (upper panel) and the corresponding chemical structures (lower panel). Source: Adapted with permission from Riss et al. [35]. Copyright 2016, Macmillan Publishers Ltd: [Nature Chemistry].

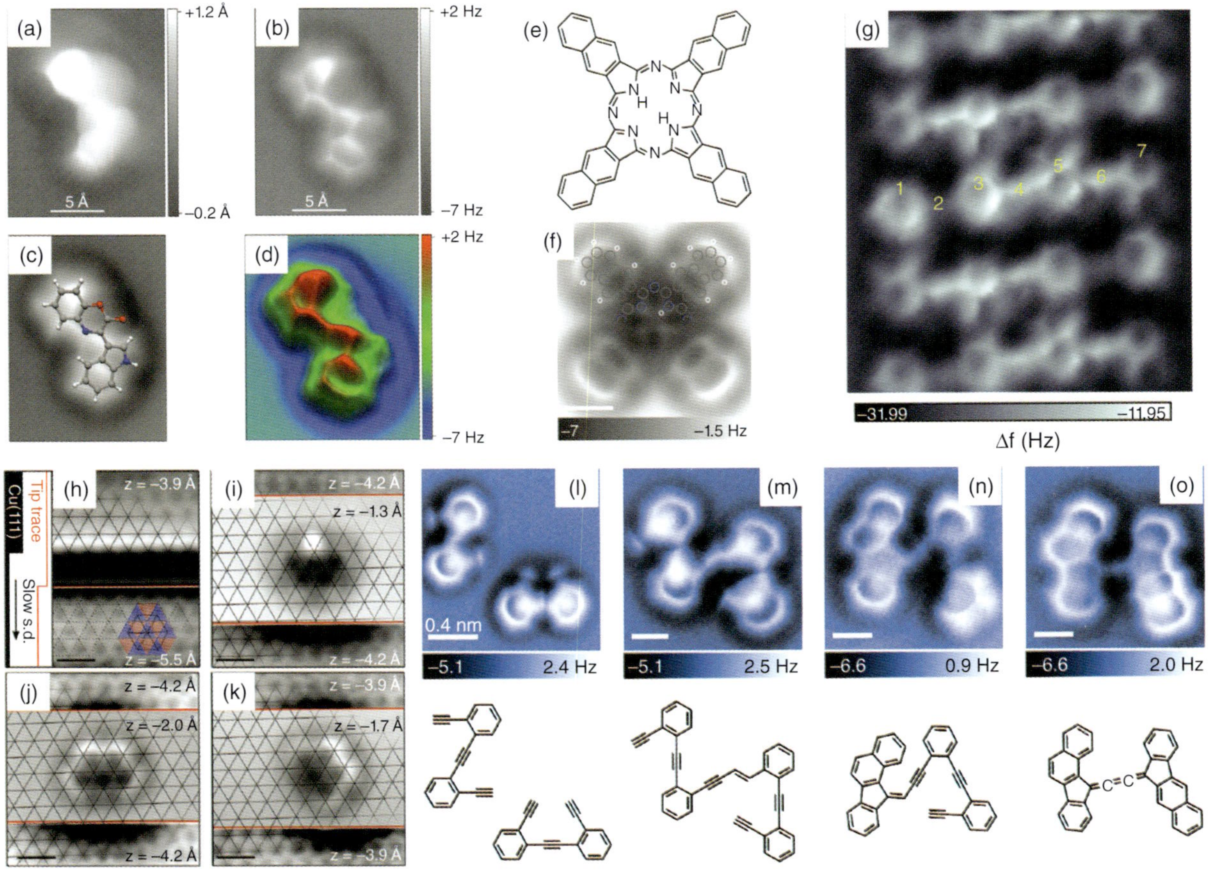

(a)
+1.2 Å
−0.2 Å
5 Å
(b)
+2 Hz
−7 Hz
5 Å
(c)
(d)
+2 Hz
−7 Hz
(e)
N
H
(f)
−7
−1.5 Hz
(g)
1
2
3
4
5
6
7
−31.99
−11.95
Δf (Hz)
(h)
Cu(111)
Tip trace
Slow s.d.
z = −3.9 Å
z = −5.5 Å
(i)
z = −4.2 Å
z = −1.3 Å
z = −4.2 Å
(j)
z = −4.2 Å
z = −2.0 Å
z = −4.2 Å
(k)
z = −3.9 Å
z = −1.7 Å
z = −3.9 Å
(l)
0.4 nm
−5.1
2.4 Hz
(m)
−5.1
2.5 Hz
(n)
−6.6
0.9 Hz
(o)
−6.6
2.0 Hz
C=C

resolution becomes high enough to resolve inner structures of molecules. Besides a CO molecule, several inert materials such as xenon and oxygen atoms as well as other small organic molecules have been used as tips [36]; yet, the CO tip is superior in the sense of resolution among them. Since the lateral stiffness of the CO tip is low [37], the CO molecule can easily be deflected by the interaction. Above the intramolecular bond, a greater charge density leads to a stronger repulsive force so that the CO tip cannot keep the initial adsorption geometry on the metal tip except exactly above the bond. Therefore, in the vicinity of the bond, the CO tip deflects away so that the resultant repulsive interaction reduces. In this way, the bond feature is drastically enhanced [38, 39]. At a small tip–sample separation, such an imaging mechanism dominates. However, the rich electronic property of the tip – the positive dipole from the metal tip and the spatially localized negative charge cloud of the frontmost oxygen atom actually – plays a role at a larger distance [40]. Nevertheless, Gross et al. successfully discriminated the bond order via a careful analysis of the observed bond lengths [41]. Similarly, AFM with a CO functionalized tip was used to resolve different extensions of π-electrons caused by fluorine substitution [42]. This direct observation of inner structures is beneficial to determine chemical structures of unknown compounds (Figure 12.3a–d) [31] and orientations of fullerene [43] as well as to sort individual compounds in petroleum [44]. In addition to structural information, Mohn et al. demonstrated that AFM with the CO tip can resolve the distribution of total charge in a naphthalocyanine molecule with Kelvin probe force microscopy (Figure 12.3e,f) [32].

Besides single molecules, molecular self-assemblies condensed by hydrogen bond interaction [33, 45–47] and halogen bond interaction (Figure 12.3g) [48] have been studied in detail. In such measurements, a bond-like feature often appears between adjacent molecules. Once it was interpreted as a hydrogen bond [45]. However, the follow-up theoretical work claimed that the bond-like feature should not be interpreted as the hydrogen bond since the increase in electron density due to the hydrogen bond interaction is not sufficiently high [39]. The tilt effect of the CO tip following the landscape of the charge density is responsible. This proposed mechanism was supported by experimental results [46, 48]. Thus, "seeing is believing" may lead to misinterpretation even in the atomic scale for the particular case. Nevertheless, seeing the chemical structure is "still" very useful for studies in on-surface chemical reaction. de Oteyza et al. for the first time employed high-resolution NC-AFM to analyze individual products synthesized via 1,2-bis((2-ethynylphenyl)ethynyl)benzene on a Ag(100) surface by heating at 90 °C [49]. Following this work, an analysis of metastable intermediates in the dimerization has been demonstrated (Figure 12.3h–k) [35]. Furthermore, the regioselectivity for aryl–aryl bond formation via dehydrogenation [50] and thermally induced three-step reaction [51] were also reported.

In the abovementioned cases, planar molecules are often observed since the imaging is conducted at a constant height mode, in which the tip–sample distance regulation is turned off. If the molecule becomes nonplanar or tilted with respect to the substrate (Figure 12.3g–j) [34], the tip–sample distance has to be adjusted for a highest part so that only the protruding part can be observed [52, 53]. Otherwise, the excessive small separation leads to the desorption of the CO molecule by the strong repulsive interaction. This issue can be solved by taking a three-dimensional (3D) force map [54] or planarization of molecules by dehydrogenation [55]. Yet, the observation of an inner structure of 3D hydrocarbons is generally challenging since the CO tip is strongly deflected by the excessive interaction between the hydrogen atom and the CO tip. On the other hand, the hydrogen atoms can be directly observed via the CO· · ·HC junction [56] so that identifying molecular structures via observed hydrogen atom positions may be a new route.

12.1.4 Scanning Probe Microscopy at Liquid/Solid Interface

Studies at the gas or liquid/solid interfaces have also progressed in on-surface chemistry. In addition to its inexpensive experimental approach, the environment allows us to investigate the dynamics of chemical reactions and structures with a local probe. Recent progress on frequency modulation AFM at the liquid/solid interfaces allows us to observe surfaces with atomic resolution. In particular, self-assemblies and dynamics of bio-related molecules in water have been intensively studied [57–59]. However, observation of on-surface chemical reactions using high-resolution AFM at the interfaces was scarcely reported. Thus, STM is mainly used to study chemical reactions. Here, a brief summary of the on-surface chemistry at the gas or liquid/solid interfaces will be provided.

Since the first observation of self-assembled monolayers of alkanes and alkylated aromatic molecules at the liquid/solid interface by Rabe and Buchholz in 1991 [3], formation of various self-assembled molecules has been reported using STM with high resolution (Figure 12.4). So far, detailed guidelines to structure and functionalize 2D molecular self-assemblies were established as described in several review articles [60–64]. In contrast, studies to produce novel organic materials, such as conductive 1D molecular wires [65] or 2D polymers [66], by controlling surficial and/or interfacial chemical reactions are ongoing.

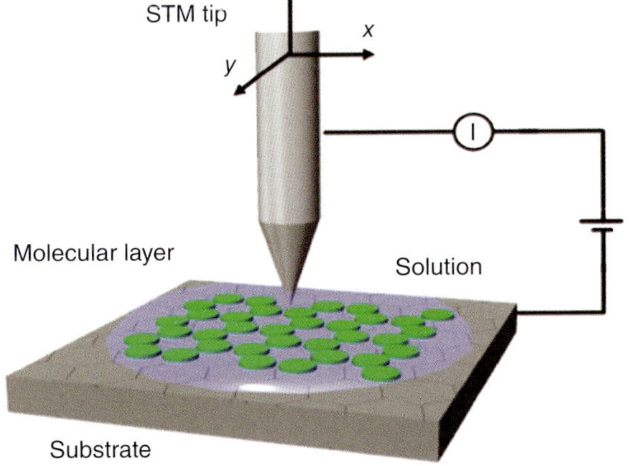

Figure 12.4 Schematic representation of STM observation at the liquid/graphite interface.

At the liquid/solid interface, the experimental approach is more straightforward than that in UHV. Since high thermal stability and sublimability of molecules were not always required, a wider range of precursors can be used. Furthermore, the environment allows us to study the dynamic nature of the system in which the supernatant liquid phase acts as a molecular reservoir. Since reactants are continuously provided to the interface, the reaction rate can be controlled by solute concentrations and solvation energy. Furthermore, defects are usually self-repaired by dynamic exchange between reactants and products by equilibrating the adsorption–desorption rate. In this way, well-ordered 2D structures can be formed. Moreover, at the interface, external physical and chemical stimuli can also be applied to initiate and/or control chemical reactions. Thus, the creation of 2D structures via surficial or interfacial reactions at the liquid/solid interfaces is of great importance in this field. Section 12.2.5 sheds light on the recent progress on the formation of novel on-surface organic structures by chemical reactions at the interfaces.

12.2 On-Surface Chemical Reaction

12.2.1 General Info About the On-Surface Chemical Reaction

Over the past decade, various kinds of on-surface chemical reactions have been reported via observations of local nanostructures by SPM. The classification of on-surface reactions is rather challenging since the reaction mechanisms may differ from those reported in a solution phase; yet, we attempt here to classify them into three types: (i) carbon–carbon, (ii) carbon–heteroatom, and (iii) heteroatom–heteroatom bond formation reactions. Figure 12.5a–d shows an overview of reported on-surface reactions. The carbon–carbon bond formation reactions include aryl–aryl coupling reactions [67–70], dehydrogenative homo-coupling reactions of sp, sp^2, and sp^3 carbon atoms [71–73], and radical cascade reactions [49, 74, 75] as well as aromatization reactions [76–79] (Figure 12.5a,b). The carbon–heteroatom bond formation reactions include reversible imine forming reactions [80, 81], amide formation reactions [82–84], 1,3-dipolar addition reaction [85], and phthalocyanine formation reactions [86] (Figure 12.5c). Figure 12.5d summarizes the heteroatom–heteroatom bond formation reactions [87, 88].

12.2.2 Ultrahigh Vacuum Environment

Metal surfaces are often used as substrates in the field of on-surface chemistry in UHV. Atomically clean and flat metal surfaces are routinely obtained by several cycles of sputtering and annealing procedure. Most of the reactions rely on the catalytic behavior of surfaces. Thus, elements as well as their crystal facets strongly affect the reaction. Density of the step edges on the surface also plays an important role since molecules can be aligned along the step edge besides the increased reactivity [71, 89–91]. Furthermore, adatoms diffusing on surfaces participate in the reaction; yet a population of the adatoms is changed by the element

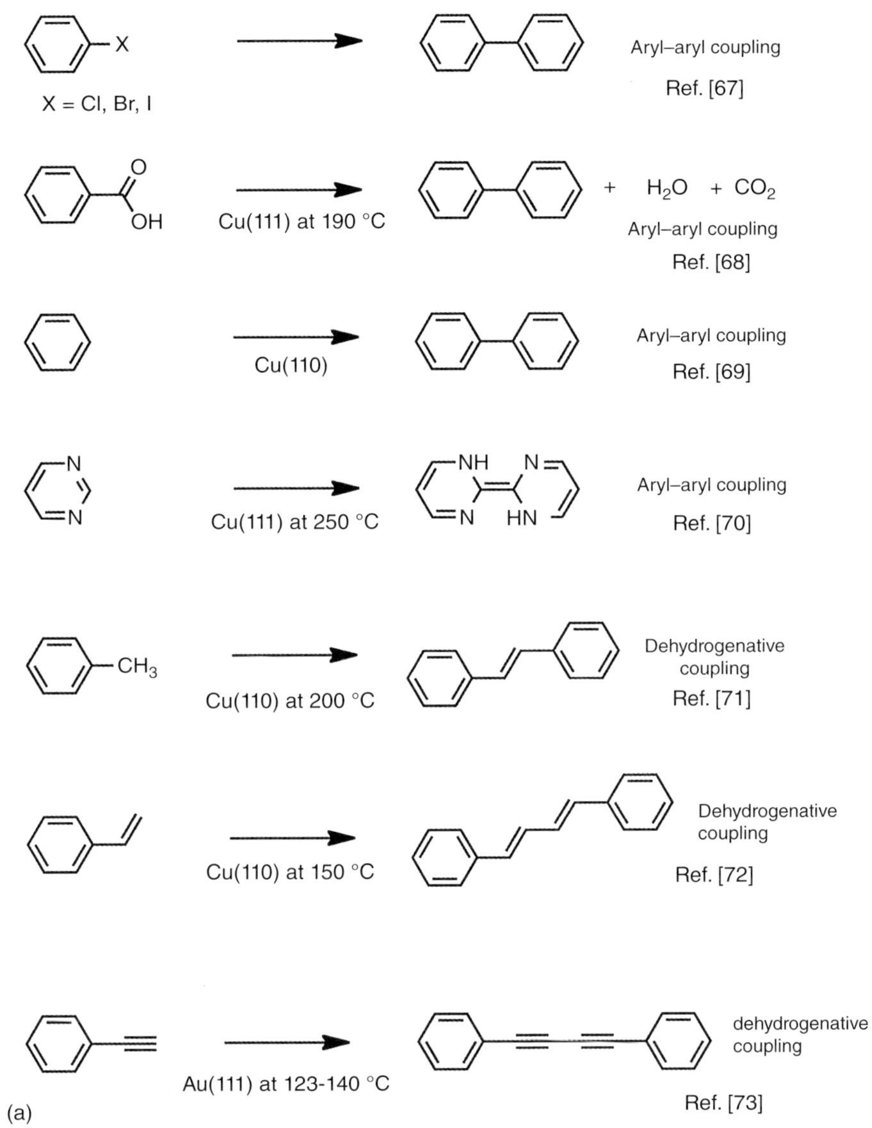

Figure 12.5 Overview of on-surface chemical reactions. (a,b) Carbon–carbon, (c) carbon–heteroatom, and (d) heteroatom–heteroatom bond formation reactions.

and facet as well as the annealing temperature [92]. For this reason, the substrate has to be judiciously selected.

After the substrate is prepared, molecules are deposited on the surface. In UHV, the thermal sublimation technique is most commonly used. First, precursor molecules are inserted into a small crucible and then are *in situ* degassed by heating for a while at below their evaporation temperature. Once molecules become clean enough, they are deposited on the surface at an elevated temperature. To maintain the surface cleanness and to control the amount

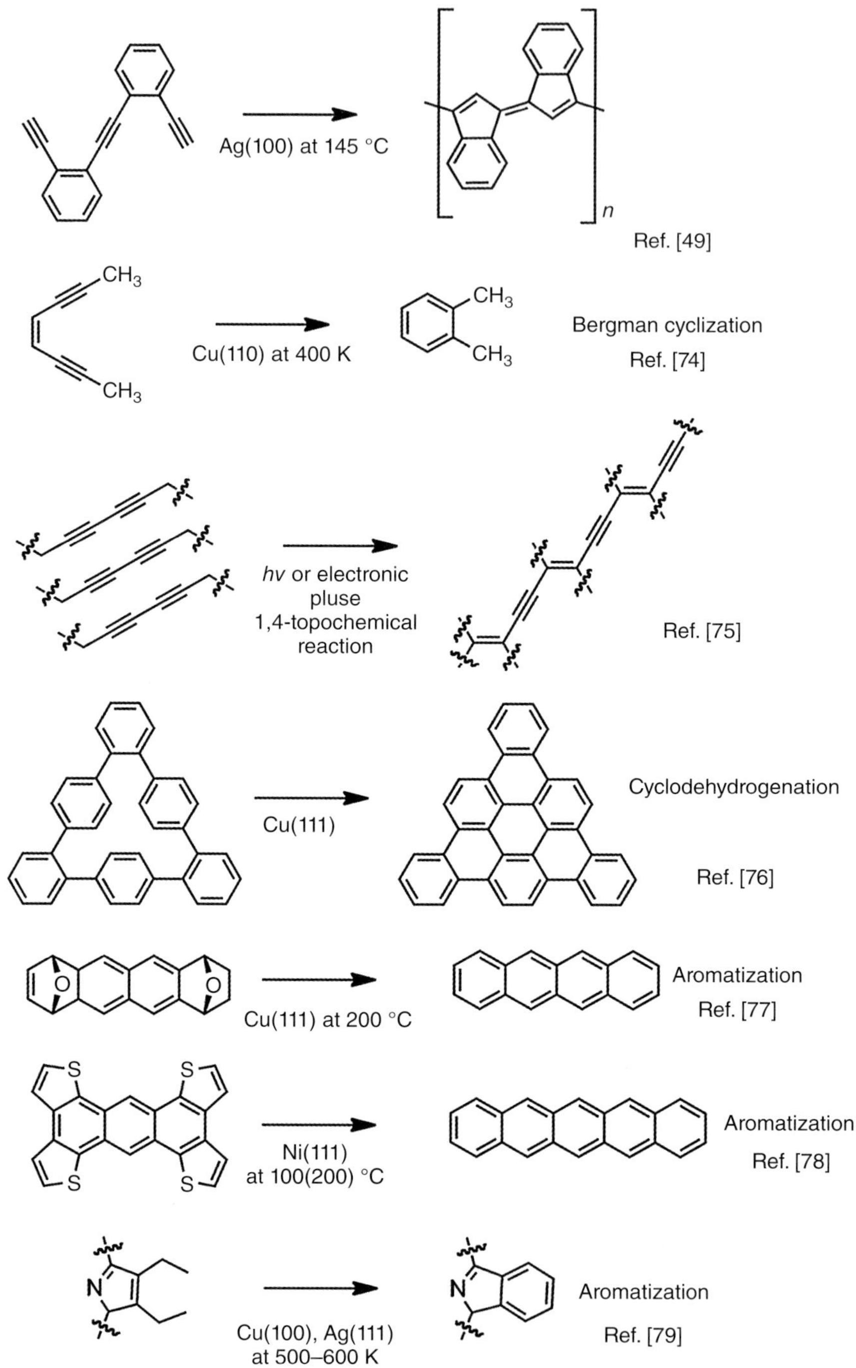

(b)

Figure 12.5 (*Continued*)

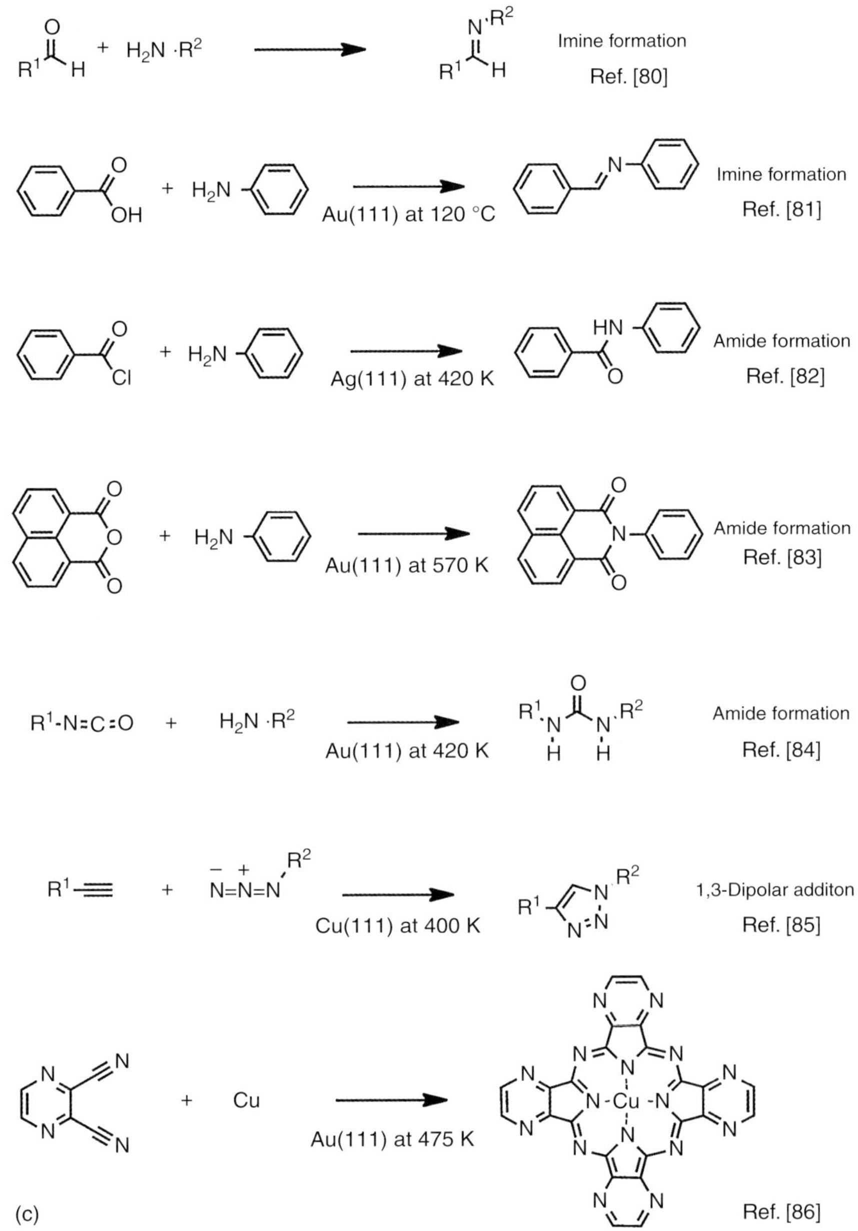

Figure 12.5 (*Continued*)

of molecules on surfaces precisely, a low evaporation rate is usually set. The evaporation temperature strongly depends on the molecular mass as well as the intermolecular interaction. Therefore, this method can be used only for a molecule with a mass in the range of about 100–800. A molecule with a lower mass is usually in a gas phase at RT so that they can be introduced via a leak

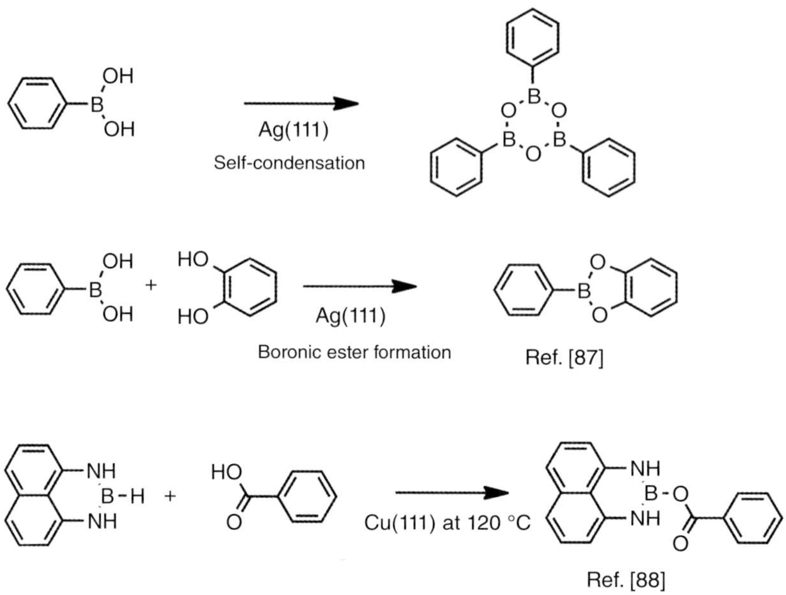

Figure 12.5 (*Continued*)

valve. However, heavier and/or fragile molecules are often decomposed during sublimation. To deposit a large mass molecule, Tanaka and Kawai employed pulse injection deposition in UHV [93]. In this method, the DNA diluted in distilled water ($0.1\,\mathrm{mM\,l^{-1}}$) was deposited by several shots, each with pulse duration of 1.5 ms. This technique is the simplest and is sufficient enough to prepare even a single DNA molecule [94]. Nevertheless, all molecules in the solvent appear on the surface. In contrast, electrospray ionization with a quadrupole mass filter is a more sophisticated approach to control and select the target molecules in solvent [95]. Since the kinetic energy of the ionized molecule can be reduced by lens, even a biomolecule can be dosed on a clean surface. An on-surface reaction with a biomolecule may lead to significant progress in life science.

Precursor molecules on surfaces can be transformed to other compounds on surfaces by various kinds of chemical reactions. Particularly, when they are linked to each other, nanoscale carbon materials with an atomic thickness can be fabricated. This bottom-up approach is of great importance since many applications for electronics and sensor devices can be realized. There are several reaction schemes, yet only thermal-assisted and local-probe-assisted reaction will be described in the following sections.

12.2.3 Thermal-Assisted On-Surface Reaction in UHV

Since the invention of STM and AFM, controlling intermolecular interaction has attracted much interest from researchers [2]. Diffusion of molecules and intermolecular bonding can be controlled by judiciously designed molecules and so that a large-scale structure is achievable. However, due to the nature of the formation, the thermal stability and the conductivity are rather low. To improve the mechanical and electronic properties, a coupling of molecules on surface is required. In 2007, Grill et al. demonstrated a first systematic Ullman-type on-surface aryl–aryl coupling reaction (Figure 12.6a–e) [67]. Tetraphenylporphyrin **1** with bromophenyl groups was employed for the reaction. When the precursor molecules are evaporated below 270 °C, more than 90% of molecules remain intact on Au(111). The ratio drops with increasing the temperature. Finally, the radical can be formed in most of the molecules by heating at the evaporation temperature of 330 °C and then the molecules are connected to each other on the surface. Alternatively, the intact precursors can be coupled to each other by annealing on surface. This reaction is categorized to the Ullmann type, since the surface metal catalyzes for cleaving the C—Br bond. Ullmann reaction was originally demonstrated in solution, in which two aryl halide molecules are transformed to a biaryl molecule by a Cu catalyst [96]. Nowadays, Ullman reactions have become one of the most popular on-surface reactions [97].

Soon after the first systematic aryl–aryl coupling reactions on surfaces [67], several kinds of on-surface chemical reactions have been reported. Imine compounds were formed from aldehydes and amines on Au(111) at RT [98]. The coupling of the methyl groups of (mesityl)porphyrins on Cu(110) was induced via dehydrogenation by annealing up to 200 °C [99]. 1,3,8,10-Tetraazaperopyrene molecules were connected via coupling of *N*-heterocyclic carbene tautomers on Cu(111) [70]. Furthermore, two types of porous 2D polymeric networks were formed on Ag(111) by dehydration of a boronic acid **2** or esterification (Figure 12.6f–h) [87]. Later in 2013, Gao et al. demonstrated Glaser-type coupling of bisalkyne **3** on Au(111) (Figure 12.6i–n) [73]. In these reactions, by-products remain on surfaces, which may give a detailed insight of the reaction path as well as the potential energy landscape. Formation of large-scale structures with these intermolecular coupling reactions is rather challenging due to an entropic point of view. In contrast, the reaction within a single molecule has generally a higher production yield. For instance, Otero et al. synthesized C_{60} and triazafullerene $C_{57}N_3$ from aromatic precursors on Pt(111) with about 100% yield [100]. Treier et al. demonstrated a systematic cyclodehydrogenation with a cyclic polyphenylene, cyclohexa-*o*-*p*-*o*-*p*-*o*-*p*-phenylene, on Cu(111) with high selectivity [76]. Kawai et al. demonstrated the three-step thermal reaction of triangular dehydrobenzo[12]annulene molecule with high reaction selectivity [51]. Furthermore, Stetsovych et al. demonstrated that a helical molecule can be transformed to a planar molecule by dehydrogenation, loosing on-surface chirality [101].

Nevertheless, among them, the Ullmann-type aryl–aryl coupling reaction is most frequently used (Figure 12.7a–j). The reaction temperature relates to the strength of the C—X bond (X = Br or I); the temperature for the C—I bond is lower than that for the C—Br bond. Two different activation temperatures allows us to conduct the two-step reaction. In this way, the two-step reaction is possible [105]. The products, up to now mainly polymers, are ideal samples to measure the conductance [106, 107] and mechanical properties by lifting from the surfaces and/or sliding on the surface (Figure 12.7k–n) [104, 108]. Furthermore, Reecht et al. demonstrated electroluminescence of polythiophene by flowing a current between the metal substrate and the tip through the polythiophene molecule [109]. Notably, the synthesis of graphene nanoribbon (GNR) is one of the most successful examples [102, 110]. In 2010, Cai et al. demonstrated the bottom-up synthesis of $n = 7$ GNR with armchair edges (n: a number of carbon atoms corresponding to the GNR width) (Figure 12.7a–c). First, 10,10′-dibromo-9,9′-bianthryl (DBBA) **4** molecules were connected by heating on an Au(111) surface and subsequent high-temperature annealing induced the dehydrogenation between the anthracene units. This synthesis inspires the creativity of researchers to produce other GNRs with different widths: $n = 5$ from 3,9-dibromoperylene (DBP), 3,10-DBP (Figure 12.7d,e) [18] or 1,4,5,8-tetrabromonaphthalene [111], $n = 6$ from fused poly(*p*-phenylene) [112], $n = 7$ from DBBA [102], $n = 9$ from 3′,6′-dibromo-1,1′:2′,1″-terphenyl [113], and $n = 13$ from 2,2′-di((1,1′-biphenyl)-2-yl)-10,10′-dibromo-9,9′-bianthracene [114]. Furthermore, boron substitution at the center of $n = 7$ GNR (Figure 12.7f–j) [103] and sulfur substitution at the edge of $n = 13$ GNR [115] have been demonstrated. In principle, wider GNRs can be synthesized by fusing GNRs [103]. Besides the straight GNR, chevron-type GNRs with armchair edges have also been synthesized [102]. Furthermore, Ruffieux et al. recently synthesized GNRs with zigzag edges [110]. In this way, the Ullmann-type reaction allows us to synthesize various types of GNRs, which may be a candidate for a key element to realize thin carbon-based electronic devices.

12.2.4 Local Probe-Assisted On-Surface Reaction in UHV

One of the most fascinating things in scanning probe microcopy is the ability to manipulate atom and/or molecule. In 1990, Eigler and Schweizer demonstrated atom manipulation, in which Xe atoms were laterally moved on Ni(110) and were repositioned at LT [116]. Furthermore, Bartels et al. manipulated a

Figure 12.6 Various on-surface chemical reactions in UHV. (a) Concept of the Ullmann-type on-surface reaction. (b) Chemical structure of tetraphenylporphyrin **1** with bromophenyl. (c) STM image of a molecular island on Au(111) after deposition at low evaporator temperatures of 550 K prior to being activated (20×20 nm^2). The inset shows the corresponding chemical structure. (d) STM image for deposition at an elevated evaporator temperature of 610 K, causing the activation and connection of the molecules to form networks (41×41 nm^2). (e) STM image of a network of eight molecular building blocks (8.5×8.5 nm^2). The chemical structure of the network is drawn in the inset. Source: Adapted with permission from Grill et al. [67]. Copyright 2007, Macmillan Publishers Ltd: [Nature Nanotechnology]. (f) Chemical structure of 1,4-benzenediboronic acid **2** and (g) the boroxine repeat unit formed from dehydration of three monomers. (h) STM image of near-complete monolayer covalent network film on Ag(111) (120×90 nm^2). The inset shows the overlaid chemical structure obtained by density functional theory (DFT) calculation. Source: Adapted with permission from Zwaneveld et al. [87]. Copyright 2008, American Chemical Society. (i) General outline of the on-surface Glaser coupling reaction and (j) its coupling for the construction of conjugated chains. (k) Chemical structure of 1,4-diethynyl-2,5-dihexylbenzene **3**. (l) High-resolution STM image as deposited on Au(111) (6×6 nm^2). (m) and (n) STM image and high-resolution STM image of the molecules after annealing (25×25 nm^2 and 5×5 nm^2). Source: Adapted with permission Gao et al. [73]. Copyright 2013, John Wiley and Sons.

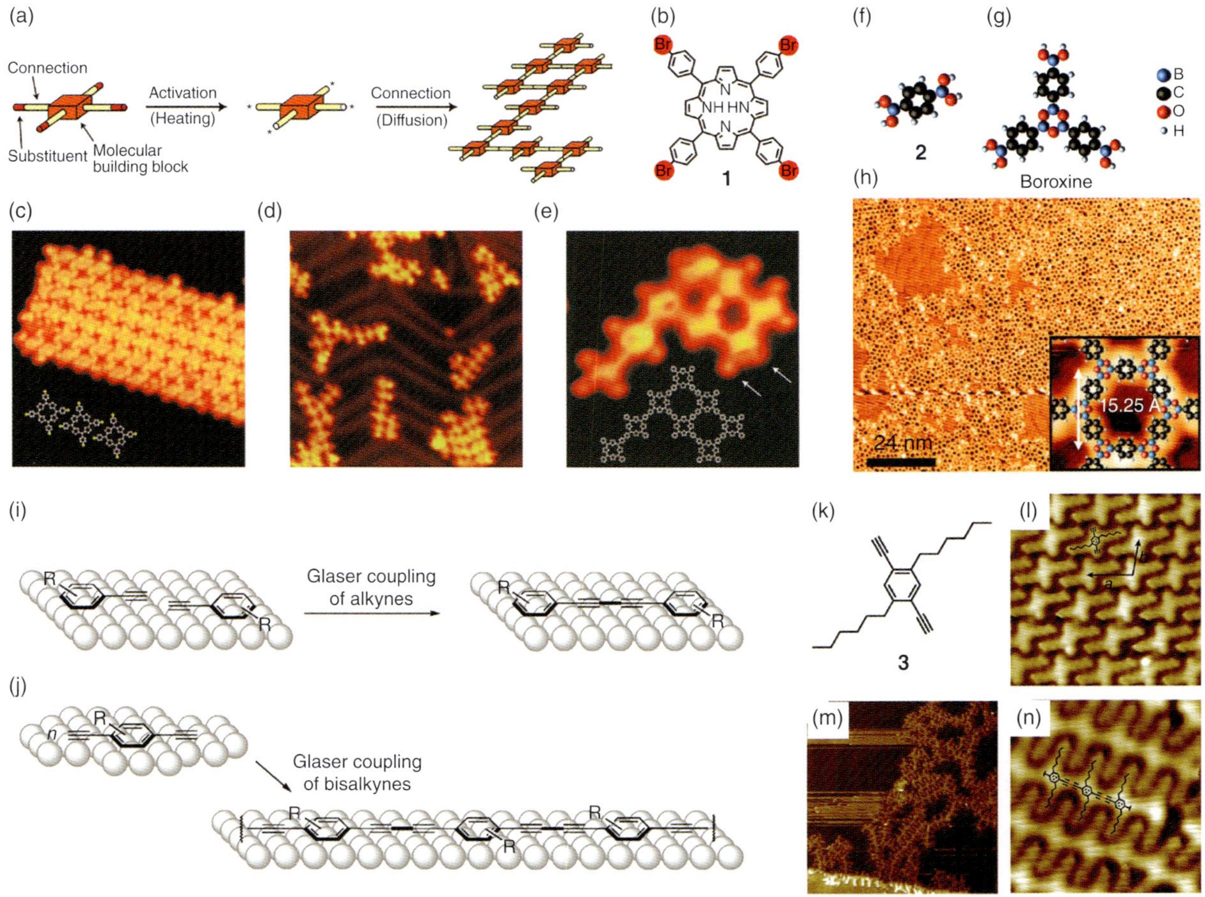
(a)
Connection
Substituent
Molecular building block
Activation
(Heating)
Connection
(Diffusion)
(b)
Br
NH HN
N
1
(c)
(d)
(e)
(f)
2
(g)
Boroxine
B
C
O
H
(h)
24 nm
15.25 Å
(i)
Glaser coupling of alkynes
R
(j)
Glaser coupling of bisalkynes
n
(k)
3
(l)
a
(m)
(n)

CO molecule on Cu(211) [117]. Such atom and molecule manipulation can be conducted with attractive and repulsive interaction forces, the so-called mechanical manipulation. In the case of molecules, besides the force-induced manipulation, the tunneling current with a certain bias voltage can excite the bond. Consequently, the molecule can be hopped [118] or rotated [119], in the so-called inelastic way. Pascual et al. demonstrated selective bond vibrations by adjusting the energy [120]. Note that the vibration energy is also affected by the substrate [121]. In addition to the repositioning, the hydrogen atom in the molecule can be desorbed from a molecule [122]. Liljeroth et al. demonstrated tautomerization in naphthalocyanine [123]. If reactive molecules are located in line on Cu(110), a relay reaction such as transferring a hydrogen atom can be demonstrated [124]. Besides hydrogen atoms, a C—Br bond can be cleaved by the inelastic way. Hla et al. demonstrated the tip-induced synthesis of biphenyl from two iodobenzene molecules at the step of Cu(111) [125]. These innovative works were performed with STM. However, once such local probe chemistry is performed with high-resolution NC-AFM, the products are readily identified. Pavliček et al. successfully formed aryne, which is generally a short-lived and highly reactive intermediate in solution but is not on a thin NaCl film formed on Cu(111) at LT, by extracting two hydrogen atoms from an aromatic compound by a local probe (Figure 12.8a–e) [126]. In their experiment, two iodine atoms in 10,11-didehydronaphtho[1,2,3,4-*g*,*h*,*i*]perylene derivative **5** were cleaved and the product was analyzed via the apparent bond lengths. They concluded that a contribution of cumulene resonance structure exists. Furthermore, Schuler et al. demonstrated controlled and reversible transformation between diyne and diradical compounds (Figure 12.8f–o). In their experiment, two bromine atoms in 9,10-dibromoanthracene (DBA) on a NaCl film formed on Cu(111) were removed by applying a pulse bias voltage of about 1.6 V for the first Br atom and of 3.3 V for the second Br atom. Then, anthracene biradical was formed. Subsequently, applying a pulse voltage of 1.7 V induces the formation of diyne. At a voltage of 1.64 V, the molecule was infrequently switched between diradical and diyne states [127]. Furthermore, triangulene was recently synthesized via

Figure 12.7 1D nanostructure and its characterization. (a) Scheme of GNR synthesis on surface from **4**. Middle, formation of linear polymers by covalent interlinking of the dehalogenated intermediates. Bottom, formation of $n = 7$ GNRs by cyclodehydrogenation. (b) Overview STM image of GNRs after cyclodehydrogenation at 400 °C. The inset shows a higher-resolution STM image. (c) High-resolution STM image with partly overlaid molecular model (blue) of the ribbon. At the bottom left is a DFT-based STM simulation of the GNR (greyscale image). Source: Adapted with permission from Cai et al. [102]. Copyright 2010, Macmillan Publishers Ltd: [Nature]. (d) Reaction scheme of the polymerization of the DBP precursor to armchair $n = 5$ GNRs. Source: Adapted from Kimouche et al. [18] under CC BY license. (e) STM image after cyclodehydrogenation at 320 °C. Scale bar, 2 nm. (f) Schematic drawing of the on-surface chemical reaction. (g) STM overview of fused B-GNR. Scale bar, 10 nm. (h) AFM image of $n = 7$ B-GNR, (i) fused $n = 14$ and (j) $n = 21$ B-GNR. Source: Adapted from Kawai et al. [103] under CC BY license. (k) Schematic drawing of the lateral manipulation procedure. (l) STM topographies before and (m) after a $n = 7$ GNR has been displaced on the Au(111) surface in the direction of the yellow arrows. (n) Frequency shifts accompanying the lateral motion at $z = 2$ nm. Source: Kawai et al. [104]. Adapted with permission from AAAS.

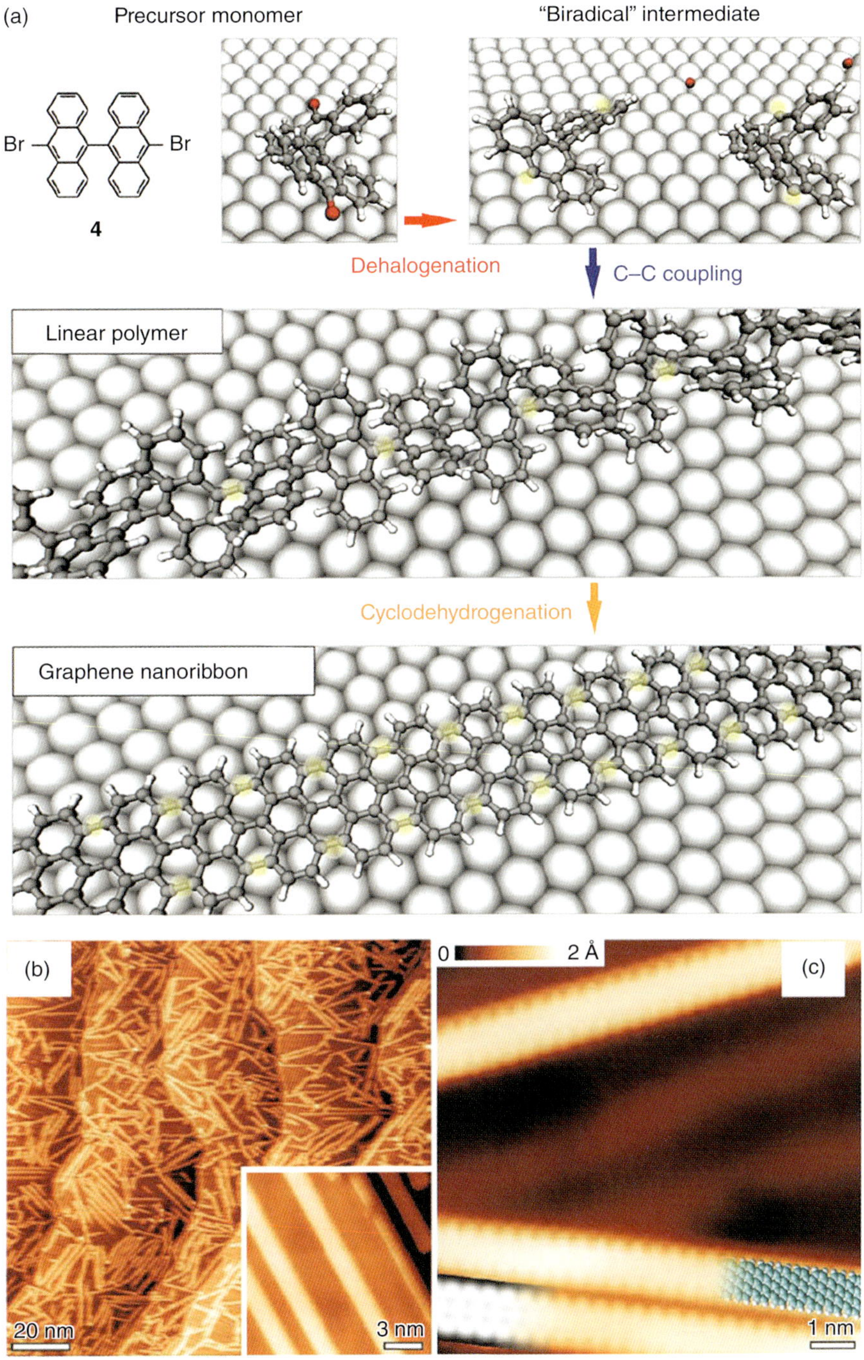
(a)
Precursor monomer
"Biradical" intermediate
Br
Br
4
Dehalogenation
C–C coupling
Linear polymer
Cyclodehydrogenation
Graphene nanoribbon
(b)
20 nm
3 nm
0
2 Å
(c)
1 nm

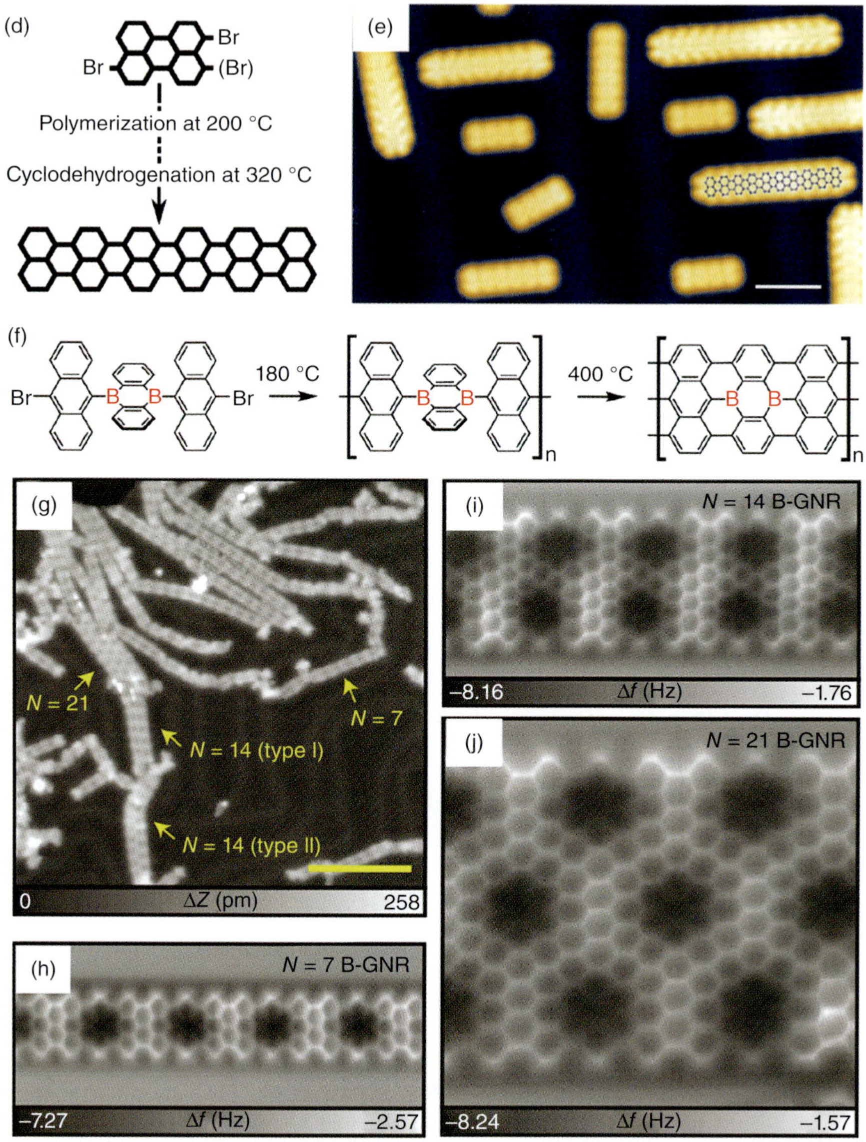

Figure 12.7 (*Continued*)

dihydrodibenzo[*cd,mn*]pyrene isomers by removing a single hydrogen atom from the CH_2 group. Since the C—H bond dissociation energy of CH_2 group is lower than that of CH group, the selective dehydrogenation was successfully performed [128]. In this way, the combination of local probe chemistry and high-resolution AFM imaging opened up single molecular chemistry.

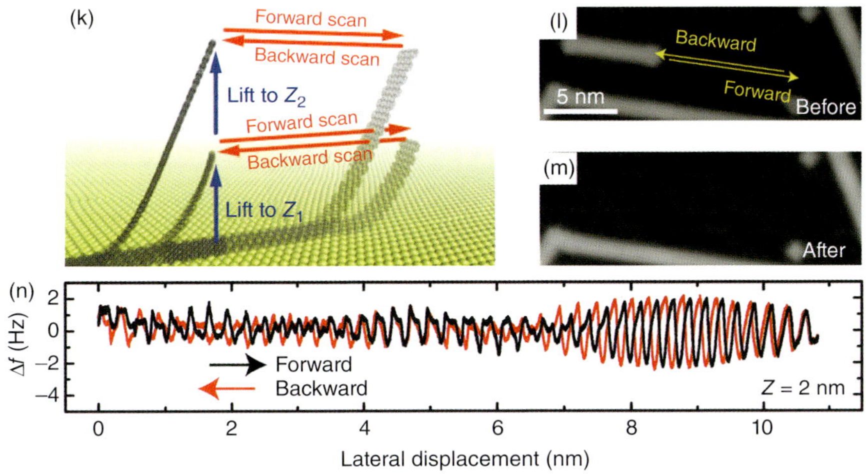

Figure 12.7 *(Continued)*

12.2.5 Chemical Reactions at the Liquid/Solid Interface

12.2.5.1 Carbon–Carbon Bond Formation Reactions

Topochemical polymerization of 1,4-diacetylene derivatives is one of the most popular reactions at the interface. De Schryver and coworkers reported that ultraviolet (UV) light irradiation or a pulsed bias voltage with an STM tip at diacetylene moieties in the self-assembled isophthalic acid derivative **16** induces formation of 1D π-conjugated oligomers (Figure 12.9) [75, 129]. Furthermore, the tip scanning could be used to desorb the oligomers from a graphite surface since the binding energy is low. Later, Aono and coworker successfully demonstrated the controlled polymerization of 1,4-butadiyne derivatives on graphite by applying a pulsed bias voltage with an STM tip [130]. In this polymerization, an arrangement of the diacetylene units by the van der Waals intermolecular interaction is indispensable.

In contrast to UHV, only a few studies on the surface-catalyzed Ullman-type aryl–aryl coupling reaction exist in ambient conditions. It may relate to the fact that a large activation energy is required to cleave the C—X bond. Nevertheless, Beton and coworkers demonstrated dimerization of tris(4-bromophenyl)benzene **17** at the liquid/solid interface. If compound **17** diluted in ethanol drop-casted onto an Au(111) surface preheated at 200 °C, dimerization can be induced; otherwise, **17** monomer self-assembles on a cold substrate (Figure 12.10a,b) [131]. Later, they conducted a similar reaction with more reactive iodine-substituted derivative **18** in nonanoic acid on Au(111) at 100 °C and formed 1D chains, closed pentagons, hexagons, and heptagons as well as irregular structures (Figure 12.10c). However, the reaction stopped at a certain point because the cleaved iodine atoms covered the entire Au(111) substrate [132], which prevent the surface-catalyzed polymerization.

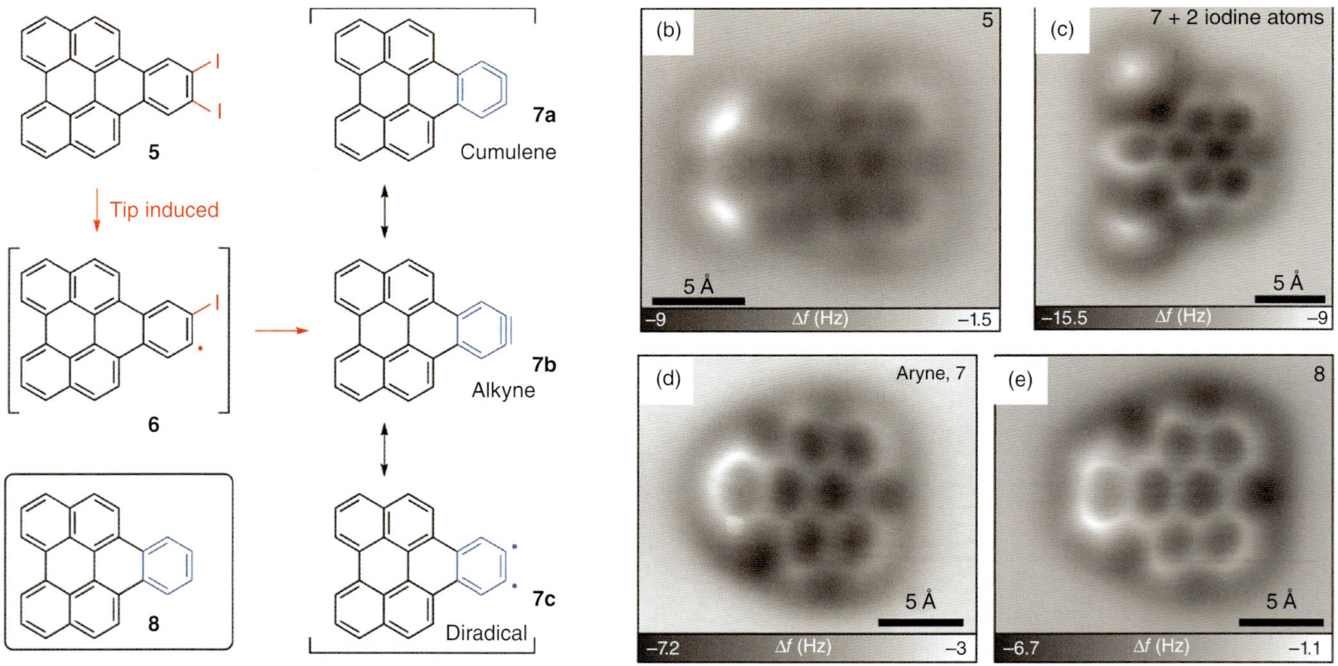

Figure 12.8 Tip-induced reactions in UHV. (a) Starting from **5**, iodine dissociation generates aryne **7**. Dehalogenation may occur either directly from **5** to **7** or sequentially via intermediate **6**. **7a**–**7c** are the three possible resonance structures of this aryne. Compound **8** is the fully hydrogenated analogue of aryne **7** for comparison. (b) AFM image of **5**. (c) AFM image of an aryne molecule close to two iodine atoms after dissociating two iodines from **5**. (d) AFM images of aryne **7** and (e) and **8**. Source: Adapted with permission from Pavliček et al. [126]. Copyright 2015, Macmillan Publishers Ltd: [Nature Chemistry]. (f) The thermal isomerization of deuterated enediynes **9** and **10** through the formation of the diradical [2,3-d_2]-1,4-didehydrobenzene **10**. (g) Bergman cyclization of the cyclic diyne 3,4-benzocyclodeca-3,7,9-triene-1,5-diyne **12** to generate the 9,10-didehydroanthracene diradical **13**. Chemical structures of the reaction products formed by STM-induced debromination of **14** in (h) and subsequent *retro*-Bergman cyclization: **14**, 9-dehydro-10-bromoanthracene (radical **15**) in (i), diradical **13** in (j) and diyne **12** in (k). (l–o) Corresponding constant-height AFM images of the molecules in (h–k), respectively, on NaCl(2 ml)/Cu(111). Source: Adapted with by permission from Schuler et al. [127]. Copyright 2016, Macmillan Publishers Ltd: [Nature Chemistry].

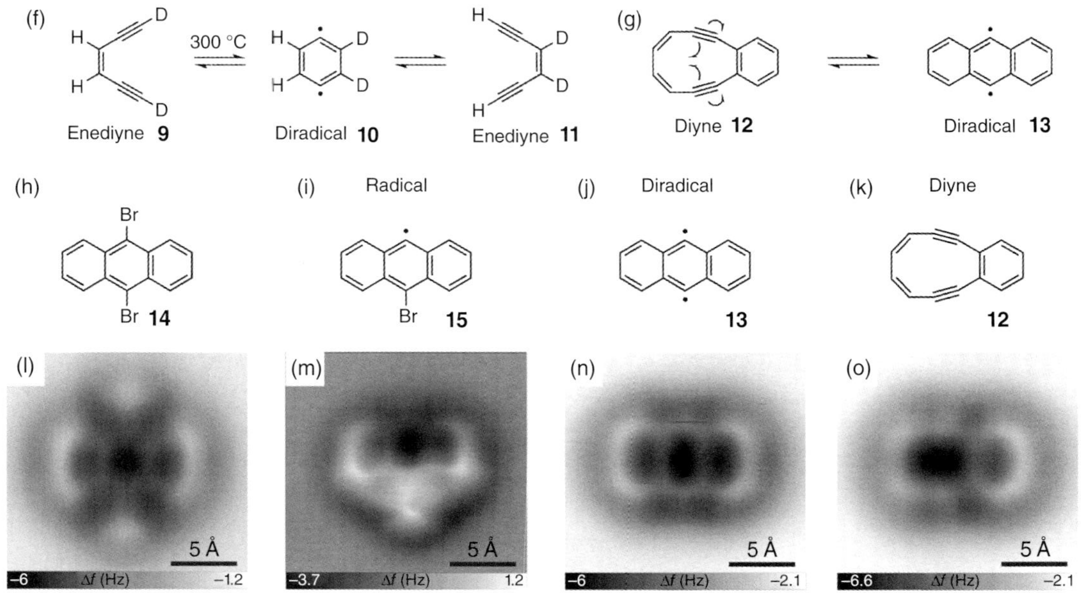

Figure 12.8 (*Continued*)

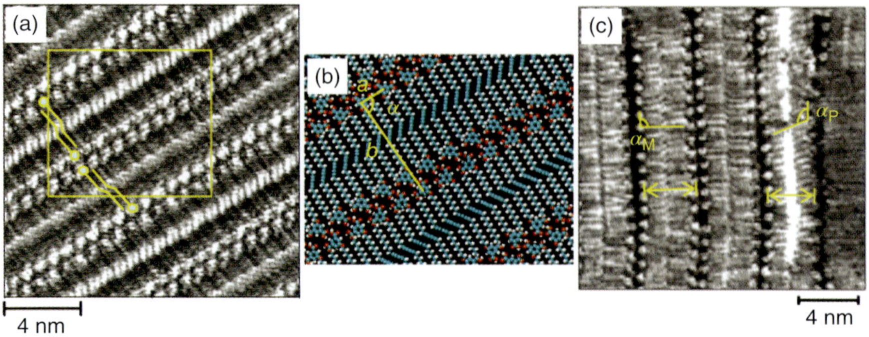

Figure 12.9 1,4-Topochemical polymerization of an isophthalic acid derivative **16** having diacetylene units at the liquid/solid interface. (a) STM image of **16** in self-assembly and (b) its corresponding structural model. (c) STM image of **16** after UV light irradiation. A bright row corresponds to the synthesized 1D π-conjugated oligomer. Source: Adapted with permission from Miura et al. [129]. Copyright 2003, American Chemical Society.

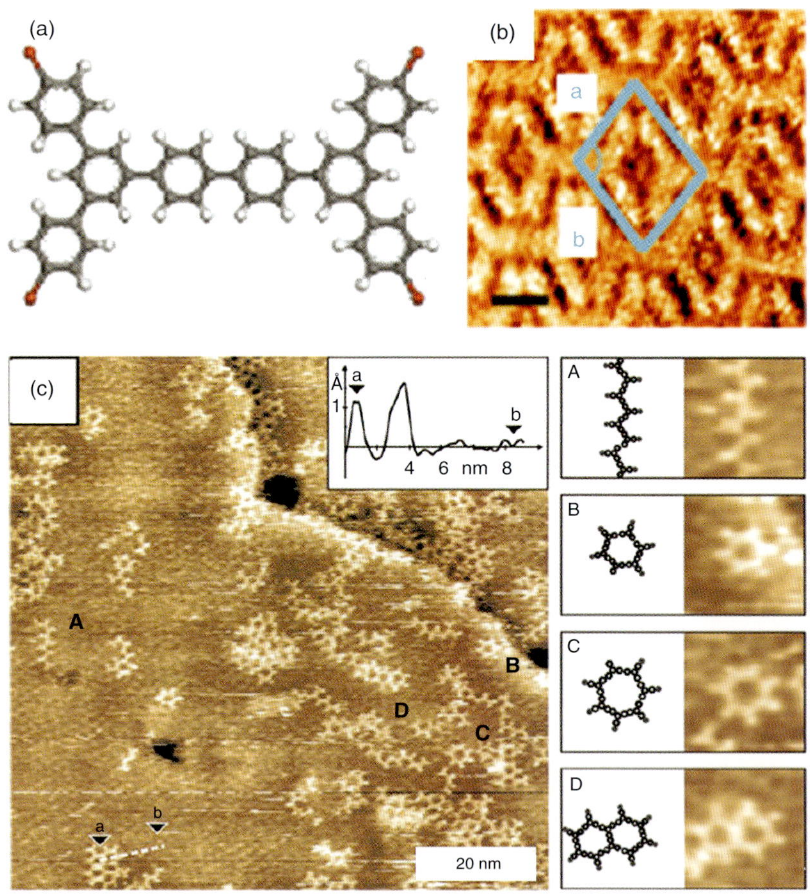

Figure 12.10 (a) Chemical structure of a dimer of **17**. (b) STM image of a monolayer of dimerized **2**. (c) STM image of on-surface coupling products of **18**. Source: Adapted with permission from Russell et al. [131] and Eder et al. [132]. Copyright 2011 and 2013, American Chemical Society.

Wang and coworkers demonstrated a Glaser-type reaction at the liquid/solid interface [133]. If compound **19** was solved in 1-phenyloctane and 1,2,4-trichlorobenzene, the resultant self-assembly on a graphite substrate became a lamella structure with co-adsorbed solvent molecules and a close-packed structure, respectively. Adding a $Cu(OAc)_2$ solution in pyridine and subsequent annealing at 60 °C for one hour induced the coupling reaction of the acetylene moieties. Then, 1D linear polymers and dimers were obtained in 1-phenyloctane and 1,2,4-trichlorobezene, respectively (Figure 12.11). Simultaneous formation of dimer, timer, tetramer, and pentamer in solution phase synthesis led to less ordered nanostructures at the interface. Therefore, they claimed that the coupling reactions happened on surface rather than in solution. Moreover, from the fact that other products were formed on different solvents, they assumed that the self-assembled monolayer also plays a role in the reaction. Recently, Colazzo et al. also reported metal-free on-surface alkyne–alkyne homocoupling reaction induced by light irradiation at the heptanoic acid/graphite interface [134]. In this case, the prearrangement of the alkyne moieties by self-assembling on graphite is indispensable for the photochemical dimerization reaction.

12.2.5.2 Dynamic Imine Formation Reactions

In contrast to the irreversible C—C bond coupling reaction, formation and decomposition of imines, the so-called Schiff-base coupling reaction, are reversible. Since the reaction can thermodynamically be controlled, it has widely been employed in the field of supramolecular chemistry, aiming to realize dynamic covalent chemistry [135, 136]. To this end, various imines such as 1D and 2D polymers were recently synthesized at the interfaces.

Kunitake and cowokers synthesized 1D aromatic polymers and 2D macoromolecular frameworks by coupling primarily amines **A1**–**A3** and aldehydes **B1**–**B4** at the electrolyte/Au(111) interface (Figures 12.12 and 12.13) [137, 138]. Imine formation reactions were promoted by adsorption of both monomers on a hydrophobic Au(111) surface covered by iodines and setting a slightly acidic condition relative to the pK_a of the amines. Moreover, adsorption–desorption dynamics of the monomers were directly probed via consecutive STM imaging.

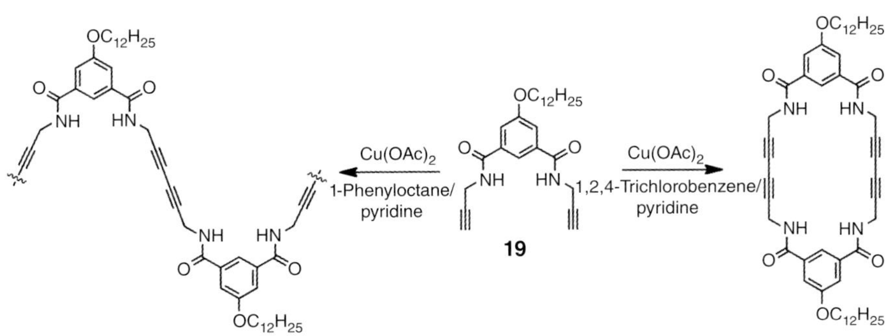

Figure 12.11 Glaser coupling reactions with **19** at the liquid/graphite interface. 1D chains were formed at the 1-phenyloctane/graphite interface, while the dimers were formed at the 1,2,4-trichlorobenzene/graphite interface.

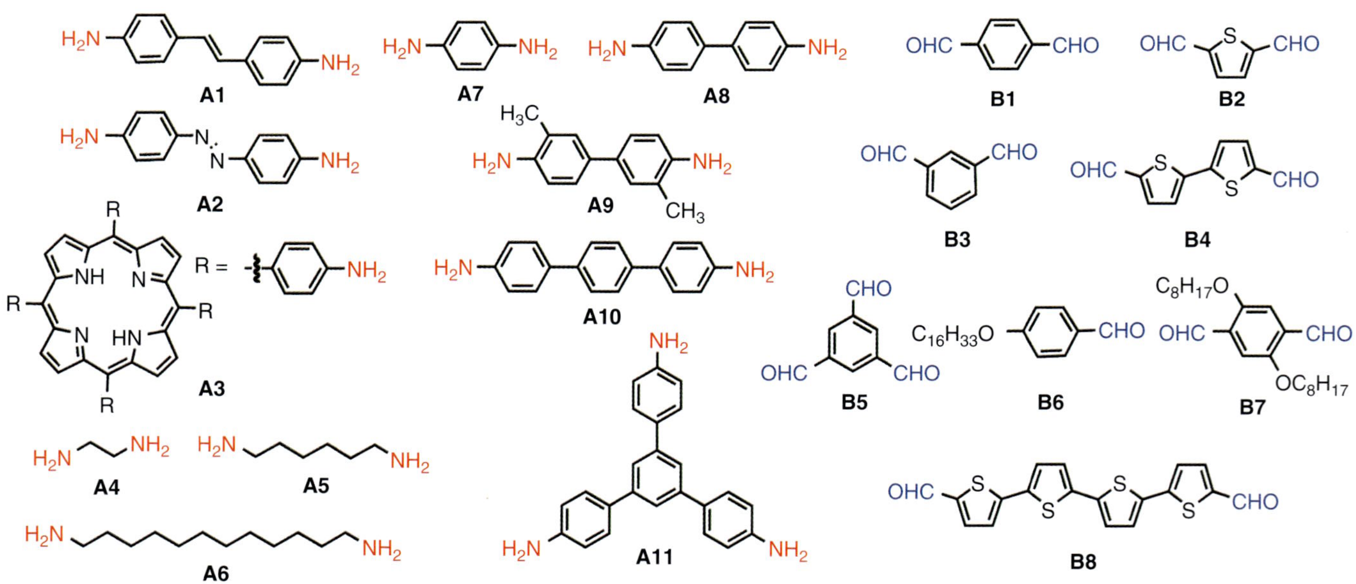

Figure 12.12 Chemical structures of amines **A1**–**A11** and aldehydes **B1**–**B8** used for on-surface imine formation reactions.

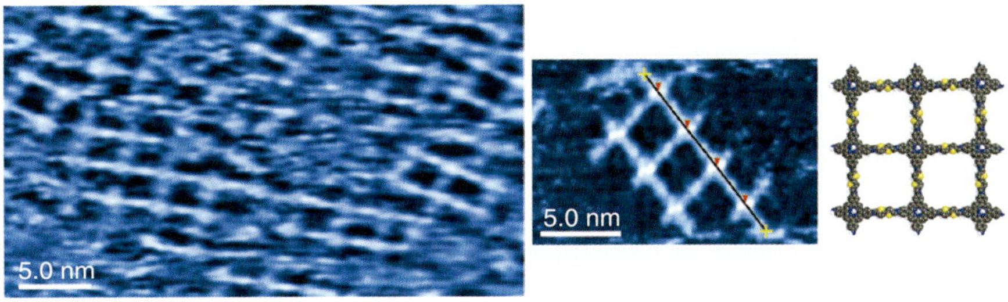

(a) (b) (c)

Figure 12.13 (a,b) STM images of small 2D lattices formed by **A3+B4**. (c) Molecular models of a small 2D lattice of **A3+B4**. Source: Adapted with permission from Tanoue et al. [137]. Copyright 2011, American Chemical Society.

Lehn and coworkers reported the surface-mediated reversible exchange process of aliphatic bisimines at the 1-phenyloctane/graphite interface [139]. To a monolayer of 4-(hexadecyloxy)benzaldehyde **B6**, a solution of one of the diamines (0.5 equiv with respect to **B6**), 1,2-diaminoethane **A4**, 1,6-diaminohexane **A5**, and 1,12-diaminododecane **A6**, in a mixture of 1-phenyloctane and pyridine (in 1 to 99 ratio) was added. High-resolution STM images confirmed that the monolayers consisted of bisimides formed by condensation between the carbonyl group of **B6** and the amino groups of diamines. The replacements of the monolayers from **B6** to bisimides were attributed to larger absorbabilities of the products. Note that the condensation reactions at the interface proceed much faster than those in solution because (i) concentrations of each reactant at the interface is higher than saturation in solution and (ii) quasi surface confinement reduces the activation barrier. Moreover, the reversible nature of C=N bond formations leads to *in situ* imination/bistransimination cycles. Adding a solution of **A5** onto the **A4+B6** monolayer, **A4** in the monolayer was replaced by **A5** on forming the **A5+6** monolayer within three minutes by replacing amines. Subsequently, when **A6** was added, **A5+B6** was transformed to **A6+B6** within one minute. The faster transformation indicates a larger free energy gain by the formation of **A6+B6**. The authors suggested that this controlled transformation relates to entropy-driven selective adsorption. Reverse *in situ* bistransimination from **A6+B6** to **A5+B6** monolayer was also accomplished by drop-casting 100 equiv of **A5** on the **A6+B6** layer while the transformation took about 430 minutes. The reported modular approach using thermodynamically controlled reversible formation of covalent architectures opened up various perspectives for the production of novel 2D molecular functional materials. Wang and coworkers also reported similar *in situ* imine formation reactions between bisaldehyde and aniline derivatives to construct chiral 2D structures [140].

Lei and coworkers also extensively synthesized 1D and 2D polymers with imine formation reactions [141]. By drop-casting a mixture of 1,3,5-triformylbenzene **B5** and *p*-diaminobenzene **A7** solved in octanoic acid onto graphite at RT, ordered honeycomb structures (domain size: tens of nanometers, pore diameter: c. 1.7 nm) were obtained within a few minutes (Figure 12.14). Interestingly,

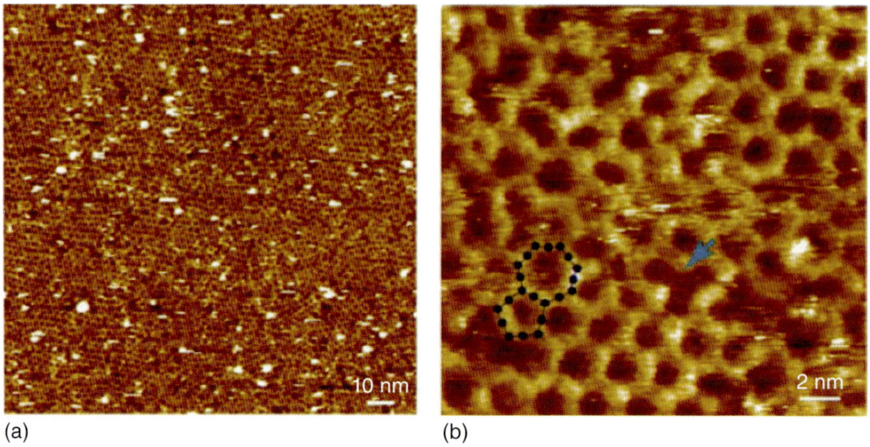

Figure 12.14 STM images of 2D structures formed by **A7+B5** at the octanoic acid/graphite interface. Source: Adapted with permission from Xu et al. [141]. Copyright 2013, American Chemical Society.

typical defects as one heptagon neighboring with a pentagon, similar to the Stone–Wales defect of graphene, were found. Simultaneous polymer growth starting from different sites may be responsible for the presence of small mis-oriented domains. With other mixtures, **B5+A8**, **B5+A9**, and **B5+A10**, amorphous polymers were observed even after days at RT or heating to 40 °C. They also demonstrated an alternative approach; the solutions of the mixtures in dimethyl sulfoxide (DMSO) were dropped onto graphite and then condensation reactions were performed in a vacuum oven (133 Pa) at 140 °C. In this condition, the higher temperature led to high reversibility of the reaction and greater diffusion of the reactants. Moreover, the unreacted monomers, water, and solvent can be removed under reduced pressure. As a result, the regularity and domain size of polymer **B5+B7** drastically improved. Moreover, well-defined honeycomb imine polymers with different pore sizes (1.7–3.5 nm) were prepared using other mixtures, **B5+A8**, **B5+A9**, and **B5+A10**. With this approach, porphyrin containing 2D square lattices with 10–30 nm size were synthesized from 5,10,15,20-*meso*-tetra(4-aminophenyl)porphyrin **A3** and two dialdehydes having dioctyloxyphenyl **B7** or quarterthiophene **B8** linker [142]. Furthermore, 1D polymers and oligomers containing quarterthiophene units were synthesized [143]. Later, they also succeeded in RT synthesis of 2D imine polymers of **B5+A8** and **B5+A10**, with honeycomb topologies at the octanoic acid/graphite interface by simply optimizing molar ratios and concentrations of the reactants [144].

At the gas/solid interface, Wan and coworkers synthesized well-ordered 2D imine polymers [145]. A graphite substrate covered by a dry film of tris(4-aminophenyl)benzene **A11** or 1,3,5-triformylbenzene **B5**, solid samples of terephthalaldehyde **B1** or *p*-diaminobenzene **A7**, and water supplier, $CuSO_4 \cdot 5H_2O$, were added to an autoclave. After annealing this closed system at 150 °C or 120 °C for three hours, highly ordered honeycomb networks were formed. In this case, reversible imine formation reactions between one monomer in the film and gaseous other monomer are promoted by water vapor supplied from $CuSO_4{\cdot}5H_2O$. Lei and coworkers further synthesized a 2D imine polymer **B5+A7** on single-layer graphene on copper. [146]. The polymers also contained a typical defect consisting of one heptagon adjacent to one pentagon as similar to the previous study on graphite [141]. Since the graphene covered the polycrystalline copper foil, no significant influence of the facets and steps of the copper on the synthesis of the polymer was observed. Yet, a strong electronic coupling existed between the 2D polymer and single-layer graphene. They further used the porous 2D imine polymer **B5+A7** on graphite as a host network for guest molecules of coronene, copper phthalocyanine (CuPc), and fluorinated copper phthalocyanine (F_{16}CuPc) (Figure 12.15) [147]. Via statistical analysis, ratios of the pore sites with fuzzy bright contrast in the STM images were >99% for coronene and ca. 13% for CuPc. Fuzzy appearance of the guest molecules relates to their lateral mobilities within the pore. For both cases, van der Waals interaction governs so that the stability relates to the size of the guest molecule. In contrast, F_{18}CuPc was highly immobilized by the C—H···F hydrogen bonding interactions.

12.2.5.3 Condensation of Boronic Acids

Reversible boronic acid condensation is often employed for the synthesis of covalent organic frameworks (COFs) in solution [148–150]. Inspired by the successful synthesis of 3D-COFs in the solution phase, the synthesis of 2D-COFs by direct on-surface polycondensation of boronic acids has been investigated, in particular under the interfacial conditions. Since hydration and dehydration reactions are reversible in the annealing process, large domains are reliably achieved.

Lackinger and coworkers formed porous honeycomb COFs on graphite with a solution of 1,4-benzenediboronic acid **2** in 1-nonanoic acid by thermal treatment under humid conditions (Figure 12.16) [151]. The domain size was limited up to $40 \times 40\ nm^2$. However, if larger aromatic diboronic acids **20**, **21**, **22**, and **23** were

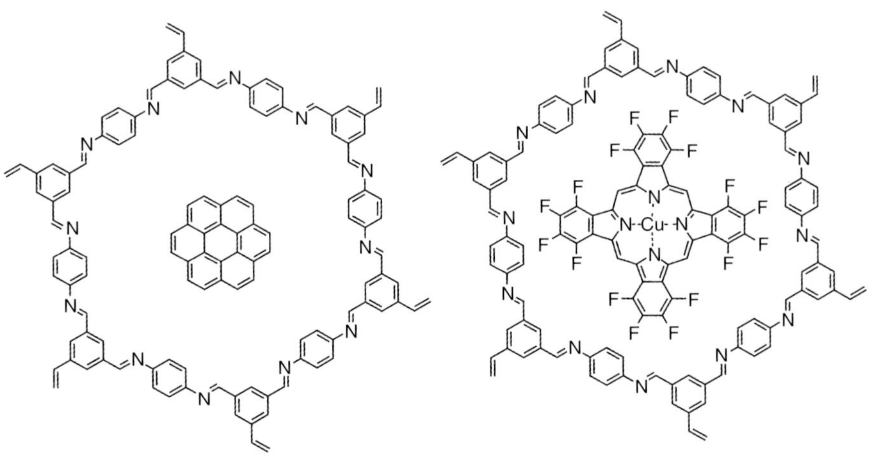

Figure 12.15 Schematic representation of 2D polymers of **A7+B5** with guest molecules of coronene (a) and F_{16}CuPc (b).

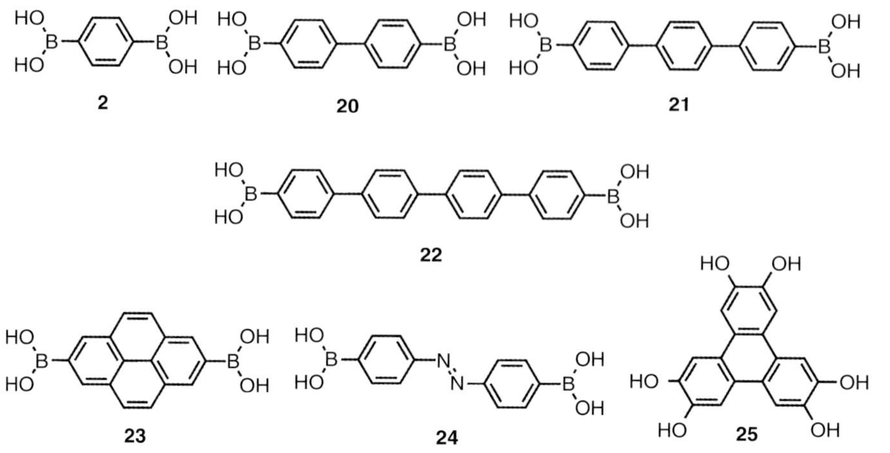

Figure 12.16 Chemical structures of boronic acids **2**, **20–24**, and hexahydroxytriphenylene **25** used for the synthesis of porous COFs.

used (Figure 12.17) [152], the domain became larger ($100 \times 100\,nm^2$). In this case, the diameters of the pores also became larger (1.0–3.2 nm). Wan et al. also synthesized a porous 2D-COF using a closed reaction vessel [153]. A dry film prepared by drop-casting a solution of **20** in tetrahydrofuran (THF) on graphite was annealed in the presence of $CuSO_4{\cdot}5H_2O$. In this system, $CuSO_4{\cdot}5H_2O$ acts as a water reservoir and releases water molecules to regulate the chemical equilibrium of the reaction. Thus, the highly ordered COF with large domains ($>200 \times 200\,nm^2$) was obtained.

Lei and coworkers synthesized COFs with boronic ester linkages from diboronic acid **24** and hexahydroxytriphenylene **25** at the octanoic acid/graphite interface at RT [154]. Depending on concentrations and molar ratios of the monomers, a series of periodic porous networks were formed. Using a solution of the mixture at molar ratio of **24** : **25** = 3 : 2, the surface was mostly covered by self-assembled monolayers of **25** at a high concentration (noncovalent). At medium concentration, various porous structures consisting of both noncovalent and covalent intermolecular interactions appeared. Finally, porous 2D-COFs were obtained at low concentration. Moreover, by comparing rates of the reversible reaction between boronic ester formation and cleavage in STM images, they stated that the boronic ester formation reaction is promoted at the interface.

De Feyter and coworkers reported a guest accommodation at the pore sites of the 2D-COF on graphite prepared by self-condensation of **2** [155] – a slightly modified protocol from Lackinger and coworkers [151]. C_{60} diluted in 1-phenyloctane was drop-casted onto the COF structures. A bright hexagonal pattern in the STM image was assigned to the presence of C_{60}. Since no stable monolayer can be formed with C_{60} alone at the interface, they concluded that C_{60} was stabilized at the pore sites. Moreover, they also reported 3D growth of the structure and removal of the host–guest architectures by an STM tip.

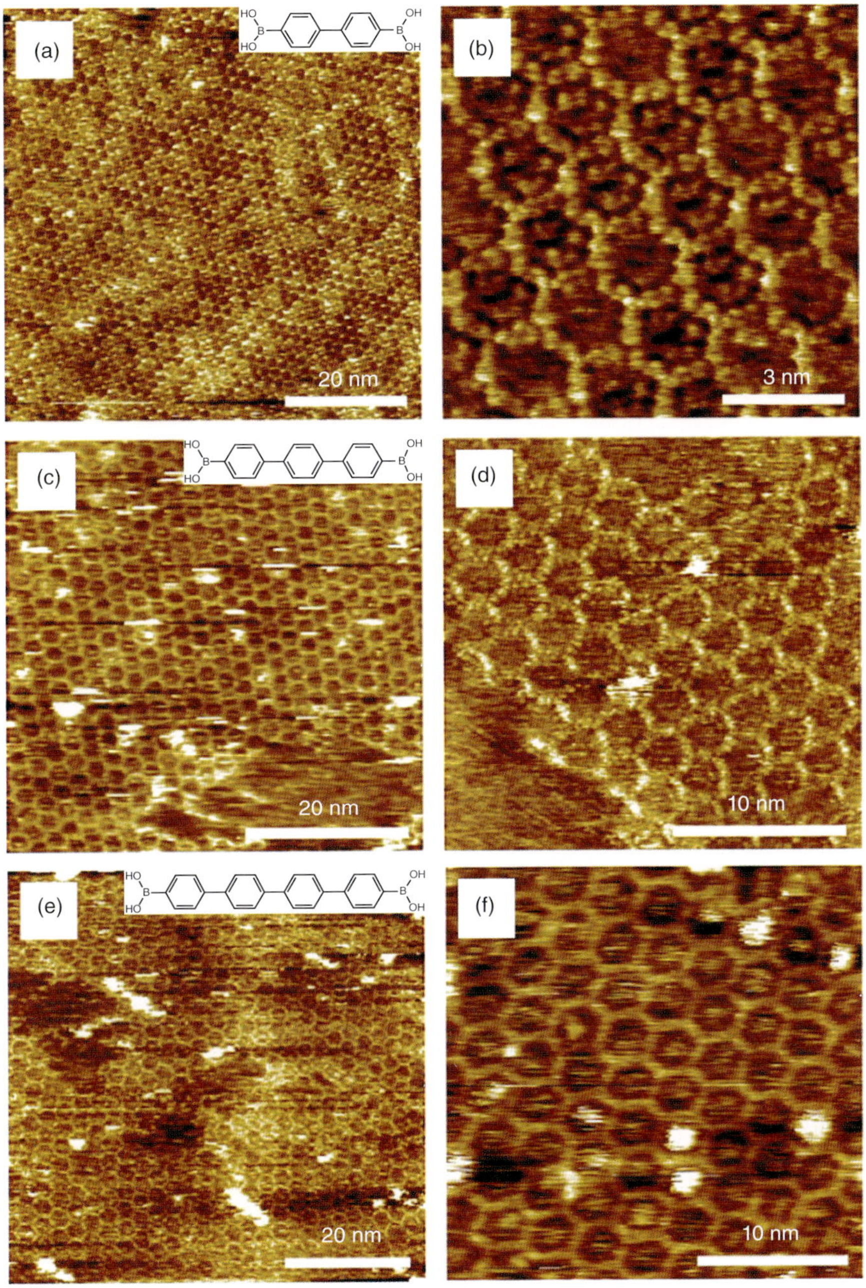

Figure 12.17 STM images of 2D-COFs formed with **20** in (a,b), **21** in (c,d), **22** in (e,f), and **23** in (g,h). Source: Adapted with permission from Dienstmaier et al. [152]. Copyright 2012, American Chemical Society.

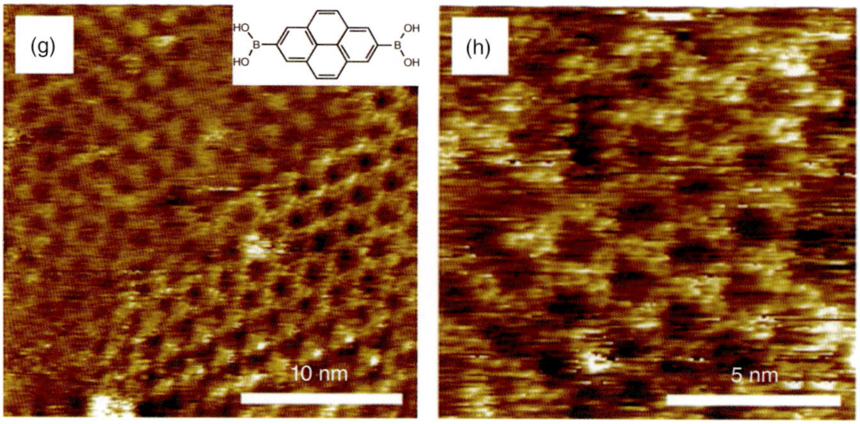

Figure 12.17 (*Continued*)

Lei and coworkers synthesized porous 2D-COFs from dry film of **24** on graphite in a closed reaction vessel containing a small amount of water [156]. Most of the graphite surface was covered with the 2D-COFs. Moreover, by irradiating with UV light for two hours, the coverage of the porous 2D-COF decreased to 11%, indicating the isomerization from trans to cis configurations by breaking B—O bonds. Note that the 2D-COF can be healed by annealing at 100 °C for three hours. At the pores of the 2D-COF, CuPc molecules were accommodated as the guest. The decomposition of the 2D-COF by UV light irradiation released the guest molecules as well. This demonstrates the possibilities to construct stimuli-responsive 2D-COFs.

12.3 Conclusion and Perspective

In this chapter, molecular technology toward functional 1D and 2D materials on surfaces is described. The key technique to investigate this field is scanning probe microscopies, STM and AFM, which allow us to observe the structures with high resolution, manipulate single molecules, and even induce chemical reactions at local part of molecules on surface. Using precursor molecules that are synthesized by solution chemistry, 1D and 2D carbon-based materials have been successfully produced in both UHV and interfacial conditions. These nanoscale materials have a great potential for various applications such as molecular electronics, coating, photonic materials, superconductor, and supercapacitor. To this end, further functional nano-carbon based materials have to be synthesized by earlier and novel chemical reactions on surfaces, which will lead to technological innovations in the near future.

Acknowledgments

This work was supported in part by the Japan Science and Technology Agency (JST) "Precursory Research for Embryonic Science and Technology (PRESTO)" for a project of "Molecular technology and creation of new function," by Japan Society for the Promotion of Science (JSPS) KAKENHI Grant Number 15K21765.

References

1 Binnig, G., Rohrer, H., Gerber, C., and Weibel, E. (1982). *Phys. Rev. Lett.* 49: 57–61.
2 Barth, J.V., Costantini, G., and Kern, K. (2005). *Nature* 437: 671–679.
3 Rabe, J.P. and Buchholz, S. (1991). *Science* 253: 424–427.
4 Tahara, K., Furukawa, S., Uji-i, H. et al. (2006). *J. Am. Chem. Soc.* 128: 16613–16625.
5 Yokoyama, T., Yokoyama, S., Kamikado, T. et al. (2001). *Nature* 413: 619–621.
6 Theobald, J.A., Oxtoby, N.S., Phillips, M.A. et al. (2003). *Nature* 424: 1029–1031.
7 Shang, J., Wang, Y., Chen, M. et al. (2015). *Nat. Chem.* 7: 389–393.
8 Repp, J. and Meyer, G. (2005). *Phys. Rev. Lett.* 94: 026803.
9 Binnig, G., Quate, C.F., and Gerber, C. (1986). *Phys. Rev. Lett.* 56: 930–933.
10 Martin, Y., Williams, C.C., and Wickramasinghe, H.K. (1987). *J. Appl. Phys.* 61: 4723–4729.
11 Albrecht, T.R., Grütter, P., Horne, D., and Rugar, D. (1991). *J. Appl. Phys.* 69: 668–673.
12 Giessibl, F.J. (1995). *Science* 267: 68–71.
13 Giessibl, F.J. (2003). *Rev. Mod. Phys.* 75: 949–983.
14 Ghasemi, S.A., Goedecker, S., Baratoff, A. et al. (2008). *Phys. Rev. Lett.* 100: 236106.
15 Sugimoto, Y., Abe, M., Hirayama, S. et al. (2005). *Nat. Mater.* 4: 156–159.
16 Kawai, S., Foster, A.S., Canova, F.F. et al. (2014). *Nat. Commun.* 5: 4403.
17 Loppacher, C., Guggisberg, M., Pfeiffer, O. et al. (2003). *Phys. Rev. Lett.* 90: 066107.
18 Kimouche, A., Ervasti, M.M., Drost, R. et al. (2015). *Nat. Commun.* 6: 10177.
19 Maier, S., Fendt, L., Zimmerli, L. et al. (2008). *Small* 4: 1115–1118.
20 Sugimoto, Y., Pou, P., Custance, O. et al. (2008). *Science* 322: 413–417.
21 Ternes, M., Lutz, C.P., Hirjibehedin, C.F. et al. (2008). *Science* 319: 1066–1069.
22 Sader, J.E. and Jarvis, S.P. (2004). *Appl. Phys. Lett.* 84: 1801–1803.
23 Lantz, M.A., Hug, H.J., Hoffmann, R. et al. (2001). *Science* 291: 2580–2583.
24 Hoffmann, R., Kantorovich, L.N., Baratoff, A. et al. (2004). *Phys. Rev. Lett.* 92: 146103.
25 Ternes, M., González, C., Lutz, C.P. et al. (2011). *Phys. Rev. Lett.* 106: 016802.
26 Kawai, S., Foster, A.S., Björkman, T. et al. (2016). *Nat. Commun.* 7: 11559.
27 Sugimoto, Y., Pou, P., Abe, M. et al. (2007). *Nature* 446: 64–67.
28 Gotsmann, B., Schmidt, C., Seidel, C., and Fuchs, H. (1998). *Eur. Phys. J. B* 4: 267–268.
29 Nony, L., Gnecco, E., Baratoff, A. et al. (2004). *Nano Lett.* 4: 2185–2189.

30 Gross, L., Mohn, F., Moll, N. et al. (2009). *Science* 325: 1110–1114.
31 Gross, L., Mohn, F., Moll, N. et al. (2010). *Nat. Chem.* 2: 821–825.
32 Mohn, F., Gross, L., Moll, N., and Meyer, G. (2012). *Nat. Nanotechnol.* 7: 227–231.
33 Kawai, S., Sadeghi, A., Feng, X. et al. (2013). *ACS Nano* 7: 9098–9105.
34 Schuler, B., Liu, W., Tkatchenko, A. et al. (2013). *Phys. Rev. Lett.* 111: 106103.
35 Riss, A., Paz, A.P., Wickenburg, S. et al. (2016). *Nat. Chem.* 8: 678–683.
36 Mohn, F., Schuler, B., Gross, L., and Meyer, G. (2013). *Appl. Phys. Lett.* 102: 073109.
37 Weymouth, A.J., Hofmann, T., and Giessibl, F.J. (2014). *Science* 343: 1120–1122.
38 Moll, N., Gross, L., Mohn, F. et al. (2012). *New J. Phys.* 14: 083023.
39 Hapala, P., Kichin, G., Wagner, C. et al. (2014). *Phys. Rev. B* 90: 085421.
40 Ellner, M., Pavliček, N., Pou, P. et al. (2016). *Nano Lett.* 16: 1974–1980.
41 Gross, L., Mohn, F., Moll, N. et al. (2012). *Science* 337: 1326–1329.
42 Moll, N., Schuler, B., Kawai, S. et al. (2014). *Nano Lett.* 14: 6127–6131.
43 Pawlak, R., Kawai, S., Fremy, S. et al. (2011). *ACS Nano* 5: 6349–6354.
44 Schuler, B., Meyer, G., Peña, D. et al. (2015). *J. Am. Chem. Soc.* 137: 9870–9876.
45 Zhang, J., Chen, P., Yuan, B. et al. (2013). *Science* 342: 611–614.
46 Hämäläinen, S.K., van der Heijden, N., van der Lit, J. et al. (2014). *Phys. Rev. Lett.* 113: 186102.
47 Sweetman, A.M., Jarvis, S.P., Sang, H. et al. (2014). *Nat. Commun.* 5: 3931.
48 Kawai, S., Sadeghi, A., Xu, F. et al. (2015). *ACS Nano* 9: 2574–2583.
49 de Oteyza, D.G., Gorman, P., Chen, Y.-C. et al. (2013). *Science* 340: 1434–1437.
50 Kocić, N., Liu, X., Chen, S. et al. (2016). *J. Am. Chem. Soc.* 138: 5585–5593.
51 Kawai, S., Haapasilta, V., Lindner, B.D. et al. (2016). *Nat. Commun.* 7: 12711.
52 Pavliček, N., Fleury, B., Neu, M. et al. (2012). *Phys. Rev. Lett.* 108: 086101.
53 Schuler, B., Liu, S., Geng, Y. et al. (2014). *Nano Lett.* 14: 3342–3346.
54 Albrecht, F., Pavliček, N., Herranz-Lancho, C. et al. (2015). *J. Am. Chem. Soc.* 137: 7424–7428.
55 Majzik, Z., Cuenca, A.B., Pavliček, N. et al. (2016). *ACS Nano* 10: 5340–5345.
56 Kawai, S., Nishiuchi, T., Kodama, T. et al. (2017). *Sci. Adv.* 3: e1603258.
57 Fukuma, T. (2010). *Sci. Technol. Adv. Mater.* 11: 033003.
58 Ido, S., Kimura, K., Oyabu, N. et al. (2013). *ACS Nano* 7: 1817–1822.
59 Ido, S., Kimiya, H., Kobayashi, K. et al. (2014). *Nat. Mater.* 13: 264–270.
60 Elemans, J.A.A.W., Lei, S., and De Feyter, S. (2009). *Angew. Chem. Int. Ed.* 48: 7298–7332.
61 Yang, Y. and Wang, C. (2009). *Chem. Soc. Rev.* 38: 2576–2589.
62 Bonifazi, D., Mohnani, S., and Llanes-Pallas, A. (2009). *Chem. Eur. J.* 15: 7004–7025.
63 Lackinger, M. and Heckl, W.M. (2009). *Langmuir* 25: 11307–11321.
64 Tobe, Y., Tahara, K., and De Feyter, S. (2016). *Bull. Chem. Soc. Jpn.* 89: 1277–1306.

65 Okawa, Y. and Aono, M. (2004). *e-J. Surf. Sci. Nanotechnol.* 2: 99–105.
66 Sakamoto, J., von Heijst, J., Lukin, O., and Schlüter, A.D. (2009). *Angew. Chem. Int. Ed.* 48: 1030–1069.
67 Grill, L., Dyer, M., Lafferentz, L. et al. (2007). *Nat. Nanotechnol.* 2: 687–691.
68 Gao, H.-Y., Held, P.A., Knor, M. et al. (2014). *J. Am. Chem. Soc.* 136: 9658–9663.
69 Haq, S., Hanke, F., Sharp, J. et al. (2014). *ACS Nano* 8: 8856–8870.
70 Matena, M., Riehm, T., Stöhr, M. et al. (2008). *Angew. Chem. Int. Ed.* 47: 2414–2417.
71 Zhong, D., Franke, J.-H., Podiyanachari, S.K. et al. (2011). *Science* 334: 213–216.
72 Sun, Q., Cai, L., Ding, Y. et al. (2015). *Angew. Chem. Int. Ed.* 54: 4549–4552.
73 Gao, H.-Y., Wagner, H., Zhong, D. et al. (2013). *Angew. Chem. Int. Ed.* 52: 4024–4028.
74 Sun, Q., Zhang, C., Li, Z. et al. (2013). *J. Am. Chem. Soc.* 135: 8448–8451.
75 Grim, P.C.M., De Feyter, S., Gesquiére, A. et al. (1997). *Angew. Chem. Int. Ed.* 36: 2601–2603.
76 Treier, M., Pignedoli, C.A., Laino, T. et al. (2011). *Nat. Chem.* 3: 61–67.
77 Krüger, J., Pavliček, N., Alonso, J.M. et al. (2016). *ACS Nano* 10: 4538–4542.
78 Dinca, L.E., Fu, C., MacLeod, J.M. et al. (2013). *ACS Nano* 7: 1652–1657.
79 Williams, C.G., Wang, M., Skomski, D. et al. (2016). *Surf. Sci.* 653: 130–137.
80 Weigelt, S., Busse, C., Bombis, C. et al. (2008). *Angew. Chem. Int. Ed.* 47: 4406–4410.
81 Gong, Z., Yang, B., Lin, H. et al. (2016). *ACS Nano* 10: 4228–4235.
82 Schmitz, C.H., Ikonomov, J., and Sokolowski, M. (2009). *J. Phys. Chem. C* 113: 11984–11987.
83 Treier, M., Richardson, N.V., and Fasel, R. (2008). *J. Am. Chem. Soc.* 130: 14054–14055.
84 Jensen, S., Früchtl, H., and Baddeley, C.J. (2009). *J. Am. Chem. Soc.* 131: 16706–16713.
85 Bebensee, F., Bombis, C., Vadapoo, S.-R. et al. (2013). *J. Am. Chem. Soc.* 135: 2136–2139.
86 Nardi, E., Chen, L., Clair, S. et al. (2014). *J. Phys. Chem. C* 118: 27549–27553.
87 Zwaneveld, N.A.A., Pawlak, R., Abel, M. et al. (2008). *J. Am. Chem. Soc.* 130: 6678–6679.
88 Shchyrba, A., Martens, S.C., Wäckerlin, C. et al. (2014). *Chem. Commun.* 50: 7628–7631.
89 Saywell, A., Schwarz, J., Hecht, S., and Grill, L. (2012). *Angew. Chem. Int. Ed.* 51: 5096–5100.
90 Denk, R., Hohage, M., Zeppenfeld, P. et al. (2014). *Nat. Commun.* 5: 4253.
91 Cirera, B., Zhang, Y.-Q., Björk, J. et al. (2014). *Nano Lett.* 14: 1891–1897.
92 Cirera, B., Giménez-Agulló, N., Björk, J. et al. (2016). *Nat. Commun.* 7: 11002.
93 Tanaka, H. and Kawai, T. (1997). *J. Vac. Sci. Technol., B* 15: 602–604.
94 Tanaka, H. and Kawai, T. (2009). *Nat. Nanotechnol.* 4: 518–522.

95 Hamann, C., Woltmann, R., Hong, I.-P. et al. (2011). *Rev. Sci. Instrum.* 82: 033903.
96 Ullmann, F. and Bielecki, J. (1901). *Ber. Dtsch. Chem. Ges.* 34: 2174–2185.
97 Sambiagio, C., Marsden, S.P., Blackera, A.J., and McGowan, P.C. (2014). *Chem. Soc. Rev.* 43: 3525–3550.
98 Weigelt, S., Busse, C., Bombis, C. et al. (2007). *Angew. Chem. Int. Ed.* 46: 9227–9230.
99 In't Veld, M., Iavicoli, P., Haq, S. et al. (2008). *Chem. Commun.* 1536–1538.
100 Otero, G., Biddau, G., Sánchez-Sánchez, C. et al. (2008). *Nature* 454: 865–868.
101 Stetsovych, O., Švec, M., Vacek, J. et al. (2017). *Nat. Chem.* 9: 213–218.
102 Cai, J., Ruffieux, P., Jaafar, R. et al. (2010). *Nature* 466: 470–473.
103 Kawai, S., Saito, S., Osumi, S. et al. (2015). *Nat. Commun.* 6: 8098.
104 Kawai, S., Benassi, A., Gnecco, E. et al. (2016). *Science* 351: 957–961.
105 Lafferentz, L., Eberhardt, V., Dri, C. et al. (2012). *Nat. Chem.* 4: 215–220.
106 Lafferentz, L., Ample, F., Yu, H. et al. (2009). *Science* 323: 1193–1197.
107 Koch, M., Ample, F., Joachim, C., and Grill, L. (2012). *Nat. Nanotechnol.* 7: 713–717.
108 Kawai, S., Koch, M., Gnecco, E. et al. (2014). *Proc. Natl. Acad. Sci. U.S.A.* 111: 3968–3972.
109 Reecht, G., Scheurer, F., Speisser, V. et al. (2014). *Phys. Rev. Lett.* 112: 047403.
110 Ruffieux, P., Wang, S., Yang, B. et al. (2016). *Nature* 531: 489–493.
111 Zhang, H., Lin, H., Sun, K. et al. (2015). *J. Am. Chem. Soc.* 137: 4022–4025.
112 Basagni, A., Sedona, F., Pignedoli, C.A. et al. (2015). *J. Am. Chem. Soc.* 137: 1802–1808.
113 Talirz, L., Söde, H., Dumslaff, T. et al. (2017). *ACS Nano* 11: 1380–1388.
114 Chen, Y.-C., de Oteyza, D.G., Pedramrazi, Z. et al. (2013). *ACS Nano* 7: 6123–6128.
115 Nguyen, G.D., Toma, F.M., Cao, T. et al. (2016). *J. Phys. Chem. C* 120: 2684–2687.
116 Eigler, D.M. and Schweizer, E.K. (1990). *Nature* 344: 524–526.
117 Bartels, L., Meyer, G., and Rieder, K.-H. (1997). *Phys. Rev. Lett.* 79: 697–700.
118 Bartels, L., Meyer, G., Rieder, K.-H. et al. (1998). *Phys. Rev. Lett.* 80: 2004–2007.
119 Stipe, B.C. and Rezaei, M.A. (1998). *Science* 279: 1907–1909.
120 Pascual, J.I., Lorente, N., Song, Z. et al. (2003). *Nature* 423: 525–528.
121 Komeda, T., Kim, Y., Kawai, M. et al. (2002). *Science* 295: 2055–2058.
122 Katano, S., Kim, Y., Hori, M. et al. (2007). *Science* 316: 1883–1886.
123 Liljeroth, P., Repp, J., and Meyer, G. (2007). *Science* 317: 1203–1206.
124 Kumagai, T., Shiotari, A., Okuyama, H. et al. (2012). *Nat. Mater.* 11: 167–172.
125 Hla, S., Bartels, L., Meyer, G., and Rieder, K. (2000). *Phys. Rev. Lett.* 85: 2777–2780.
126 Pavliček, N., Schuler, B., Collazos, S. et al. (2015). *Nat. Chem.* 7: 623–628.
127 Schuler, B., Fatayer, S., Mohn, F. et al. (2016). *Nat. Chem.* 8: 220–224.

128 Pavliček, N., Mistry, A., Majzik, Z. et al. (2017). *Nat. Nanotechnol.* https://doi.org/10.1038/nnano.2016.305.
129 Miura, A., De Feyter, S., Abdel-Mottaleb, M.M.S. et al. (2003). *Langmuir* 19: 6474–6482.
130 Okawa, Y. and Aono, M. (2001). *Nature* 409: 683–684.
131 Russell, J.C., Blunt, M.O., Garfitt, J.M. et al. (2011). *J. Am. Chem. Soc.* 133: 4220–4223.
132 Eder, G., Smith, E.F., Cebula, I. et al. (2013). *ACS Nano* 7: 3014–3021.
133 Zhang, X., Liao, L., Wang, S. et al. (2014). *Sci. Rep.* 4: 3899.
134 Colazzo, L., Sedona, F., Moretto, A. et al. (2016). *J. Am. Chem. Soc.* 138: 10151–10156.
135 Rowan, S.J., Cantrill, S.J., Cousins, G.R.L. et al. (2002). *Angew. Chem. Int. Ed.* 41: 898–952.
136 Jin, Y., Yu, C., Denman, R.J., and Zhang, W. (2013). *Chem. Soc. Rev.* 42: 6634–6654.
137 Tanoue, R., Higuchi, R., Enoki, N. et al. (2011). *ACS Nano* 5: 3923–3929.
138 Tanoue, R., Higuchi, R., Ikebe, K. et al. (2012). *Langmuir* 28: 13844–13851.
139 Ciesielski, A., Garah, M.E., Haar, S. et al. (2014). *Nat. Chem.* 6: 1017–1023.
140 Hu, F.-Y., Zhang, X.-M., Wang, X.-C. et al. (2013). *ACS Appl. Mater. Interfaces* 5: 1583–1587.
141 Xu, L., Zhou, X., Yu, Y. et al. (2013). *ACS Nano* 7: 8066–8073.
142 Sun, X., Fan, L., Zhou, X. et al. (2015). *Chem. Commun.* 51: 5864–5867.
143 Sun, X.-L., Fan, L.-X., Yang, Y.-J. et al. (2015). *Chem. Eur. J.* 21: 6898–6905.
144 Yu, Y., Lin, J., Wang, Y. et al. (2016). *Chem. Commun.* 52: 6609–6612.
145 Liu, X.-H., Guan, C.-Z., Ding, S.-Y. et al. (2013). *J. Am. Chem. Soc.* 135: 10470–10474.
146 Xu, L., Zhou, X., Tian, W.Q. et al. (2014). *Angew. Chem. Int. Ed.* 53: 9564–9568.
147 Sun, J., Zhou, X., and Lei, S. (2016). *Chem. Commun.* 52: 8691–8694.
148 Côté, A.P., Benin, A.I., Ockwig, N.W. et al. (2005). *Science* 310: 1166–1170.
149 Feng, X., Ding, X., and Jiang, D. (2012). *Chem. Soc. Rev.* 41: 6010–6022.
150 Thomas, A. (2010). *Angew. Chem. Int. Ed.* 49: 8328–8344.
151 Dienstmaier, J.F., Gigler, A.M., Goetz, A.J. et al. (2011). *ACS Nano* 5: 9737–9745.
152 Dienstmaier, J.F., Medina, D.D., Dogru, M. et al. (2012). *ACS Nano* 6: 7234–7242.
153 Guan, C.-Z., Wang, D., and Wan, L.-J. (2012). *Chem. Commun.* 48: 2943–2945.
154 Liu, C., Yu, Y., Zhang, W. et al. (2016). *Chem. Eur. J.* 22: 18412–18418.
155 Plas, J., Ivasenko, O., Martsinovich, N. et al. (2016). *Chem. Commun.* 52: 68–71.
156 Liu, C., Zhang, W., Zeng, Q., and Lei, S. (2016). *Chem. Eur. J.* 22: 6768–6773.

Index

Molecular Technology: Materials Innovation, Volume 3, First Edition.
Edited by Hisashi Yamamoto and Takashi Kato.

p

t